Engineering Mechanics: STATICS AND STRENGTH OF MATERIALS

R. D. SNYDER / E. F. BYARS
West Virginia University

McGRAW-HILL BOOK COMPANY
New York St. Louis San Francisco Düsseldorf Johannesburg
Kuala Lumpur London Mexico Montreal New Delhi
Panama Rio de Janeiro Singapore Sydney Toronto

Library of Congress Cataloging in Publication Data

Snyder, Robert D
Engineering mechanics.

1. Strength of materials. 2. Statics.
I. Byars, Edward Ford, 1925- joint author.
II. Title.
TA405.S65 620.1'12 72-6568
ISBN 0-07-059530-5

ENGINEERING MECHANICS: STATICS AND STRENGTH OF MATERIALS

6789-MAMM-76543

This book was set in Century Expanded by Progressive Typographers. The editors were B. J. Clark and Michael Gardner; the designer was Merrill Haber; and the production supervisor was Thomas J. Lo Pinto. The drawings were done by Vantage Art, Inc.
The printer and binder was Kingsport Press, Inc.

The cover design was done by Hartmuth Bender. The sculpture was made by Bernard Rosenthal; it is on the premises of a building in New York City owned by Jack and Burton Resnick, Arthur Belfer, and Lawrence Ruben and managed by Jack Resnick & Sons, Inc.

CONTENTS

PREFACE

This text represents our attempt to write an undergraduate engineering mechanics text covering much of the subject material often included in the traditional statics and strength of materials courses. We were motivated in this endeavor by the changing emphasis in engineering education in which curricula are becoming more flexible to meet the needs of broader educational objectives.

The book is not intended to cover "all" of statics or strength but includes those topics which we feel are essential to a cohesive development of the subject and necessary for the education of any professional engineer. We believe the text to be quite adaptable to most formats of three-, four-, or five-semester-credit-hour courses in mechanics. For example, it can be used in a conventional three-hour statics course by covering the material in Chaps. 1 to 4 and parts of 5 and 10, and Appendix A without loss in continuity. On the other hand, a conventional three-hour strength course can be made up of Chaps. 5, 6, 7, 8, 9, and parts of 10. At West Virginia University we have used this text in our two four-credit-hour course sequence in mechanics in the sophomore year, covering most of the text in the first course. With modern dynamics courses leaning more toward systems rather than the traditional rigid-body analysis, we feel that statics and strength form a compatible combination in attempting to consolidate coverage.

An introductory course in engineering mechanics should provide a meaningful tie between the student's previous background in formal mathematics and his native desire to solve useful problems. To this end we have striven to achieve an optimum blend of engineering rationalizations and approximations with mathematical abstraction and rigor. We have tried to make the student acutely aware of the relative roles played by physical intuition and formal mathematical manipulation in the solution of real problems.

There are certain features which may strike some veteran mechanics teachers as unusual and unconventional. In particular, while some of our symbolism and notation is somewhat unconventional, we think it is quite transparent and uncomplicated, particularly for the student who is not already familiar with some other convention.

Also, we have not included a separate chapter on friction but have discussed this topic at various places throughout the text, most noticeably in Chap. 10 on stability.

We would like to thank our colleagues who made numerous suggestions and criticisms, and our several secretaries who painstakingly typed and retyped the manuscript.

R. D. Snyder and E. F. Byars

TO THE STUDENT

Before you embark on studying the contents of this book, we would like to acquaint you with our philosophy of what engineering is all about. We hope that this will help you to better understand and appreciate the objective of this course.

We believe that engineering is an art! The artist engineer uses his creative talent and technical skills to help man utilize nature in his continuing quest for a better life. The basic ingredients of good engineering are reliable technical knowledge, careful rational analysis, and prudent judgment. In addition, a good engineer is invariably enthusiastic about his work and takes great pride in his profession and its accomplishments.

In all probability, up to now your career as a student engineer has been concerned with acquiring certain technical knowledge in the math and science areas. To be sure, in this course you will continue to acquire technical knowledge, but that is only one objective of this course. An equally or more important objective is to apply your acquired technical knowledge to solving real problems involving real things. The "artistic" aspect of engineering will begin to develop as you learn to relate real problems to the paper-and-pencil-type problems to which you can apply your technical skills. In the final analysis, the engineer must possess the creative talent to relate the knowledge in his books and head to the surrounding world in which he lives.

We hope you enjoy this course as much as we enjoyed writing the book. Good luck.

R. D. SNYDER AND E. F. BYARS

chapter 1 FORCES

1-1 INTRODUCTION

Mechanics is the oldest of the physical sciences in that throughout all recorded history man has utilized mechanics in some way or other to aid him in his daily chores for comfort and survival. The wheel, bow and arrow, automobile jack, and rocket engine are all examples of useful applications of mechanics by man. But what is mechanics? Certainly the first wheel was used by man long before the first book on mechanics was ever written. Every student in this course was familiar with the behavior of a jack or a bow and arrow before reading this text. Most of you no doubt understand how a rocket engine is able to operate even in a vacuum like outer space. All of these ideas you have accumulated via your experiences, along with occasional explanations from other sources.

But man cannot always rely on his experiences alone. He must often try to extrapolate his experience to another situation with which he is unfamiliar. For example, with only his experience on earth man has attempted to predict occurrences and behavior on the moon and in outer space. This extrapolation is accomplished by abstracting physical experiences into theorems and mathematical representations, or models, which enable man to solve problems without first experiencing their solutions. How good the solution is, that is, how well it agrees with physical reality, greatly depends upon how good the mathematical model of the physical situation is.

Mechanics is a reasonable and rational interpretation of the physical processes which we observe in our daily life. It is an ever-increasing collection of theorems, postulates, and principles most of which are unproved and unprovable but which seem to work in enabling us to predict the physical behavior of the world around us. For example, by applying certain principles of mechanics we can predict the orbital motion of satellites, the lift on the wings of an airplane, and the force necessary to slide a crate across a warehouse floor. Make no mistake, mechanics does not attempt to explain the mechanisms or intricate processes of nature but merely attempts to interpret the phenomenological results of these mechanisms in a reasonable and useful manner; e.g., we do not know "why" a body falls toward the earth, but we know that it *will* fall, and we can predict from the "law of gravity" how far and fast it will fall in a given time. In summary, mechanics is a creation of man's thoughts based on his experiences.

In this course we do not attempt to study all of mechanics since that includes electrical, thermal, chemical, and atomic, as well as mechanical, phenomena. You will encounter many of these branches in subsequent courses. In this course we confine our attention to the primitive mechanical concepts of force and displacement. You will soon see how the abstracting of the physical idea of a force leads to a systematic development of postulates and theorems which enable us to solve many useful problems.

1-2 FORCE AS A VECTOR

Probably our most primitive notion of a force is that of a push or pull, such as those illustrated in Fig. 1-1. In Fig. 1-1(*a*) the man pushes the crate up the incline, in Fig. 1-1(*b*) the earth pulls the command module toward the ground, and in Fig. 1-1(*c*) the screw jack supports the sagging floor beam.

In each of these examples, one body exerts a force on another body in producing or trying to produce some mechanical effect. Thus we might define a force as the action of one body on another body that tends to produce a mechanical effect of motion or deformation. Such a definition suffices as a general description of a force, but it is too imprecise to be of much value in solving problems involving forces.

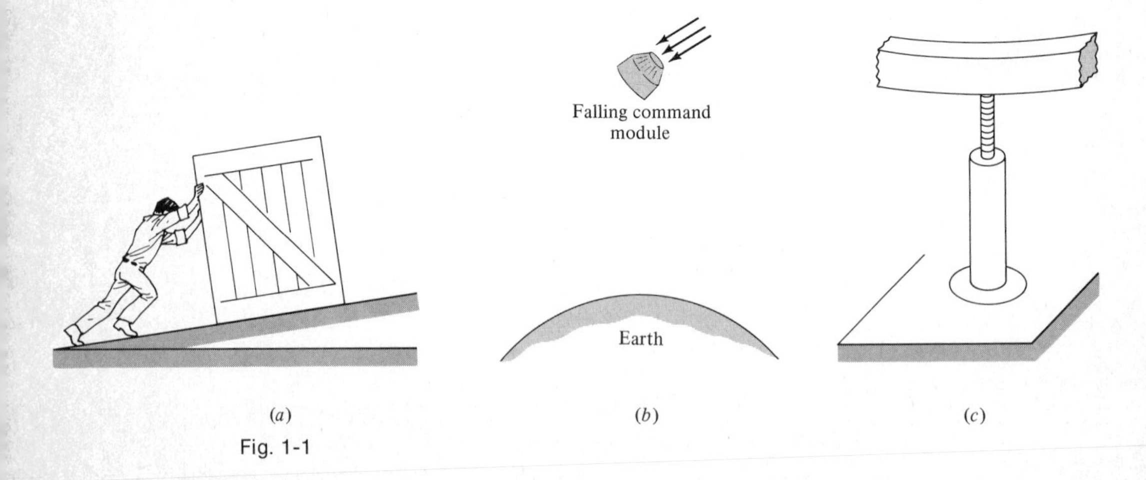

Fig. 1-1

Going back to the illustrations, each of the forces described has a direction and location as well as a size, e.g., a force of 20 lb *on* the left side of the crate and *up* the plane. Thus, in describing or representing forces, we must take into consideration their size, location, and direction.

Before abstracting these notions of forces, we make one important additional observation of our experience in dealing with forces. Suppose the large box in Fig. 1-2(*a*) is being simultaneously pushed by two boys. Our experience tells us that the box will tend to move diagonally across the floor in the same manner as if the box were being pushed by a single man exerting a single diagonal force as in Fig. 1-2(*b*). With this important additional observation, we are now ready to abstract these primitive notions about forces in a precise and useful manner.

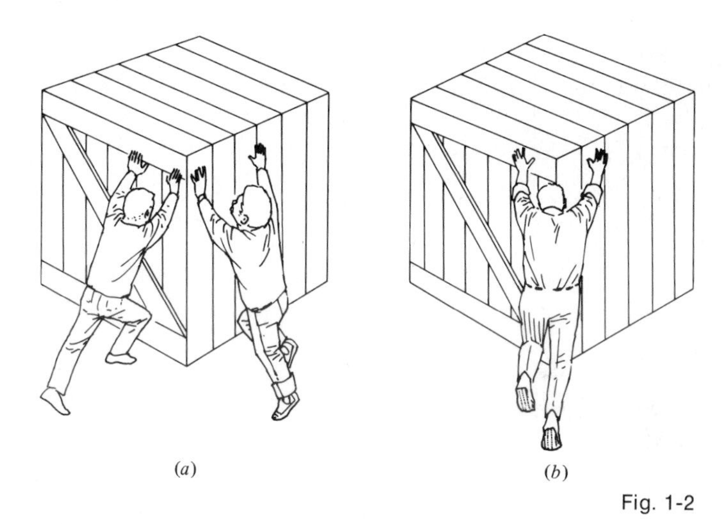

Fig. 1-2

Definition 1-1 Quantities which can be described by real numbers and which add according to the usual rules of algebraic addition are called *scalars*.

For example, the total volume of liquid in five different bottles can be found by adding together the volume of liquid in each bottle. In an electrical dc series circuit, the voltage across any two contact points is the input voltage minus the voltage drops throughout the other parts of the circuit. You can no doubt think of numerous other physical quantities which can be treated as scalars. However, from the previous discussion, it can be seen that not all physical quantities can be completely described by real numbers, and thus we are led to define a second type of quantity.

Definition 1-2 Quantities which have magnitude and direction and which add according to the parallelogram law are called *vectors.*

In a euclidean 3-space, such as the space in which we live, vectors are usually represented by directed line segments with an arrow indicating the *sense* of the vector, as shown by the vectors $\boldsymbol{\alpha}$ and $\boldsymbol{\beta}$ in Fig. 1-3(*a*). (Throughout the sequel we will "name" vectors with boldface letters.) Usually some scale is chosen so that the larger the vector, the longer is the directed line segment representing that vector. For example, using the same scale, a line segment representing a force vector of 10 lb would be twice as long as a line segment representing a force vector of 80 oz, since 16 oz $=1$ lb. The sum $\boldsymbol{\alpha}+\boldsymbol{\beta}$ is the vector given by the construction shown in Fig. 1-3(*b*), where the dashed lines are the other two sides of the parallelogram. This is called addition according to the parallelogram law, or tip-to-tail addition.

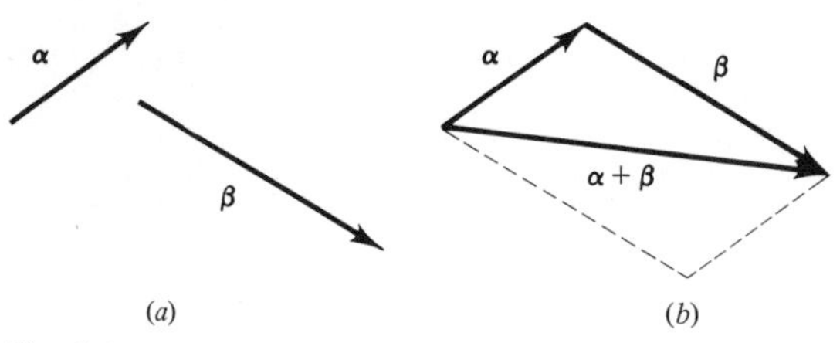

Fig. 1-3

Forces are often treated as vectors and are thereby represented by directed line segments. You can probably think of other physical quantities which we can represent in a similar manner. For example, on a map of the United States the position of Boston relative to New York City can be represented by a straight line drawn on the map from New York to Boston. This line might be several inches in length, whereas the actual distance between New York and Boston is approximately 200 mi in a north-easterly direction. Thus, relative position is a vector quantity which we represent with a directed line segment.

The essential feature of this discussion of scalars and vectors is that in dealing with physical quantities which have the characteristics of scalars and/or vectors, we need not deal with the physical quantities themselves but rather with their representations as numbers or directed line segments. This is a simple example of what is known as abstracting or modeling physical notions into mathematical or geometrical concepts and is the backbone of good engineering. An engineer must be able to construct good models and be able to correctly interpret their physical meaning.

EXAMPLE 1-1

Represent the forces illustrated in Fig. 1-1 as directed line segments if (1) the man *pushes* the crate with a force of 100 lb, (2) the command module *weighs* 3 tons, and (3) the jack *exerts* a force of 4 kips, a kip being equal to 1,000 lb.

Solutions: (1) Assuming the man is pushing parallel to the inclined plane and letting 1 in of length represent 100 lb of force, we would have the representation shown in Fig. 1-4(*a*), where the line segment is 1 in long. It should be mentioned that location of this force is somewhat imprecise since the palm of the human hand covers roughly 20 to 30 in^2. However, if the crate is much larger than the hand, the exact location is relatively unimportant as a physical consideration. Note also that the drawing in Fig. 1-4 is two dimensional even though the physical situation is three dimensional.

(2) The earth's gravitational field exerts a downward force equal to the weight of the module. Thus, letting $\frac{1}{4}$ in represent 1 ton, a downward line segment of $\frac{3}{4}$ in length can represent the force of the earth on the module, as shown in Fig. 1-4(*b*). Again, its exact location is imprecise, but since the module appears as a point in the huge vastness of space, putting the vector at the "middle" of the module is quite satisfactory for most considerations. In the delicate maneuvers of reentry navigation a much more precise location of the middle, or center of gravity, would be needed.

(3) Letting $\frac{1}{4}$ in represent 1 kip (1,000 lb), a line segment 1 in long is used to represent the force of the jack on the sagging beam, as shown in Fig. 1-4(*c*). As to its exact location, the same comment as in (1) applies to this situation.

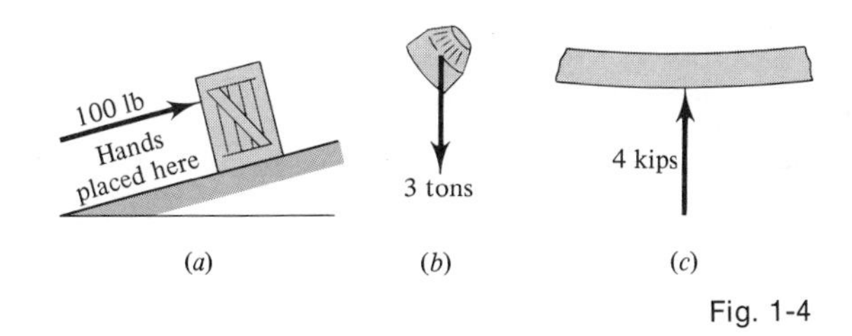

Fig. 1-4

Important Remark: Compare Fig. 1-1 with Fig. 1-4. Note that in each case in Fig. 1-4 the agent which exerts the force in Fig. 1-1 is missing (the man, the earth, and the jack), but its effect has been represented by a force vector.

EXAMPLE 1-2

A small boat is being towed upstream by two boys, one on each side of the stream, pulling on ropes attached to the boat, as in-

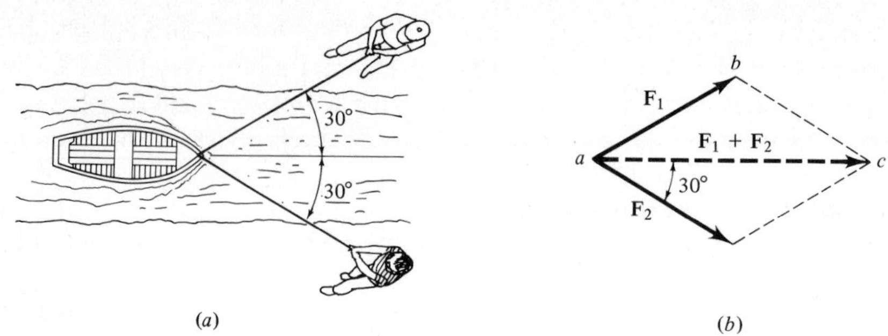

Fig. 1-5

dicated in Fig. 1-5(a). If each boy pulls with about 7.5 lb of force, what is the sum of their forces?

Solution: From Fig. 1-5(b) the sum of $\mathbf{F}_1 + \mathbf{F}_2$ will be the third side of an isosceles triangle abc. Hence, the magnitude ac will be

$$ac = 2ab \cos 30° = 15.0 \frac{\sqrt{3}}{2} = 7.5\sqrt{3}$$

Thus

$$\mathbf{F}_1 + \mathbf{F}_2 = \xrightarrow{7.5\sqrt{3}\text{ lb}}$$

which we interpret physically as an upstream force of approximately 13 lb.

EXERCISE PROBLEMS

1-1 Give five examples of physical quantities that can be treated as scalar quantities.

1-2 Give five examples of physical quantities that can be treated as vector quantities.

PROB. 1-3 PROB. 1-5

1-3 A 160-lb man is standing on the ladder as shown. Represent the force(s) exerted by the man on the ladder. Represent the force(s) which the ladder exerts on the man.

1-4 In Prob-1.3, without worrying about their sizes, represent the force(s) which the wall exerts on the ladder, and which the ground exerts on the ladder. What keeps the ladder from sliding out from under the man?

1-5 A man pushes down and sideways at a 45° angle on the table with a total force of 60 lb on which is resting a box weighing 35 lb. Using some scale, represent the forces exerted on the table by the man and the box.

1-6 In Prob. 1-5, without worrying about their sizes, represent the force which the floor exerts on the table. What keeps the table from sliding across the floor?

1-7 If Morgantown is 60 mi due east of New Martinsville and Pittsburgh is 80 mi due north of Morgantown, what is the position of Pittsburgh relative to New Martinsville? Give your answer as a single vector.

1-8 A car is traveling west at 50 mph on U.S. 30 while another car is traveling north on U.S. 11 at 60 mph. What is the relative velocity between the two cars; that is, how does the first car appear to be moving as seen by the driver of the second car?

1-9 to 1-12 Find the sum of the vectors shown by two methods: (*a*) graphically by drawing them to scale; (*b*) analytically, using trigonometry.

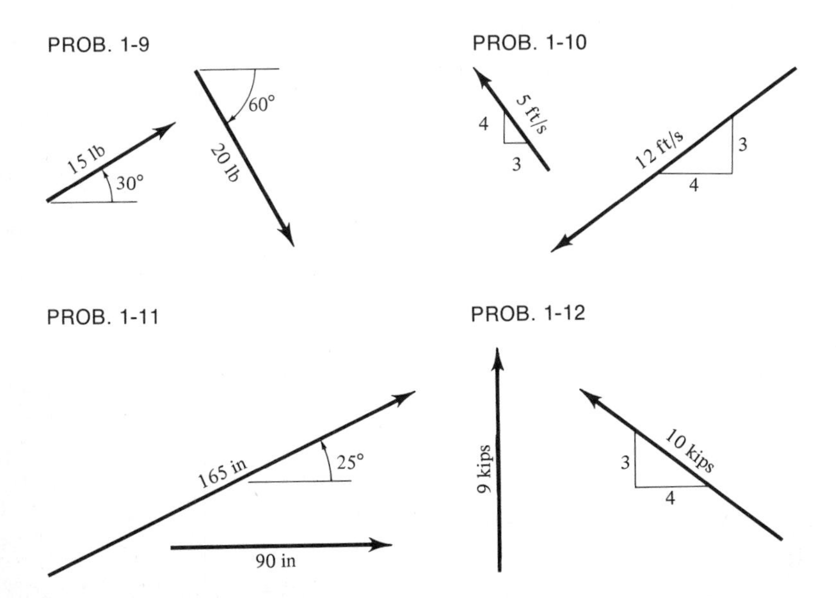

1-3 FURTHER PROPERTIES OF VECTORS

The physical notions associated with forces were observed long before the mathematics of vectors was developed by Euler, Lagrange, Gibbs, and others. However, the subsequent development of vector mathematics has greatly clarified and made more precise the implications and interpretations of our physical observations. While Definition 1-2 suffices for a physical interpretation of a vector, we need a more formal and complete mathematical concept of a vector. Most of what we do in this section is formal mathematics and is not motivated by physical laws or observations, although, later in the course, we will use this mathematics to solve physical problems.

Let $\mathscr{V}$ be the set or collection of all vectors in the euclidean 3-space. Then

(1) If $\boldsymbol{\alpha}$ and $\boldsymbol{\beta}$ are in $\mathscr{V}$, their *sum* $\boldsymbol{\alpha} + \boldsymbol{\beta}$ as given by the parallelogram law is also in $\mathscr{V}$. Furthermore, this addition is commutative and associative in that

$$\boldsymbol{\alpha} + \boldsymbol{\beta} = \boldsymbol{\beta} + \boldsymbol{\alpha}$$
$$(\boldsymbol{\alpha} + \boldsymbol{\beta}) + \boldsymbol{\gamma} = \boldsymbol{\alpha} + (\boldsymbol{\beta} + \boldsymbol{\gamma}) = \boldsymbol{\alpha} + \boldsymbol{\beta} + \boldsymbol{\gamma}$$

See Fig. 1-6.

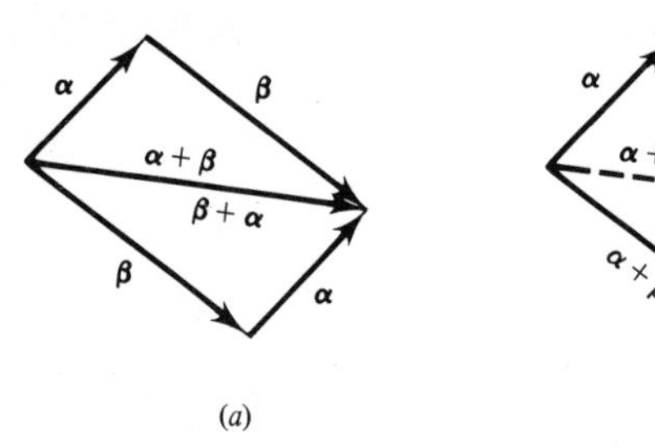

Fig. 1-6

Remark: Although there are many types of physical vector quantities, such as forces, positions, heat flux, etc., it usually makes physical sense to add only vectors of the same species, i.e., forces add to forces, positions to positions, etc.

(2) For mathematical reasons we define the *null* (or zero) vector $\mathbf{0}$ in $\mathscr{V}$ such that for all $\boldsymbol{\alpha}$

$$\boldsymbol{\alpha} + \mathbf{0} = \boldsymbol{\alpha}$$

Physically, $\mathbf{0}$ is a vector of zero length (magnitude). You are cautioned not to think of $\mathbf{0}$ as the number 0 (zero) since numbers cannot be added to vectors by means of the parallelogram law.

(3) The existence of the null vector $\mathbf{0}$ implies that for every $\boldsymbol{\alpha}$ there exists a unique vector $\boldsymbol{\beta}$ called the *additive inverse* or *opposite* of $\boldsymbol{\alpha}$ such that

$$\boldsymbol{\alpha} + \boldsymbol{\beta} = \mathbf{0}$$

Formally

$$\boldsymbol{\beta} = \mathbf{0} - \boldsymbol{\alpha} = -\boldsymbol{\alpha}$$

Thus the opposite of $\boldsymbol{\alpha}$ is written as $-\boldsymbol{\alpha}$, which we interpret physically as the reverse of $\boldsymbol{\alpha}$ as in Fig. 1-7.

Fig. 1-7

(4) Let $\boldsymbol{\alpha}$ be a vector and c be any real number. Then we define the "c multiple of $\boldsymbol{\alpha}$" as the vector $c\boldsymbol{\alpha}$ whose direction is the same as $\boldsymbol{\alpha}$ if c is positive and whose magnitude is c times the magnitude of $\boldsymbol{\alpha}$. If c is negative, the direction of $c\boldsymbol{\alpha}$ is opposite to that of $|c|\boldsymbol{\alpha}$, so that $-1\boldsymbol{\alpha} = -\boldsymbol{\alpha}$. Intuitively we expect that $0\boldsymbol{\alpha} = \mathbf{0}$. See Fig. 1-8 for an example of scalar multiples. In this figure, $\boldsymbol{\alpha}$ could be any type of physical quantity represented by a vector. For example, on a straight highway, $\boldsymbol{\alpha}$ might represent the velocity of a car traveling at 40 mph west while $-2\boldsymbol{\alpha}$ would then represent the velocity of a car traveling at 80 mph east.

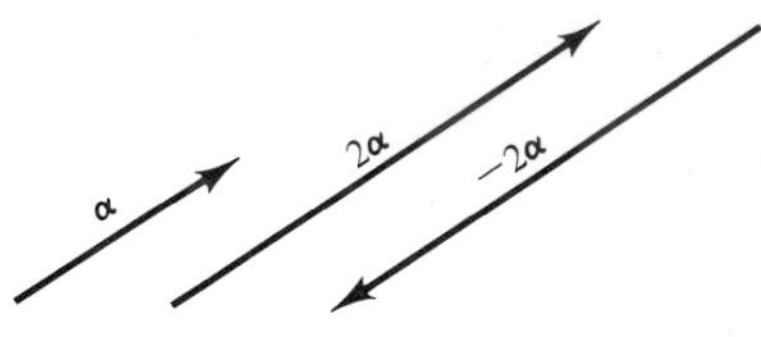

Fig. 1-8

We pause in our development to observe that all of the formalities listed above are entirely consistent with our physical notions associated with vector-type quantities such as forces and positions. We now lay down some additional vector concepts.

Definition 1-3 A *unit vector* is a dimensionless vector whose magnitude is unity. If $\boldsymbol{\alpha}$ is any nonzero vector and we denote its magnitude by $\|\boldsymbol{\alpha}\|$, then the vector

$$\boldsymbol{\mu}_\alpha = \frac{\boldsymbol{\alpha}}{\|\boldsymbol{\alpha}\|}$$

is a unit vector in the direction of $\boldsymbol{\alpha}$. Physically, unit vectors are used to identify some particular direction in euclidean 3-space.

(5) From (4) and Definition 1-3 we see that any vector $\boldsymbol{\alpha}$ can be written in the form

$$\boldsymbol{\alpha} = \alpha\boldsymbol{\mu}_\alpha$$

where $\boldsymbol{\mu}_\alpha$ is a unit vector in the direction of $\boldsymbol{\alpha}$ and the scalar α is the magnitude of the vector $\boldsymbol{\alpha}$, $\alpha = \|\boldsymbol{\alpha}\|$.

Definition 1-4 Let $\boldsymbol{\alpha}$ be any vector and $\boldsymbol{\mu}$ be a unit vector in some prescribed direction. Then the *projection of* $\boldsymbol{\alpha}$ onto $\boldsymbol{\mu}$ is

$$\text{Proj}_{\boldsymbol{\mu}}\boldsymbol{\alpha} \overset{\text{def}}{=} \|\boldsymbol{\alpha}\| \cos\theta$$

where θ is the angle formed by $\boldsymbol{\alpha}$ and $\boldsymbol{\mu}$ with $0 \leqslant \theta \leqslant \pi$. See Fig. 1-9.

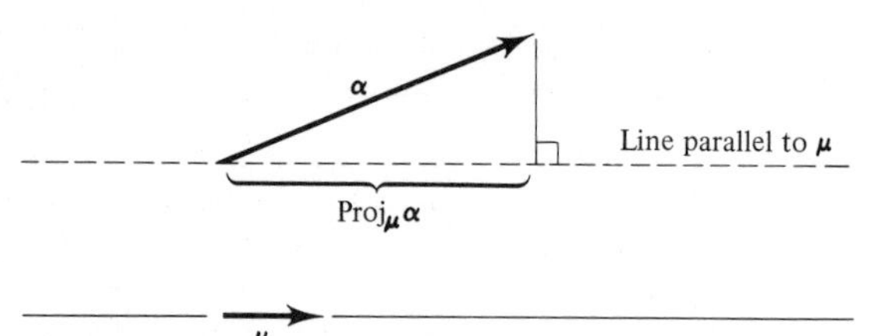

Fig. 1-9

Physically, $\text{Proj}_{\boldsymbol{\mu}}\boldsymbol{\alpha}$ is merely the perpendicular projection of $\boldsymbol{\alpha}$ onto a line parallel to $\boldsymbol{\mu}$. For convenience we write

$$\text{Proj}_{\boldsymbol{\mu}}\boldsymbol{\alpha} = \boldsymbol{\mu}\cdot\boldsymbol{\alpha} = \|\boldsymbol{\alpha}\| \cos(\boldsymbol{\mu},\boldsymbol{\alpha})$$

and call the $\cdot$ operation the inner product, dot product, or scalar product between the unit vector $\boldsymbol{\mu}$ and the vector $\boldsymbol{\alpha}$. We observe that a projection can be a positive or negative number depending upon the angle $(\boldsymbol{\mu},\boldsymbol{\alpha})$, for example, $\cos 135° = -0.707$.

(6) Let $\boldsymbol{\alpha}$ and $\boldsymbol{\beta}$ be two arbitrary vectors and write

$$\boldsymbol{\alpha} = \alpha\boldsymbol{\mu}_\alpha \qquad \text{and} \qquad \boldsymbol{\beta} = \beta\boldsymbol{\mu}_\beta$$

where $\boldsymbol{\mu}_\alpha$ and $\boldsymbol{\mu}_\beta$ are unit vectors in the directions of $\boldsymbol{\alpha}$ and $\boldsymbol{\beta}$, respectively. Then, from Definition 1-4

$$\begin{aligned}\alpha\,\text{Proj}_{\boldsymbol{\mu}_\alpha}\boldsymbol{\beta} &= \alpha\|\boldsymbol{\beta}\| \cos(\boldsymbol{\mu}_\alpha,\boldsymbol{\beta})\\ &= \alpha\beta\cos(\boldsymbol{\alpha},\boldsymbol{\beta}) = \beta\alpha\cos(\boldsymbol{\beta},\boldsymbol{\alpha}) = \beta\alpha\cos(\boldsymbol{\mu}_\beta,\boldsymbol{\alpha})\\ &= \beta\,\text{Proj}_{\boldsymbol{\mu}_\beta}\boldsymbol{\alpha}\end{aligned}$$

since $\alpha\beta = \beta\alpha$, $\cos(\boldsymbol{\alpha},\boldsymbol{\beta}) = \cos(\boldsymbol{\beta},\boldsymbol{\alpha})$, and angle $(\boldsymbol{\mu}_\alpha,\boldsymbol{\beta}) =$ angle $(\boldsymbol{\alpha},\boldsymbol{\beta})$. Thus we can define the commutative inner product of $\boldsymbol{\alpha}$ and $\boldsymbol{\beta}$ by

$$\boldsymbol{\alpha}\cdot\boldsymbol{\beta} = \boldsymbol{\beta}\cdot\boldsymbol{\alpha} = \alpha\beta\cos(\boldsymbol{\alpha},\boldsymbol{\beta}) \tag{1-1}$$

and geometrically interpret it as the magnitude of $\boldsymbol{\alpha}$ times the projection of $\boldsymbol{\beta}$ onto $\boldsymbol{\mu}_\alpha$ or as the magnitude of $\boldsymbol{\beta}$ times the projection of $\boldsymbol{\alpha}$ onto $\boldsymbol{\mu}_\beta$. We observe the important result that

$$\boldsymbol{\alpha}\cdot\boldsymbol{\alpha} = \alpha\alpha(1) = \|\boldsymbol{\alpha}\|^2 = \alpha^2$$

(7) If $\boldsymbol{\alpha} \cdot \boldsymbol{\beta} = 0$, $\boldsymbol{\alpha}$ and $\boldsymbol{\beta}$ are said to be *orthogonal*, or perpendicular, or normal, to each other. Geometrically, if $\boldsymbol{\alpha}$ and $\boldsymbol{\beta}$ are orthogonal, the projection of $\boldsymbol{\alpha}$ onto $\boldsymbol{\beta}$ (and vice versa) is zero.

Remark: When forming the inner product of two vectors, the vectors need not be of the same species; that is, one vector could be a force vector while the other might be a position vector.

The following example problems more fully illustrate some of the ideas presented in this section.

EXAMPLE 1-3

Suppose $\boldsymbol{\alpha}$ and $\boldsymbol{\gamma}$ are vectors as shown in Fig. 1-10(a). If $\boldsymbol{\alpha} + \boldsymbol{\beta} = \boldsymbol{\gamma}$, determine the vector $\boldsymbol{\beta}$.

Solution: According to the parallelogram law for addition of vectors, if $\boldsymbol{\alpha} + \boldsymbol{\beta} = \boldsymbol{\gamma}$, $\boldsymbol{\beta}$ must be the dashed vector in Fig. 1-10(b). By the law of cosines

$$\begin{aligned}\beta^2 &= \alpha^2 + \gamma^2 - 2\alpha\gamma \cos 135^\circ \\ &= (25)^2 + (40)^2 - 2(25)(40)\left(-\frac{\sqrt{2}}{2}\right) = 3{,}639 \text{ in}^2 \\ \beta &= 60.3 \text{ in}\end{aligned}$$

Also, by the law of sines

$$\frac{\sin\theta}{25} = \frac{\sin 135^\circ}{60.3}$$

$$\therefore \sin\theta = 0.293 \Rightarrow \theta = 17.0^\circ$$

Thus $\boldsymbol{\beta}$ is a vector with magnitude 60.3 in directed downward and to the right at an angle of 17.0° with the horizontal.

Alternate Solution: Formally, since

$$\boldsymbol{\alpha} + \boldsymbol{\beta} = \boldsymbol{\gamma}$$

we can "solve" for $\boldsymbol{\beta}$ by

$$\boldsymbol{\beta} = \boldsymbol{\gamma} - \boldsymbol{\alpha} = \boldsymbol{\gamma} + (-\boldsymbol{\alpha})$$

This last equation is illustrated by the vectors in Fig. 1-10(c). We will now solve for $\boldsymbol{\beta}$ in a different manner simply to illustrate

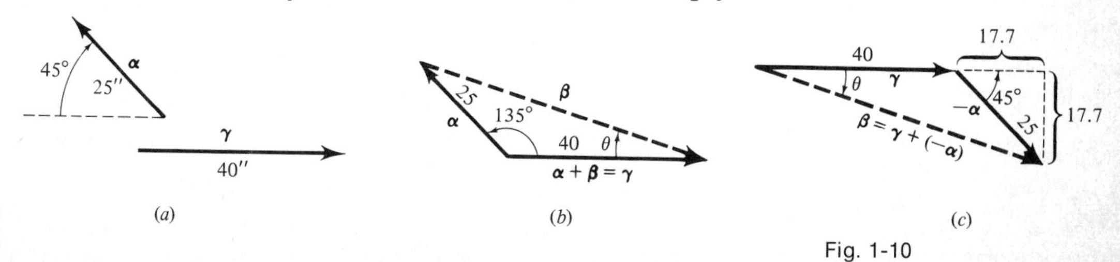

Fig. 1-10

another method of handling such problems. By the pythagorean theorem

$$\beta = \sqrt{(40 + 17.7)^2 + (17.7)^2}$$
$$= 60.3 \text{ in}$$

while
$$\theta = \arctan \frac{17.7}{57.7} = 17.1°$$

which agrees with the first solution to within the limits of slide-rule accuracy.

EXAMPLE 1-4

Given the two vectors shown in Fig. 1-11(*a*) with $\|\boldsymbol{\alpha}\| = 30$ in, $\|\boldsymbol{\beta}\| = 50$ lb. Determine (*a*) the projection of $\boldsymbol{\alpha}$ onto $\boldsymbol{\beta}$, (*b*) the projection of $\boldsymbol{\beta}$ onto $\boldsymbol{\alpha}$, and (*c*) the inner product between $\boldsymbol{\alpha}$ and $\boldsymbol{\beta}$.

Solution: By definition,

(*a*) $\text{Proj}_{\boldsymbol{\mu}_\beta} \boldsymbol{\alpha} = \|\boldsymbol{\alpha}\| \cos(\boldsymbol{\beta},\boldsymbol{\alpha}) = 30 \cos 105° = -7.76$ in

(*b*) $\text{Proj}_{\boldsymbol{\mu}_\alpha} \boldsymbol{\beta} = \|\boldsymbol{\beta}\| \cos(\boldsymbol{\alpha},\boldsymbol{\beta}) = 50 \cos 105° = -12.9$ lb

(*c*) $\boldsymbol{\alpha} \cdot \boldsymbol{\beta} = \alpha\beta \cos(\boldsymbol{\alpha},\boldsymbol{\beta}) = (30)(50) \cos 105° = -388$ in-lb

The negative projections are indicated in Fig. 1-11(*b*).

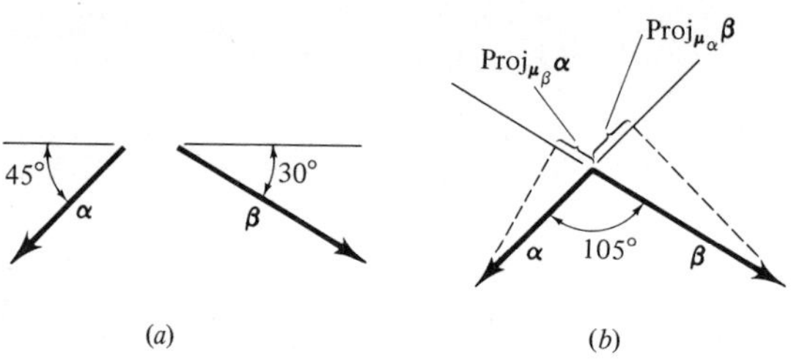

Fig. 1-11

EXERCISE PROBLEMS

1-13 Given the distributive law for scalar multiplication

$$(c + d)\boldsymbol{\alpha} = c\boldsymbol{\alpha} + d\boldsymbol{\alpha}$$

where c and d are scalars and $\boldsymbol{\alpha}$ a vector, first prove that $0\boldsymbol{\alpha} = 0$ and then prove that $(-1)\boldsymbol{\alpha} = -\boldsymbol{\alpha}$.

Hint: Use the definition of the null vector $\mathbf{0}$.

1-14 Let $\boldsymbol{\alpha}$ and $\boldsymbol{\beta}$ be two arbitrary vectors and $\boldsymbol{\mu}$ some arbitrary unit vector. Geometrically show that

$$\text{Proj}_{\boldsymbol{\mu}}(\boldsymbol{\alpha} + \boldsymbol{\beta}) = \text{Proj}_{\boldsymbol{\mu}}\boldsymbol{\alpha} + \text{Proj}_{\boldsymbol{\mu}}\boldsymbol{\beta}$$

or
$$\boldsymbol{\mu} \cdot (\boldsymbol{\alpha} + \boldsymbol{\beta}) = \boldsymbol{\mu} \cdot \boldsymbol{\alpha} + \boldsymbol{\mu} \cdot \boldsymbol{\beta}$$

This is a form of the distributive law for the inner product.

1-15 to 1-18 For the vectors $\boldsymbol{\alpha}$ and $\boldsymbol{\beta}$ shown, find (*a*) $\boldsymbol{\alpha} + \boldsymbol{\beta}$, (*b*) $\boldsymbol{\alpha} - \boldsymbol{\beta}$, and (*c*) $\boldsymbol{\beta} - \boldsymbol{\alpha}$.

PROBS. 1-15 AND 1-19

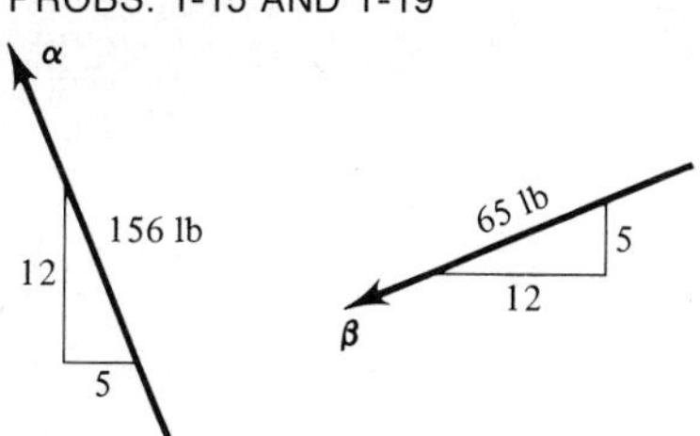

PROBS. 1-16 AND 1-20

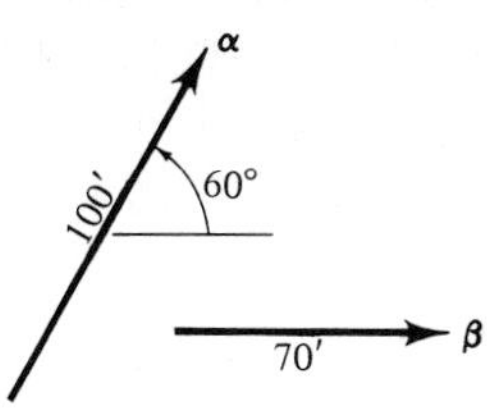

PROBS. 1-17 AND 1-21

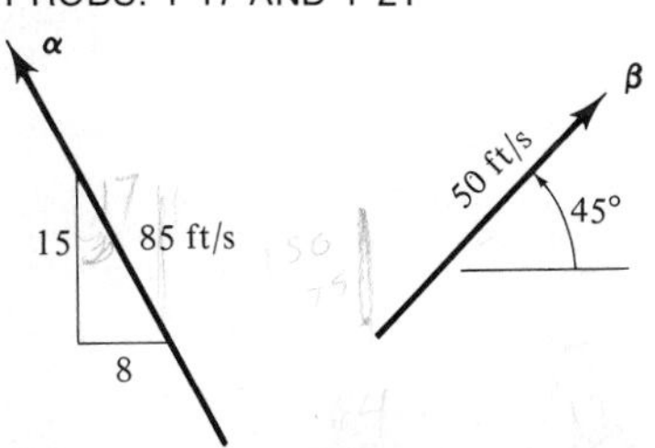

PROBS. 1-18 AND 1-22

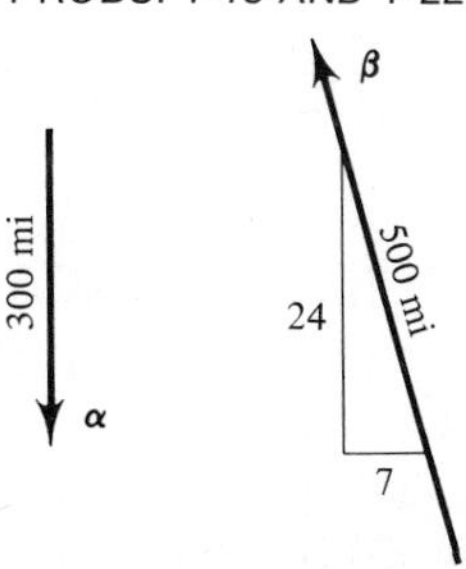

1-19 to 1-22 For the vectors $\boldsymbol{\alpha}$ and $\boldsymbol{\beta}$ shown, find (*a*) the projection of $\boldsymbol{\alpha}$ onto $\boldsymbol{\beta}$, (*b*) the projection of $\boldsymbol{\beta}$ onto $\boldsymbol{\alpha}$, and (*c*) the inner product $\boldsymbol{\alpha} \cdot \boldsymbol{\beta}$.

1-23 and 1-24 The vector $\boldsymbol{\gamma}$ is the sum of two vectors $\boldsymbol{\alpha}$ and $\boldsymbol{\beta}$, where $\boldsymbol{\alpha}$ is directed along line *a–a* and $\boldsymbol{\beta}$ along line *b–b*. Determine $\boldsymbol{\alpha}$ and $\boldsymbol{\beta}$.

PROB. 1-23

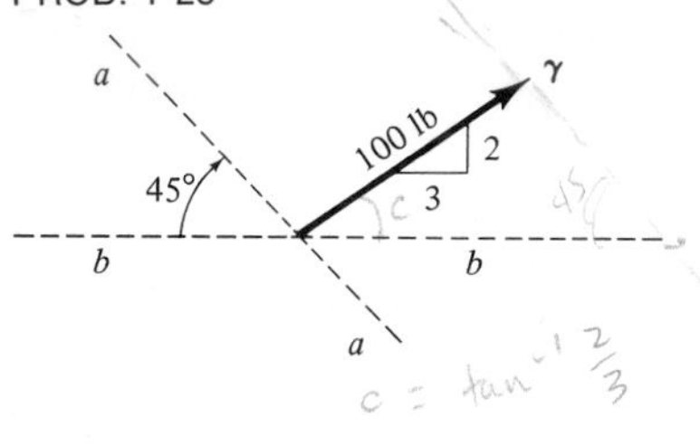

PROB. 1-24

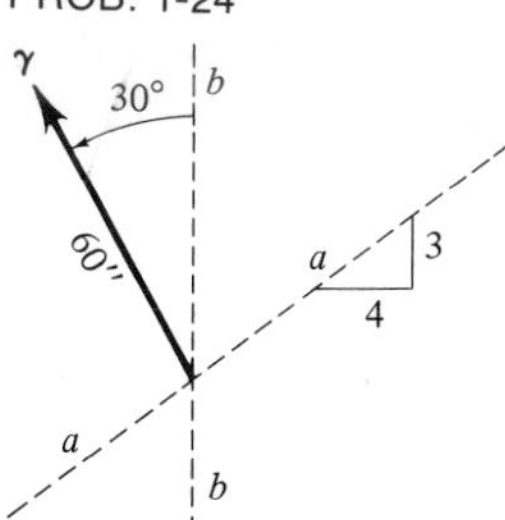

1-25 and 1-26 The inner product of two vectors $\boldsymbol{\alpha}$ and $\boldsymbol{\beta}$ is known to be −400 in-lb. If $\boldsymbol{\alpha}$ is known to be a 100-lb force as shown, and $\boldsymbol{\beta}$ is directed along *b–b*, determine $\boldsymbol{\beta}$.

PROB. 1-25

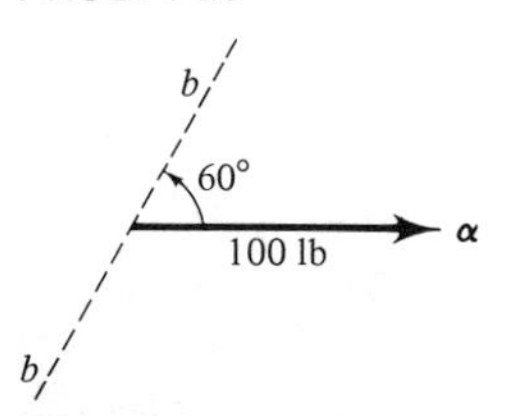

PROB. 1-26

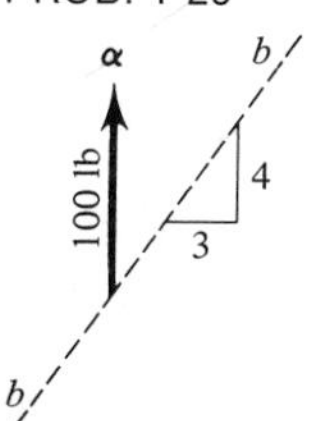

1-27 and 1-28 The inner product of two vectors $\boldsymbol{\alpha}$ and $\boldsymbol{\beta}$ is known to be 3,000 ft-lb. If $\boldsymbol{\alpha}$ is as shown and the magnitude of $\boldsymbol{\beta}$ is 10 ft, determine the vector $\boldsymbol{\beta}$.

1-4 ORTHOGONAL COMPONENTS

In the previous sections we have been representing vector quantities by directed line segments in space. Specifying the magnitude of a vector quantity is done by giving a number and a unit of measure, for example, 10 lb, 50 fps, etc., but it is much more difficult to *clearly* specify the direction of a vector. Of course, we could specify the direction in terms of north-south, east-west, and up-down, as long as we knew where the north pole was and which way was up. Actually, all references like the "north pole," a "fixed star," or the "horizontal ground" are quite arbitrary. The "best" reference is most often simply the most convenient one. There is no such thing as an absolute direction.

Suppose you walk into a large room and look at something located on one of the walls, as in Fig. 1-12. Your line of sight A is a directed line from your eye to the object.

To specify the direction of this line, probably the most convenient reference would be the walls of the room without worrying about which way was north. By measuring along the walls and floor of the room we are able to tell how far our line of sight goes to

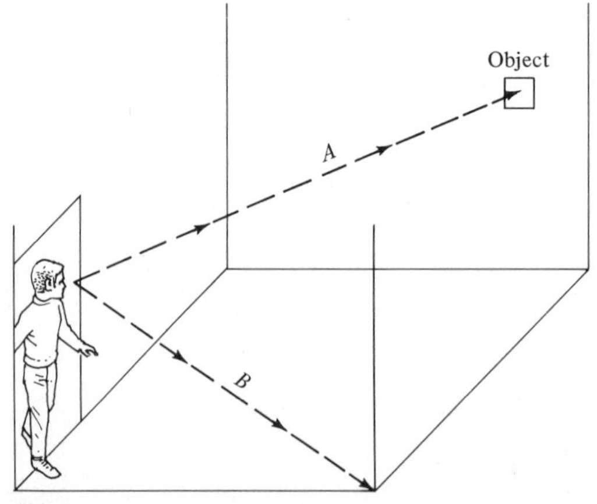

Fig. 1-12

the left or right, how far above the floor, and how deep into the room the line travels; that is, the room itself provides us with a reference for measuring left-right, up-down, and in-out. Thus we can easily distinguish between two directions, such as A and B, just by using measurements within the room and specifying different heights, widths, and depths. We are now going to use this room or box idea in working with vectors in a general way.

Let $\boldsymbol{\alpha}$ be an arbitrary nonzero vector, as shown in Fig. 1-13(*a*). Construct a rectangular "box" with $\boldsymbol{\alpha}$ as the diagonal of the box, as indicated in Fig. 1-13(*b*). Clearly, such a construction can be made for any vector with, of course, a different box as the result. Although the sides of our box in Fig. 1-13(*b*) are horizontal and vertical, this was merely a convenience and we could just as well have constructed the box in Fig. 1-13(*c*) or a multitude of others. Note that $\boldsymbol{\alpha}$ is exactly the same vector in Fig. 1-13(*a*) to (*c*).

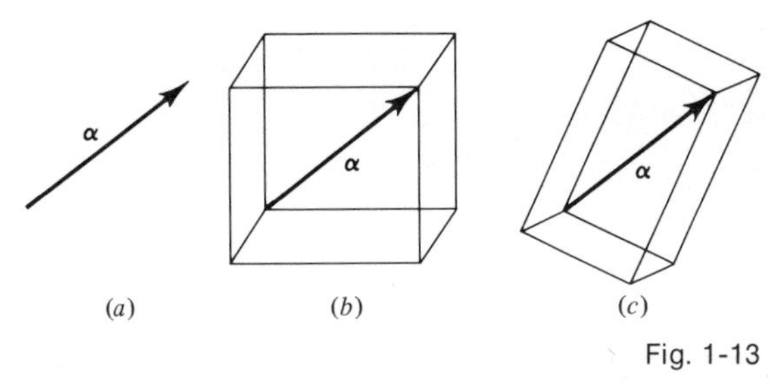

Fig. 1-13

Let us look at this box a little more closely. We first observe from Fig. 1-14(*a*) that side o–b has a length which is equal to the projection of $\boldsymbol{\alpha}$ onto the horizontal line o–b, and similarly for lengths o–c and o–d. Secondly, we observe from Fig. 1-14(*b*) that $\boldsymbol{\alpha}$ is the sum of the three vectors $\boldsymbol{\beta}$, $\boldsymbol{\gamma}$, and $\boldsymbol{\delta}$.

$$\boldsymbol{\alpha} = \boldsymbol{\beta} + \boldsymbol{\gamma} + \boldsymbol{\delta}$$

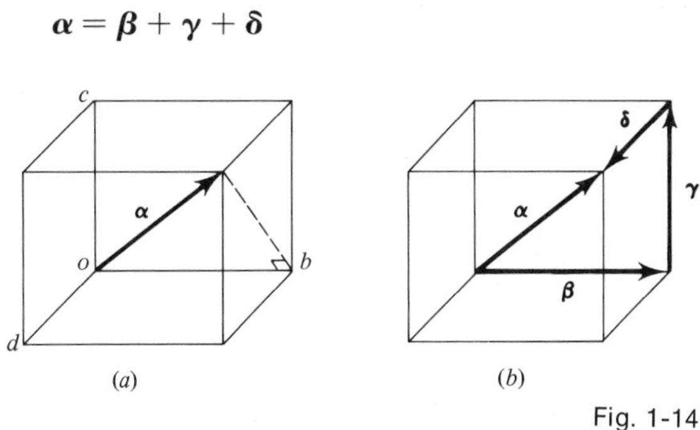

Fig. 1-14

We are now ready to formalize these observations.

Let **i**, **j**, and **k** denote three *unit* vectors mutually perpendicular to each other, that is, an *orthonormal triad.* In most situations it is convenient to choose these to be horizontal and vertical vectors, like those shown in Fig. 1-15(*a*), although any orthogonal triad of

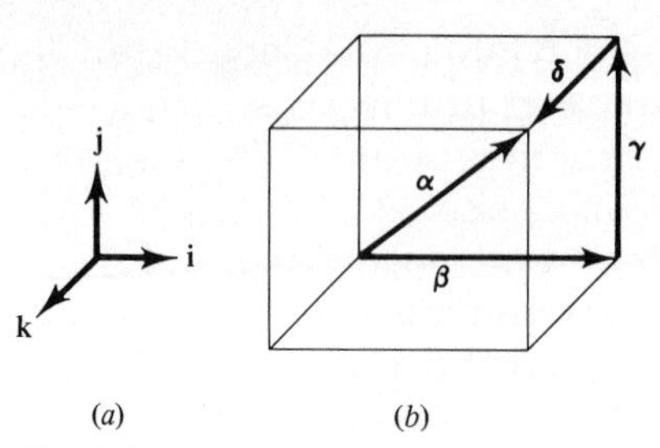

Fig. 1-15

unit vectors will suffice. Now let $\boldsymbol{\alpha}$ be some vector and let us construct the box with sides parallel to the triad, as shown in Fig. 1-15(*b*). Just as before, we have

$$\boldsymbol{\alpha} = \boldsymbol{\beta} + \boldsymbol{\gamma} + \boldsymbol{\delta}$$

but, since $\mathbf{i}$ is a unit vector parallel to $\boldsymbol{\beta}$, we can write

$$\boldsymbol{\beta} = \beta\mathbf{i}$$

and similarly

$$\boldsymbol{\gamma} = \gamma\mathbf{j}$$
$$\boldsymbol{\delta} = \delta\mathbf{k}$$

Hence we can write

$$\boldsymbol{\alpha} = \beta\mathbf{i} + \gamma\mathbf{j} + \delta\mathbf{k}$$

where β, γ, and δ are the projections of $\boldsymbol{\alpha}$ onto the $\mathbf{i}$, $\mathbf{j}$, and $\mathbf{k}$ unit vectors, respectively; that is,

$$\begin{aligned} \text{Proj}_{\mathbf{i}}\boldsymbol{\alpha} &= \mathbf{i} \cdot \boldsymbol{\alpha} = \mathbf{i} \cdot (\beta\mathbf{i} + \gamma\mathbf{j} + \delta\mathbf{k}) \\ &= \beta 1 + \gamma 0 + \delta 0 \\ &= \beta \end{aligned}$$

since $\mathbf{i} \cdot \mathbf{i} = 1$, $\mathbf{i} \cdot \mathbf{j} = 0$, and $\mathbf{i} \cdot \mathbf{k} = 0$. Similarly

$$\text{Proj}_{\mathbf{j}}\boldsymbol{\alpha} = \gamma$$
$$\text{Proj}_{\mathbf{k}}\boldsymbol{\alpha} = \delta$$

Definition 1-5 Let $\boldsymbol{\alpha}$ be any vector and $\mathbf{i}, \mathbf{j}, \mathbf{k}$ be an orthogonal unit triad. Then, $\boldsymbol{\alpha}$ can be written as a sum

$$\boldsymbol{\alpha} = \boldsymbol{\alpha}_1 + \boldsymbol{\alpha}_2 + \boldsymbol{\alpha}_3$$

where
$$\begin{aligned} \boldsymbol{\alpha}_1 &= \alpha_1\mathbf{i} = (\text{Proj}_{\mathbf{i}}\boldsymbol{\alpha})\mathbf{i} \\ \boldsymbol{\alpha}_2 &= \alpha_2\mathbf{j} = (\text{Proj}_{\mathbf{j}}\boldsymbol{\alpha})\mathbf{j} \\ \boldsymbol{\alpha}_3 &= \alpha_3\mathbf{k} = (\text{Proj}_{\mathbf{k}}\boldsymbol{\alpha})\mathbf{k} \end{aligned}$$

are called the *orthogonal vector components* of $\boldsymbol{\alpha}$ relative to the triad $\mathbf{i}, \mathbf{j}, \mathbf{k}$ and the projections α_1, α_2, and α_3 are called the *scalar components* of $\boldsymbol{\alpha}$.

Obviously, from the definition, the orthogonal components depend upon the choice of the unit triad, although in most situations we will choose the **i**, **j**, **k** triad as in Fig. 1-15(*a*) with one unit vector directed to the right, one upward, and the third toward you. Hereafter we will refer to such a triad as a *natural* triad.

The example in Fig. 1-15(*b*) is somewhat a special case since all the projections are positive and the orthogonal components are in the same direction as the unit vectors. As a different illustration, consider the 70-lb force acting on the cable in the direction shown in Fig. 1-16(*a*). The "box" serves to specify the direction of the cable. The diagonal of the box has a length of

$$\ell = \sqrt{2^2 + 3^2 + 6^2} = 7$$

Hence, we can let a 7-in line segment correspond to a force of 70 lb, or, equivalently, a 1-in segment correspond to a force of 10 lb. With this scaling factor it is easily seen that the 70-lb force can be written in terms of orthogonal components relative to the natural triad as

$$\mathbf{F} = -30\mathbf{i} - 20\mathbf{j} + 60\mathbf{k}$$

These components are represented by line segments in Fig. 1-16(*b*), recalling that minus signs indicate opposites, so that $-20\mathbf{j}$ means 20 lb in the direction opposite to **j**.

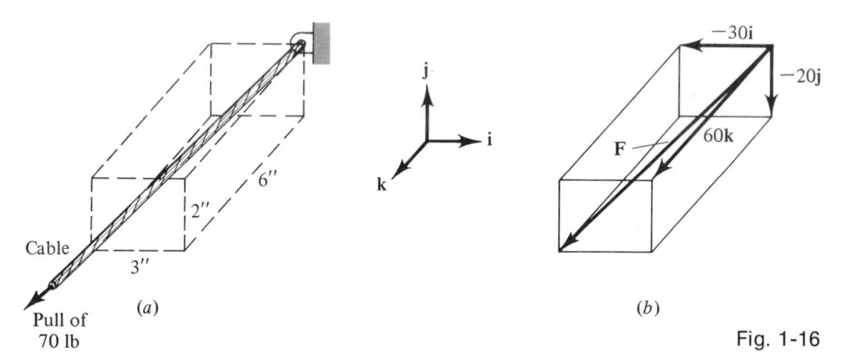

Fig. 1-16

We close this section with one additional idea. Let $\boldsymbol{\mu}$ be a unit vector in some arbitrary direction and **i**, **j**, **k** be a natural triad. From the previous discussion in this section we can write

$$\boldsymbol{\mu} = \mu_1\mathbf{i} + \mu_2\mathbf{j} + \mu_3\mathbf{k}$$

where

$$\mu_1 = \text{Proj}_{\mathbf{i}}\boldsymbol{\mu} = \mathbf{i} \cdot \boldsymbol{\mu} = \cos\theta_1$$
$$\mu_2 = \text{Proj}_{\mathbf{j}}\boldsymbol{\mu} = \mathbf{j} \cdot \boldsymbol{\mu} = \cos\theta_2$$
$$\mu_3 = \text{Proj}_{\mathbf{k}}\boldsymbol{\mu} = \mathbf{k} \cdot \boldsymbol{\mu} = \cos\theta_3$$

in which θ_1, θ_2, and θ_3 are the angles between the $\boldsymbol{\mu}$ direction and the **i**, **j**, and **k** directions, respectively, as shown in Fig. 1-17. The numbers μ_1, μ_2, and μ_3 are referred to as the *direction cosines* for

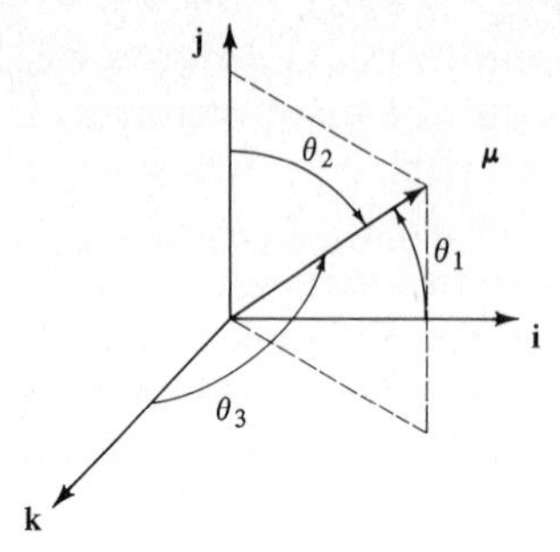

Fig. 1-17

the unit vector $\boldsymbol{\mu}$. Hence, specifying a set of direction cosines is a convenient way to specify a particular direction in space.

EXERCISE PROBLEMS

1-29 to 1-36 Using a natural triad, determine the orthogonal components of each of the appropriate vectors.

PROB. 1-29

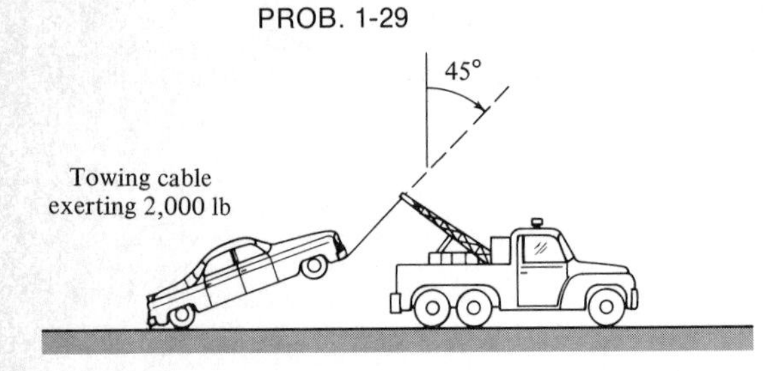

PROB. 1-30

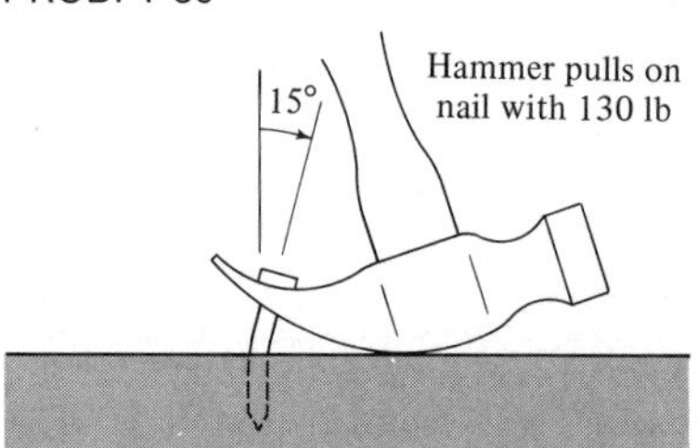

PROB. 1-31

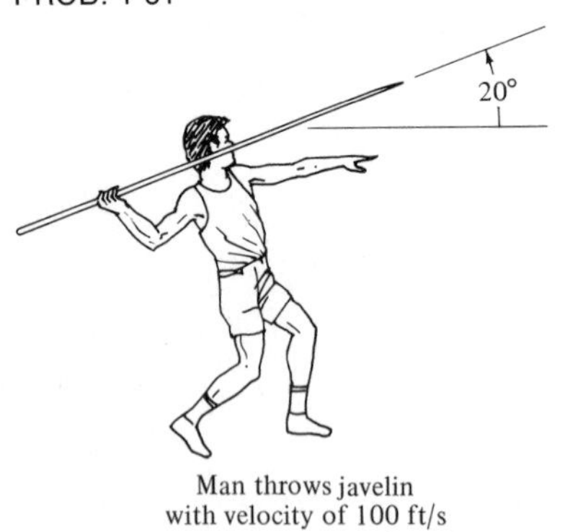

PROB. 1-32

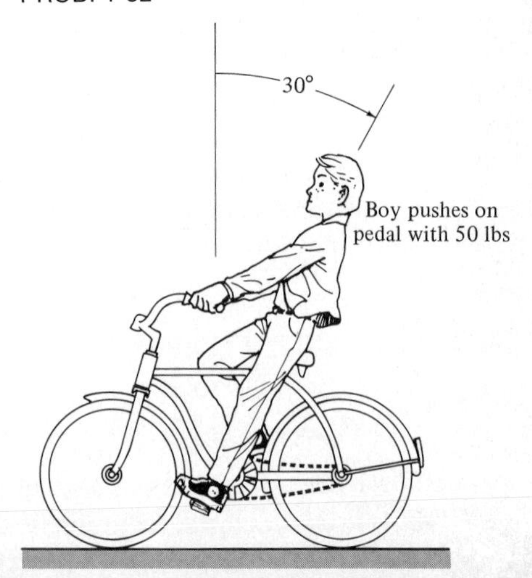

PROB. 1-33

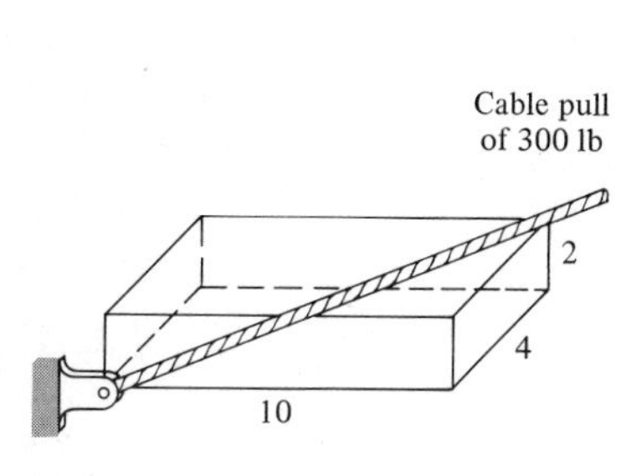

PROB. 1-34

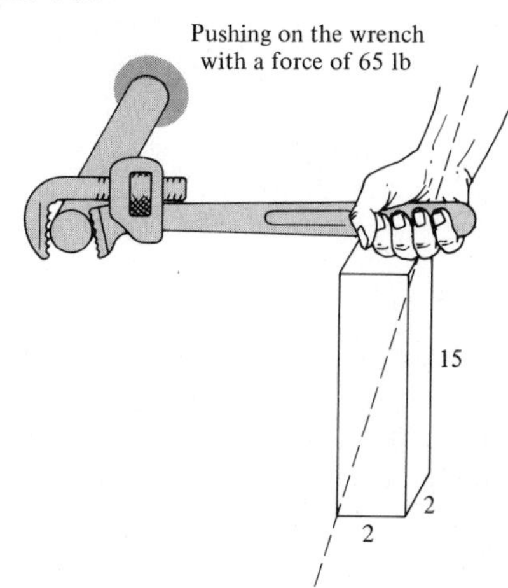

PROB. 1-35

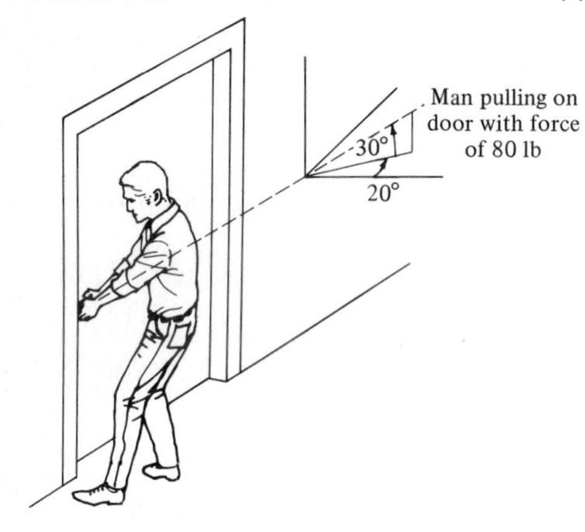

PROB. 1-36

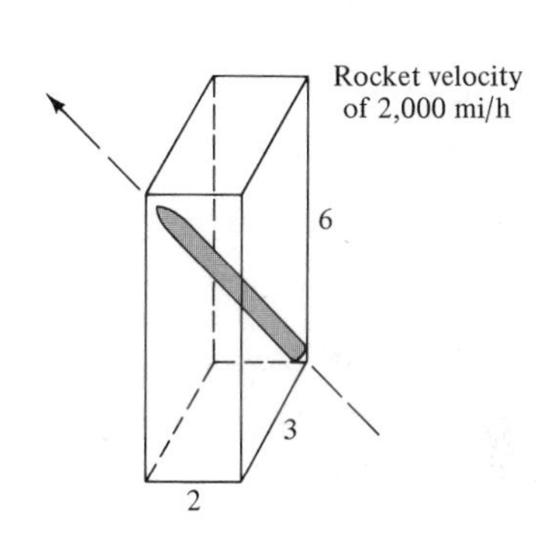

1-37 to 1-40 Determine a unit vector for the directions indicated.

PROB. 1-37

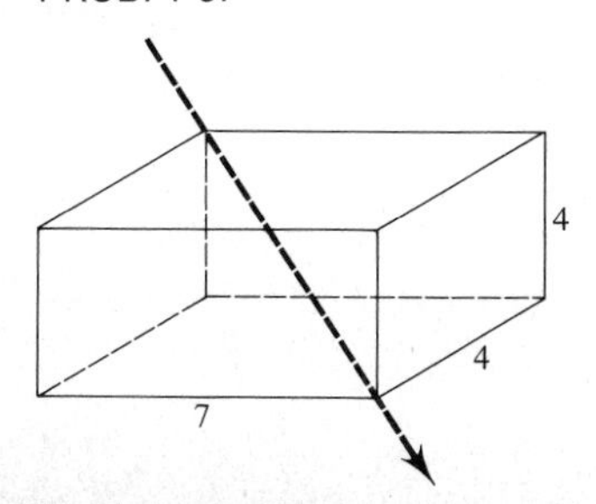

PROB. 1-38

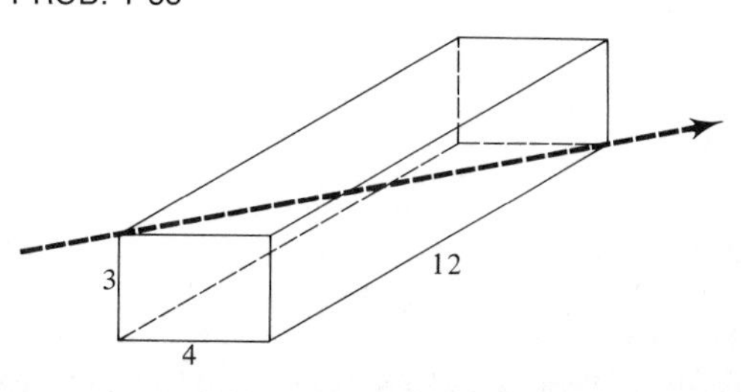

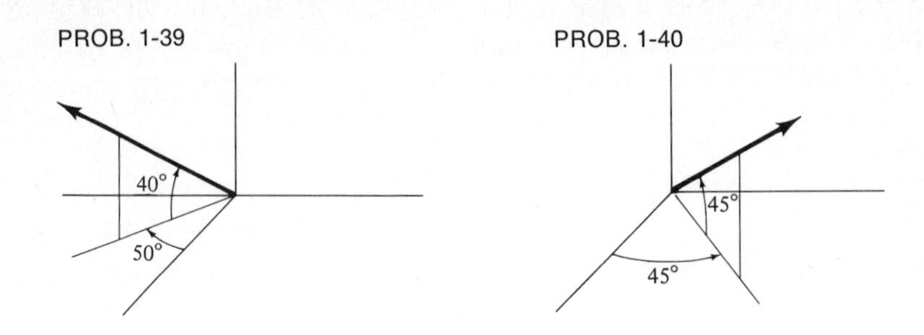

1-5 SUMS AND INNER PRODUCTS WITH ORTHOGONAL COMPONENTS

The procedure just discussed of decomposing any vector into three orthogonal components can be carried out for any vector in the euclidean 3-space. Thus, if $\boldsymbol{\alpha}$ and $\boldsymbol{\beta}$ are two vectors and **i, j, k** are an orthogonal triad, we can write

$$\boldsymbol{\alpha} = \alpha_1\mathbf{i} + \alpha_2\mathbf{j} + \alpha_3\mathbf{k}$$
$$\boldsymbol{\beta} = \beta_1\mathbf{i} + \beta_2\mathbf{j} + \beta_3\mathbf{k}$$

Suppose we wish to form their sum, $\boldsymbol{\alpha} + \boldsymbol{\beta}$. Formally

$$\boldsymbol{\alpha} + \boldsymbol{\beta} = \alpha_1\mathbf{i} + \alpha_2\mathbf{j} + \alpha_3\mathbf{k} + \beta_1\mathbf{i} + \beta_2\mathbf{j} + \beta_3\mathbf{k}$$

Now, using the commutativity of addition and then regrouping

$$\boldsymbol{\alpha} + \boldsymbol{\beta} = (\alpha_1 + \beta_1)\mathbf{i} + (\alpha_2 + \beta_2)\mathbf{j} + (\alpha_3 + \beta_3)\mathbf{k}$$

Thus $(\alpha_1 + \beta_1)\mathbf{i}$, $(\alpha_2 + \beta_2)\mathbf{j}$, and $(\alpha_3 + \beta_3)\mathbf{k}$ are the orthogonal components of the sum $\boldsymbol{\alpha} + \boldsymbol{\beta}$, that is, the orthogonal components of a sum are the sums of the orthogonal components of the vectors being summed.

We utilize this result most effectively when we wish to add three or more vectors together to form their sum. The parallelogram law is quite laborious to use since only two vectors can be added together with one parallelogram. Thus, if we wanted to add five vectors together we would have to construct four individual parallelograms.

However, with the result just obtained we can perform this sum of five vectors much more efficiently. We would first write *each* vector in terms of its orthogonal components relative to the same unit triad **i, j, k**. Then we would add all five **i** components to form the **i** component of the sum, and similarly for the **j** and **k** components. In this manner we would obtain the orthogonal components of the sum and thus the sum itself. Symbolically, for n vectors,

$$\overbrace{\boldsymbol{\alpha} + \boldsymbol{\beta} + \cdots + \boldsymbol{\gamma}}^{n \text{ vectors}} = (\alpha_1 + \beta_1 + \cdots + \gamma_1)\mathbf{i} + (\alpha_2 + \beta_2 + \cdots + \gamma_2)\mathbf{j} + (\alpha_3 + \beta_3 + \cdots + \gamma_3)\mathbf{k}$$

We remark that since the numbers α_1, β_1, α_2, γ_3, etc., are projections, they can be positive or negative, and you are cautioned not to get careless about "dropping signs." Careless bookkeeping is a source of much difficulty. You should discipline yourself to develop good bookkeeping techniques.

Suppose now that instead of adding $\boldsymbol{\alpha}$ and $\boldsymbol{\beta}$ we wish to form their inner product. While we could use Eq. (1-1) of Sec. 1-3.

$$\boldsymbol{\alpha} \cdot \boldsymbol{\beta} = \alpha\beta \cos(\boldsymbol{\alpha},\boldsymbol{\beta})$$

as a practical matter it is often very difficult to geometrically construct and determine the angle between two arbitrary vectors in 3-space. Therefore, it is often desirable to proceed in the following manner.

First write $\boldsymbol{\alpha}$ and $\boldsymbol{\beta}$ in terms of their orthogonal components relative to some unit triad **i**, **j**, **k**. Then formally

$$\boldsymbol{\alpha} \cdot \boldsymbol{\beta} = (\alpha_1\mathbf{i} + \alpha_2\mathbf{j} + \alpha_3\mathbf{k}) \cdot (\beta_1\mathbf{i} + \beta_2\mathbf{j} + \beta_3\mathbf{k})$$

Now, using the distributive law for inner products (see Prob. 1-14) we have

$$\boldsymbol{\alpha} \cdot \boldsymbol{\beta} = \alpha_1\beta_1 + \alpha_2\beta_2 + \alpha_3\beta_3 \tag{1-2}$$

since all terms like

$$\alpha_1\mathbf{i} \cdot \beta_2\mathbf{j} = \alpha_1\beta_2(\mathbf{i} \cdot \mathbf{j}) = \alpha_1\beta_2(0)$$

and

$$\alpha_1\mathbf{i} \cdot \beta_1\mathbf{i} = \alpha_1\beta_1(\mathbf{i} \cdot \mathbf{i}) = \alpha_1\beta_1(1)$$

since the **i**, **j**, **k** are an orthogonal unit triad. Hence we can form the inner product of two vectors without actually measuring the angle between them. We simply write each vector in terms of its orthogonal components and add their products according to Eq. (1-2).

Once again you are cautioned about bookkeeping errors which can and should be avoided. We now illustrate the use of orthogonal components by a few examples.

EXAMPLE 1-5

Determine the natural orthogonal components of the velocity vector of 80 fps shown in Fig. 1-18.

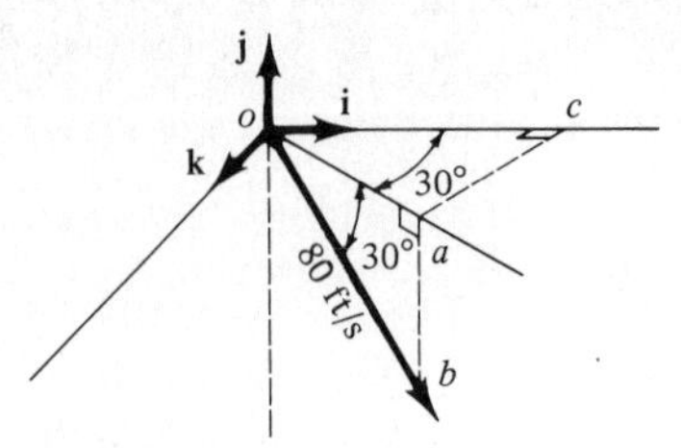

Fig. 1-18

Solution: From the figure

$$o\text{–}a = 80 \cos 30° = 69 \text{ fps}$$
$$a\text{–}b = 80 \sin 30° = 40 \text{ fps}$$
$$o\text{–}c = o\text{–}a \cos 30° = 60 \text{ fps}$$
$$a\text{–}c = o\text{–}a \sin 30° = 35 \text{ fps}$$

Thus, o–c is the **i** projection, a–b is a negative **j** projection, and a–c is the **k** projection. Hence

$$\mathbf{v} = 60\mathbf{i} - 40\mathbf{j} + 35\mathbf{k}$$

Check: Since $\|\mathbf{v}\|^2 = \mathbf{v} \cdot \mathbf{v}$ we see that

$$\|\mathbf{v}\| = \sqrt{(60)^2 + (-40)^2 + (35)^2} = 80 \text{ fps}$$

EXAMPLE 1-6

Find the sum and inner product of the two force vectors shown in Fig. 1-19(*a*). The 40-lb force lies in the vertical plane and the 39-lb force lies in the horizontal plane. What is the angle between the two vectors?

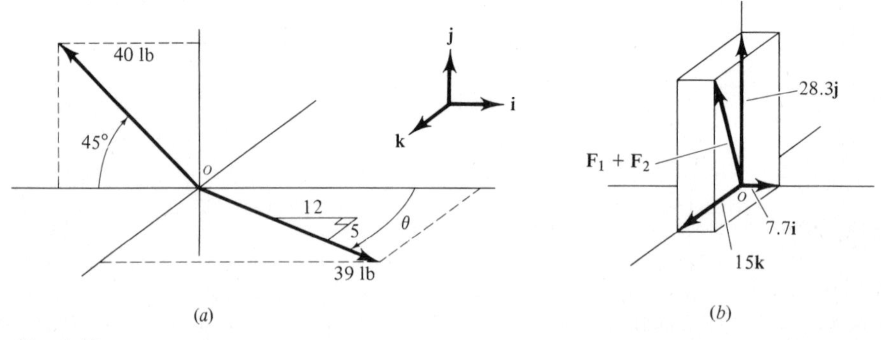

Fig. 1-19

Solution: The first thing we do is express each force in terms of orthogonal components relative to the natural triad **i**, **j**, **k**. Letting $\mathbf{F}_1$ and $\mathbf{F}_2$ be the 40-lb and 39-lb force, respectively, we have

$$\begin{aligned}\mathbf{F}_1 &= -(40 \cos 45°)\mathbf{i} + (40 \sin 45°)\mathbf{j} + 0\mathbf{k} \\ &= -28.3\mathbf{i} + 28.3\mathbf{j} + 0\mathbf{k} \\ \mathbf{F}_2 &= \tfrac{12}{13}(39)\mathbf{i} + 0\mathbf{j} + \tfrac{5}{13}(39)\mathbf{k} \\ &= 36\mathbf{i} + 0\mathbf{j} + 15\mathbf{k}\end{aligned}$$

since $\cos \theta = \frac{12}{13}$ and $\sin \theta = \frac{5}{13}$. Now, the sum is given by

$$\begin{aligned}\mathbf{F}_2 + \mathbf{F}_1 &= (36 - 28.3)\mathbf{i} + (0 + 28.3)\mathbf{j} + (15 + 0)\mathbf{k} \\ &= 7.7\mathbf{i} + 28.3\mathbf{j} + 15\mathbf{k}\end{aligned}$$

The magnitude of this vector is

$$\begin{aligned}\|\mathbf{F}_1 + \mathbf{F}_2\| &= \sqrt{(7.7)^2 + (28.3)^2 + (15)^2} \\ &= 32.9 \text{ lb}\end{aligned}$$

and the vector and its orthogonal components are shown in Fig. 1-19(*b*).

The inner product is given by Eq. (1-2) and is

$$\begin{aligned}\mathbf{F}_1 \cdot \mathbf{F}_2 &= (-28.3)(36) + (28.3)(0) + (0)(15) \\ &= -1{,}019 \text{ lb}^2\end{aligned}$$

This last result has no meaningful physical interpretation and was included simply to demonstrate the procedure for evaluating the inner product of two vectors. Finally, since $\mathbf{F}_1 \cdot \mathbf{F}_2 = \|\mathbf{F}_1\| \, \|\mathbf{F}_2\| \cos (\mathbf{F}_1,\mathbf{F}_2)$, we have

$$\angle(\mathbf{F}_1,\mathbf{F}_2) = \arccos \frac{-1{,}019}{(40)(39)} = 130.8°$$

1-6 CLOSURE

The main idea introduced in this chapter has been the representation of forces by means of directed line segments. This device helps us to draw a picture describing the physical situation as we believe it to be.

Most of the examples in this chapter deal with relatively simple situations in which the force involved was easily identified as a push or pull. However, probably you can think of many physical situations involving forces where it would be an unreasonable stretch of the imagination to represent the forces involved with a simple directed line segment, for example, the force that a high wind exerts on a roof or tree, or the torque exerted by the magnetic field on the armature of an electric motor. As we progress in this course we will broaden our basic ideas about forces so that eventually we will be able to satisfactorily model these and many other more complex physical situations involving forces and their effects. But before moving on you must develop facility and confidence in using the simple ideas presented in this chapter.

EXERCISE PROBLEMS

1-41 Let $\boldsymbol{\alpha}$ be an arbitrary vector and **i**, **j**, **k** an orthonormal triad, so that

$$\boldsymbol{\alpha} = \alpha_1\mathbf{i} + \alpha_2\mathbf{j} + \alpha_3\mathbf{k}$$

Show that

$$\|\boldsymbol{\alpha}\|^2 = \alpha_1{}^2 + \alpha_2{}^2 + \alpha_3{}^2$$

1-42 Let $\boldsymbol{\mu}$ and $\boldsymbol{\nu}$ be arbitrary unit vectors and **i**, **j**, **k** an orthonormal triad. Show that

$$\cos(\boldsymbol{\mu},\boldsymbol{\nu}) = \mu_1\nu_1 + \mu_2\nu_2 + \mu_3\nu_3$$

where μ_1, μ_2, μ_3 and ν_1, ν_2, ν_3 are the direction cosines of $\boldsymbol{\mu}$ and $\boldsymbol{\nu}$, respectively.

1-43 Let $\boldsymbol{\alpha}$ and $\boldsymbol{\beta}$ be two-dimensional vectors lying in a plane. The direction of $\boldsymbol{\alpha}$ is given by the angle θ_1 and the direction of $\boldsymbol{\beta}$ by the angle θ_2. Using orthogonal components, prove the cosine identity

$$\cos(\theta_1 - \theta_2) = \cos\theta_1 \cos\theta_2 + \sin\theta_1 \sin\theta_2$$

PROB. 1-43

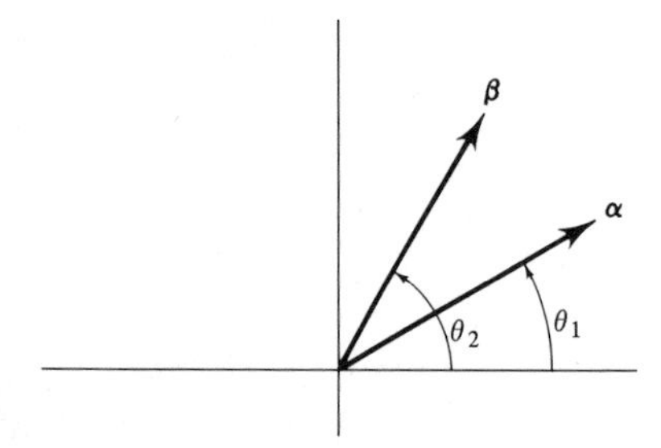

1-44 Let $\boldsymbol{\alpha}$, $\boldsymbol{\beta}$, and $\boldsymbol{\gamma}$ be three nonzero vectors. Suppose

$$\boldsymbol{\alpha} \cdot \boldsymbol{\beta} = \boldsymbol{\alpha} \cdot \boldsymbol{\gamma}$$

Does this imply that $\boldsymbol{\beta} = \boldsymbol{\gamma}$? Why?

1-45 to 1-56 Rework Probs. 1-15 to 1-26, using orthogonal components.

1-57 to 1-62 For the vectors shown, find (*a*) their sum, (*b*) their inner product, and (*c*) the angle between them.

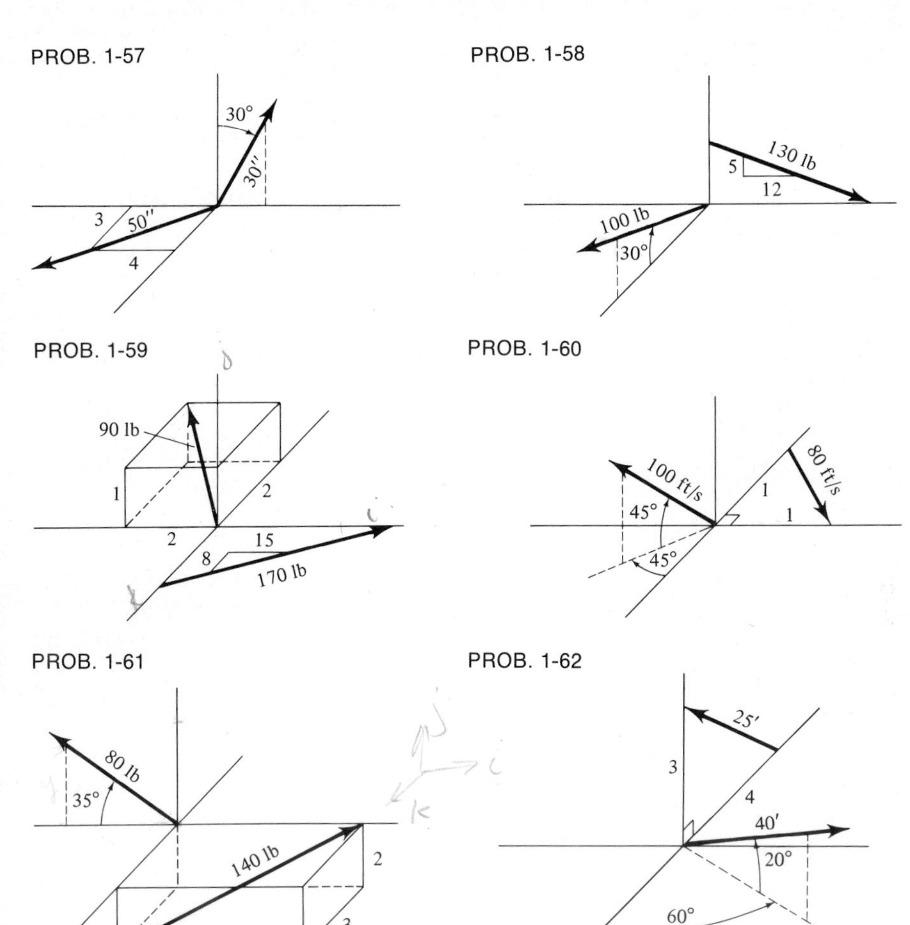

1-63 and 1-64 The mast is supported by the cables as shown. Write an expression for a unit vector parallel to cable *a-b* and one parallel to cable *a-c*. What is the angle between these cables?

PROBS. 1-63 AND 1-65

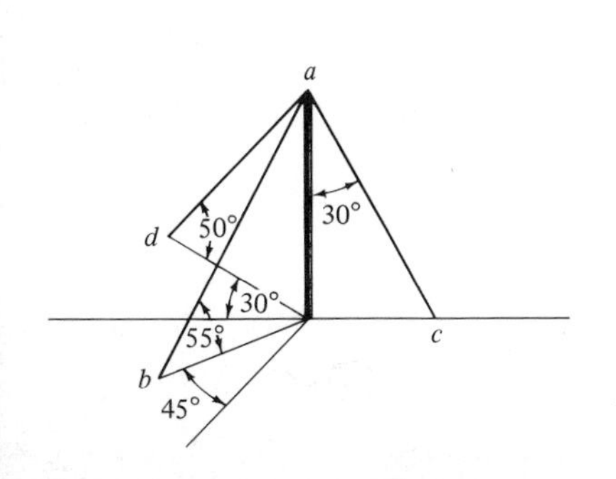

PROBS. 1-64 AND 1-66

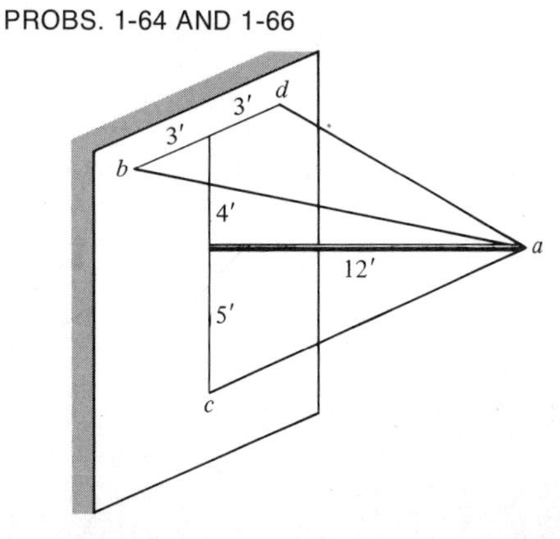

1-65 Same as Prob. 1-63 except replace *a-c* with *a-d*.

1-66 Same as Prob. 1-64 except replace *a-c* with *a-d*.

1-67 and 1-68 In the figure, $\boldsymbol{\alpha}$ is a force vector and $\boldsymbol{\mu}$ is a unit vector. Find the projection of $\boldsymbol{\alpha}$ onto $\boldsymbol{\mu}$. Write $\boldsymbol{\alpha}$ as the sum of two vectors $\boldsymbol{\gamma}$ and $\boldsymbol{\delta}$, $\boldsymbol{\gamma}$ being parallel to $\boldsymbol{\mu}$ and $\boldsymbol{\delta}$ being perpendicular to $\boldsymbol{\mu}$. Find $\boldsymbol{\gamma}$ and $\boldsymbol{\delta}$.

PROB. 1-67

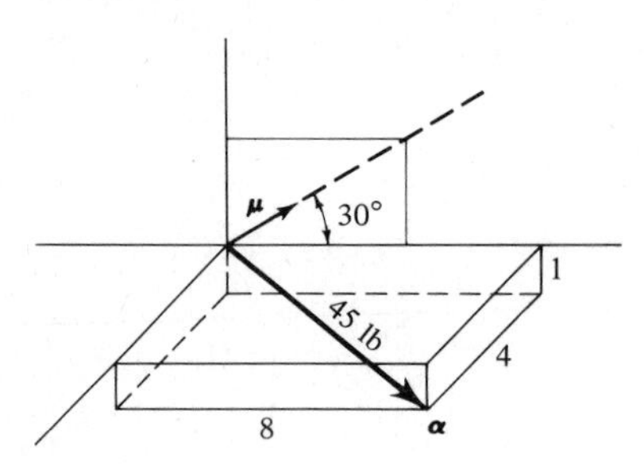

PROB. 1-68

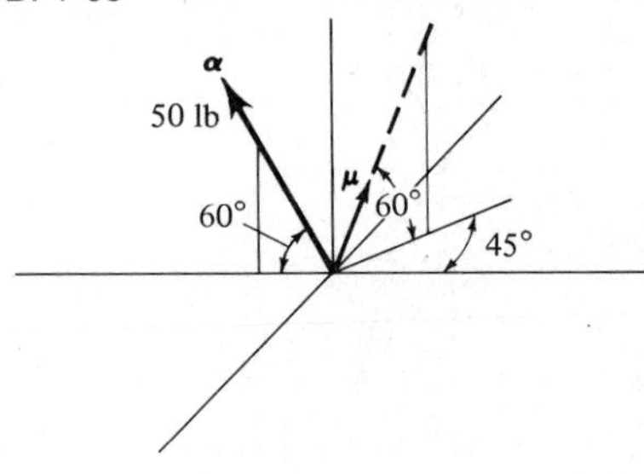

chapter 2 EQUIVALENT FORCES AND SYSTEMS

2-1 INTRODUCTION

The dictionary defines equivalent as "having the same value." This definition suffices for quantities which can be completely described by a number, but even in these cases the definition is often insufficient. For example, is a dozen small eggs equivalent to a dozen extra large ones? Maybe, maybe not, depending upon other factors such as price, use, quality, etc. Hence, equivalence is more often a matter of comparison of various factors rather than just equating numbers.

The previous chapter demonstrated that, when comparing two forces, one has to take into consideration their directions as well as magnitudes. However, two forces of equal magnitude and direction need not be "physically equivalent," as demonstrated by the situation in Fig. 2-1, where a 20-lb downward force on the board produces two distinctly different physical effects.

In this chapter we will attempt to define equivalence of forces in terms of certain physical effects. As a matter of fact, much of what we do in the remainder of this course will deal with equivalence of forces and force systems.

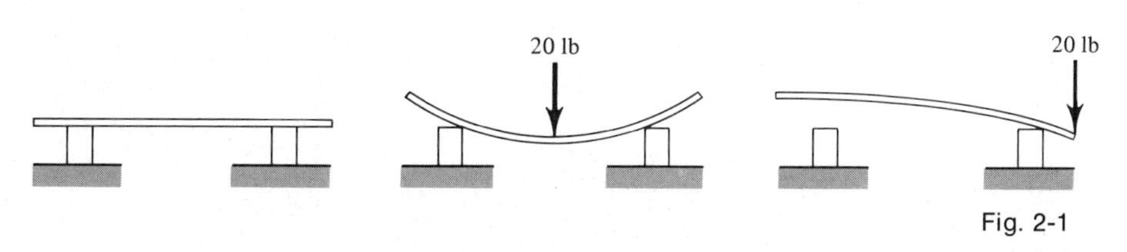

Fig. 2-1

2-2 MOMENT OF A FORCE

Consider the physical situation depicted in Fig. 2-2. Assuming the men are of equal strength and each exerting, say, 50 lb, these 50-lb forces are not equivalent in that the man in Fig. 2-2(*c*) has the best chance of opening the valve on the fire hydrant. Why? Because he has the best "leverage" on the valve. This leverage is not just a matter of how much force he can exert but is also governed by the length of the handle on the pipe wrench. Thus the ability of a force to produce a turning or twisting effect at a given point is governed by the force itself and its location relative to the given point. In the example just cited, the point in question is the valve of the hydrant, and the location of the force is where the force is applied on the wrench handle. We now formalize this idea with the following definition.

Fig. 2-2

Definition 2-1 The *position* vector of a force relative to a given point in space is the directed line segment **p** *from* the point *to* the location where the force is applied.

Looking down on the top of the hydrant we can schematically represent the physical situation as in Fig. 2-3.

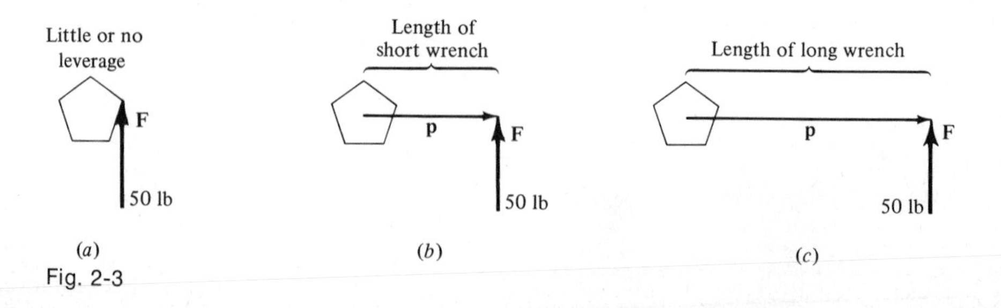

Fig. 2-3

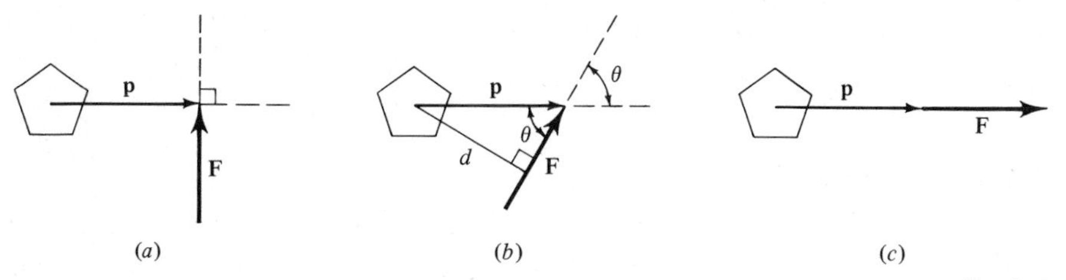

Fig. 2-4

We now pose the question: What is the relation between the force vector, its position vector relative to some point, and the turning or twisting effect it produces at that point? To answer this question, suppose the 50-lb force in Fig. 2-3 had not been directed perpendicular to the wrench. Would the turning effect be the same for the two situations in Fig. 2-4(*a*) and (*b*)? Our experience tells us that the answer would be no, and that the perpendicularity is a vital factor in the ability of a force to produce a turning effect. In fact, our experience tells us that the situation in Fig. 2-4(*c*) will produce no turning effect. Returning to Fig. 2-4(*b*) we see that the length d of the perpendicular distance between the valve and the direct line segment representing the force is given by

$$d = \|\mathbf{p}\| \sin \theta$$

where θ is the angle between $\mathbf{p}$ and $\mathbf{F}$, with $0 \leq \theta \leq 180°$. Hence, guided by our experience we can calculate the turning effect by

$$T = Fd = \|\mathbf{F}\| \, \|\mathbf{p}\| \sin \theta \qquad (2\text{-}1)$$

which agrees with the observations made previously in that when $\theta = 90°$ (perpendicular), as in Fig. 2-4(*a*), the turning effect is maximum, and when $\theta = 0$ (parallel), as in Fig. 2-4(*c*), the turning effect is zero.

While it might seem that Eq. (2-1) is quite satisfactory when discussing turning or rotational effects, it does have some deficiencies and ambiguities. First of all, suppose the force is reversed as in Fig. 2-5(*a*). Then, physically the turning effect is reversed

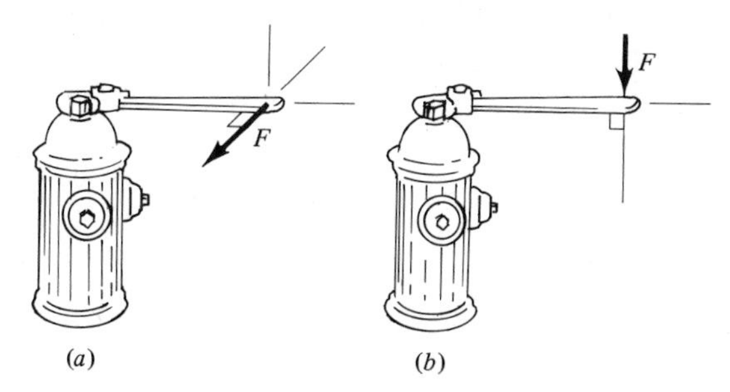

Fig. 2-5

although the perpendicularity is maintained; that is, the man now tends to tighten the valve rather than loosen it. As a second situation, suppose in Fig. 2-5(*b*) the man pushes down on the wrench toward the ground rather than parallel to the ground. Obviously, the force and position vector are still perpendicular, but, just as obvious, this effect will not open or close the valve. Hence, Eq. (2-1) will not suffice to accurately describe these situations, but as we shall soon see, the following formal definition is physically reasonable and yet avoids the previously observed difficulties.

Definition 2-2 Let $\mathbf{p}$ be the position vector of a force $\mathbf{F}$ relative to some given point q. The moment of the force $\mathbf{F}$ with respect to the point q is the *vector* $\mathbf{M}_q(\mathbf{F})$ whose magnitude is given by

$$\|\mathbf{M}_q(\mathbf{F})\| = Fd = \|\mathbf{p}\|\ \|\mathbf{F}\| \sin(\mathbf{p},\mathbf{F}) \tag{2-2}$$

and whose direction is such that $\mathbf{M}_q(\mathbf{F})$ is perpendicular to both $\mathbf{p}$ and $\mathbf{F}$, so that $\mathbf{p}$, $\mathbf{F}$, $\mathbf{M}_q(\mathbf{F})$ is a right-hand triple. The angle $(\mathbf{p},\mathbf{F})$ is always taken to be less than 180°. Formally, we write

$$\mathbf{M}_q(\mathbf{F}) = \mathbf{p} \times \mathbf{F} \tag{2-3}$$

where the $\times$ denotes the cross or vector product of the two vectors $\mathbf{p}$ and $\mathbf{F}$.

Before delving into a discussion of the vector product, let us briefly return to the hydrant example. Figure 2-6 illustrates three situations previously discussed. In Fig. 2-6(*a*) the double-headed vector represents the direction of $\mathbf{p} \times \mathbf{F}$ since $\mathbf{p}$, $\mathbf{F}$, $\mathbf{M}_q(\mathbf{F})$ form a right-hand triple. The phrase "right-hand triple" can be interpreted in several ways. One way is to visually rotate a vector pointing in the direction of $\mathbf{p}$ through the angle $(\mathbf{p},\mathbf{F})$ less than 180° until it points in the direction of $\mathbf{F}$. The direction of $\mathbf{M}_q(\mathbf{F})$ is that

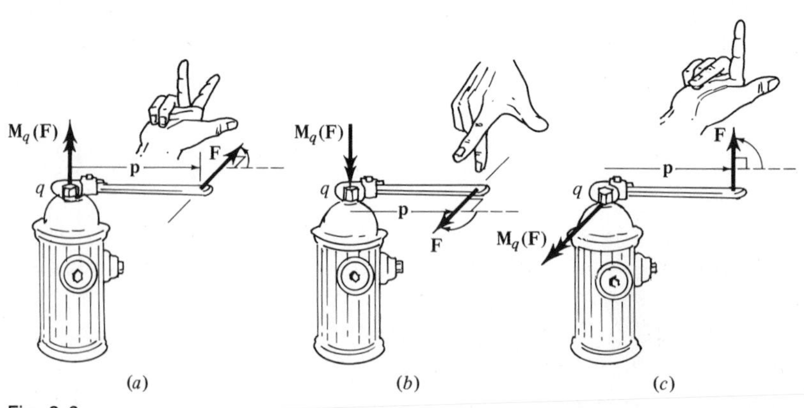

Fig. 2-6

which a *right-hand* screw would advance if rotated in this manner. A second method is to point the thumb of your *right hand* in the direction of **p**, your index finger in the direction of **F**, and your middle finger perpendicular to your thumb and index finger. Your middle finger then points in the direction of **p** × **F**. Thus, in Fig. 2-6(*a*) the moment vector tends to loosen the valve (if it has a right-handed thread), while in Fig. 2-6(*b*) it tends to tighten the valve; in Fig. 2-6(*c*) it tends to bend the valve rather than tighten or loosen it.

Remark: Although, in the above examples, **p** and **F** are perpendicular, in general they need not be, as in Fig. 2-4(*b*). However, $\mathbf{M}_q(\mathbf{F})$ will *always* be perpendicular to both **p** and **F**.

Although Definition 2-2 formally defines the moment of a force as a vector, physically you should think of this vector as a tendency to rotate or cause rotation just as you think of a force as a tendency to push or pull. The direction of the moment vector specifies the axis of rotation, and the arrow on the moment vector tells us whether the rotational tendency is clockwise or counterclockwise as we look along the axis from one of its ends.

Study Fig. 2-6 carefully to be sure you understand the physical significance of the moment vectors.

EXAMPLE 2-1

Determine the moment with respect to the left end of the beam of each of the forces shown in Fig. 2-7.

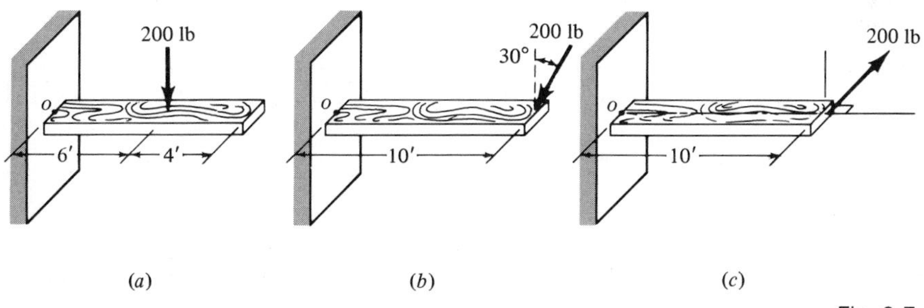

Fig. 2-7

Solutions: (*a*) For this case, p, **F**, and $\mathbf{M}_o(\mathbf{F})$ are as shown in Fig. 2-8(*a*), in which **p** is directed to the right, **F** is directed downward, and $\mathbf{M}_o(\mathbf{F})$ is directed inward ($-\mathbf{k}$ direction). By Eqs. (2-2) and (2-3),

$$\|\mathbf{M}_o(\mathbf{F})\| = \|\mathbf{p}\|\,\|\mathbf{F}\| \sin 90^\circ$$
$$= (6)(200)(1) = 1{,}200 \text{ ft-lb}$$
$$\mathbf{M}_o(\mathbf{F}) = \mathbf{p} \times \mathbf{F} = -1{,}200\mathbf{k} \text{ ft-lb}$$

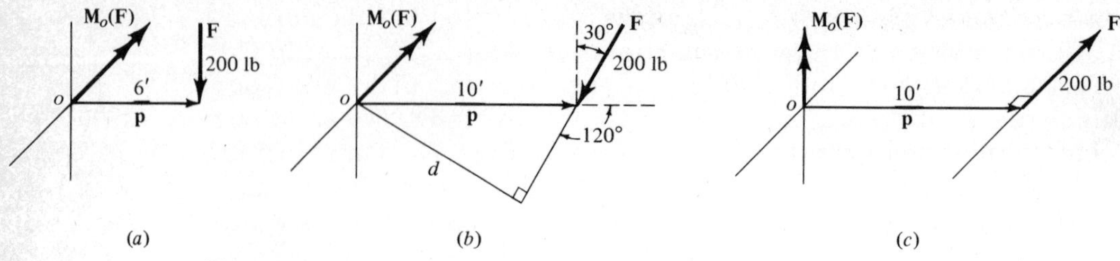

Fig. 2-8

(*b*) This case is shown in Fig. 2-8(*b*), where $\mathbf{M}_o(\mathbf{F})$ is again a vector directed inward ($-\mathbf{k}$ direction). Note that perpendicular distance d lies in the vertical (i–j) plane and the angle from $\mathbf{p}$ to $\mathbf{F}$ is 120°.

$$\|\mathbf{M}_o(\mathbf{F})\| = (10)(200)\sin 120° = 1{,}732 \text{ ft-lb}$$

$$\mathbf{M}_o(\mathbf{F}) = -1{,}732\mathbf{k} \text{ ft-lb}$$

(*c*) This case is shown in Fig. 2-8(*c*), in which $\mathbf{F}$ is directed inward ($-\mathbf{k}$ direction) and $\mathbf{M}_o(\mathbf{F})$ is directed upward.

$$\|\mathbf{M}_o(\mathbf{F})\| = (10)(200)\sin 90° = 2{,}000 \text{ ft-lb}$$

$$\mathbf{M}_o(\mathbf{F}) = 2{,}000\mathbf{j} \text{ ft-lb}$$

Satisfy yourself on the directions of $\mathbf{M}_o(\mathbf{F})$ by using the right-hand rule.

EXERCISE PROBLEMS

2-1 to 2-10 For each of the situations shown, determine the moment vector of the force with respect to point *o*.

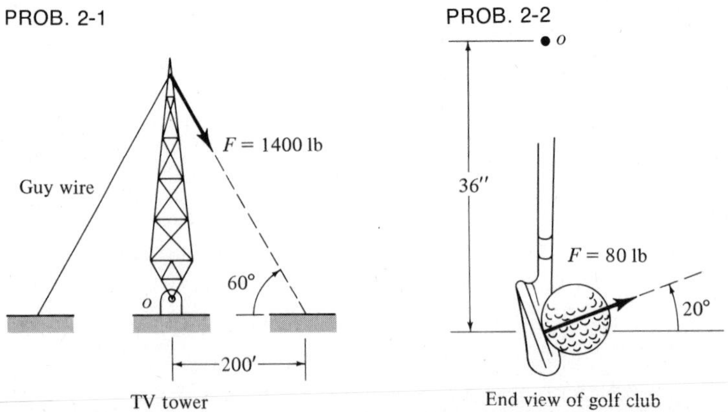

PROB. 2-3

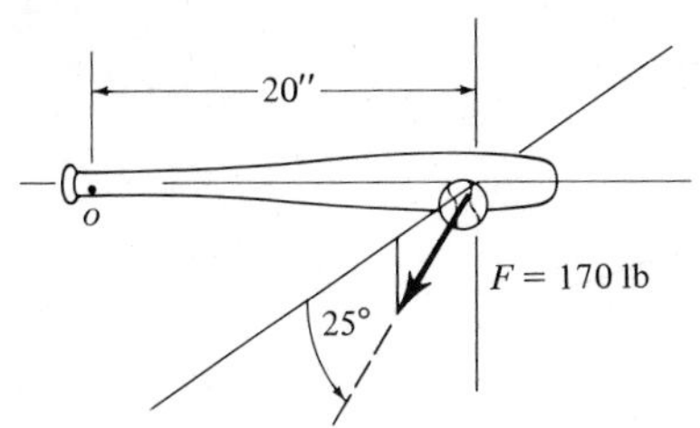

Bat hitting ground ball

PROB. 2-4

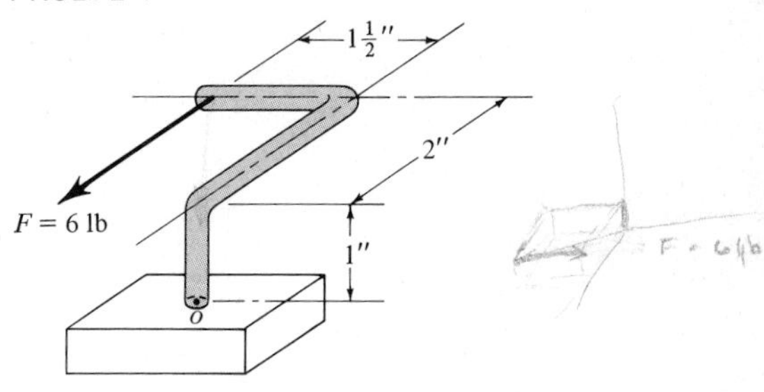

PROB. 2-5

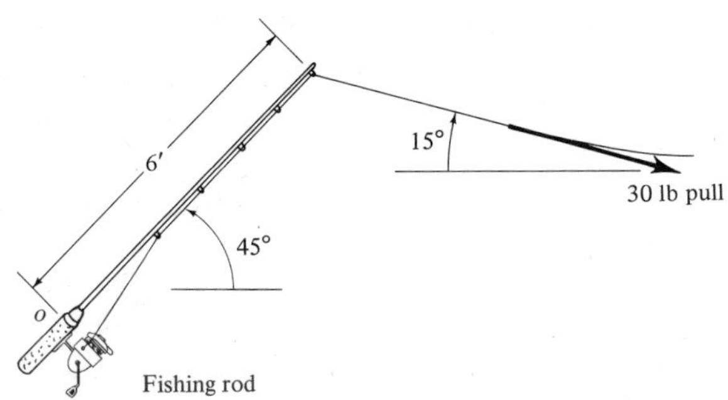

Fishing rod

PROB. 2-6

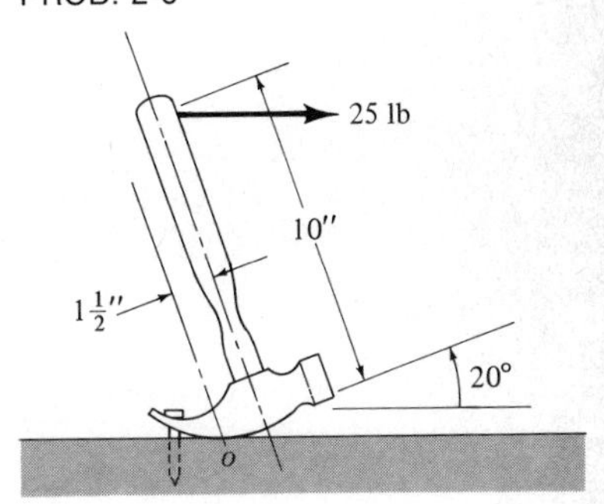

Pulling a nail

PROB. 2-7

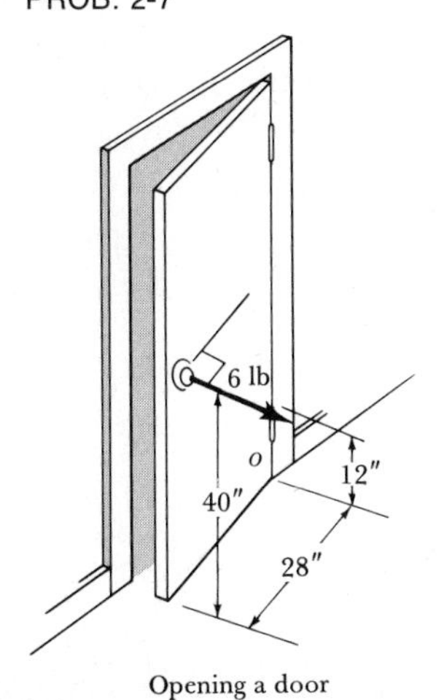

Opening a door

PROB. 2-8

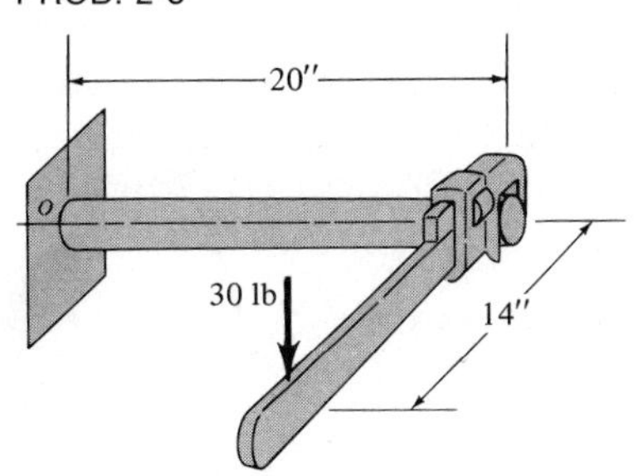

Pushing on pipe wrench

PROB. 2-9

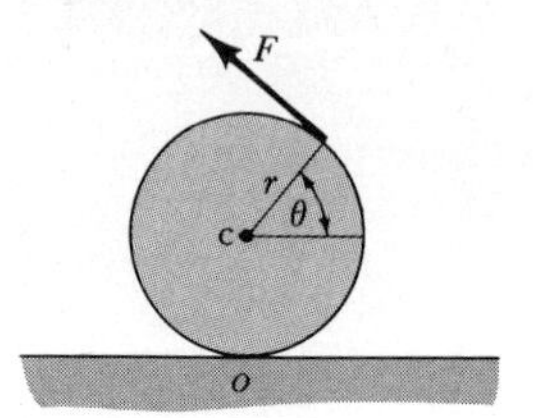

Rolling cylinder

PROB. 2-10

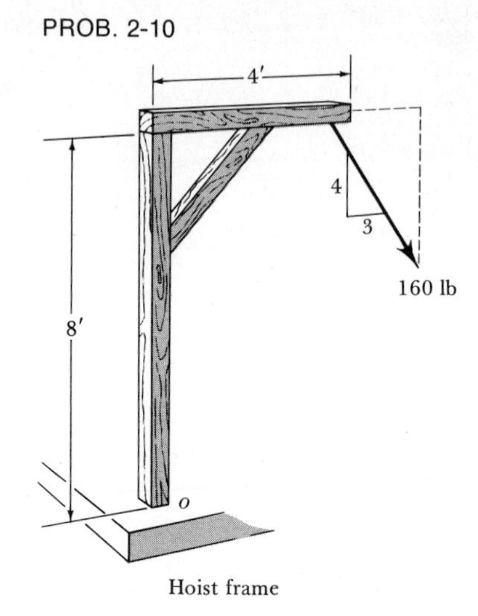

Hoist frame

2-3 PROPERTIES OF THE CROSS PRODUCT

The cross or vector product was defined in Definition 2-2 in terms of the moment of a force. However, using an analogous definition, one can define the cross product of any two vectors $\boldsymbol{\alpha}$ and $\boldsymbol{\beta}$, say, as

$$\boldsymbol{\alpha} \times \boldsymbol{\beta} = \boldsymbol{\gamma}$$

where

$$\|\boldsymbol{\gamma}\| = \|\boldsymbol{\alpha}\| \, \|\boldsymbol{\beta}\| \sin(\boldsymbol{\alpha},\boldsymbol{\beta})$$

and $\boldsymbol{\alpha}$, $\boldsymbol{\beta}$, $\boldsymbol{\gamma}$ form a right-hand triple. We now make some important observations.

1. If $\boldsymbol{\alpha} \times \boldsymbol{\beta} = \boldsymbol{\gamma}$, then $\boldsymbol{\beta} \times \boldsymbol{\alpha} = -\boldsymbol{\gamma}$, that is, the vector product is noncommutative. This follows from the fact that if $\boldsymbol{\alpha}$, $\boldsymbol{\beta}$, $\boldsymbol{\gamma}$ is a right-hand triple, $\boldsymbol{\beta}$, $\boldsymbol{\alpha}$, $\boldsymbol{\gamma}$ will be a left-hand triple. Satisfy yourself on this fact by using one or both of the right-hand devices mentioned previously.

2. If $\boldsymbol{\alpha}$ and $\boldsymbol{\beta}$ have parallel directions

$$\boldsymbol{\alpha} \times \boldsymbol{\beta} = \mathbf{0}$$

since the angle between parallel vectors is 0 or 180°, and by Eq. (2-2)

$$\|\boldsymbol{\alpha} \times \boldsymbol{\beta}\| = \|\boldsymbol{\alpha}\| \, \|\boldsymbol{\beta}\| \sin(\boldsymbol{\alpha},\boldsymbol{\beta}) = 0$$

3. As a special case of (2) we observe that

$$\boldsymbol{\alpha} \times \boldsymbol{\alpha} = \mathbf{0}$$

4 If c is a scalar

$$c(\boldsymbol{\alpha} \times \boldsymbol{\beta}) = c\boldsymbol{\alpha} \times \boldsymbol{\beta} = \boldsymbol{\alpha} \times c\boldsymbol{\beta}$$

5 If $\boldsymbol{\alpha}$, $\boldsymbol{\beta}$, and $\boldsymbol{\gamma}$ are vectors, then

$$\boldsymbol{\alpha} \times (\boldsymbol{\beta} + \boldsymbol{\gamma}) = \boldsymbol{\alpha} \times \boldsymbol{\beta} + \boldsymbol{\alpha} \times \boldsymbol{\gamma}$$
$$(\boldsymbol{\alpha} + \boldsymbol{\beta}) \times \boldsymbol{\gamma} = \boldsymbol{\alpha} \times \boldsymbol{\gamma} + \boldsymbol{\beta} \times \boldsymbol{\gamma}$$

which are the distribution laws for the vector product. Geometric proofs of (4) and (5) are left as exercises (Probs. 2-11 and 2-12).

Although the formal definition of the cross product of two vectors is quite useful in solving physical problems, the actual determination of the vector $\boldsymbol{\alpha} \times \boldsymbol{\beta}$ can be quite difficult, particularly for two arbitrarily directed vectors in 3-space. To alleviate this difficulty, let $\boldsymbol{\alpha}$ and $\boldsymbol{\beta}$ be written in terms of their orthogonal components relative to some **i, j, k** right-handed orthogonal triad of unit vectors, so that

$$\boldsymbol{\alpha} = \alpha_1\mathbf{i} + \alpha_2\mathbf{j} + \alpha_3\mathbf{k}$$
$$\boldsymbol{\beta} = \beta_1\mathbf{i} + \beta_2\mathbf{j} + \beta_3\mathbf{k}$$

Then

$$\boldsymbol{\alpha} \times \boldsymbol{\beta} = (\alpha_1\mathbf{i} + \alpha_2\mathbf{j} + \alpha_3\mathbf{k}) \times (\beta_1\mathbf{i} + \beta_2\mathbf{j} + \beta_3\mathbf{k})$$

Recall that **i, j, k** is a right-hand orthogonal unit triad. Then, by virtue of observation (3)

$$\mathbf{i} \times \mathbf{i} = \mathbf{j} \times \mathbf{j} = \mathbf{k} \times \mathbf{k} = \mathbf{0}$$

but

$$\mathbf{i} \times \mathbf{j} = \mathbf{k} \qquad \mathbf{j} \times \mathbf{k} = \mathbf{i} \qquad \mathbf{k} \times \mathbf{i} = \mathbf{j}$$

as illustrated in Fig. 2-9. Using these and the distributive laws for vector products we now have

$$\boldsymbol{\alpha} \times \boldsymbol{\beta} = (\alpha_2\beta_3 - \alpha_3\beta_2)\mathbf{i} + (\alpha_3\beta_1 - \alpha_1\beta_3)\mathbf{j} + (\alpha_1\beta_2 - \alpha_2\beta_1)\mathbf{k} \qquad (2\text{-}4)$$

At first you might think that Eq. (2-4) is more difficult to use than the original definition of the vector product. However, this difficulty is often overcome by using the following mnemonic device. Write the 3×3 matrix having **i, j, k** as the first row; α_1, α_2, α_3 as the

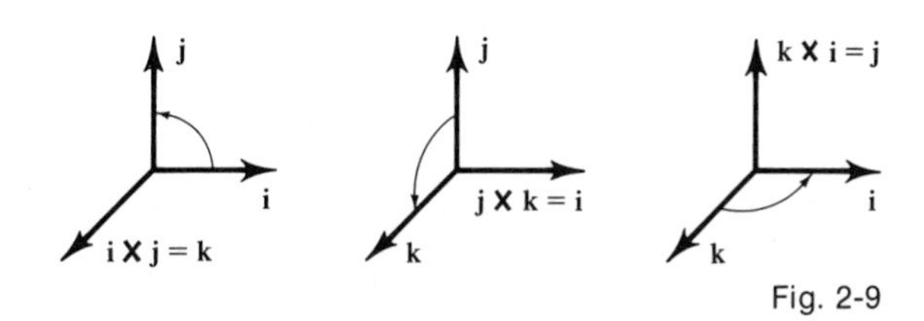

Fig. 2-9

second row; and $\beta_1, \beta_2, \beta_3$ as the third row, and then perform a *first-row* determinant expansion. In this manner, we obtain

$$\det \begin{vmatrix} \mathbf{i} & \mathbf{j} & \mathbf{k} \\ \alpha_1 & \alpha_2 & \alpha_3 \\ \beta_1 & \beta_2 & \beta_3 \end{vmatrix} = \mathbf{i}(\alpha_2\beta_3 - \alpha_3\beta_2) + \mathbf{j}(\alpha_3\beta_1 - \alpha_1\beta_3) + \mathbf{k}(\alpha_1\beta_2 - \alpha_2\beta_1) \tag{2-5}$$

which is precisely Eq. (2-4). Hence Eq. (2-5) is a very useful algorithm for calculating the cross product of two vectors.

Scalar Triple Product

We close this section with another useful algorithm. Since $\boldsymbol{\alpha} \times \boldsymbol{\beta}$ is a vector, we can take its inner product with another vector $\boldsymbol{\gamma}$, say, to obtain

$$\boldsymbol{\gamma} \cdot (\boldsymbol{\alpha} \times \boldsymbol{\beta})$$

which is called the scalar triple product of the vectors $\boldsymbol{\alpha}$, $\boldsymbol{\beta}$, and $\boldsymbol{\gamma}$. According to Eq. (1-2), this can be written in terms of orthogonal components as

$$\begin{aligned} \boldsymbol{\gamma} \cdot (\boldsymbol{\alpha} \times \boldsymbol{\beta}) &= \gamma_1(\boldsymbol{\alpha} \times \boldsymbol{\beta})_1 + \gamma_2(\boldsymbol{\alpha} \times \boldsymbol{\beta})_2 + \gamma_3(\boldsymbol{\alpha} \times \boldsymbol{\beta})_3 \\ &= \gamma_1(\alpha_2\beta_3 - \alpha_3\beta_2) + \gamma_2(\alpha_3\beta_1 - \alpha_1\beta_3) + \gamma_3(\alpha_1\beta_2 - \alpha_2\beta_1) \end{aligned}$$

But, if we expand the determinant, we obtain

$$\det \begin{vmatrix} \gamma_1 & \gamma_2 & \gamma_3 \\ \alpha_1 & \alpha_2 & \alpha_3 \\ \beta_1 & \beta_2 & \beta_3 \end{vmatrix} = \gamma_1(\alpha_2\beta_3 - \alpha_3\beta_2) + \gamma_2(\alpha_3\beta_1 - \alpha_1\beta_3) + \gamma_3(\alpha_1\beta_2 - \alpha_2\beta_1)$$

Hence we arrive at the algorithm

$$\boldsymbol{\gamma} \cdot (\boldsymbol{\alpha} \times \boldsymbol{\beta}) = \det \begin{vmatrix} \gamma_1 & \gamma_2 & \gamma_3 \\ \alpha_1 & \alpha_2 & \alpha_3 \\ \beta_1 & \beta_2 & \beta_3 \end{vmatrix} \tag{2-6}$$

for the scalar triple product of three vectors. By utilizing the properties of a determinant, it is a relatively simple matter to show that

(1) $$\boldsymbol{\gamma} \cdot (\boldsymbol{\alpha} \times \boldsymbol{\beta}) = (\boldsymbol{\gamma} \times \boldsymbol{\alpha}) \cdot \boldsymbol{\beta} \tag{2-7}$$

that is, the dot and cross operators may be interchanged, and

(2) $$\begin{aligned} \boldsymbol{\gamma} \cdot (\boldsymbol{\alpha} \times \boldsymbol{\beta}) &= \boldsymbol{\alpha} \cdot (\boldsymbol{\beta} \times \boldsymbol{\gamma}) = \boldsymbol{\beta} \cdot (\boldsymbol{\gamma} \times \boldsymbol{\alpha}) \\ &= -\boldsymbol{\gamma} \cdot (\boldsymbol{\beta} \times \boldsymbol{\alpha}) = -\boldsymbol{\alpha} \cdot (\boldsymbol{\gamma} \times \boldsymbol{\beta}) = -\boldsymbol{\beta} \cdot (\boldsymbol{\alpha} \times \boldsymbol{\gamma}) \end{aligned} \tag{2-8}$$

that is, preserving the cyclic order of $\boldsymbol{\alpha}$, $\boldsymbol{\beta}$, $\boldsymbol{\gamma}$ preserves the sign of their scalar triple product, but changing their cyclic order changes the sign of their scalar triple product.

As a final observation we see that

$$\boldsymbol{\alpha} \cdot (\boldsymbol{\alpha} \times \boldsymbol{\beta}) = 0 \tag{2-9}$$

since $\boldsymbol{\alpha} \times \boldsymbol{\beta}$ must be perpendicular to both $\boldsymbol{\alpha}$ and $\boldsymbol{\beta}$ by the definition of the cross product. This same result is obvious from Eq. (2-6) since two rows of the matrix would be identical, thus making its determinant vanish.

This concludes our formal discussion of the cross product. In subsequent articles these results will be used in obtaining answers to physical problems; but first you must gain experience in the manipulations involved in using cross products.

EXAMPLE 2-2

For the situation shown in Fig. 2-10(*a*), find (*a*) the moment of the force with respect to point *o*, (*b*) the moment with respect to point *q*, and (*c*) the perpendicular distance between *o* and the line of action of the force.

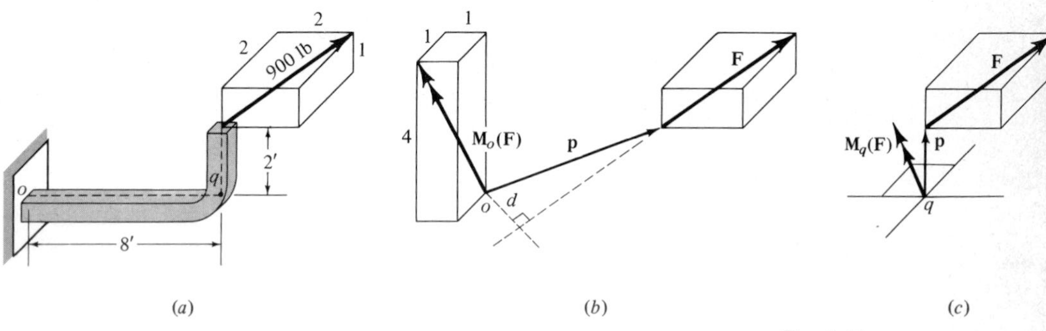

Fig. 2-10

Solution: First, we write **F** in terms of orthogonal components relative to a natural triad. Since the diagonal of the box is

$$\sqrt{2^2 + 2^2 + 1^2} = 3$$

we have
$$\begin{aligned} \mathbf{F} &= \tfrac{2}{3}\,900\mathbf{i} + \tfrac{1}{3}\,900\mathbf{j} - \tfrac{2}{3}\,900\mathbf{k} \\ &= 600 \text{ lb } \mathbf{i} + 300 \text{ lb } \mathbf{j} - 600 \text{ lb } \mathbf{k} \end{aligned}$$

(*a*) The position of **F** relative to point *o* is

$$\mathbf{p} = 8 \text{ ft } \mathbf{i} + 2 \text{ ft } \mathbf{j} + 0 \text{ ft } \mathbf{k}$$

Hence

$$\mathbf{p} \times \mathbf{F} = \det \begin{vmatrix} \mathbf{i} & \mathbf{j} & \mathbf{k} \\ 8 & 2 & 0 \\ 600 & 300 & -600 \end{vmatrix} = -1{,}200\mathbf{i} + 4{,}800\mathbf{j} + 1{,}200\mathbf{k}$$

$$\|\mathbf{p} \times \mathbf{F}\| = 3{,}600\sqrt{2} \text{ ft-lb}$$

This vector is shown in Fig. 2-10(*b*).

(*b*) The position of **F** relative to point *q* is

$$\mathbf{p} = 0 \text{ ft } \mathbf{i} + 2 \text{ ft } \mathbf{j} + 0 \text{ ft } \mathbf{k}$$

Hence

$$\mathbf{p} \times \mathbf{F} = \det \begin{vmatrix} \mathbf{i} & \mathbf{j} & \mathbf{k} \\ 0 & 2 & 0 \\ 600 & 300 & -600 \end{vmatrix} = -1{,}200\mathbf{i} + 0\mathbf{j} - 1{,}200\mathbf{k}$$

$$\|\mathbf{p} \times \mathbf{F}\| = 1{,}200\sqrt{2} \text{ ft-lb}$$

This vector is shown in Fig. 2-10(*c*)

(*c*) By Eq. (2-1) the perpendicular distance is related to the magnitude of the moment by

$$T = \|\mathbf{M}_o(\mathbf{F})\| = Fd$$

Now, for point *o*, by (*a*)

$$\mathbf{M}_o(\mathbf{F}) = -1{,}200\mathbf{i} + 4{,}800\mathbf{j} + 1{,}200\mathbf{k}$$

$$\|\mathbf{M}_o(\mathbf{F})\| = 1{,}200\sqrt{(-1)^2 + (4)^2 + (1)^2}$$

$$= 3{,}600\sqrt{2} \text{ ft-lb}$$

thus

$$d = \frac{3{,}600\sqrt{2}}{900} = 4\sqrt{2} \text{ ft}$$

and is shown in Fig. 2-10(*b*).

In the example just completed, the moment vectors were given in terms of orthogonal vector components. Since the moment vector itself has a tendency to produce rotation, its vector components can also be physically interpreted as tendencies to rotate. Thus, the moment vector in (*a*) of Example 2-2 can be interpreted in terms of the rotational effects indicated in Fig. 2-11. Physically, this moment is tending to twist the member about its longitudinal axis, while at the same time it is tending to bend the member as shown in Fig. 2-11(*c*).

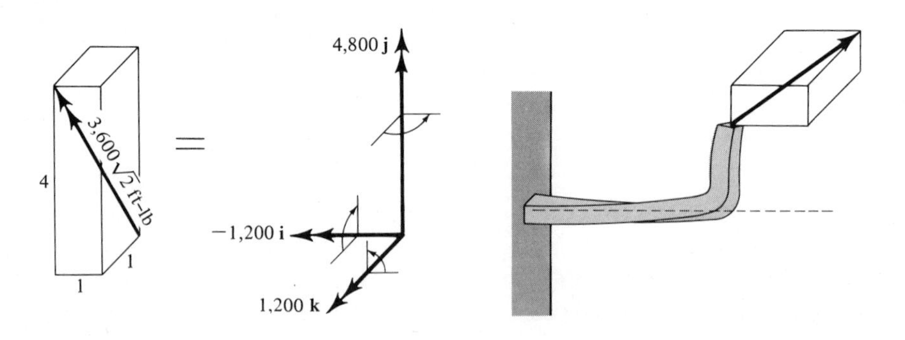

Fig. 2-11

EXERCISE PROBLEMS

2-11 Give a geometric proof of the equation

$$c(\boldsymbol{\alpha} \times \boldsymbol{\beta}) = c\boldsymbol{\alpha} \times \boldsymbol{\beta} = \boldsymbol{\alpha} \times c\boldsymbol{\beta}$$

where c is a scalar.

2-12 Give a geometric proof of the distributive law

$$\boldsymbol{\alpha} \times (\boldsymbol{\beta} + \boldsymbol{\gamma}) = \boldsymbol{\alpha} \times \boldsymbol{\beta} + \boldsymbol{\alpha} \times \boldsymbol{\gamma}$$

2-13 Using properties of determinants prove Eq. (2-7).

2-14 Using properties of determinants prove Eq. (2-8).

2-15 to 2-20 For the given $\boldsymbol{\alpha}$, $\boldsymbol{\beta}$, $\boldsymbol{\gamma}$, determine (*a*) $\boldsymbol{\alpha} \times \boldsymbol{\beta}$, (*b*) $\boldsymbol{\alpha} \times \boldsymbol{\gamma}$, and (*c*) the scalar triple product $(\boldsymbol{\alpha} \times \boldsymbol{\beta}) \cdot \boldsymbol{\gamma}$.

2-15 $\boldsymbol{\alpha} = 2\mathbf{i} + 2\mathbf{j} - \mathbf{k}$, $\boldsymbol{\beta} = 3\mathbf{i} + 0\mathbf{j} + 2\mathbf{k}$, $\boldsymbol{\gamma} = 0\mathbf{i} + \mathbf{j} - \mathbf{k}$.

2-16 $\boldsymbol{\alpha} = 12\mathbf{i} - 3\mathbf{j} - 4\mathbf{k}$, $\boldsymbol{\beta} = 2\mathbf{i} + 2\mathbf{j} + 0\mathbf{k}$, $\boldsymbol{\gamma} = 2\mathbf{i} + 0\mathbf{j} + 2\mathbf{k}$.

2-17 $\boldsymbol{\alpha} = 3\mathbf{i} + \mathbf{j} - \mathbf{k}$, $\boldsymbol{\beta} = -\mathbf{i} - \mathbf{j} - \mathbf{k}$, $\boldsymbol{\gamma} = -6\mathbf{i} - 2\mathbf{j} + 2\mathbf{k}$.

2-18 $\boldsymbol{\alpha} = 4\mathbf{i} - 4\mathbf{j} - 2\mathbf{k}$, $\boldsymbol{\beta} = 3\mathbf{i} - \mathbf{j} + \mathbf{k}$, $\boldsymbol{\gamma} = 8\mathbf{i} - 8\mathbf{j} - 4\mathbf{k}$.

2-19 $\boldsymbol{\alpha} = -2\mathbf{i} + 3\mathbf{j} + 6\mathbf{k}$, $\boldsymbol{\beta} = 5\mathbf{i} + 4\mathbf{j} + 3\mathbf{k}$, $\boldsymbol{\gamma} = -2\mathbf{i} - 6\mathbf{j} + 3\mathbf{k}$.

2-20 $\boldsymbol{\alpha} = -\mathbf{i} + 2\mathbf{j} + 3\mathbf{k}$, $\boldsymbol{\beta} = 2\mathbf{i} + 2\mathbf{j} - 2\mathbf{k}$, $\boldsymbol{\gamma} = -\mathbf{i} - \mathbf{j} + \mathbf{k}$.

2-21 to 2-30 Using orthogonal components, rework Probs. 2-1 to 2-10.

2-31 to 2-40 For each of the situations shown, determine (*a*) the moment of the force with respect to point *o*, and (*b*) the perpendicular distance between point *o* and the line of action of the force.

PROB. 2-31

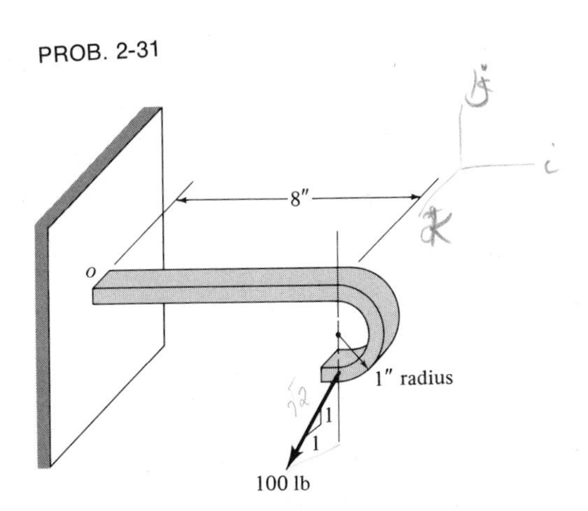

PROB. 2-32

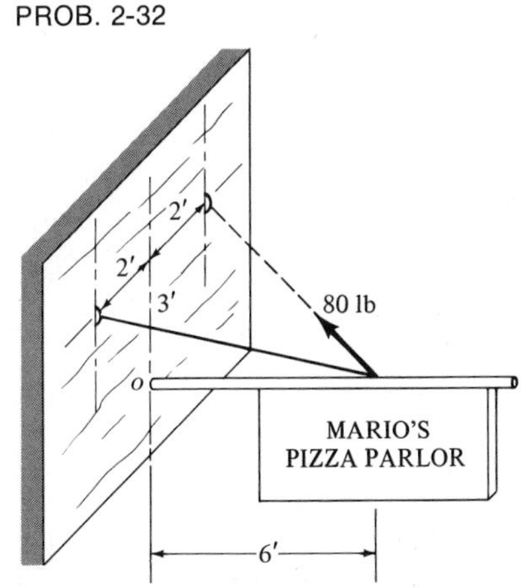

PROB. 2-33

Fishing rod

PROB. 2-34

Bat hitting foul ball

PROB. 2-35

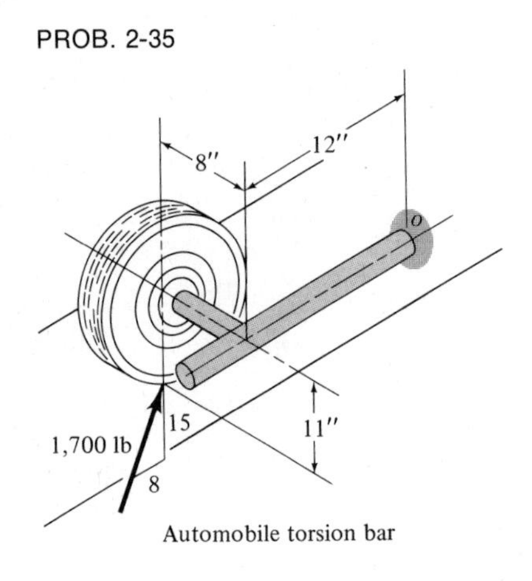

Automobile torsion bar

PROB. 2-36

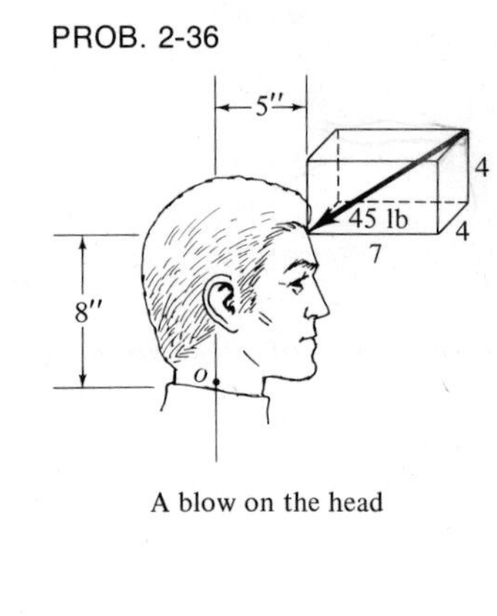

A blow on the head

PROB. 2-37

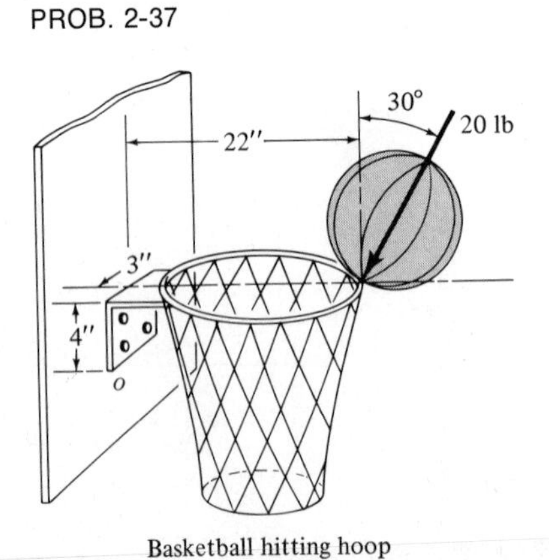

Basketball hitting hoop

PROB. 2-38

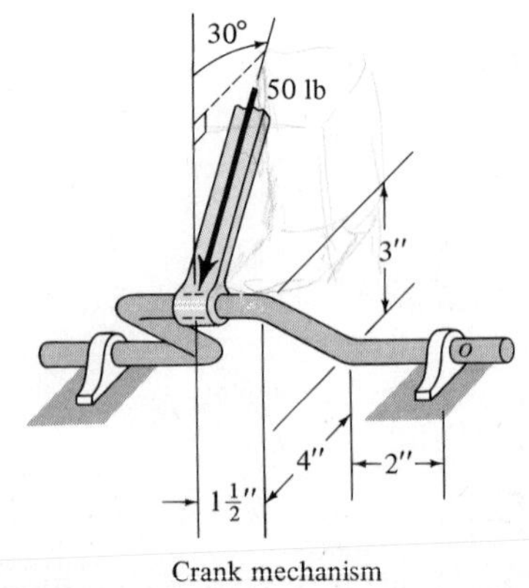

Crank mechanism

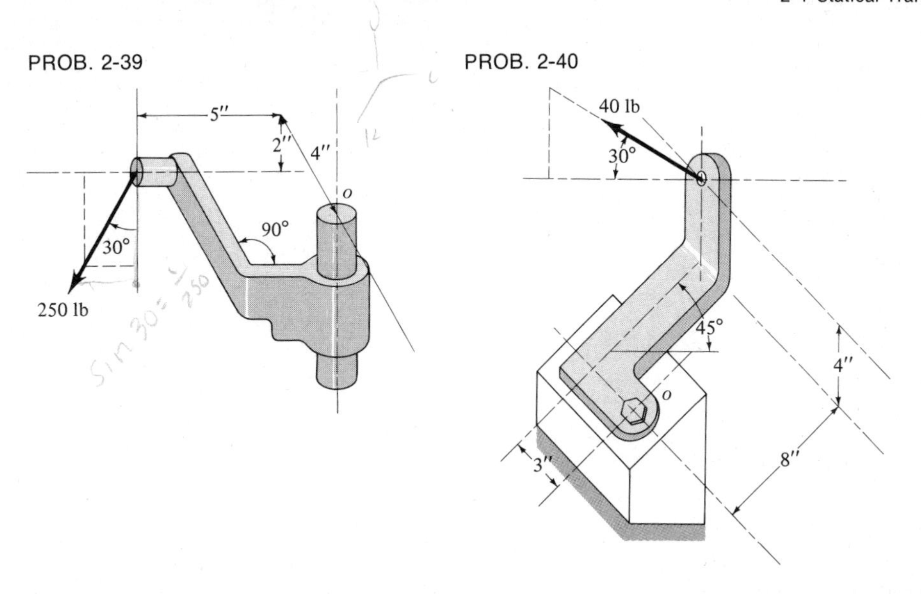

2-4 STATICAL TRANSFORMATIONS

Before we formally define what is meant by a statical transformation, let us make some simple but important observations. Previously in this chapter we have defined the moment of a force $\mathbf{F}$ with respect to an arbitrary point q as the vector product $\mathbf{p} \times \mathbf{F}$, where $\mathbf{p}$ is the position vector of $\mathbf{F}$ relative to point q. Thus, in general, changing the location of a force will change its moment with respect to a given point q since the position vector $\mathbf{p}$ will change, as illustrated in Fig. 2-12.

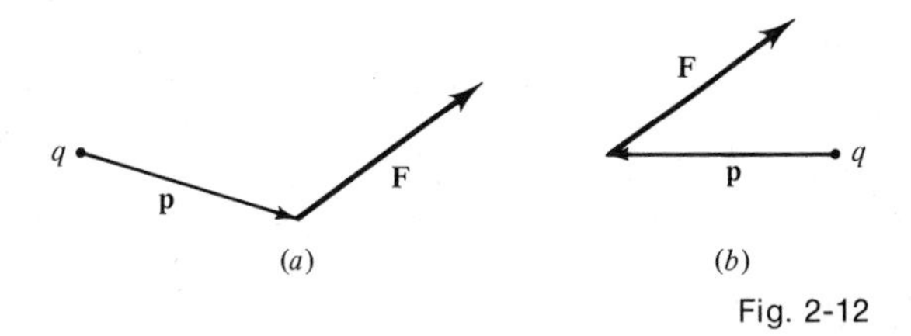

Fig. 2-12

However, consider the rather special situation depicted in Fig. 2-13(a) to (c). In each of these cases, with $\mathbf{F}$ located at a, b, and c, respectively, the moment of $\mathbf{F}$ with respect to q will be the same with

$$\|\mathbf{M}_q(\mathbf{F})\| = Fd$$

Hence, we make the following important observations.

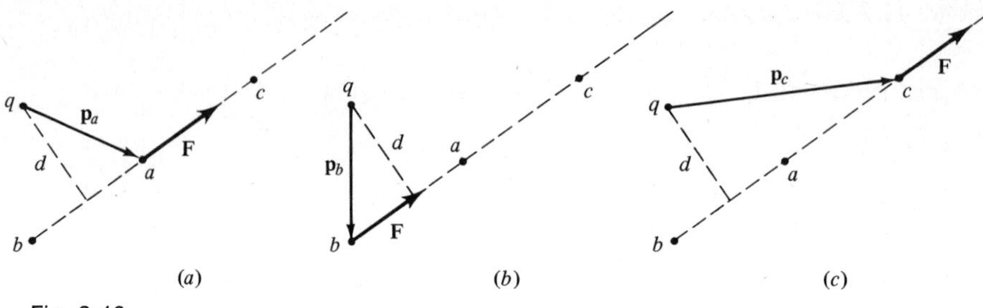

Fig. 2-13

> ***Observation 1:*** The location of a force can be translated along its line of action without changing its moment with respect to any point q.

As a second consideration, suppose that **F**, located at point p, is known to be the sum of two or more other forces, such as $\mathbf{F}'$ and $\mathbf{F}''$, as illustrated in Fig. 2-14.

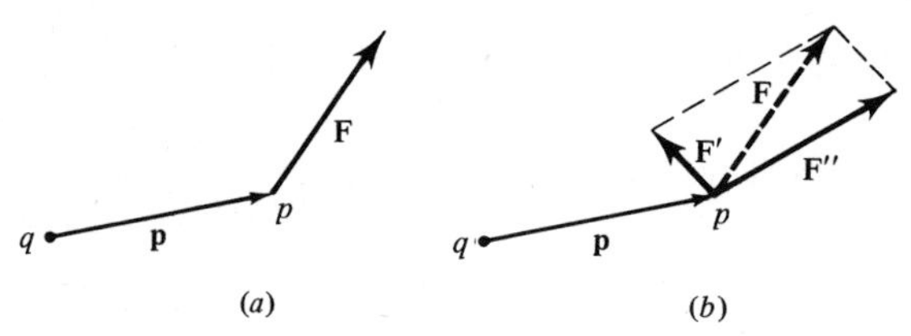

Fig. 2-14

Now,

$$\begin{aligned}\mathbf{M}_q(\mathbf{F}) &= \mathbf{p} \times \mathbf{F} = \mathbf{p} \times (\mathbf{F}' + \mathbf{F}'') \\ &= \mathbf{p} \times \mathbf{F}' + \mathbf{p} \times \mathbf{F}'' \\ &= \mathbf{M}_q(\mathbf{F}') + \mathbf{M}_q(\mathbf{F}'')\end{aligned}$$

where we have utilized the distributive law for vector products. Thus, the moment of **F** is the same as the sum of the moments of $\mathbf{F}'$ and $\mathbf{F}''$. In particular, suppose we write **F** as the sum of its orthogonal components relative to a natural triad, so that

$$\mathbf{F} = F_1\mathbf{i} + F_2\mathbf{j} + F_3\mathbf{k}$$

Then

$$\mathbf{M}_q(\mathbf{F}) = \mathbf{M}_q(F_1\mathbf{i}) + \mathbf{M}_q(F_2\mathbf{j}) + \mathbf{M}_q(F_3\mathbf{k})$$

Thus we are led to the following:

> ***Observation 2:*** The moment of a force **F** located at a point p is equal to the sum of the moments of its vector components, also located at p.

We now formalize these two observations with the following:

Definition 2-3 A *statical transformation* of $\mathbf{F}$ located at a point p is

(1) A statical translation—translation of $\mathbf{F}$ along its line of action.

(2) Resolution and composition—replacement of $\mathbf{F}$ at p by components $\mathbf{F}_1, \mathbf{F}_2, \ldots, \mathbf{F}_n$ (not necessarily orthogonal) at p, or vice versa, such that $\mathbf{F} = \mathbf{F}_1 + \mathbf{F}_2 + \cdots + \mathbf{F}_n$.

(3) A combination of (1) and (2).

Such transformations preserve the moments with respect to any point q. Physically this means that a statical transformation of a force does not change its turning effect.

Figure 2-15 illustrates various statical transformations of a given force. Satisfy yourself that each of these situations can be obtained from any of the others by a sequence or chain of statical transformations and that in all cases the moment with respect to point q remains unchanged.

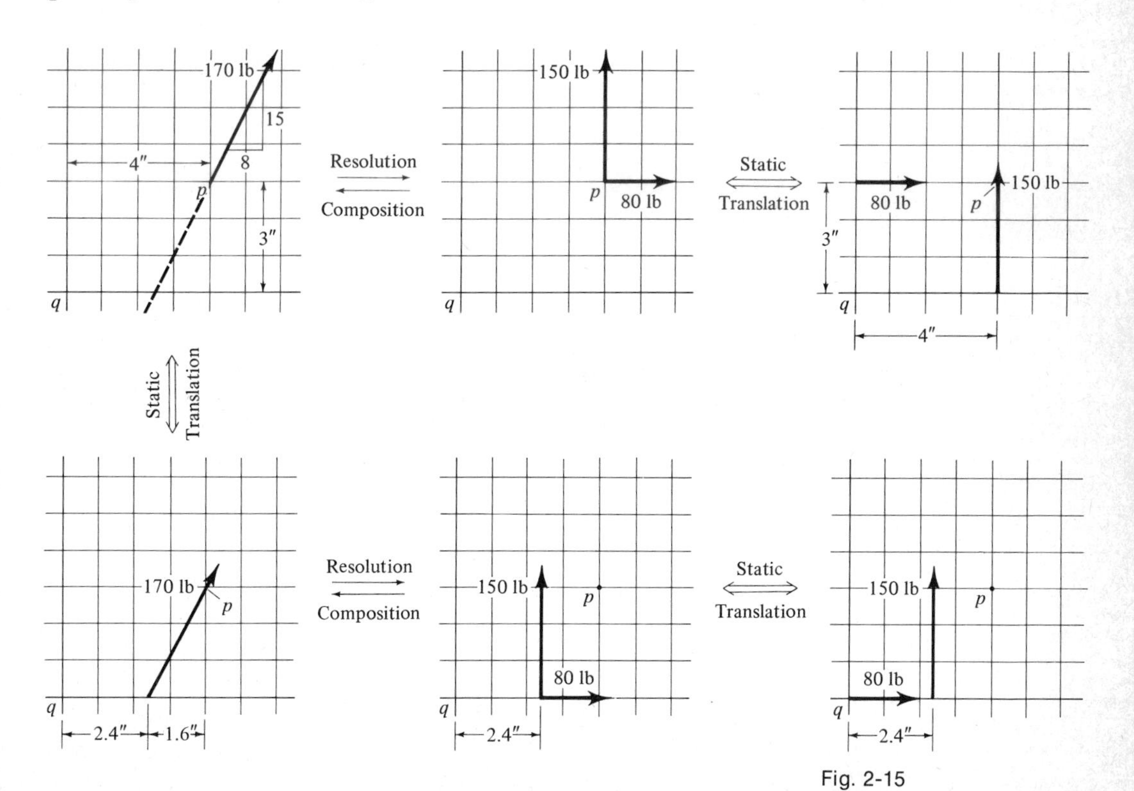

Fig. 2-15

In real situations, a force is applied to some body at some unique location, and we seldom actually change the location of a force. However, in determining the rotational effect of a given force with respect to some point q, it is often convenient to visually resolve the force into components and/or relocate the force at some other point. For example, in Fig. 2-15, to determine the moment of the given 170-lb force with respect to point q, the last figure in the sequence shows that

$$\mathbf{M}_q(\mathbf{F}) = (2.4)(150)\mathbf{k} \text{ in-lb}$$

whereas this result is not immediately apparent from the first figure of the sequence.

2-5 STATICALLY EQUIVALENT SYSTEMS

Up to now, we have been discussing moments of a force and statical transformations of a force, but in actual physical situations we usually have to deal with several forces at one time; that is, we must actually deal with a *system of forces*. As a very simple illustration, the crate being lifted in Fig. 2-16(a) is actually being acted upon by three forces, as indicated by the representation in Fig. 2-16(b) where each force $\mathbf{T}_1$, $\mathbf{T}_2$, and $\mathbf{W}$ has a definite location p_1, p_2, and p_3. We denote such a system by

$$\mathscr{S} = \{\mathbf{T}_1, \mathbf{T}_2, \mathbf{W} | p_1, p_2, p_3\}$$

and define its *sum* to be

$$\mathbf{S}(\mathscr{S}) = \mathbf{T}_1 + \mathbf{T}_2 + \mathbf{W}$$

and its moment with respect to some point q to be

$$\mathbf{M}_q(\mathscr{S}) = \mathbf{qp}_1 \times \mathbf{T}_1 + \mathbf{qp}_2 \times \mathbf{T}_2 + \mathbf{qp}_3 \times \mathbf{W}$$

which is the vector sum of the moments of the individual forces, where $\mathbf{qp}_1$ is the position vector drawn from q to p_1, etc.

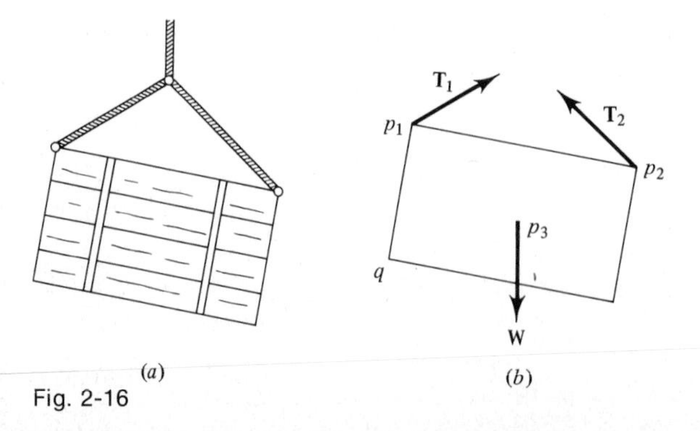

Fig. 2-16

Let us now choose our terminology very carefully. In Chap. 1 we treated force as a vector, and, as such, two forces are "equal" if and only if they have the same magnitude and the same direction. But, as we have just seen in this chapter, two "equal" forces are not necessarily equivalent in a statical sense in that they can produce different moments depending upon their locations. However, if the first force can be statically transformed into the second force, then both forces produce the same moment with respect to any point q, since moments are preserved under statical transformations. The two forces are then said to be *statically equivalent* to each other.

We now pose the question: Under what conditions are two force systems statically equivalent? Since two forces are statically equivalent if and only if they are equal and produce the same moment, we give the following definition:

Definition 2-4 Two systems of forces

$$\mathscr{S} = \{\mathbf{F}_1,\mathbf{F}_2, \ldots ,\mathbf{F}_n|p_1,p_2, \ldots ,p_n\}$$

and

$$\mathscr{S}' = \{\mathbf{F}'_1,\mathbf{F}'_2, \ldots ,\mathbf{F}'_{n'}|p'_1,p'_2, \ldots ,p'_{n'}\}$$

are *statically equivalent* (or *equipollent*) if and only if their sums are equal and their moments with respect to *any* point q are equal; that is,

$$\mathbf{S}(\mathscr{S}) = \mathbf{S}(\mathscr{S}') \tag{2-10}$$

or

$$\sum_{i=1}^{n} \mathbf{F}_i = \sum_{i=1}^{n'} \mathbf{F}'_i$$

and

$$\mathbf{M}_q(\mathscr{S}) = \mathbf{M}_q(\mathscr{S}') \tag{2-11}$$

or

$$\sum_{i=1}^{n} \mathbf{qp}_i \times \mathbf{F}_i = \sum_{i=1}^{n'} \mathbf{qp}'_i \times \mathbf{F}'_i$$

Hence, to determine whether or not two systems of forces are statically equivalent, we need to compare their sums and their moments with respect to some point. Physically, for two systems to be statically equivalent they must produce the same *net* force and the same *net* moment. Otherwise, the systems are statically different.

Before looking at an example problem, we pose a question that perhaps has already occurred to you. Suppose two systems $\mathscr{S}$ and $\mathscr{S}'$ have the same sum and the same moment with respect to one particular point q. Do the systems have the same moment with respect to any other arbitrary point, say, o? To answer this question, consider the two systems $\mathscr{S}$ and $\mathscr{S}'$ shown in Fig. 2-17(*a*) and (*b*), respectively. Suppose we have found that Eqs. (2-10) and (2-11)

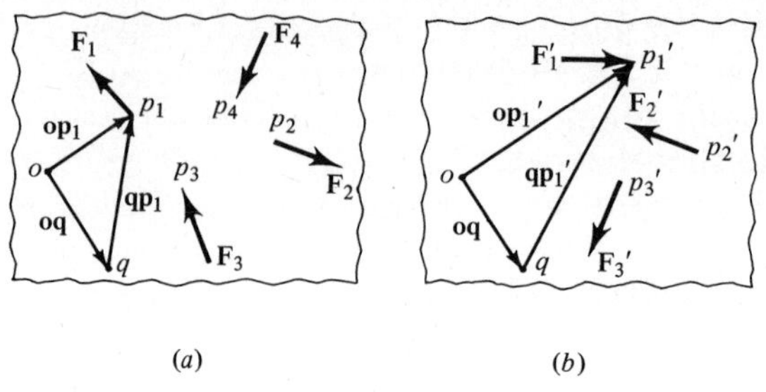

Fig. 2-17

are satisfied. Then consider, for system $\mathscr{S}$, moments with respect to point o.

$$
\begin{aligned}
\mathbf{M}_0(\mathscr{S}) &= \sum_{i=1}^{4} \mathbf{op}_i \times \mathbf{F}_i \\
&= \sum_{i=1}^{4} (\mathbf{oq} + \mathbf{qp}_i) \times \mathbf{F}_i \\
&= \sum_{i=1}^{4} \mathbf{oq} \times \mathbf{F}_i + \sum_{i=1}^{4} \mathbf{qp}_i \times \mathbf{F}_i \\
&= \mathbf{oq} \times \sum_{i=1}^{4} \mathbf{F}_i + \sum_{i=1}^{4} \mathbf{qp}_i \times \mathbf{F}_i \\
&= \mathbf{oq} \times \mathbf{S}(\mathscr{S}) + \mathbf{M}_q(\mathscr{S}) \qquad (2\text{-}12)
\end{aligned}
$$

Similarly, for system $\mathscr{S}'$ we would obtain

$$\mathbf{M}_o(\mathscr{S}') = \mathbf{oq} \times \mathbf{S}(\mathscr{S}') + \mathbf{M}_q(\mathscr{S}')$$

Hence, by virtue of Eqs. (2-10) and (2-11)

$$\mathbf{M}_o(\mathscr{S}') = \mathbf{M}_o(\mathscr{S})$$

and thus two systems which have the same sum and the same moment with respect to one point produce the same moment with respect to *any* point.

EXAMPLE 2-3

Determine whether or not the systems shown in Fig. 2-18(a) and (b) are statically equivalent. If they are not, can you add another force to (b) so that it is equivalent to (a)?

Solutions: (1) To check for statical equivalence we could try to statically transform Fig. 2-18(a) into Fig. 2-18(b), or vice versa. However, rather than do that, we will use Eqs. (2-10) and (2-11) to check for statical equivalence.

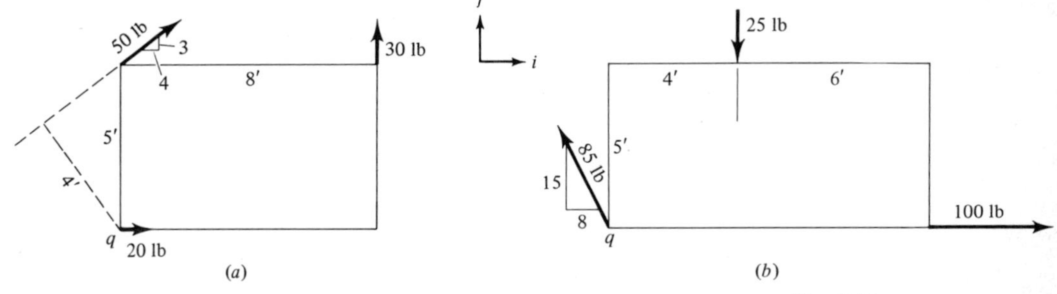

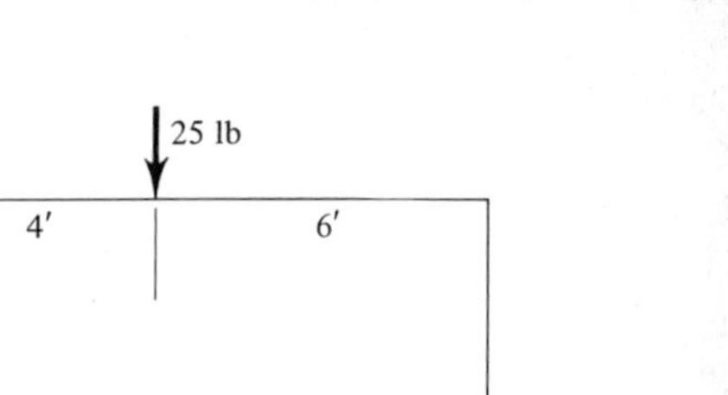

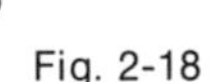

Fig. 2-18

Since all of the vectors lie in the plane of the paper, we will write each force vector in terms of **i** and **j** components. For convenience, we construct the following tables.

	System (*a*)	
Force, lb	**i Component**	**j Component**
50	+40	+30
30	0	+30
20	+20	0
Sum	+60	+60

$\mathbf{S}(\mathscr{S}_a) = 60\mathbf{i} + 60\mathbf{j}$

	System (*b*)	
Force, lb	**i Component**	**j Component**
85	−40	+75
25	0	−25
100	+100	0
Sum	+60	+50

$\mathbf{S}(\mathscr{S}_b) = 60\mathbf{i} + 50\mathbf{j}$

Thus, even without taking moments we see that the given systems cannot be equivalent.

(2) If we hope to make the systems equivalent we must first make their sums equal. We can do this by adding a force of 10**j** lb to system (*b*).

Let us now find the moments of each system with respect to point *q*. Again, we use a table. Physically, the moments are either clockwise or counterclockwise as we look at *q*, and, rather than use the cross product, we will use the fact that the magnitude of the moment of a force is equal to the magnitude of the force times the perpendicular distance *d*.

	System (*a*)	
Force, lb	***d*, ft**	**$\mathbf{M}_q$, ft-lb**
50	4	200 ↻
30	8	240 ↺
20	0	0
$\mathbf{M}_q(\mathscr{S}_a)$		40 ↺

	System (*b*)	
Force, lb	***d*, ft**	**$\mathbf{M}_q$, ft-lb**
85	0	0
25	4	100 ↻
100	0	0
$\mathbf{M}_q(\mathscr{S}_b)$		100 ↻

Thus, we see once again that the given systems are not statically equivalent.

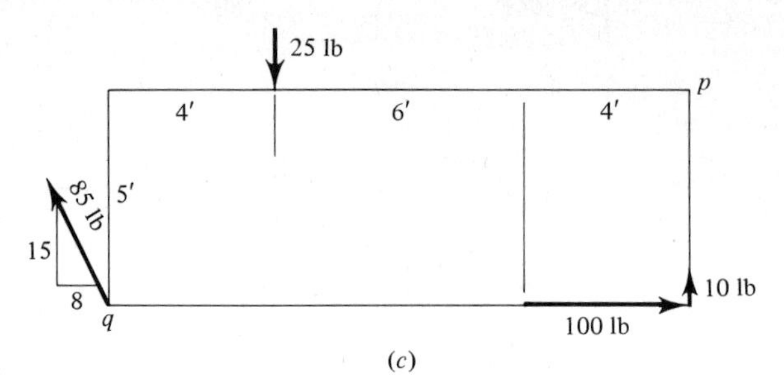

Fig. 2-18

If we add a force of 10**j** to system (*b*) as suggested earlier, where should we place it? In order to make $\mathbf{M}_q(\mathcal{S}_a)$ equal to $\mathbf{M}_q(\mathcal{S}_b)$, system (*b*) needs an additional moment of 140 ft-lb ↺. Hence the 10-lb upward force would have to be placed 14 ft to the right of point *q*. Then the system in Fig. 2-18(*c*) would be equivalent to system (*a*).

Remark: The vertical 10-lb force in Fig. 2-18(*c*) could have just as well been placed at point *p* or any other point along the vertical line through *p*.

Before ending this discussion of statically equivalent systems, we wish to reemphasize the physical meaning of statical equivalence. The system in Fig. 2-18(*a*) is *statically equivalent* to that in Fig. 2-18(*c*) in that both systems have the same sum and produce the same turning effect with respect to any given point. Whether or not these two systems are "equivalent" in any other physical sense is an open question whose answer depends upon what type of equivalence we mean. For example, suppose these systems are each acting upon some easily deformed body such as a rubber plate. Would the systems produce the same deformation of the plates. Intuitively, we expect the answer to be no, so that as far as deformation effects are concerned, the systems are *not* equivalent.

In later chapters, we will be concerned about other types of equivalence, but for now, we are solely concerned with statical equivalence.

EXERCISE PROBLEMS

2-41 to 2-48 Statically transform the given force into a system composed of a force $\mathbf{F}_1$ at *p* parallel to *op* and a force $\mathbf{F}_2$ at *q* parallel to *oq*.

PROB. 2-41

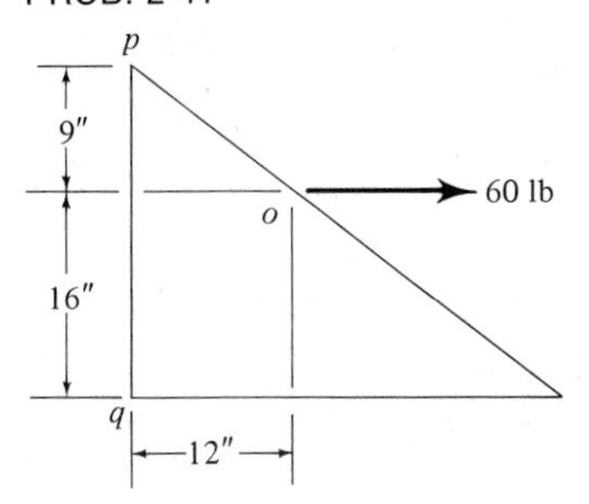

PROB. 2-42

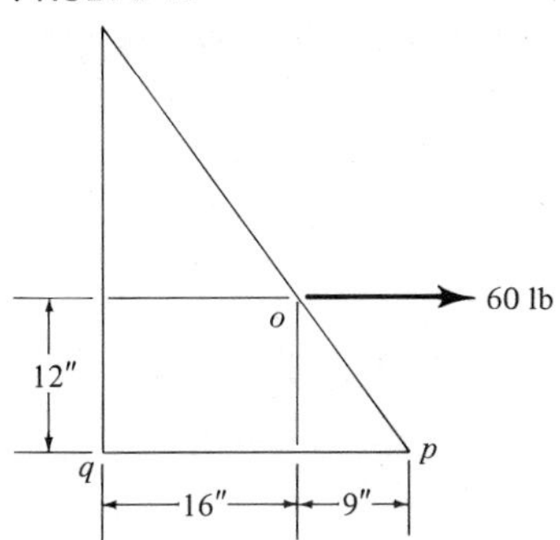

PROB. 2-43

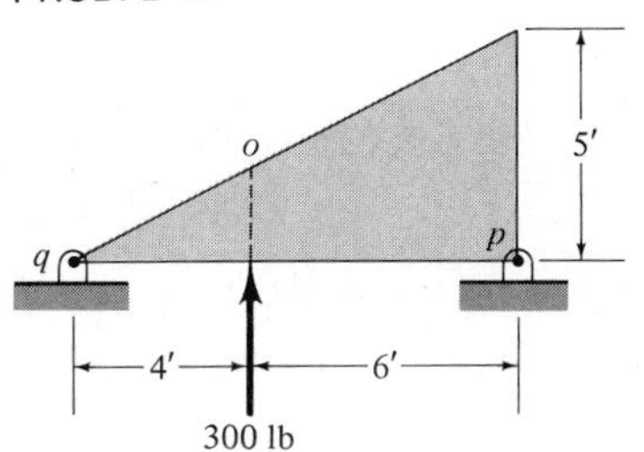

PROB. 2-44

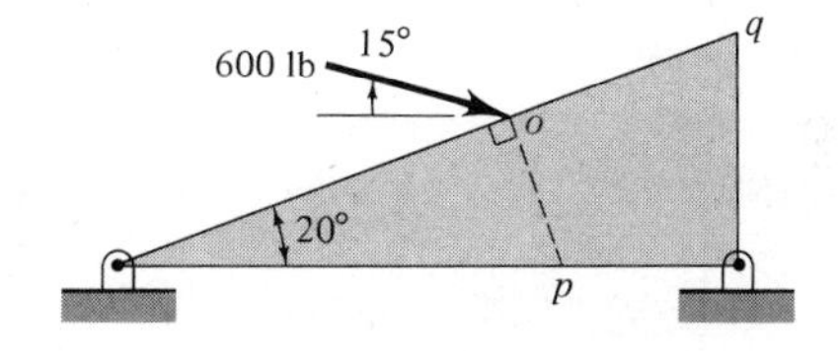

PROB. 2-45

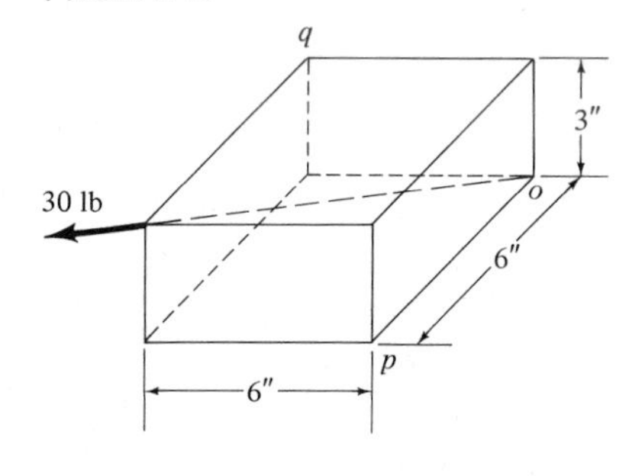

PROB. 2-46

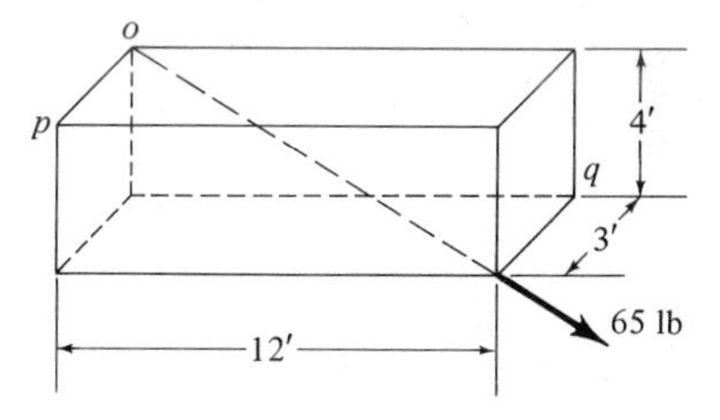

PROB. 2-47

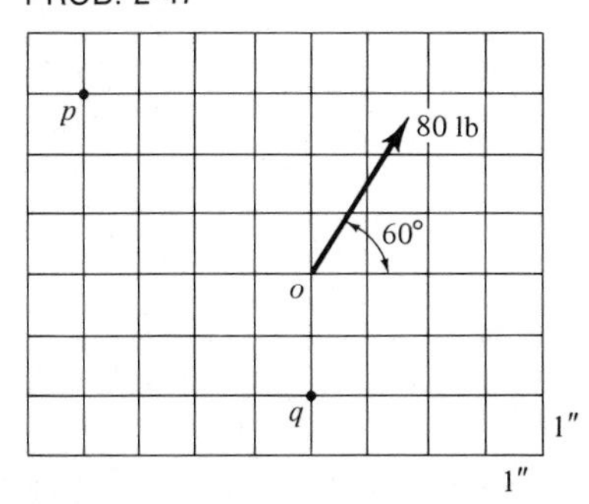

PROB. 2-48

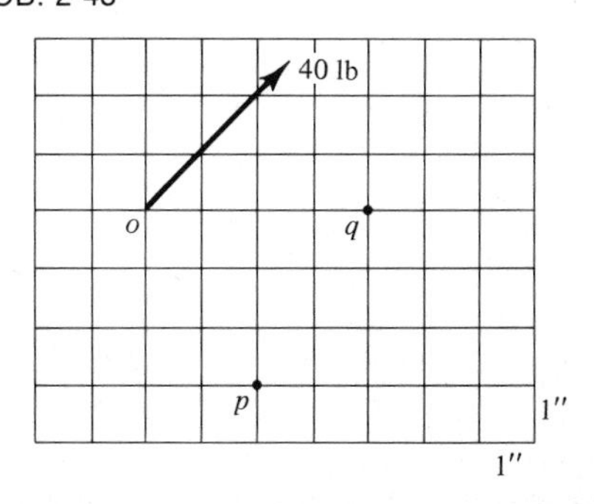

2-49 to 2-55 Determine whether or not System (*a*) is statically equivalent to System (*b*). If not, can you make them statically equivalent by adding another force to (*b*)?

PROB. 2-49

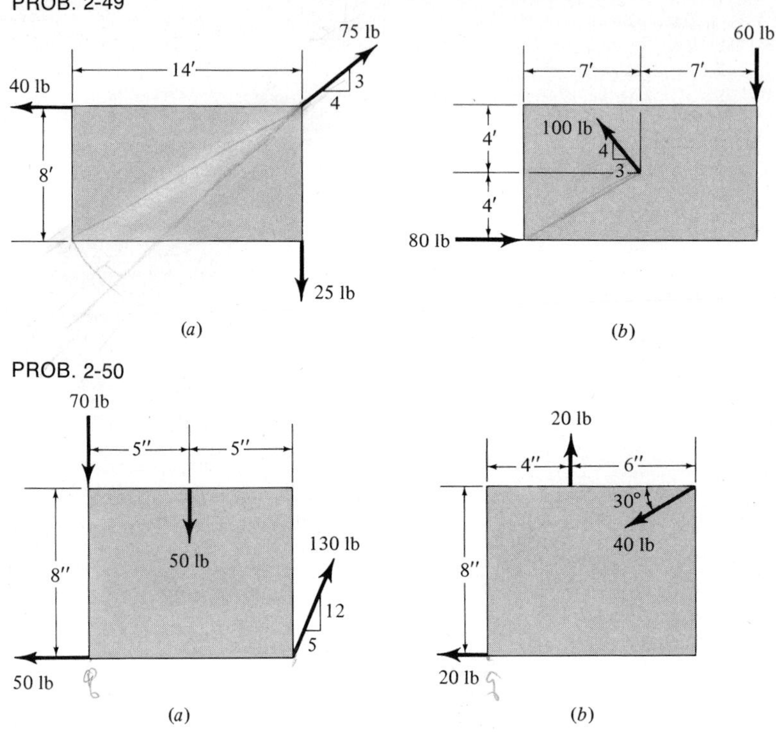

PROB. 2-51

All pulleys are 6 in diameter

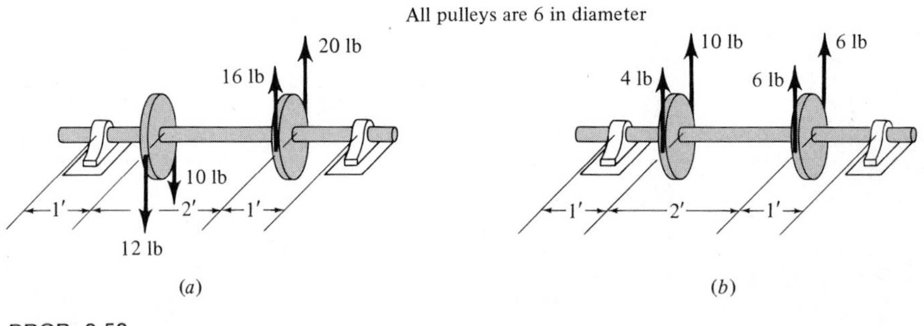

(*a*) (*b*)

PROB. 2-52

85 lb
15
8
35 lb
4′
40 lb
70 lb
8′
40 lb
50 lb
4
3
4′
8′

(*a*) (*b*)

PROB. 2-53

PROB. 2-54

PROB. 2-55

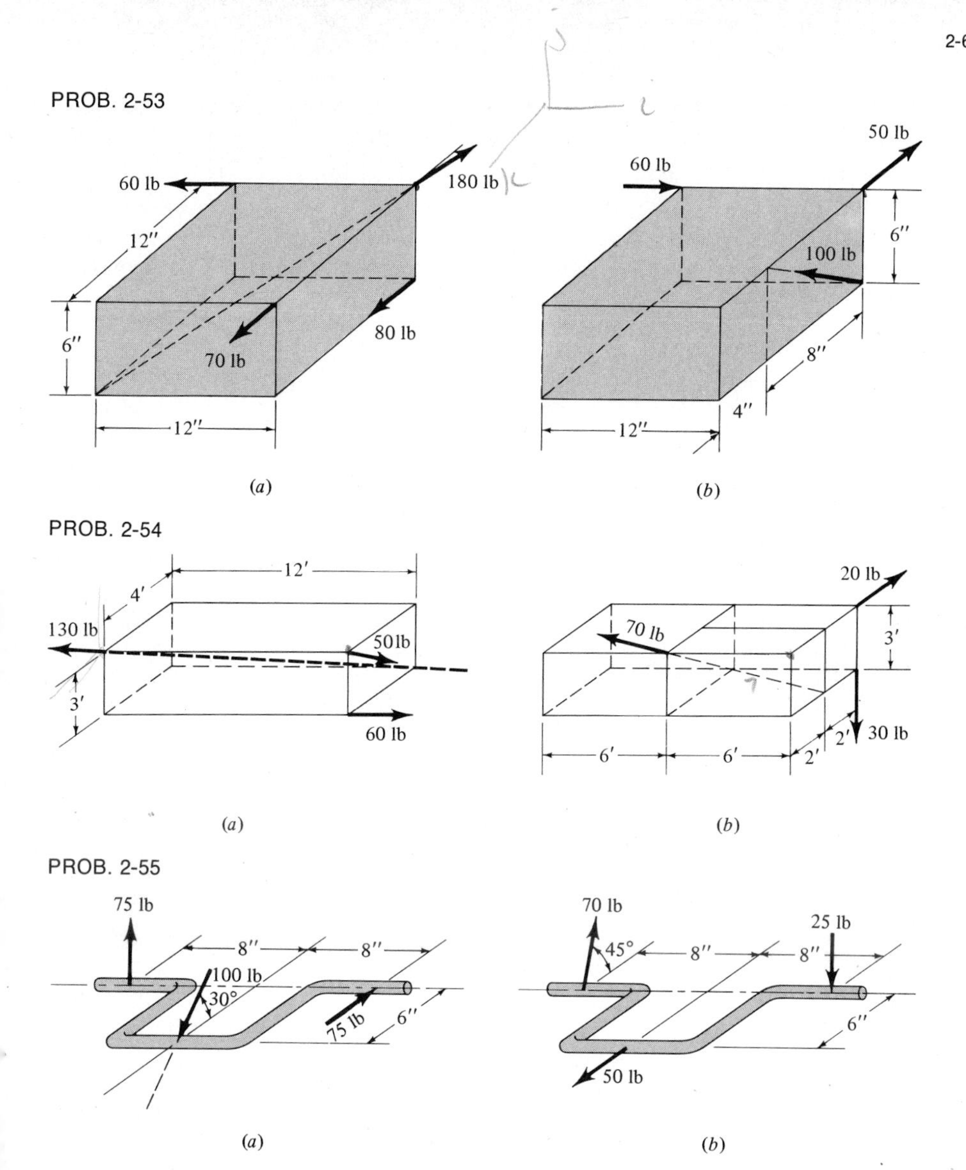

2-6 COUPLES

In general, it is highly improbable that two arbitrary force systems will be statically equivalent. However, given one force system, it is usually a relatively simple matter to construct a second force system that is statically equivalent to the first. For example, if we merely perform statical transformations on the first system, we can obtain a multitude of statically equivalent systems (see Fig. 2-15).

We now pose the question: Of all the systems that are statically equivalent to a given system, what is the most simple equivalent system? The simplest system that is statically equivalent to the original system is called the *statical resultant* of the system or simply the *resultant*. Intuitively, one would probably expect that the most simple system would be a single force properly located so as to produce the required moments. Indeed, this is quite often the case. Unfortunately, however, this is not always the case, as illustrated in Fig. 2-19(a).

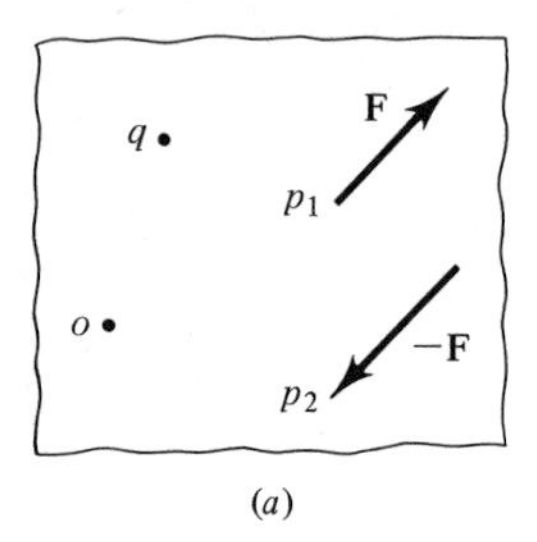

(a)

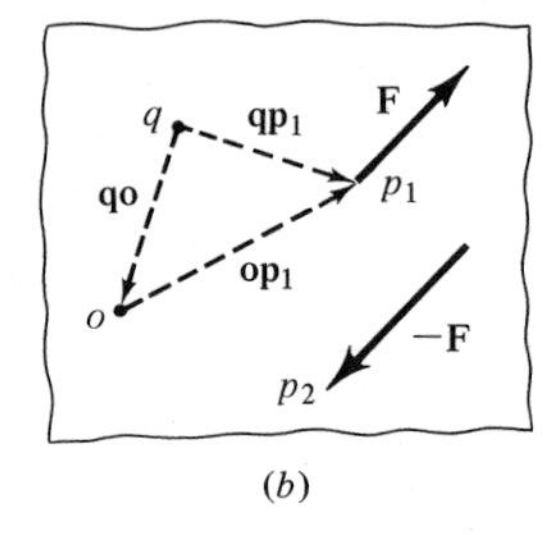

(b)

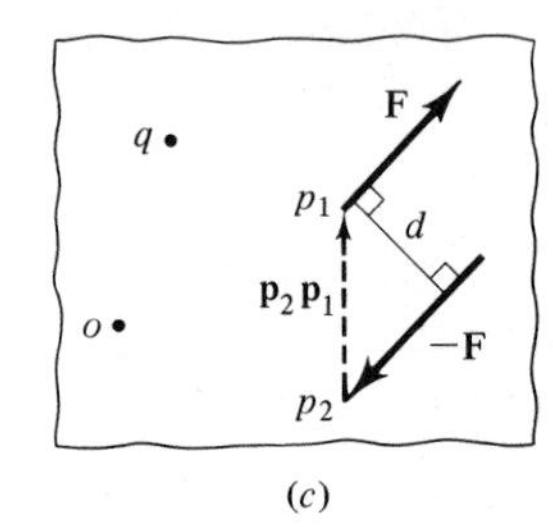

(c)

Fig. 2-19

For such a system

$$\mathbf{S}(\mathscr{S}) = \mathbf{F} + (-\mathbf{F}) = \mathbf{0}$$

while

$$\mathbf{M}_o(\mathscr{S}) = \mathbf{op}_1 \times \mathbf{F} + \mathbf{op}_2 \times (-\mathbf{F})$$

and

$$\begin{aligned} \mathbf{M}_q(\mathscr{S}) &= \mathbf{qp}_1 \times \mathbf{F} + \mathbf{qp}_2 \times (-\mathbf{F}) \\ &= (\mathbf{qo} + \mathbf{op}_1) \times \mathbf{F} + (\mathbf{qo} + \mathbf{op}_2) \times (-\mathbf{F}) \\ &= \mathbf{qo} \times \mathbf{F} + \mathbf{op}_1 \times \mathbf{F} + \mathbf{qo} \times (-\mathbf{F}) + \mathbf{op}_2 \times (-\mathbf{F}) \\ &= \mathbf{op}_1 \times \mathbf{F} + \mathbf{op}_2 \times (-\mathbf{F}) \end{aligned}$$

That is, the moment with respect to point o of this system of two forces is equal to its moment with respect to point q. In fact, since o and q were quite arbitrary, the moment of this system with respect to any point is the same. Apparently, then, we need no subscript o or q, and the moment of this system is

$$\mathbf{M}(\mathscr{S}) = (\mathbf{op}_1 - \mathbf{op}_2) \times \mathbf{F} = \mathbf{p}_2\mathbf{p}_1 \times \mathbf{F}$$

with

$$\|\mathbf{M}(\mathscr{S})\| = Fd$$

where d is the perpendicular distance between $\mathbf{F}$ and $(-\mathbf{F})$, as shown in Fig. 2-19(c). For the system in Fig. 2-19, the vector $\mathbf{M}(\mathscr{S})$ is directed in toward the paper and represents a clockwise rotational effect.

Although this type of system might appear to be a very special one, such systems do, in fact, often occur in physical problems, and they are quite useful in analyzing force systems in general. In all probability, at some time or other, you have produced such a system yourself—perhaps in removing a wheel from an automobile, as in Fig. 2-20. For these reasons, we now give the following definition:

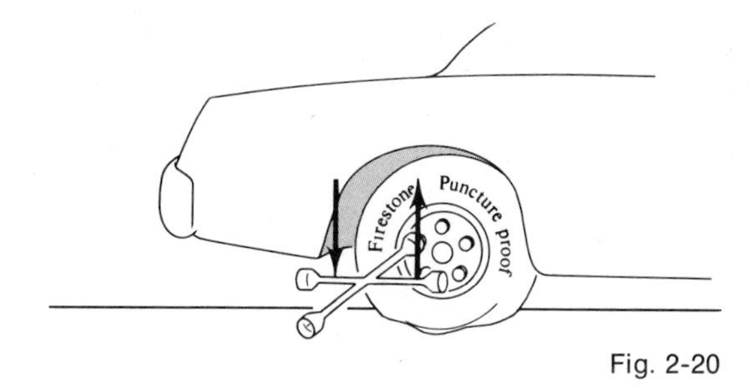

Fig. 2-20

Definition 2-5 A force system whose sum is the zero vector but whose net moment is a nonzero vector is called a *couple.* Physically, a couple is a turning effect whose net force is zero.

Since the net force (sum) of a couple is zero, from Eq. (2-12) its moment is the same with respect to any point, as our example of Fig. 2-19 indicated. Hence, we write

$$\mathbf{M}_o(\mathscr{S}) = \mathbf{M}_q(\mathscr{S}) = \mathbf{M}(\mathscr{S})$$

Also, since the sum of a couple is zero, the only systems $\mathscr{S}'$ which can be statically equivalent to a given couple $\mathscr{S}$ are other couples with the same net moment. Hence, we write

$$\mathbf{M}(\mathscr{S}) = \mathbf{M}(\mathscr{S}') = \mathbf{C}$$

where $\mathbf{C}$ denotes all couples having the same moment $\mathbf{C}$.

Since couples are, in fact, moment vectors, they can be represented by double-arrowed directed line segments, just as we have done throughout this chapter. Also, two or more couples can be added vectorially, just as we added other vectors in Chap. 1. However, because a couple physically produces a turning effect rather than a push or pull, a number of different schemes are often used to represent a couple. For example, a couple of 10 in-lb tending to rotate your book counterclockwise can be represented in each of the ways shown in Fig. 2-21. All of them are acceptable ways of representing a couple. Notice that the exact location of the couple is not distinctly specified since a couple produces the same moment with respect to any point.

Couples are quite useful in transforming a given system composed of numerous forces into a more simplified statically equivalent system, often called a *reduced system.* The following example illustrates how one system can be "reduced" to other statically equivalent systems each composed of a single force and a couple.

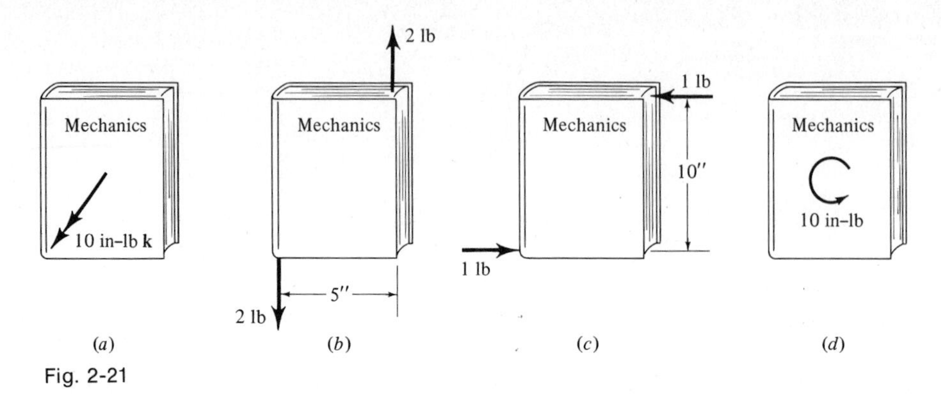

Fig. 2-21

EXAMPLE 2-4

Given the force system in Fig. 2-22(*a*), statically transform this system into (*a*) an equivalent system composed of a single force at *q* and a couple, and (*b*) an equivalent system composed of a single force at *p* and a couple.

Solutions: We first simplify the system by using resolution and statical translations, as indicated by the sequence in Fig. 2-22(*b*) and (*c*). The two 50-lb forces in (*c*) then form a 150-ft-lb couple, and thus the original system is statically equivalent to the system shown in (*d*).

(*a*) To end up with a single force at *q* we must "remove" the 10-lb force from *p* by means of statical transformations. To do this, let us add "nothing" to the system by placing two "equal-but opposite" 10-lb forces at *q*, as shown in (*e*). Now, the upward 10-lb force at *q* and the downward one at *p* form a clockwise 30-ft-lb couple. Hence the original system is statically equivalent to the one shown in (*f*) having a single 48-lb force at *q* and a 180-ft-lb couple.

(*b*) Returning to (*d*), we must remove the 47-lb force from *q*. Once again we add "nothing" in the form of two 47-lb forces at *p*, as in (*g*). The 47-lb force at *q* and the one to the left at *p* form a 47-ft-lb couple, and the final system is shown in (*h*).

> ***Remark:*** Note that the final force in (*f*) and (*h*) is always 48 lb, which is the sum of the original system in (*a*) as required by statical transformations. However, changing the final location of the force from *q* to *p* results in a different couple in order to maintain rotational equivalence.

Alternate Solution: The equivalent system is to be composed of a force at *q* and a couple. Accordingly, let us construct such a system, as shown in Fig. 2-22(*i*), with a force **F** at *q* and a couple **C**.

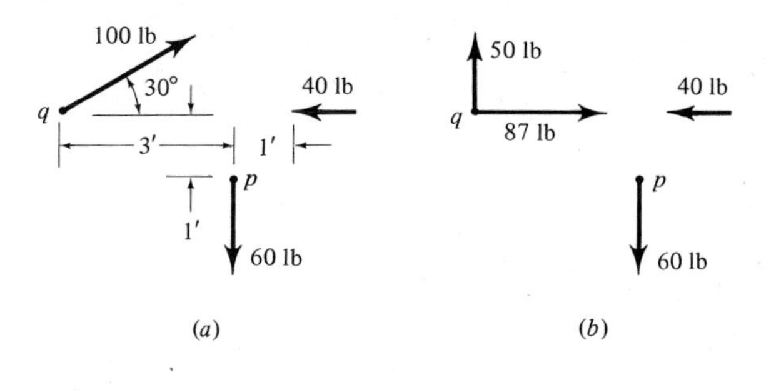
100 lb
30°
q
3′
1′
40 lb
1′
p
60 lb
(a)
50 lb
q
87 lb
40 lb
p
60 lb
(b)

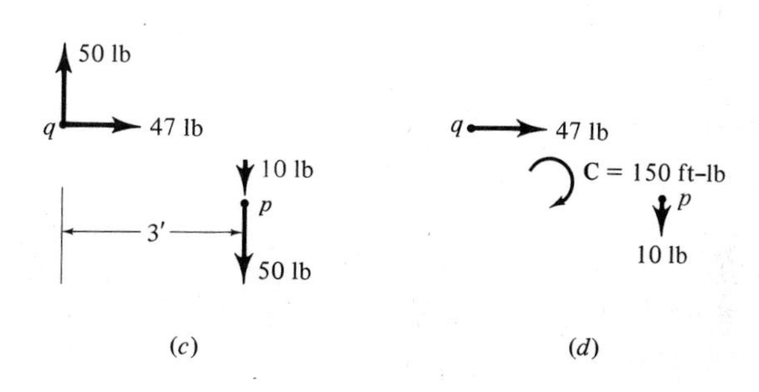
50 lb
q
47 lb
10 lb
p
3′
50 lb
(c)
q
47 lb
C = 150 ft–lb
p
10 lb
(d)

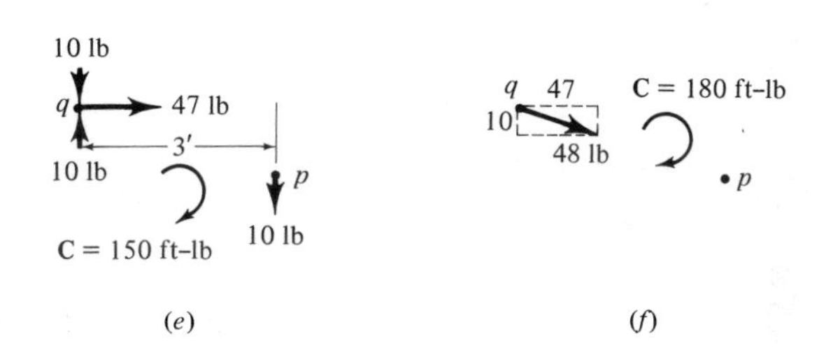
10 lb
q
47 lb
3′
10 lb
p
10 lb
C = 150 ft–lb
(e)
q
47
C = 180 ft–lb
10
48 lb
p
(f)

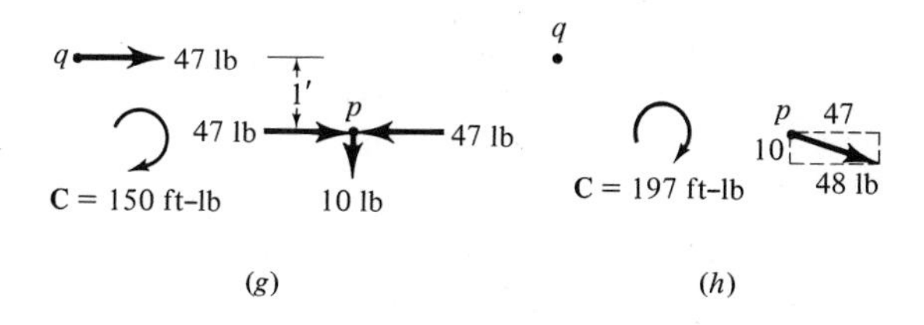
q
47 lb
1′
47 lb
p
47 lb
C = 150 ft–lb
10 lb
(g)
q
p
47
10
C = 197 ft–lb
48 lb
(h)

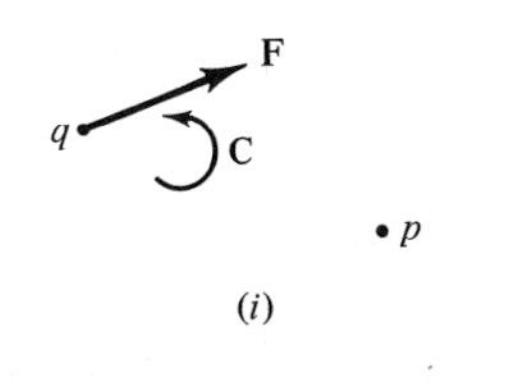
F
q
C
p
(i)

Fig. 2-22

Since this system is to be statically equivalent to the original system in Fig. 2-22(a), by Eqs. (2-10) and (2-11) we have

$$\mathbf{S}[\text{System } (a)] = \mathbf{S}[\text{System } (i)] = \mathbf{F}$$
$$\mathbf{M}_q[\text{System } (a)] = \mathbf{M}_q[\text{System } (i)] = \mathbf{C}$$

That is, if the reduced system is composed of a force at q and a couple, the *couple* must be equal to the *moment with respect to point q of the original system.*

Thus, referring to Fig. 2-22(b) and using the usual unit vectors **i**, **j**, and **k**, we obtain

$$\mathbf{F} = 87\mathbf{i} + 50\mathbf{j} - 40\mathbf{i} - 60\mathbf{j} = 47\mathbf{i} - 10\mathbf{j}$$
$$\mathbf{C} = (3\mathbf{i} - 1\mathbf{j}) \times (-60\mathbf{j}) = -180\mathbf{k} \text{ ft-lb}$$

Hence the system in Fig. 2.22(i), with **C** and **F** given above, is exactly the system obtained previously in Fig. 2-22(f).

This same alternative method can also be used to obtain the system in Fig. 2-22(h) composed of a force at p and a couple. We leave this for you to complete.

EXERCISE PROBLEMS

2-56 to 2-67 Transform the given systems into statically equivalent ones: (a) with a single force at q and a couple; (b) with a single force at p and a couple.

PROB. 2-56

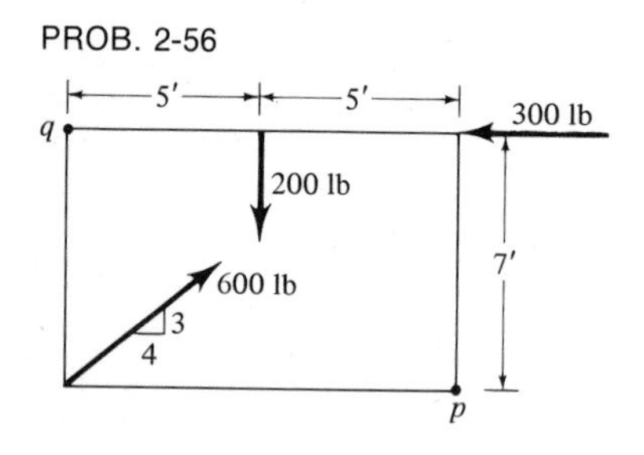

PROB. 2-57

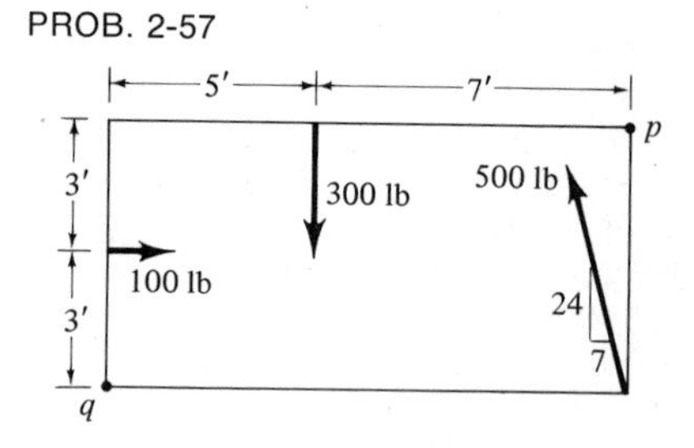

PROB. 2-58

PROB. 2-59

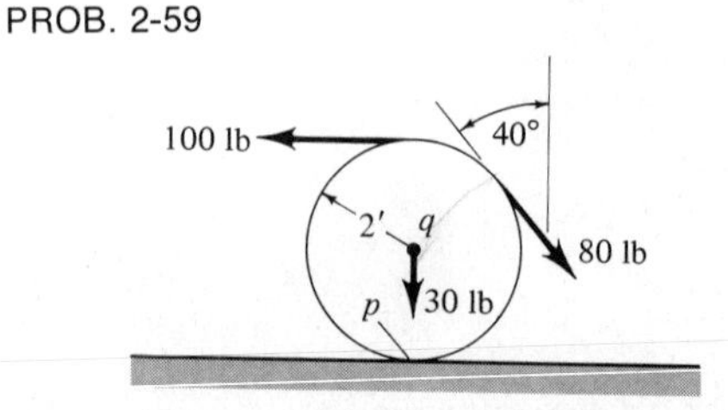

PROB. 2-60

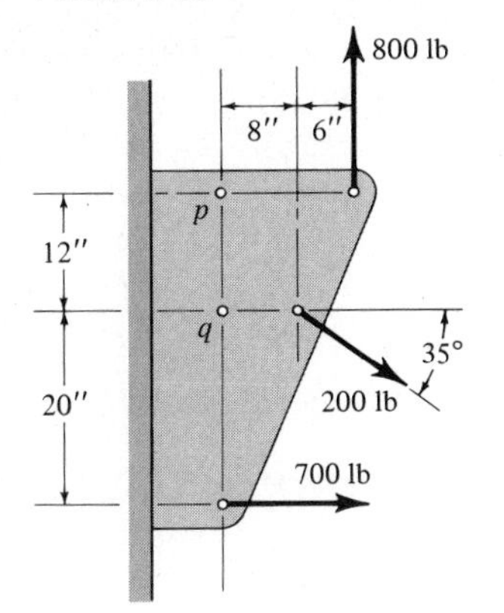

PROB. 2-61

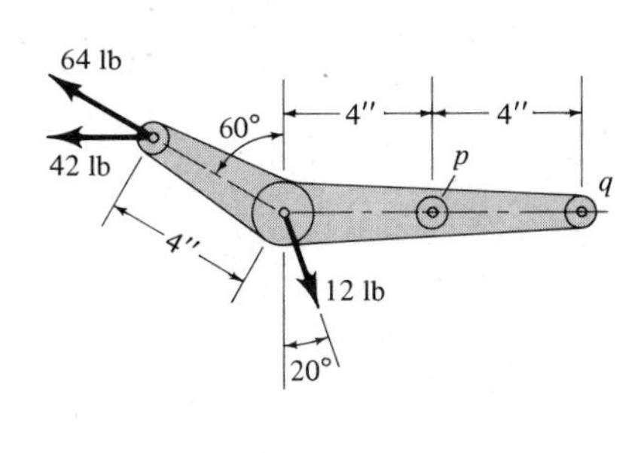

PROB. 2-62

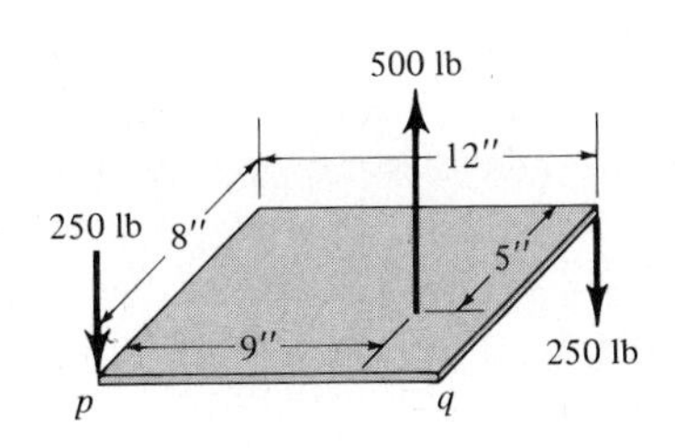

PROB. 2-63

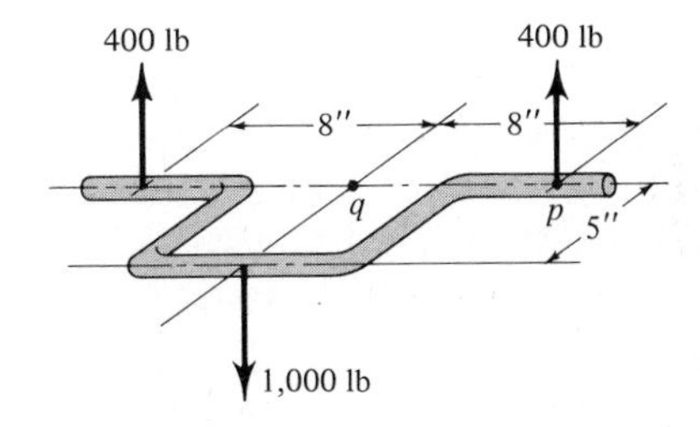

PROB. 2-64

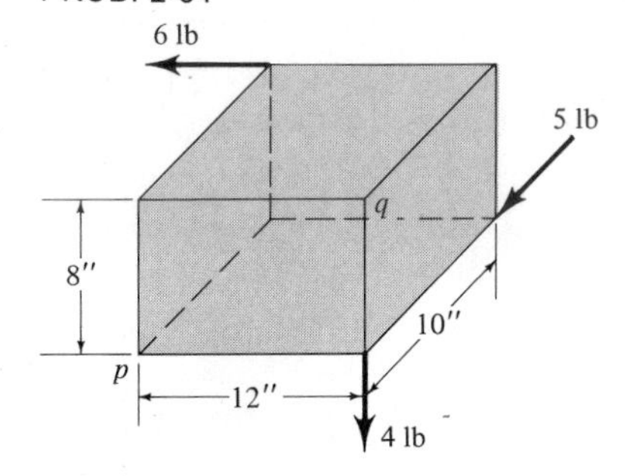

PROB. 2-65

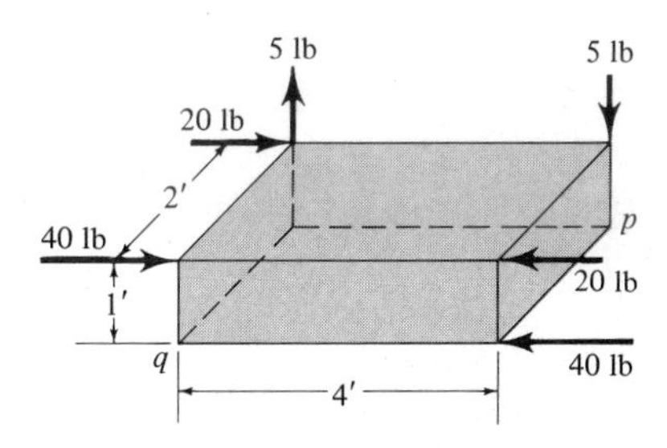

PROB. 2-66

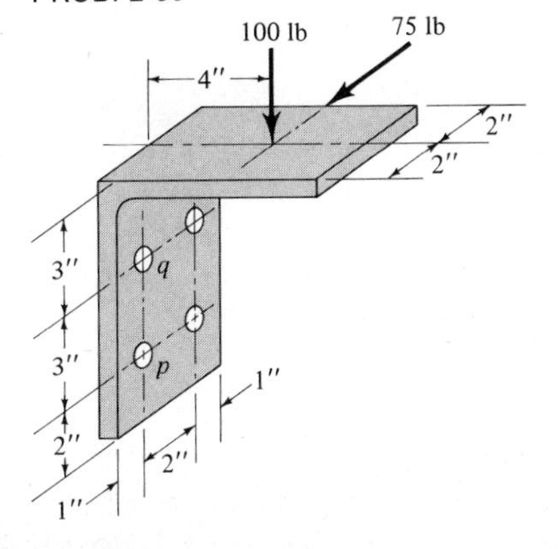

PROB. 2-67

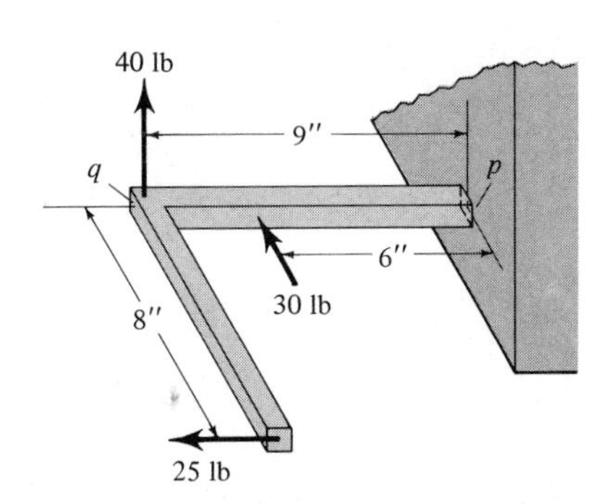

2-7 RESULTANTS

We now return to the questions posed in the previous article: Given a general system of forces, what is the *simplest* equivalent system? Intuitively, we have speculated that the resultant might be a force or possibly a couple. But rather than rely on our not-always-reliable intuition, let us set down our known facts and logically deduce the answer to our question.

First, any given system $\mathscr{S}$ will have a sum (maybe zero) and a moment (maybe zero) with respect to any arbitrary point q. For our statically equivalent resultant system $\mathscr{R}$ we must have

$$\mathbf{S}(\mathscr{R}) = \mathbf{S}(\mathscr{S}) \tag{2-13}$$

$$\mathbf{M}_q(\mathscr{R}) = \mathbf{M}_q(\mathscr{S}) \tag{2-14}$$

that is, the resultant system must have the same sum as $\mathscr{S}$ and the same moment. There are only four possible situations:

$$\begin{array}{llll} (1) & \mathbf{S}(\mathscr{S}) = \mathbf{0} & \text{and} & \mathbf{M}_q(\mathscr{S}) = \mathbf{0} \\ (2) & \mathbf{S}(\mathscr{S}) = \mathbf{0} & \text{and} & \mathbf{M}_q(\mathscr{S}) \neq \mathbf{0} \\ (3) & \mathbf{S}(\mathscr{S}) \neq \mathbf{0} & \text{and} & \mathbf{M}_q(\mathscr{S}) = \mathbf{0} \\ (4) & \mathbf{S}(\mathscr{S}) \neq \mathbf{0} & \text{and} & \mathbf{M}_q(\mathscr{S}) \neq \mathbf{0} \end{array}$$

We will consider these cases in order. For case (1), since $\mathbf{S}(\mathscr{S}) = 0$, according to the previous article the resultant *must* be a couple $\mathbf{C}$. But then $\mathbf{C} = \mathbf{M}_q(\mathscr{S}) = \mathbf{0}$, and thus $\mathbf{C} = \mathbf{0}$. Hence the resultant is said to be the *null* or *zero* resultant $\mathscr{O}$. There is no net force or turning effect for such a system. This situation plays an important role in mechanics and will be treated in greater detail in later chapters.

For case (2), $\mathbf{S}(\mathscr{S}) = \mathbf{0}$ but $\mathbf{M}_q(\mathscr{S}) \neq \mathbf{0}$. Thus, by Definition 2-5 the resultant is a couple $\mathbf{C}$ given by

$$\mathbf{C} = \mathbf{M}_q(\mathscr{S})$$

For case (3) the sum is nonzero, so the resultant system must contain a force $\mathbf{R}$ equal to the sum $\mathbf{S}(\mathscr{S})$. Also, since $\mathbf{M}_q(\mathscr{S}) = \mathbf{0}$, this means that the line of action of the resultant force $\mathbf{R}$ must pass through the point q. Thus the resultant for this case is a single force

$$\mathbf{R} = \mathbf{S}(\mathscr{S})$$

located in such a manner that its line of action passes through q.

Each of the possible foregoing cases (1) to (3) is illustrated in Fig. 2-23(a) to (c), respectively, for the two-dimensional force system shown in Fig. 2-23(e).

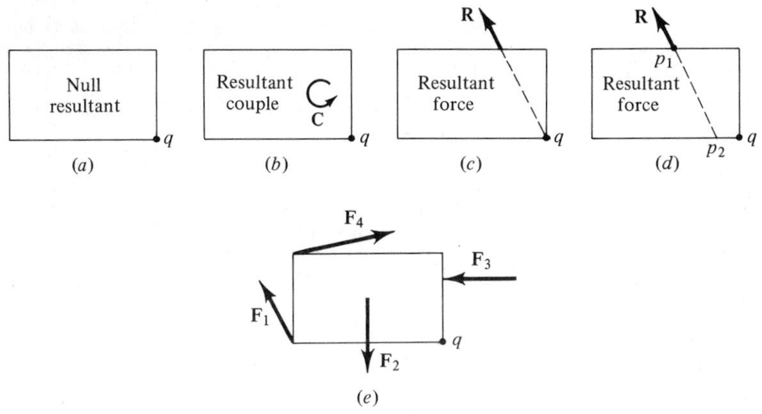

Fig. 2-23

Finally, for case (4), the resultant system must include a nonzero force $\mathbf{R} = \mathbf{S}(\mathscr{S})$ and must be capable of producing the proper moment $\mathbf{M}_q(\mathscr{S}) \neq \mathbf{0}$. The only question is whether or not **R** can be properly located to meet this requirement. Assuming it is possible, this situation is depicted in Fig. 2-23(*d*).

Notice that **R** could be located anywhere along its line of action, for example, at p_1 or p_2, just so long as

$$\mathbf{qp}_1 \times \mathbf{R} = \mathbf{qp}_2 \times \mathbf{R} = \mathbf{M}_q(\mathscr{S})$$

But we must still ask the question: Is this always possible? Formally, we are posing the question: Does a **p** exist such that

$$\mathbf{p} \times \mathbf{R} = \mathbf{M}_q(\mathscr{S}) \tag{2-15}$$

when $\mathbf{R} = S(\mathscr{S})$ and $\mathbf{M}_q(\mathscr{S})$ are known? The answer is sometimes yes, sometimes no. If there is a **p** which satisfies Eq. (2-15), then

$$\mathbf{R} \cdot (\mathbf{p} \times \mathbf{R}) = \mathbf{R} \cdot \mathbf{M}_q(\mathscr{S})$$

which, by Eq. (2-9), implies

$$\mathbf{R} \cdot \mathbf{M}_q(F) = 0$$

so that **R** and $\mathbf{M}_q(\mathscr{S})$ must be perpendicular. Then, if $\mathbf{M}_q(\mathscr{S})$ is perpendicular to $\mathbf{R} = \mathbf{S}(\mathscr{S})$, a **p** will exist such that $\mathbf{M}_q(\mathscr{S}) = \mathbf{p} \times \mathbf{R}$. In fact, infinitely many **p**'s will exist by virtue of statical translations (see Fig. 2-13).

On the other hand, if **R** and $\mathbf{M}_q(\mathscr{S})$ are not perpendicular, Eq. (2-15) cannot be satisfied by *any* **p**, and, therefore, **R** cannot be positioned to produce the moment $\mathbf{M}_q(\mathscr{S})$. However, we can do the following:

Let us decompose $\mathbf{M}_q(\mathscr{S})$ into the sum of two vectors, one parallel to **R**, $\mathbf{M}_q^{\mathbf{R}}(\mathscr{S})$, and the other perpendicular to **R**, $\mathbf{M}_q^{\perp}(\mathscr{S})$, so that

$$\mathbf{M}_q(\mathscr{S}) = \mathbf{M}_q^{\mathbf{R}}(\mathscr{S}) + \mathbf{M}_q^{\perp}(\mathscr{S})$$

From the previous paragraphs we know that a $\mathbf{p}$ exists such that

$$\mathbf{p} \times \mathbf{R} = \mathbf{M}_q^{\perp}(\mathscr{S})$$

That is, we can locate $\mathbf{R}$ so as to produce $\mathbf{M}_q^{\perp}(\mathscr{S})$. But the equivalent resultant system $\mathscr{R}$ must also produce $\mathbf{M}_q^{\mathrm{R}}(\mathscr{S})$. The only way this moment may be produced is by a couple $\mathbf{C} = \mathbf{M}_q^{\mathrm{R}}(\mathscr{S})$. Thus, in general, for case (4) the resultant system $\mathscr{R}$ will be composed of a force

$$\mathbf{R} = \mathbf{S}(\mathscr{S})$$

located so that

$$\mathbf{p} \times \mathbf{R} = \mathbf{M}_q^{\perp}(\mathscr{S})$$

and a couple

$$\mathbf{C} = \mathbf{M}_q^{\mathrm{R}}(\mathscr{S})$$

Such an equivalent resultant system is called a *wrench*, although a screwdriver might be a more descriptive name, since the effect of such a system is a force $\mathbf{R}$ and a couple $\mathbf{C}$ parallel to $\mathbf{R}$, as depicted in Fig. 2-24.

The ideas presented in this section will be examined more closely by considering certain special but important types of force systems.

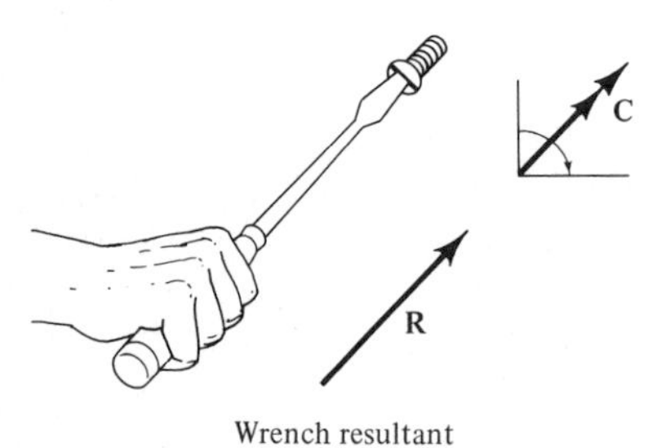

Fig. 2-24

Concurrent Force System

A force system for which the lines of action of all the forces intersect at one common point is called a concurrent force system, and the point of intersection is called the *point of concurrency.* Figure 2-25(a) is an example of such a system, the point of concurrency being o. Clearly, $\mathbf{M}_o(\mathscr{S}) = \mathbf{0}$ and thus, if $\mathbf{S}(\mathscr{S}) \neq \mathbf{0}$, the resultant of this system must be a force

$$\mathbf{R} = \mathbf{S}(\mathscr{S}) = \mathbf{F}_1 + \mathbf{F}_2 + \mathbf{F}_3 + \mathbf{F}_4$$

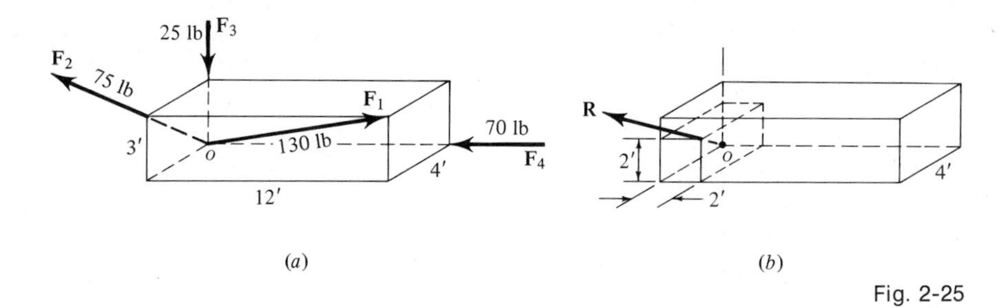

Fig. 2-25

whose line of action must pass through o as indicated in Fig. 2-25(b). If $\mathbf{S}(\mathscr{S}) = \mathbf{0}$, then the resultant would be the null system $\mathscr{O}$.

EXAMPLE 2-5

Determine the resultant of the system shown in Fig. 2-25(a).

Solution: First of all, we will write each of the forces in terms of its orthogonal components. For convenience, we construct the following table:

Force, lb	i Component	j Component	k Component
130	$\frac{12}{13}130 = +120$	$\frac{3}{13}130 = +30$	$\frac{4}{13}130 = +40$
75	0	$\frac{3}{5}75 = +45$	$\frac{4}{5}75 = +60$
25	0	-25	0
70	-70	0	0
Sum	$+50$	$+50$	$+100$

Thus, this system has a resultant force

$$\mathbf{R} = (50\mathbf{i} + 50\mathbf{j} + 100\mathbf{k})\ \text{lb}$$

By inspection, the moments of the given system with respect to point o will be zero:

$$\mathbf{M}_o(\mathscr{S}) = \mathbf{0}$$

Thus, the resultant system must also produce zero $\mathbf{M}_q(\mathscr{R})$. Hence, the line of action of the result force $\mathbf{R}$ must pass through o, as indicated in Fig. 2-25(b). If we wish, we can locate $\mathbf{R}$ at o.

Coplanar Force System

If all of the forces lie in a common plane so that the problem is basically two dimensional, the system is called coplanar. Such a system is shown in Fig. 2-26(a). Clearly, the sum of such a system will also lie in the same plane.

Also, the moment of each force with respect to, say, point q will be either clockwise or counterclockwise as we view the plane; that

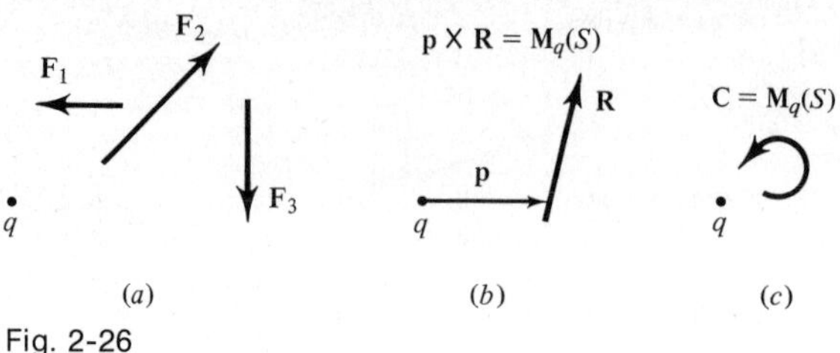

Fig. 2-26

is, the moment vector will either be pointing in toward the plane or out from the plane. In either case, the sum

$$\mathbf{R} = \mathbf{S}(\mathscr{S})$$

will be perpendicular to the moment vector

$$\mathbf{M}_q(\mathscr{S}) = \text{sum of the moments of } \mathbf{F}_1, \mathbf{F}_2, \text{ and } \mathbf{F}_3$$

If both $\mathbf{R}$ and $\mathbf{M}_q(\mathscr{S})$ are nonzero, then the resultant of this coplanar system will be a single force $\mathbf{R}$ located so as to give the moment $\mathbf{M}_q(\mathscr{S})$, as indicated in Fig. 2-26(*b*). If $\mathbf{R} = \mathbf{0}$ and $\mathbf{M}_q(\mathscr{S}) \neq \mathbf{0}$, then the resultant must be a couple $\mathbf{C} = \mathbf{M}_q(\mathscr{S})$, as shown in Fig. 2-26(*c*). In any case, there is *no possibility* that the resultant is a wrench.

EXAMPLE 2-6

Determine completely the resultant of the system shown in Fig. 2-27(*a*).

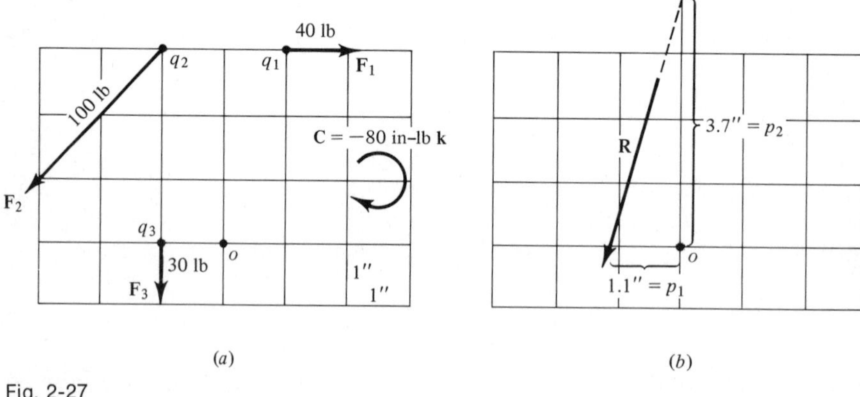

Fig. 2-27

Solution: The system is composed of three forces and a couple. We first determine the sum of this system. Once again we will use a table.

Force, lb	i Component	j Component
40	40	0
100	−71	−71
30	0	−30
Sum	−31	−101

Remark: Notice that the couple **C** does not enter into the determination of the sum of the forces. Thus, this system has a resultant force given by

$$\mathbf{R} = (-31\mathbf{i} - 101\mathbf{j})\ \text{lb}$$

We must now determine a location for **R** to make it statically equivalent to the given system. To this end we construct the following table:

Force F	p Relative to o	$\mathbf{p} \times \mathbf{F} = \mathbf{M}_o$
$40\mathbf{i}$	$1\mathbf{i} + 3\mathbf{j}$	$-120\mathbf{k}$
$-71\mathbf{i} - 71\mathbf{j}$	$-1\mathbf{i} + 3\mathbf{j}$	$284\mathbf{k}$
$-30\mathbf{j}$	$-1\mathbf{i}$	$30\mathbf{k}$

Hence

$$\begin{aligned}\mathbf{M}_o(\mathscr{S}) &= \mathbf{M}_o(\mathbf{F}_1) + \mathbf{M}_o(\mathbf{F}_2) + \mathbf{M}_o(\mathbf{F}_3) + \mathbf{C}\\ &= -120\mathbf{k} + 284\mathbf{k} + 30\mathbf{k} - 80\mathbf{k}\\ &= +114\mathbf{k}\ \text{in-lb}\end{aligned}$$

Note that in determining $\mathbf{M}_o(\mathscr{S})$ we had to include the moment of the couple **C** as well as those of the forces.

The resultant force $\mathbf{R} = -31\mathbf{i} - 101\mathbf{j}$ must be located in such a manner as to produce the moment with respect to o of $114\mathbf{k}$ in-lb.

In general, the position vector of the resultant force can be written as

$$\mathbf{p} = p_1\mathbf{i} + p_2\mathbf{j}$$

However, from our knowledge of static translations, we know that an infinity of such **p**'s will exist. In our case

$$(p_1\mathbf{i} + p_2\mathbf{j}) \times (-31\mathbf{i} - 101\mathbf{j}) = 114\mathbf{k}$$

$$-101p_1\mathbf{k} + 31p_2\mathbf{k} = 114\mathbf{k}$$

or

$$-101p_1 + 31p_2 = 114$$

In particular, for $p_1 = 0$, p_2 will be 3.7 in, while for $p_2 = 0$, p_1 will be −1.1 in. Thus, the resultant force **R** will have the line of action indicated in Fig. 2-27(b).

Before leaving this example (or any other problem for that matter), you should satisfy yourself that the obtained answer makes physical sense. In this particular problem, the original system produces a counterclockwise moment of 114 in-lb with respect to point o. The resultant force must also produce such a moment. Physically, we see that it does if its line of action is above and to the left of point o. Thus, the location of **R** seems physically reasonable for this situation.

We also make the hindsight observation that this problem would have been much simpler if we had chosen point q_2 instead of o for our reference point. Why? What is the moment $\mathbf{M}_{q_2}(\mathscr{S})$? Using foresight instead of hindsight can often save you a lot of work.

Parallel Force System

A parallel force system is one for which all the forces are parallel to one another. For such a system all the forces can be written as a multiple of some unit vector. For example, each of the forces in the system shown in Fig. 2-28(a) can be written as a multiple of the **j** unit vector of a natural triad. Clearly, the sum of such a system will also be a force parallel to the **j** vector, so that we can write

$$\mathbf{S}(\mathscr{S}) = \mathbf{R} = R\mathbf{j}$$

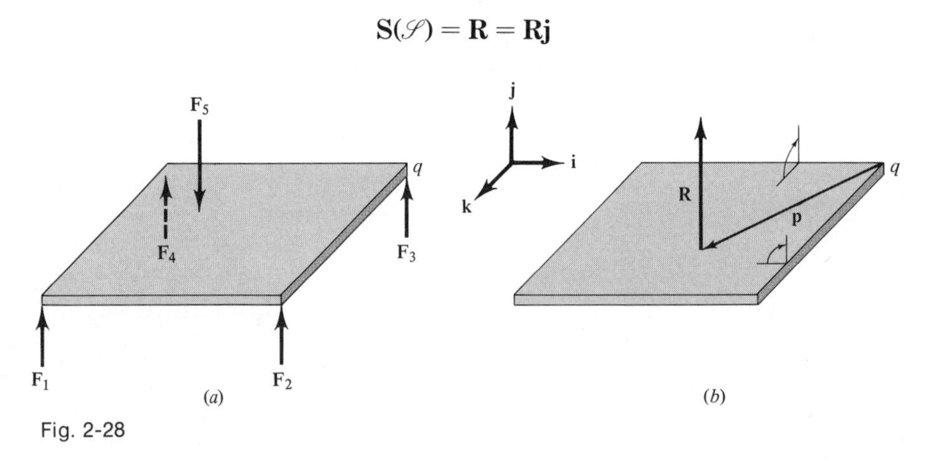

Fig. 2-28

In considering moments, observe that the line of action of each force intersects the horizontal **i**–**k** plane so that the moment of each force with respect to q can be written in the form

$$\begin{aligned}\mathbf{M}_q(\mathbf{F}) &= \mathbf{p} \times \mathbf{F} = (p_1\mathbf{i} + p_2\mathbf{k}) \times (F\mathbf{j}) \\ &= p_1F\mathbf{k} - p_2F\mathbf{i}\end{aligned}$$

so that the moment vector will always lie in the **i**–**k** plane. Physically, the forces are tending to produce rotation about the **i** and **k** axes, but not about the **j** axis. Hence, the moment $\mathbf{M}_q(\mathscr{S})$ of such a parallel system will always be perpendicular to the sum $\mathbf{R} = \mathbf{S}(\mathscr{S})$. Therefore, if $\mathbf{R} \neq \mathbf{0}$, it is always possible to position **R** in such a manner as to give the desired moments. For example, in Fig.

2-28(b) **R** is positioned so that it produces the type of rotation indicated by the circular arrows. If $\mathbf{R} = \mathbf{0}$ and $\mathbf{M}_q(\mathscr{S}) \neq \mathbf{0}$, then the resultant is a couple lying in the plane perpendicular to the forces. In either case, the resultant *cannot* be a wrench.

EXAMPLE 2-7

Determine the resultant of the parallel force system acting on the flat plate shown in Fig. 2-29(a).

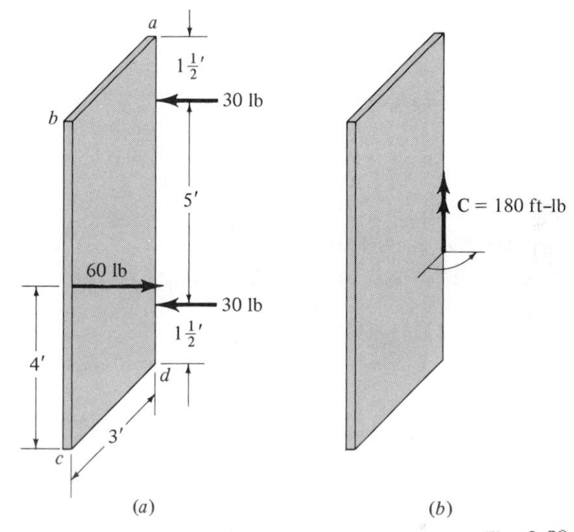

Fig. 2-29

Solution: By inspection, the sum of this system is zero:

$$\mathbf{S}(\mathscr{S}) = \mathbf{0}$$

However, we are not finished because we must check for rotational effects. Instead of using the formalism of the vector cross products, we will examine the situation in a more physically intuitive manner.

The 60-lb force tends to rotate the plate about the vertical a–d axis with a moment of

$$\mathbf{M}_{ad} = (60)(3) = 180 \text{ ft-lb}$$

On the other hand, the net moment about the horizontal c–d axis is given by

$$\mathbf{M}_{cd} = (60)(4) + (30)(1\tfrac{1}{2}) + (30)(6\tfrac{1}{2})$$

$$= 240 + 45 + 195 = \mathbf{0}$$

Thus, the overall effect of the given system of forces is a net zero force with a net turning effect of 180 ft-lb about the a–d axis. In more formal terminology the resultant of this system is a couple of

$$\mathbf{C} = 180\mathbf{j} \text{ ft-lb}$$

as shown in Fig. 2-29(b).

Before closing this discussion of resultants, we will look at one example of a noncurrent, noncoplanar, nonparallel force system. The inherent complexity of such a system will probably make you appreciate the simplifications which are possible when the system is concurrent, coplanar, or parallel as in the previous examples.

EXAMPLE 2-8

During a certain maneuver the forces on an airplane are approximately those shown in Fig. 2-30(a). Find the resultant of this system. (k stands for kip = 1,000 lb.)

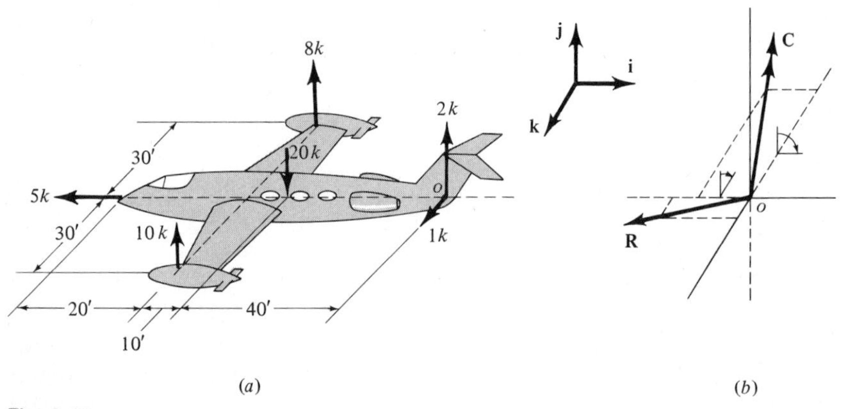

Fig. 2-30

Solution: Using a natural triad, we see that the sum of this system is

$$\begin{aligned} S(\mathscr{S}) &= -5\mathbf{i} + 8\mathbf{j} + 10\mathbf{j} + 2\mathbf{j} - 20\mathbf{j} + 1\mathbf{k} \\ &= -5\mathbf{i} + 1\mathbf{k} \end{aligned}$$

Hence, the net force is

$$\mathbf{R} = -5 \text{ kips } \mathbf{i} + 1 \text{ kip } \mathbf{k}$$
$$\|\mathbf{R}\| = \sqrt{26} \text{ kips}$$

In considering moments, probably the most convenient point to choose is the tail of the plane since three of the forces are concurrent at this point. Then, the moment of the system with respect to point o is given by

$$\begin{aligned}\mathbf{M}_o(\mathscr{S}) &= (-50\mathbf{i} + 30\mathbf{k}) \times (10\mathbf{j}) + (-50\mathbf{i} - 30\mathbf{k}) \times (8\mathbf{j}) + (-40\mathbf{i}) \times (-20\mathbf{j}) \\ &= -500\mathbf{k} - 300\mathbf{i} - 400\mathbf{k} + 240\mathbf{i} + 800\mathbf{k} \\ &= (-60\mathbf{i} - 100\mathbf{k}) \text{ kip-ft}\end{aligned}$$

Thus, we have now reduced the original system to a force

$$\mathbf{R} = (5\mathbf{i} + 1\mathbf{k}) \text{ kip}$$

which, if located at point o, has to be accompanied by a couple

$$\mathbf{C} = \mathbf{M}_o(\mathscr{S}) = (-60\mathbf{i} - 100\mathbf{k}) \text{ kip-ft}$$

This system, shown in Fig. 2-30(b), is a more simple system than the original one in that we can see that the net effect is a force in the direction of $\mathbf{R}$ and a rotational effect about the moment axis of $\mathbf{C}$ tending to pitch and roll the plane.

However, we can further reduce this system into a wrench. To do this, we must express $\mathbf{M}_o$ in terms of two components, one parallel and one perpendicular to $\mathbf{R}$. Thus, we write

$$\mathbf{M}_o = \mathbf{M}_o^{\mathbf{R}} + \mathbf{M}_o^{\perp}$$

with

$$\mathbf{M}_o^{\mathbf{R}} = (\mathbf{M}_o \cdot \boldsymbol{\mu}_R)\boldsymbol{\mu}_R$$

and

$$\mathbf{M}_o^{\perp} = \mathbf{M}_o - \mathbf{M}_o^{\mathbf{R}}$$

where $\boldsymbol{\mu}_R$ is a unit vector parallel to $\mathbf{R}$. Now

$$\boldsymbol{\mu}_R = \frac{\mathbf{R}}{\|\mathbf{R}\|} = \frac{-5}{\sqrt{26}}\mathbf{i} + \frac{1}{\sqrt{26}}\mathbf{k}$$

Hence

$$\begin{aligned}\mathbf{M}_o \cdot \boldsymbol{\mu}_R = \mathbf{M}_o^{R} &= (-60\mathbf{i} - 100\mathbf{k}) \cdot \left(\frac{-5}{\sqrt{26}}\mathbf{i} + \frac{1}{\sqrt{26}}\mathbf{k}\right) \\ &= \frac{300}{\sqrt{26}} - \frac{100}{\sqrt{26}} = \frac{200}{\sqrt{26}}\end{aligned}$$

$$\mathbf{M}_o^{\mathbf{R}} = \frac{200}{\sqrt{26}}\left(-\frac{5}{\sqrt{26}}\,\mathbf{i} + \frac{1}{\sqrt{26}}\mathbf{k}\right) \text{ kip-ft}$$

and

$$\begin{aligned}\mathbf{M}_o^{\perp} &= (-60\mathbf{i} - 100\mathbf{k}) - \frac{200}{\sqrt{26}}\left(\frac{-5}{\sqrt{26}}\mathbf{i} + \frac{1}{\sqrt{26}}\mathbf{k}\right) \\ &= \frac{560}{\sqrt{26}}\left(-\frac{1}{\sqrt{26}}\mathbf{i} - \frac{5}{\sqrt{26}}\mathbf{k}\right) \text{ kip-ft}\end{aligned}$$

$$M_o^{\perp} = \frac{560}{\sqrt{26}} \text{ kip-ft}$$

These components are shown in Fig. 2-31(a).

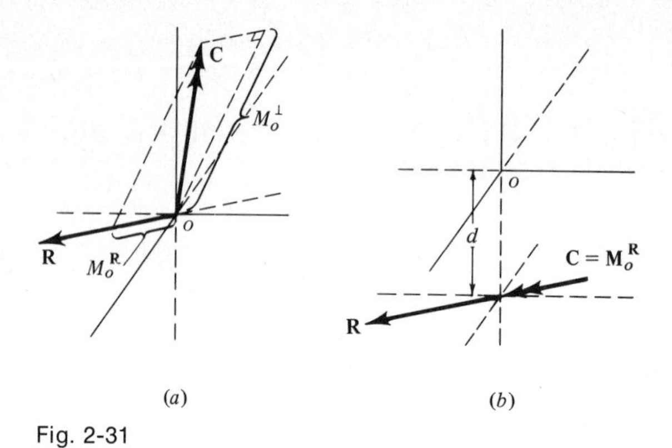

Fig. 2-31

Now, **R** cannot produce $\mathbf{M}_o^R$ no matter where it is positioned, but it can produce $\mathbf{M}_o^\perp$. In fact, if **R** is positioned below o by an amount d, where

$$Rd = M_o^\perp$$

R will produce the moment $\mathbf{M}_o^\perp$. Thus

$$d = \frac{560/\sqrt{26}}{\sqrt{26}} = \frac{560}{26} = 21.5 \text{ ft}$$

and the resultant wrench is shown in Fig. 2-31(*b*).

> ***Remark:*** It is quite debatable as to whether it is always worth the effort to reduce a system such as the one in Fig. 2-30(*b*) to a wrench as in Fig. 2-31(*b*). One must not lose sight of the fact that the purpose of simplifying any complex system of forces is to better understand the physical effect of the system. To this end, the wrench in Fig. 2-31(*b*) is only slightly more useful than the force-couple system of Fig. 2-30(*b*).

The ideas in this section are extremely important and it would be worth your time to restudy this section before proceeding to the exercise problems which follow.

EXERCISE PROBLEMS

2-68 In the structural joint shown, determine the resultant of the three member forces.

PROB. 2-68

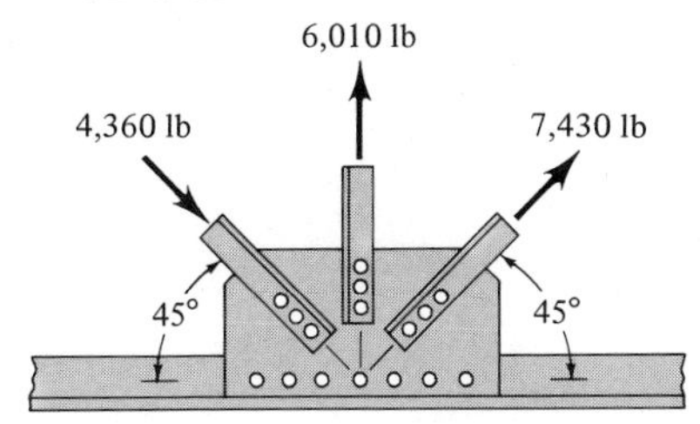

2-69 Find the resultant of the three forces shown.

PROB. 2-69

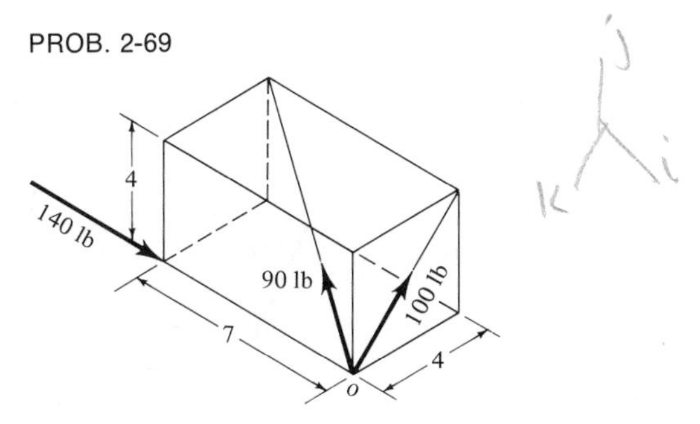

2-70 The 120-lb force is the resultant of a four-force system, three of which are shown. Determine the missing fourth force.

PROB. 2-70

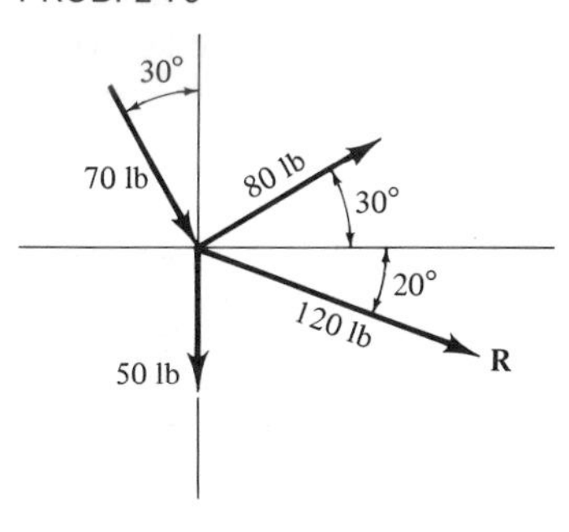

PROB. 2-71

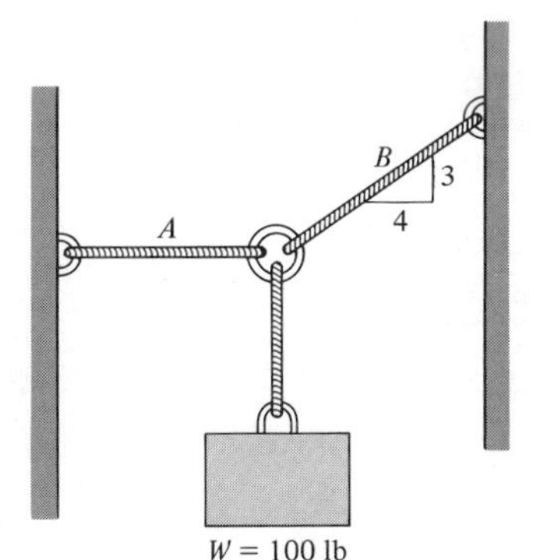

2-71 The cables A and B and the 100-lb weight are fastened to the steel ring as shown. The resultant of these three forces on the ring is the null resultant. Determine the tension of cables A and B.

2-72 and 2-73 The four cables are attached to the fixed eyebolt as shown. Replace these four cables with one cable which is statically equivalent to them.

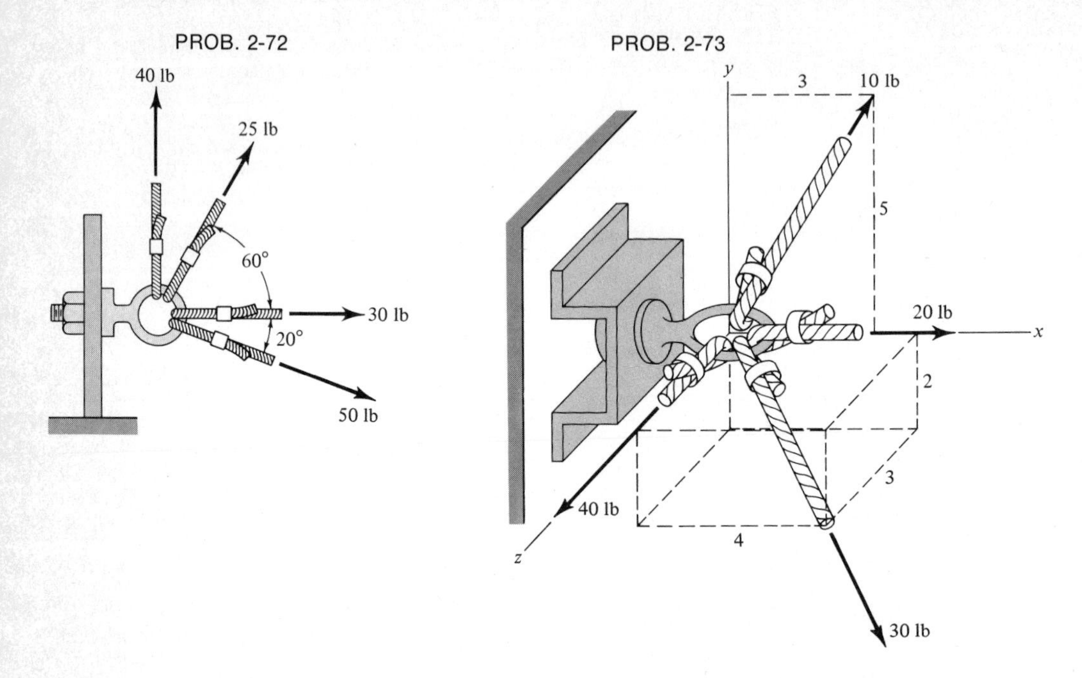

2-74 The three forces $\mathbf{T}_1$, $\mathbf{T}_2$, and $\mathbf{T}_3$ are pulling on the ring as shown. If their resultant is the null resultant, find $\mathbf{T}_3$.

2-75 The boat is resting in the water in the middle of the channel. Each man is pulling with a force of 50 lb. Find their resultant pull on the boat.

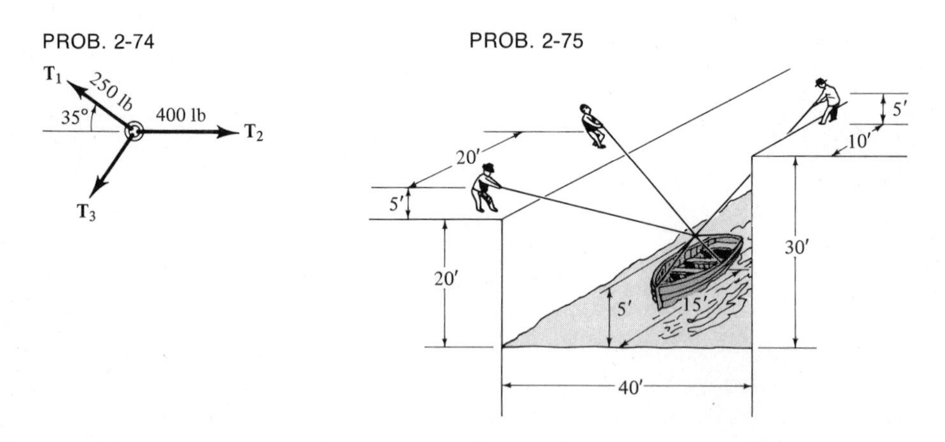

2-76 The three guy wires are supporting the mast as shown. If their resultant is a downward vertical force of 85 lb, what is the tension in each of the wires?

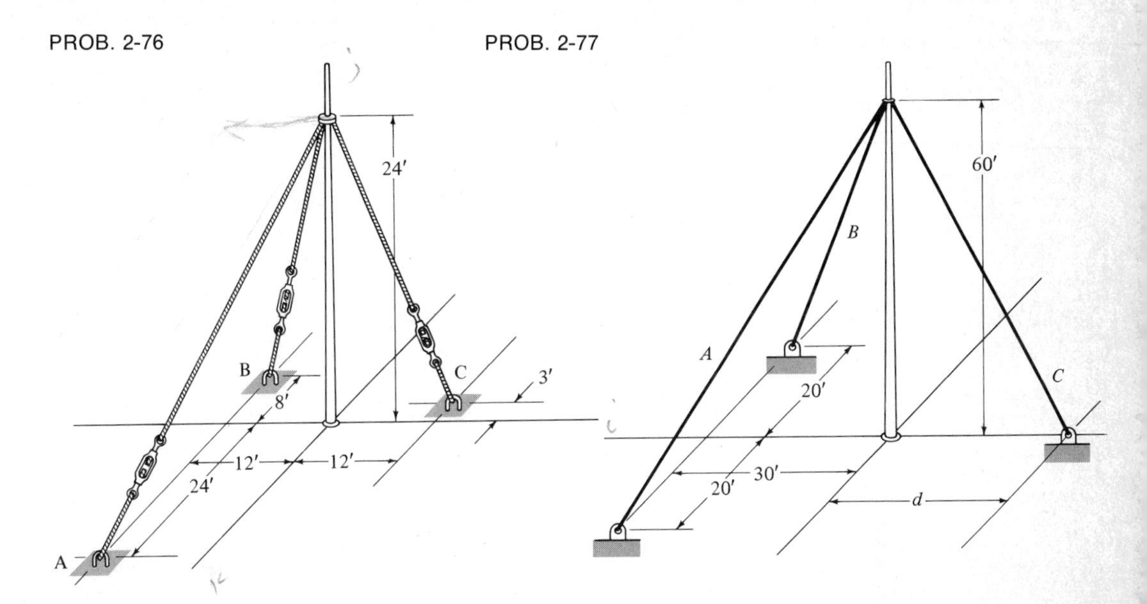

2-77 The resultant of the cables A, B, and C acting on the mast is a downward vertical force of 200 lb. If the cables A and B each carry a tension of 70 lb, determine the tension in C and the distance d.

2-78 Find the resultant of the two forces acting on the cantilever beam shown.

2-79 Three tugboats pushing in the directions shown are used to maneuver a freighter from its dock berth into the main channel. Each tug pushes with a force of 4,000 lb. Find an equivalent force-couple system with the force located at point o. Find the resultant of the three tugboat forces.

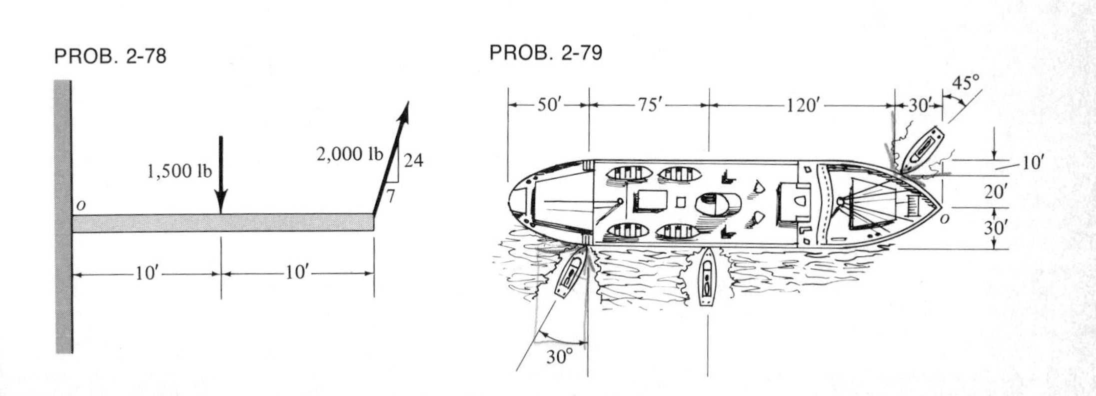

2-80 and 2-81 Determine the resultant of the forces shown.

PROB. 2-80

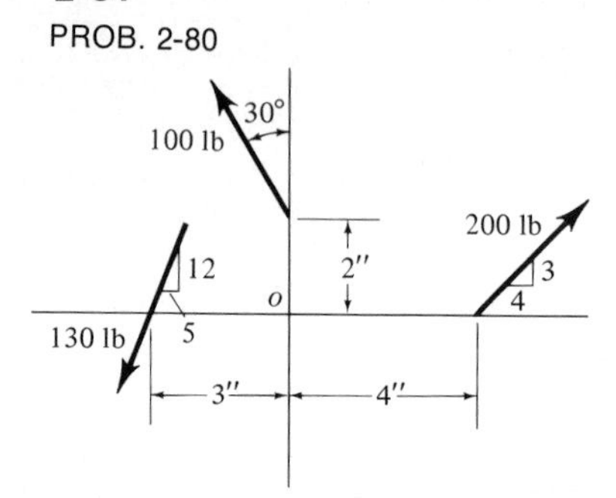

PROB. 2-81

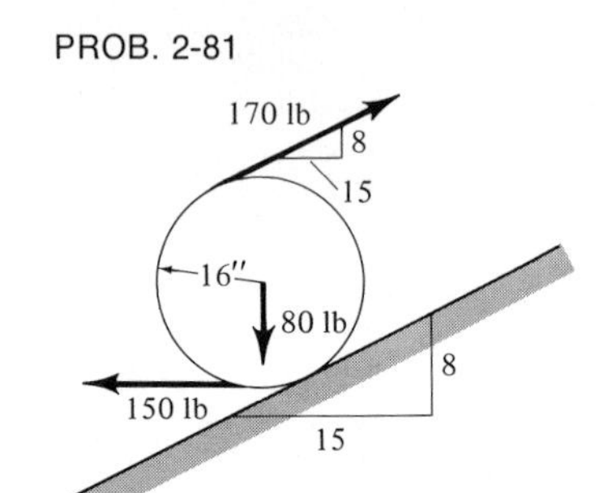

2-82 to 2-84 Replace the force-couple system shown by a single force.

PROB. 2-82

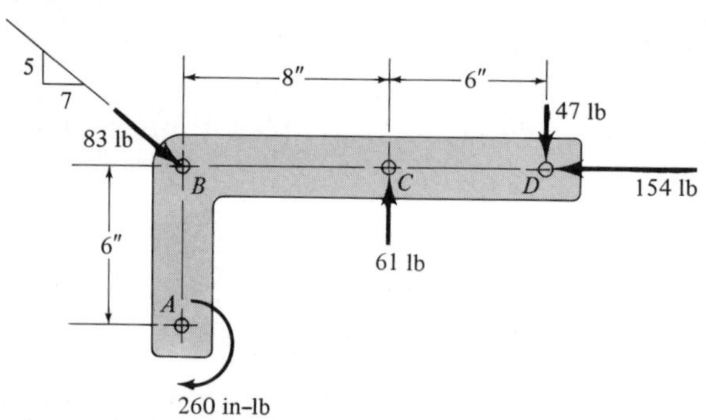

PROB. 2-83

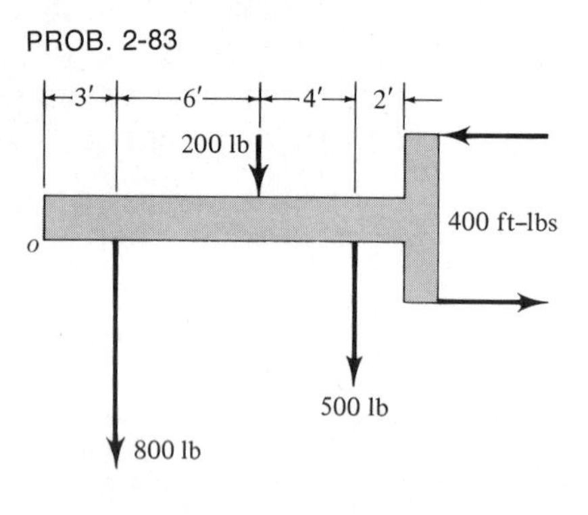

PROB. 2-84

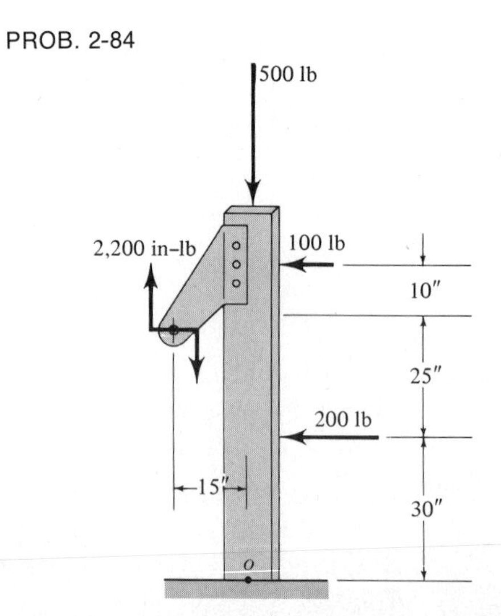

2-85 The 500-lb force is the resultant of a three-force system, two of which are shown. Determine the missing third force.

PROB. 2-85

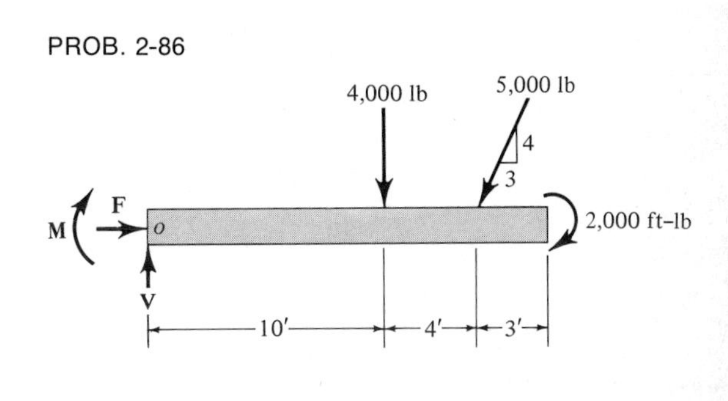

PROB. 2-86

2-86 Determine **V**, **F**, and **M** such that the resultant of all the forces on the beam shown is the null resultant.

2-87 Find the resultant of the force system shown.

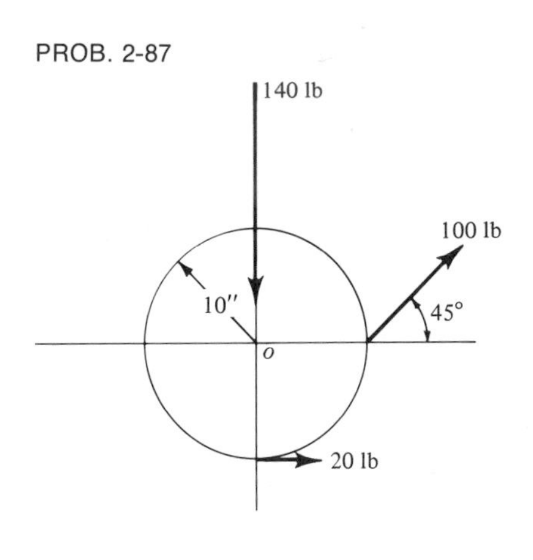

PROB. 2-87

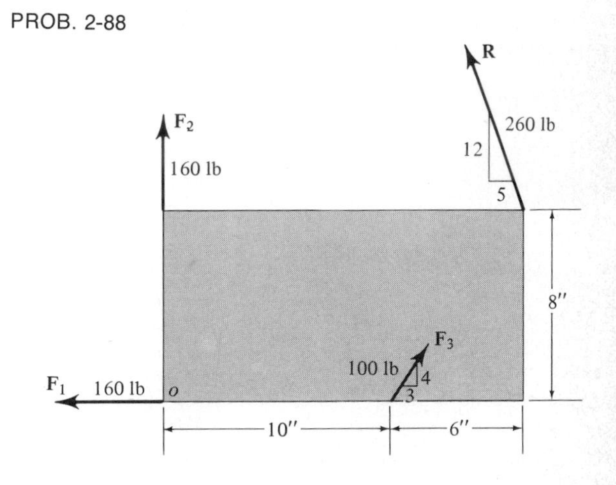

PROB. 2-88

2-88 A force-couple system has the resultant **R** shown. The three forces $\mathbf{F}_1$, $\mathbf{F}_2$, and $\mathbf{F}_3$ are part of the system. What is the complete system?

2-89 **R** is the resultant of a four-force system. Three of the forces are shown and the fourth passes through point *q*. Determine the missing force and the resultant **R** whose direction is indicated in the figure.

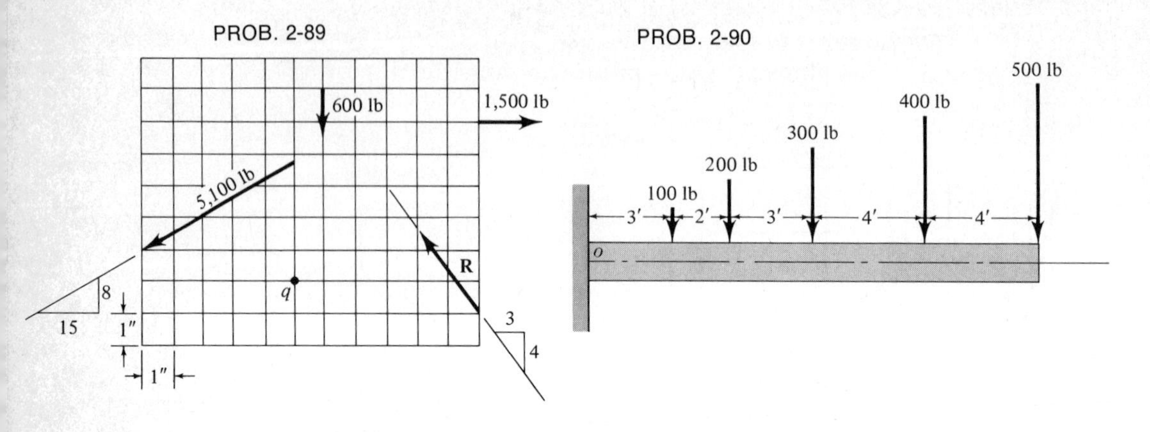

2-90 and 2-91 Find the resultant of the force system shown and locate it relative to point *o*.

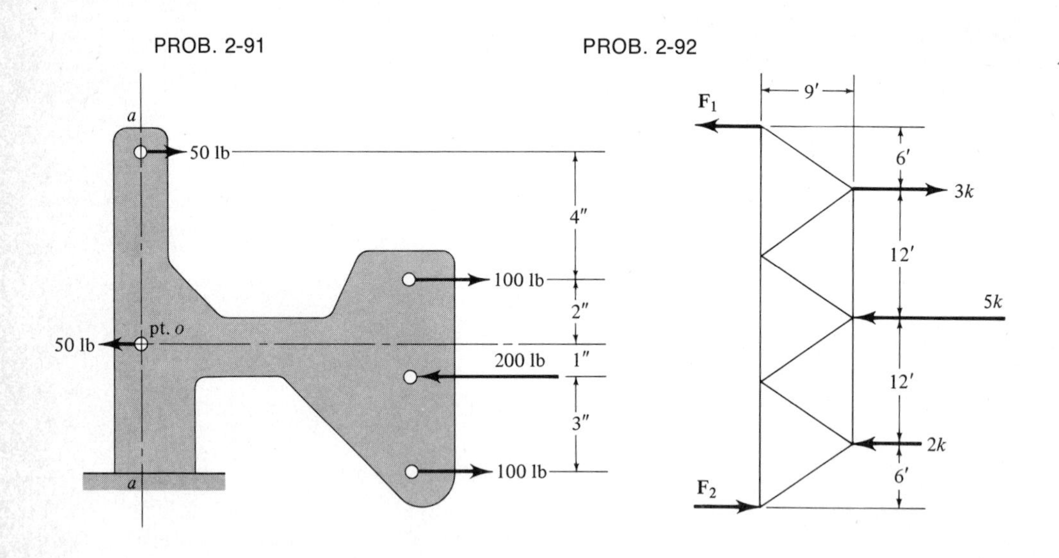

2-92 The resultant of the forces acting on the truss shown is the null resultant. Determine $\mathbf{F}_1$ and $\mathbf{F}_2$.

2-93 to 2-96 Determine the resultant of the force system shown.

PROB. 2-93

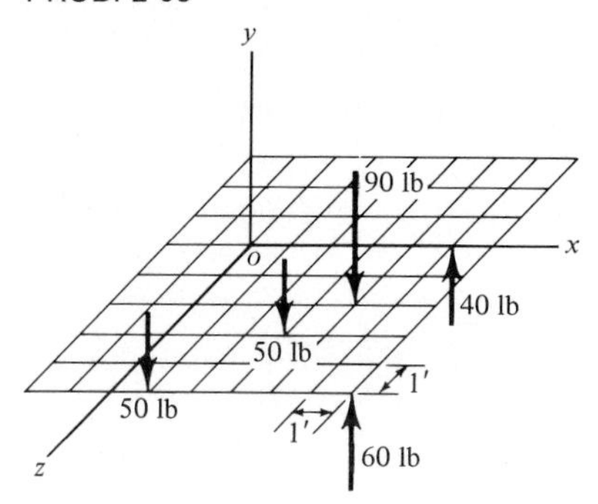

PROB. 2-94

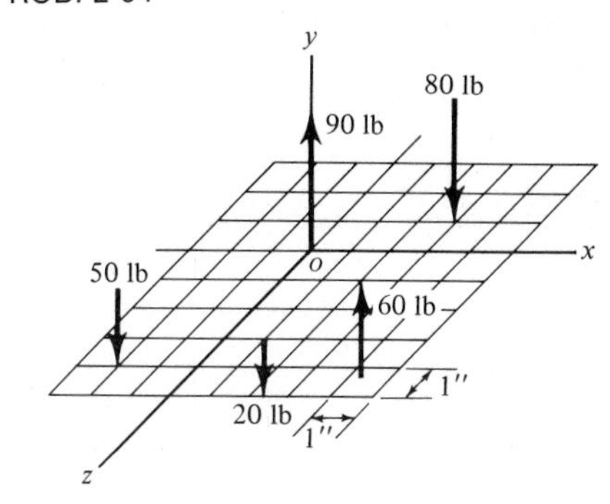

PROB. 2-95

PROB. 2-96

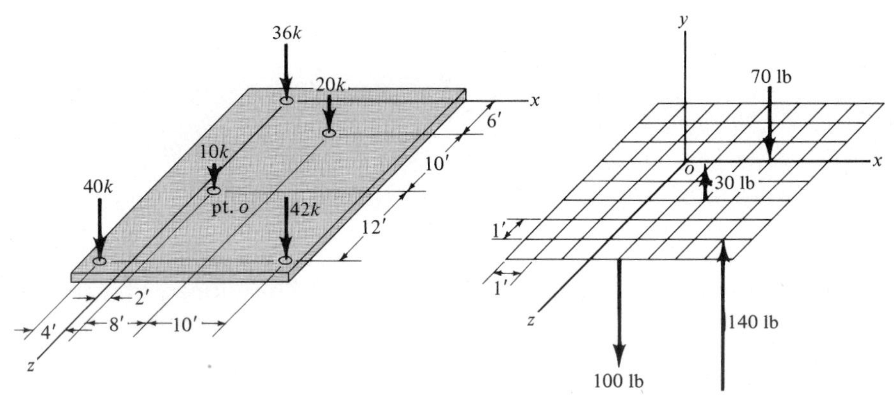

2-97 For the four-force system shown, add another force to this system so that its resultant is the null resultant.

PROB. 2-97

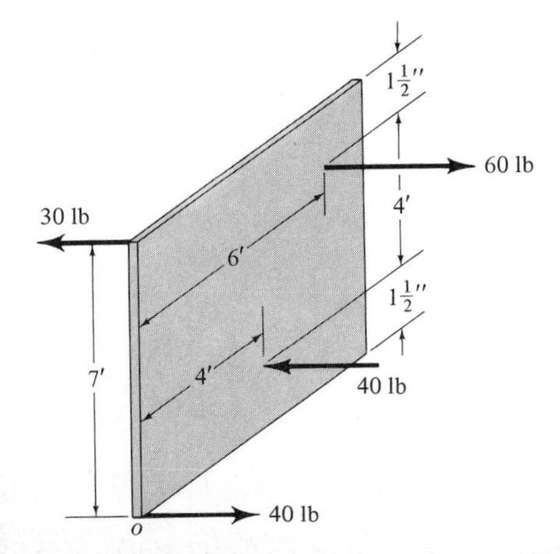

PROB. 2-98

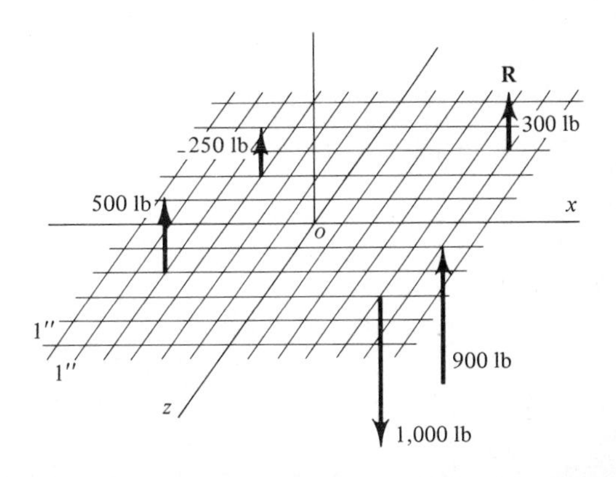

2-98 **R** is the resultant of a five-force system, four of which are shown. Determine the missing force and its location.

2-99 During flight the loads on an airplane are as shown. Find the resultant of these forces.

PROB. 2-99

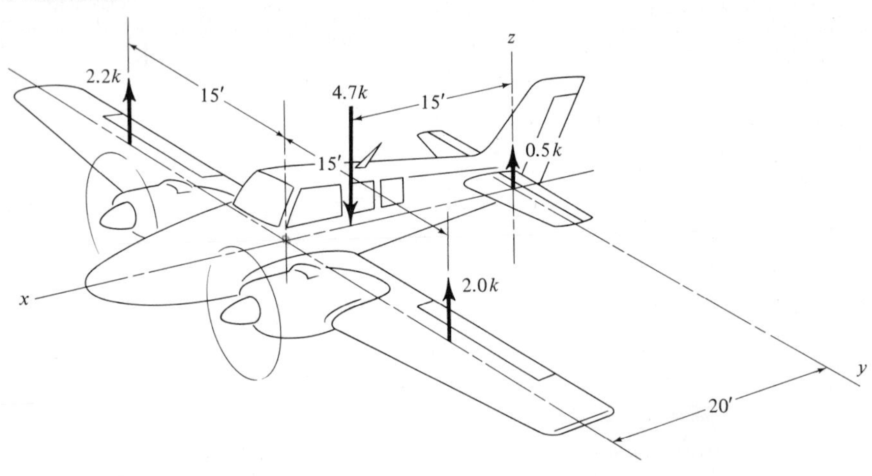

2-100 to 2-105 Reduce the given force system to a single force located at point o and a couple. Can this system be reduced to a single force?

PROBS. 2-100 AND 2-106

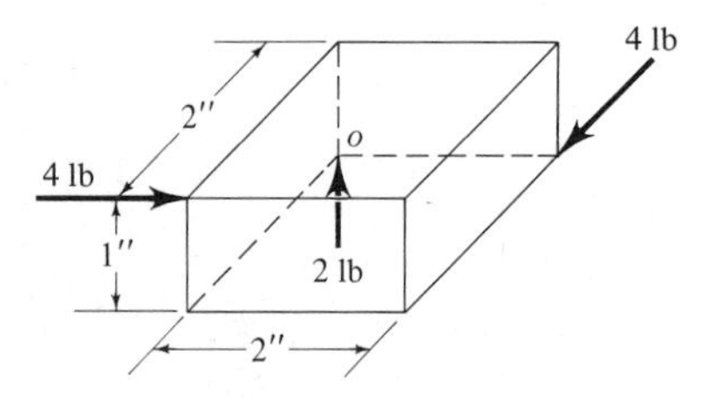

PROB. 2-101

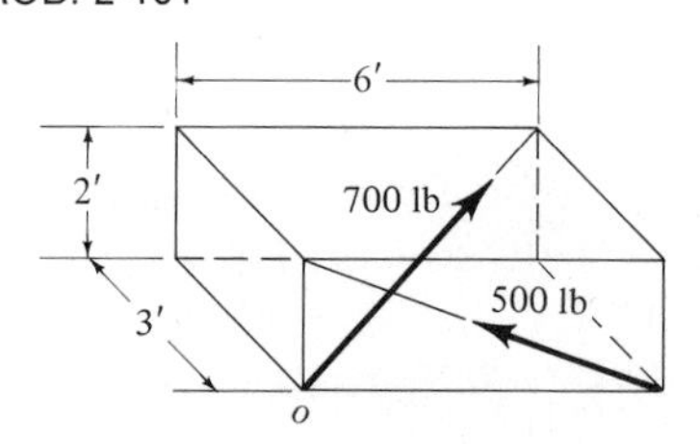

PROB. 2-102

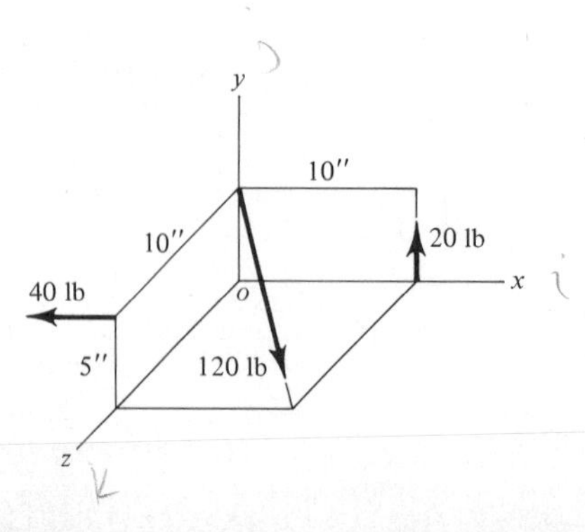

PROB. 2-103

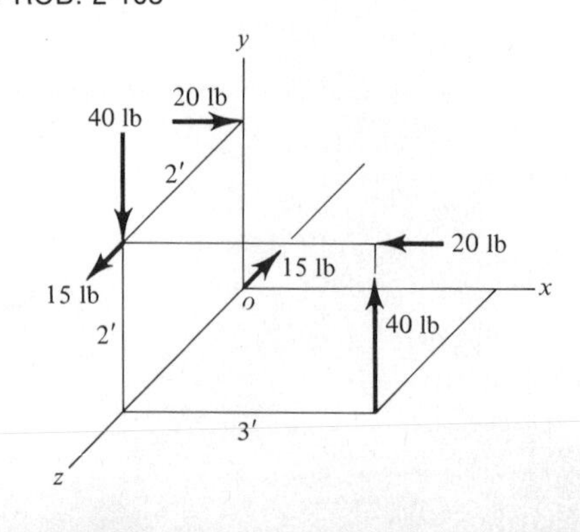

PROBS. 2-105 AND 2-107

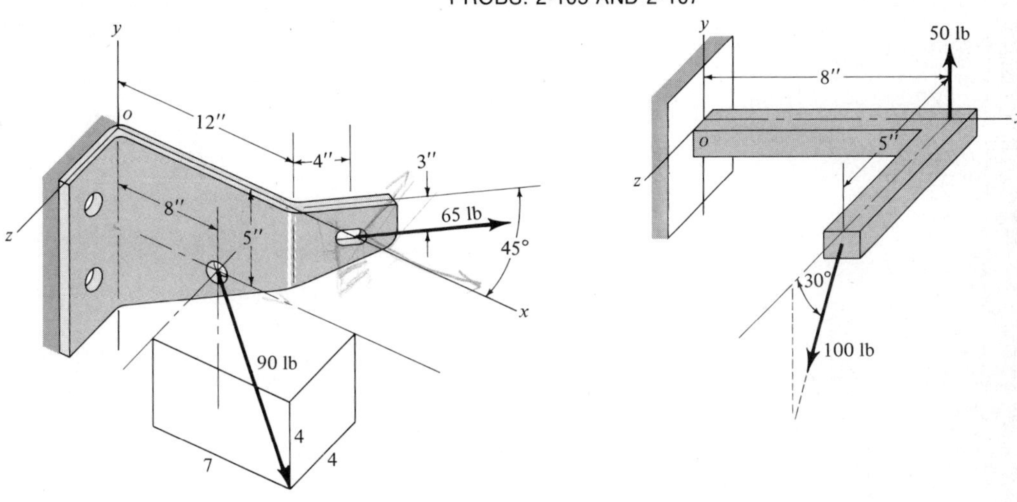

2-106 to 2-109 Find the resultant of the forces shown.

PROB. 2-108

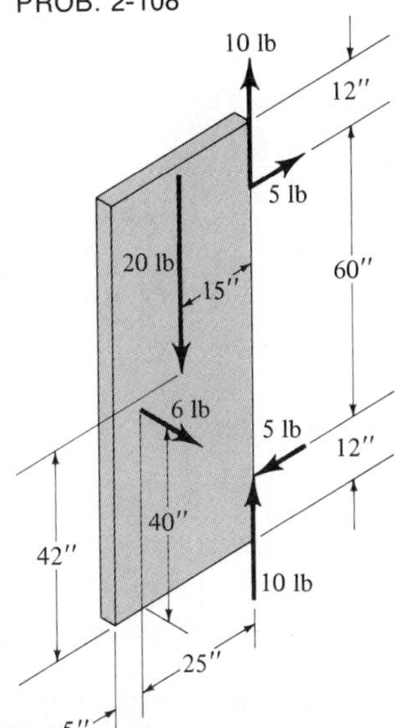

PROB. 2-109

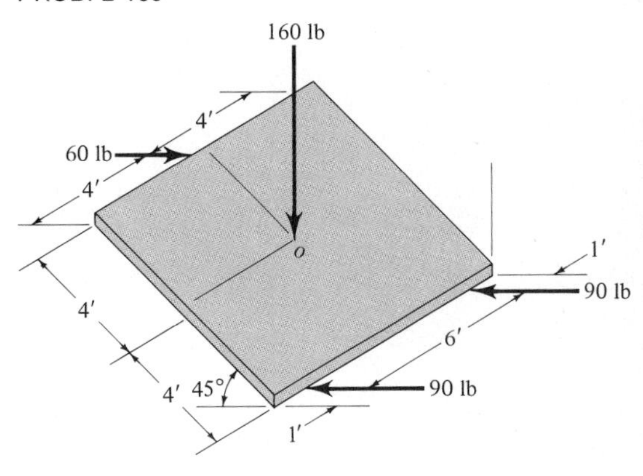

2-110 and 2-111 Find **F** and **C** if the resultant of the system is the null resultant.

PROB. 2-110

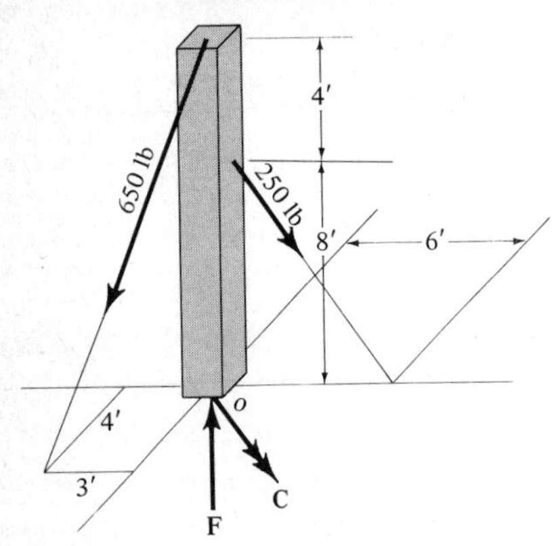

PROB. 2-111

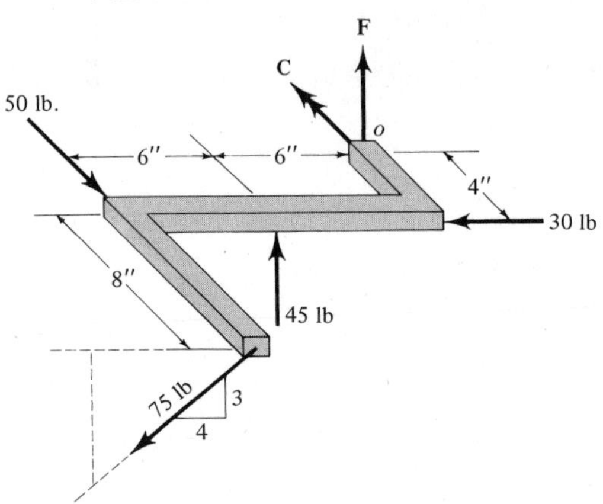

chapter 3 DISTRIBUTED FORCES

3-1 INTRODUCTION

In the first two chapters, every force, regardless of how it was produced, was represented by a single directed line segment called the force vector. This vector was located at what we called the point of application of the force. However, it was mentioned in the closure of Chap. 2 that there are many physical situations in which the forces involved cannot be modeled or represented in a realistic manner by a single vector line segment acting at a particular location. Some simple examples of such situations might be the pressure force which keeps an automobile tire inflated, the wind force blowing against the roof of an A-frame cabin, or the weight of a large building on its soil foundation.

In each of these situations, the fact that the forces involved are actually *distributed* over a surface or volume must be taken into consideration in any attempt to model or represent such physical situations in a realistic manner. This chapter will introduce some ideas which can be quite useful in dealing with situations involving distributed forces.

3-2 BODY FORCES AND SURFACE TRACTIONS

It is true that practically all forces are distributed over some surface or volume. For example, your weight is actually distributed over your entire body although you might represent your weight by a single downward vector located, say, at your stomach. Or, when you push against a door with your hand, the force is distributed over the contact area between your hand and the door, although this force would probably be represented by a single vector located at the "middle" of your hand. Such simplified representations are usually quite adequate for many, if not most, situations. But, let

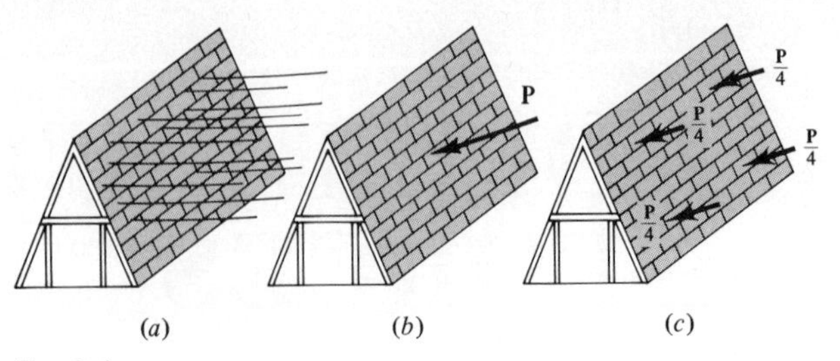

Fig. 3-1

us now return to the example of the wind blowing against the A-frame-cabin roof.

A crude first-approximate representation of such a distributed force might be a single force vector equal to the total wind load **P** located at the "middle" of the roof, as illustrated in Fig. 3-1(*b*). Perhaps a better approximate representation of the situation might be accomplished by using several smaller vectors appropriately located on the roof, as, for example, in Fig. 3-1(*c*). Of course, since the wind load is actually distributed over the entire surface of the roof, the most accurate way to represent this situation would be by a distribution of infinitely many force vectors over the entire surface. The basic requirement of such a distribution is that its *sum* must be equal to the total load **P**, that is,

$$\text{Total load} = \text{summation of the distribution}$$

In the case of the wind-roof example, the summation must be carried out over the surface area of the roof. In the case of the weight of a large building, the summation must be carried out over the volume of the building.

If the representation of the distributed load consists of a large but finite number of discrete forces each located at some particular point, then this representation can be handled as an ordinary force system, such as those in Chap. 2; that is, its statical resultant can be found by vectorially summing the discrete forces and their moments. The use of a computer can often simplify and speed up the actual work involved and permits the use of a large number of discrete forces in representing a distributed load. On the other hand, it may be more desirable for the engineer to represent the distributed load by a *continuous* distribution of forces rather than a discrete distribution. In such cases the summing operation formally becomes an integration, and it becomes necessary to introduce the concepts of surface tractions and body forces.

Definition 3-1 When a force is distributed over a surface area A, we define the *surface-traction distribution* **t** such that

$$\mathbf{F} = \iint_{\text{area } A} \mathbf{t}\, dA \tag{3-1}$$

where **F** is the total net force. Since **t** can vary from point to point over the surface A, both **t** and dA must usually be written in terms of some surface coordinates in order to perform the indicated integration.

Definition 3-2 When a force is distributed over a volume V, we define the *body-force density* **b** such that

$$\mathbf{F} = \iiint_{\text{vol } V} \mathbf{b}\, dV \tag{3-2}$$

where **F** is the total net force. Since **b** can vary from point to point throughout the volume V, both **b** and dV must usually be written in terms of some volume coordinates in order to perform the indicated integration.

> ***Remark:*** In these formal definitions, the integral signs should be interpreted simply as summing operations over areas or volumes. In most situations the actual integration operations will reduce to the ordinary Riemann integration of elementary calculus.

We usually interpret **t** to be a force per unit area, such as pounds per square inch. If the traction is a pushing force, we often call its magnitude the *pressure*, such as air pressure, water pressure, etc.

In a similar manner we usually interpret **b** to be a force per unit volume, such as pounds per cubic inch. The most common example of such a body force is that due to gravity. In this case we interpret its magnitude as the *specific weight* of the body, which is usually calculated by dividing the total weight of a body by the volume of the body.

While pressure and specific weight are by far the most common examples of surface tractions and body forces, respectively, there are many other types of both. For example, the magnetic effect of a ferrous body is a body force, and the frictional effect between two rubbing surfaces is a surface traction.

We pause now in our development to give some examples dealing with distributed forces. Please study these carefully as they provide some very useful ideas in working with distributed forces.

EXAMPLE 3-1 HYDROSTATIC PRESSURE

A 20 × 20 × 10-ft tank is partially filled to a depth of 8 ft with water weighing 62.4 lb/ft^3, as indicated in Fig. 3-2(a). What is the total weight of the water in the tank? What is the pressure of the water against the floor of the tank? What is the pressure of the water against the sides of the tank?

Solutions: The first question is essentially a body-force problem with the magnitude of the body-force density corresponding to the specific weight w of the water. Hence, by Definition 3-2, with the total weight W corresponding to the magnitude of **F** and w corresponding to the magnitude of **b**,

$$W = \iiint_{\substack{\text{vol}\\\text{water}}} w\, dV = w \iiint_V dV = wV$$

since w is a constant 62.4 throughout the entire volume V of the water. Hence

$$\begin{aligned} W &= 62.4 \text{ lb/ft}^3 \,[(20)(20)(8)] \text{ ft}^3 \\ &= 200{,}000 \text{ lb} \end{aligned}$$

The second question is a matter of finding the force of the water against a unit of area of the floor of the tank. To this end, consider a column of water 1 ft square and 8 ft high, as in Fig. 3-2(b). Since the water is the same height throughout the tank, this column represents the amount of water supported by each square foot of floor. Thus the weight supported by each square foot of floor is

$$\begin{aligned} \text{Weight of column} &= w \times \text{area} \times \text{height} \\ &= (62.4)[(1)(1)(8)] = 500 \text{ lb} \end{aligned} \tag{a}$$

Hence the pressure of the water against the floor of the tank can be given as 500 lb for each square foot, or 500 lb/ft^2.

> ***Remark:*** This same result of 500 lb/ft^2 could have been obtained by making the following observation: Since the water is a uniform depth throughout the tank, its total weight is *evenly* supported by the entire floor of the tank. This means that the surface traction (or pressure) is uniform over the entire floor. Then, by Definition 3-1, with the weight W corresponding to the magnitude of **F** and the pressure p corresponding to the magnitude of **t**, we have
>
> $$\begin{aligned} W &= \iint_{\substack{\text{floor}\\\text{area}}} p\, dA = p \iint_A dA \\ &= pA \end{aligned}$$

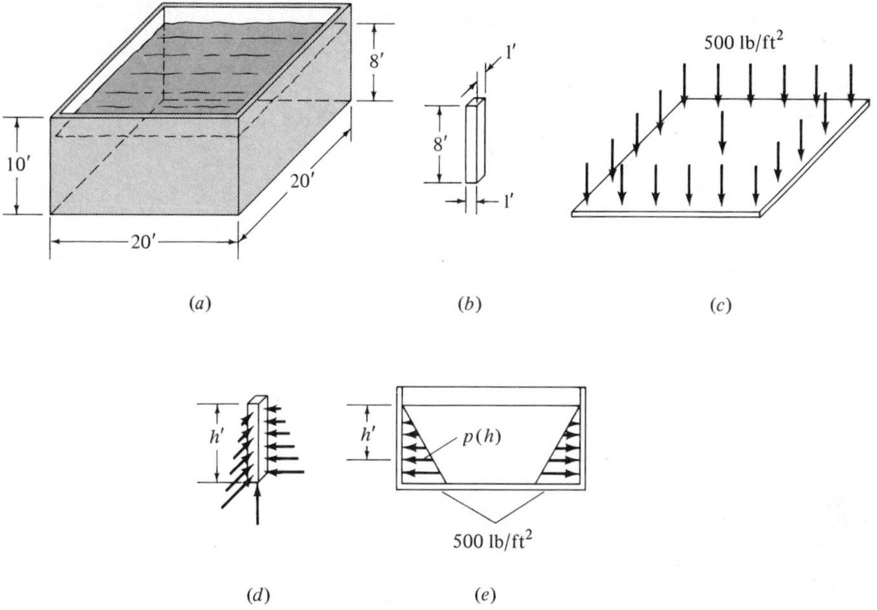

Fig. 3-2

Therefore,

$$p = \frac{W}{A} = \frac{200{,}000\ \text{lb}}{(20\ \text{ft})(20\ \text{ft})} = 500\ \text{lb/ft}^2$$

This uniform pressure is illustrated in Fig. 3-2(*c*).

The third question in the problem is to find the pressure (force per unit area) of the water against the sides of the tank. Returning to the column of water in Fig. 3-2(*b*), obviously such a column of water could not stand by itself since the water would tend to flow. Furthermore, a fluid such as water tends to flow equally in all directions (except up) and, therefore, has to be constrained by equal tractions (pressure) in *all* directions to prevent its flowing, as indicated in Fig. 3-2(*d*). Such tractions are called fluid pressure or *hydrostatic pressure.*

We have seen from Eq. (*a*) how the pressure on the bottom of such a column is directly related to the height of the column as measured from the top surface of the water. Consequently, the water pressure at the top would be zero (since the height would be zero) but would increase linearly with the depth (height) of the water. Hence, the water pressure varies linearly from zero at the top to its maximum at the bottom. Accordingly, for a given depth h from the *top* surface of the water, the pressure $p(h)$ is given by

$$\begin{aligned} p(h) &= 62.4\ \text{lb/ft}^3 \times h\ \text{ft} \\ &= 62.4\,h\ \text{lb/ft}^2 \end{aligned}$$

Therefore, the hydrostatic water pressure on the sides of the tank varies linearly with the depth from the water surface. This variation is illustrated in Fig. 3-2(*e*).

EXAMPLE 3-2

A gas-storage tank has the shape shown in Fig. 3-3(*a*) and stores gas at a gage pressure of p psi. What is the net total force on the semicircular portion A–B of the tank?

Solution: Realizing that the tank is L ft long, consider the two-dimensional figure in Fig. 3-3(*b*). The pressure is always perpendicular to the surface, and, therefore, the traction **t** is given by

$$\mathbf{t} = p\boldsymbol{\mu}_r$$

where $\boldsymbol{\mu}_r$ is a unit vector parallel to the radius. In terms of the *fixed* **i** and **j** unit vectors we have

$$\mathbf{t} = p(\mathbf{i}\cos\theta + \mathbf{j}\sin\theta)$$

Now, by Definition 3-1,

$$\mathbf{F} = \iint_{\substack{\text{semicircular}\\ \text{area}}} \mathbf{t}\, da$$

$$= \iint_{\text{area}} p(\mathbf{i}\cos\theta + \mathbf{j}\sin\theta)\, dA$$

To carry out the indicated integration over the simicircular area, we must express dA in terms of θ. Recalling that the length of the tank is L, we take as dA the small strip of width $r\,d\theta$ running the length L of the tank, so that

$$dA = Lr\,d\theta$$

as shown in Fig. 3-3(*c*). Thus, since p, r, **i**, and **j** are all constant,

$$\mathbf{F} = \int_{-\pi/2}^{\pi/2} p(\mathbf{i}\cos\theta + \mathbf{j}\sin\theta)Lr\,d\theta$$

$$= prL\left[\mathbf{i}\int_{-\pi/2}^{\pi/2}\cos\theta\,d\theta + \mathbf{j}\int_{-\pi/2}^{\pi/2}\sin\theta\,d\theta\right]$$

$$= prL\,[\mathbf{i}2 + \mathbf{j}0]$$

$$= \mathbf{i}p\,(2rL)$$

That is, the net force is a horizontal force with a magnitude equal to the pressure p times the flat vertical projected area $2rL$.

Remark: This result is indicative of problems involving hydrostatic fluid pressures in that the net force in a given direction is governed by the *projected* area perpendicular to that direction,

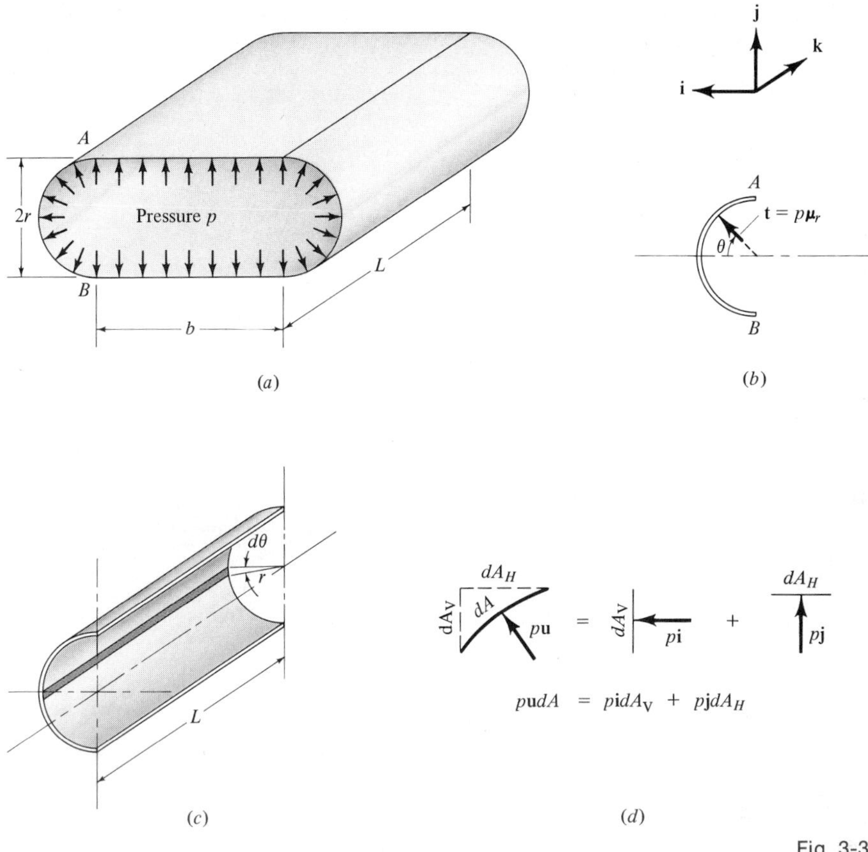

Fig. 3-3

i.e., horizontal force obtained from vertical projected area and vertical force from horizontal area. This is illustrated in Fig. 3-3(d), where **u** is a unit vector perpendicular to the area dA.

Remark: The gage pressure of the gas inside the tank is used rather than the absolute pressure since the gage pressure represents the difference between the outside and inside pressures and hence is the net effect of the gas.

EXAMPLE 3-3

The bin shown in Fig. 3-4(a) is filled level with loosely packed dry sand weighing 100 lb/ft^3. What is the net force of the sand against the inclined door in the bin floor?

Solution: Although the sand is not a fluid, for dry sand it is common practice to estimate the sand pressure in the same manner as we did the hydrostatic fluid pressure in Example 3-1. Thus,

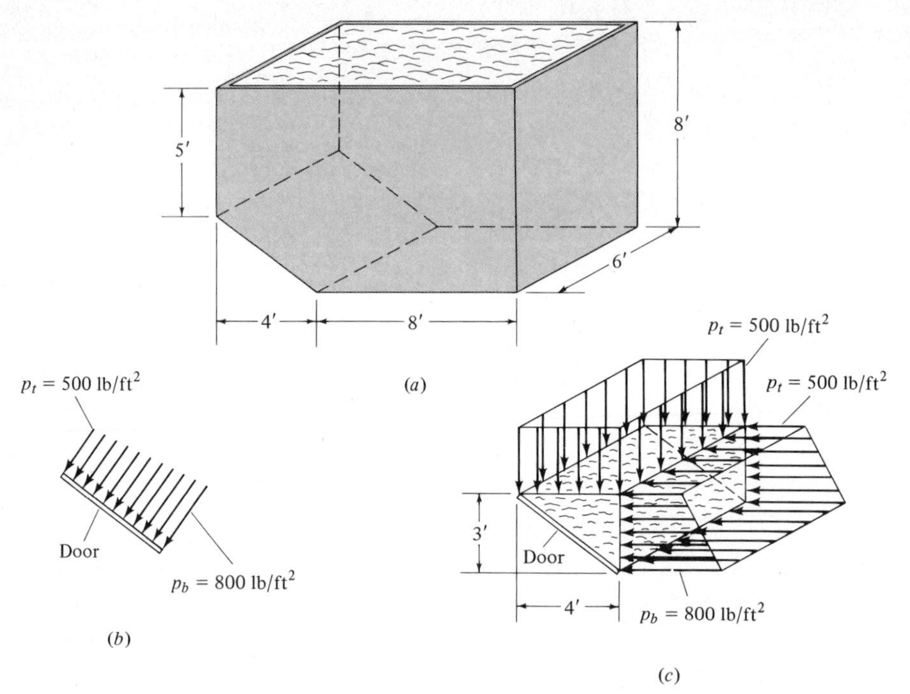

Fig. 3-4

treating the sand as a fluid, the pressure at the top of the door would be

$$
\begin{aligned}
p_t &= (100\ \text{lb/ft}^3)(5\ \text{ft}) \\
&= 500\ \text{lb/ft}^2
\end{aligned}
$$

while at the bottom of the door the pressure would be

$$
\begin{aligned}
p_b &= (100\ \text{lb/ft}^3)(8\ \text{ft}) \\
&= 800\ \text{lb/ft}^2
\end{aligned}
$$

The pressure variation on the door is indicated in the two-dimensional diagram of Fig. 3-4(*b*). If we wish, we could integrate this pressure variation over the area of the door to determine the total force, just as we did in Example 3-2. However, we will solve this problem in another manner, which is often helpful in handling problems of this type.

Recall that the *hydrostatic* pressure at a point is the same in *all* directions. With this in mind, let us visualize the forces acting on the door by the scheme shown in Fig. 3-4(*c*). In this figure we have shown the door, a triangular volume of sand resting on the door, and the pressure exerted on this volume of sand by the rest of the sand in the bin. The pressure across the top of this triangular pile of sand is a uniform 500 lb/ft², while the pressure on the side of this pile varies linearly from 500 lb/ft² at the top to 800 lb/ft² at the bottom.

The total vertical force on the door will be the weight of the triangular volume of sand plus the downward force due to the pressure of 500 lb/ft² on the top of this pile. Hence

$$\begin{aligned} F_v &= w(\text{volume}) + p_t(\text{area}) \\ &= (100)(\tfrac{1}{2})(3)(4)(6) + 500(4)(6) \\ &= 3{,}600 + 12{,}000 = 15{,}600 \text{ lb downward} \end{aligned}$$

To find the horizontal force on the door, by the remark at the end of Example 3-2, the net *horizontal* force is due to the pressure of the sand pushing against the *vertical* 3 × 6-ft area. Hence

$$F_h = \iint_{\text{area}} p \, dA$$

where the pressure p varies linearly with the depth. While we could formally integrate this integral to obtain the net horizontal force, instead let us make a useful observation. Since the pressure varies *linearly* from 500 lb/ft² at the top of the area in question to 800 lb/ft² at the bottom of the area, the *average* pressure would be

$$p_{\text{ave}} = \frac{800 + 500}{2} = 650 \text{ lb/ft}^2$$

Therefore, the net horizontal force will be the *average horizontal* pressure multiplied by the *vertical* area. Thus

$$\begin{aligned} F_h &= p_{\text{ave}}(\text{area}) \\ &= 650(6)(3) \\ &= 11{,}700 \text{ lb to the left} \end{aligned}$$

Hence, the net total force on the door will be

$$\begin{aligned} \mathbf{F}_{\text{total}} &= \mathbf{i}F_h + \mathbf{j}F_v \\ &= -11{,}700\mathbf{i} - 15{,}600\mathbf{j} \\ &= 19{,}500 \text{ lb} \end{aligned}$$

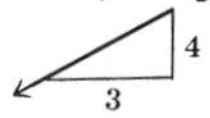

That is, the net force will be perpendicular to the door, as we anticipated from the diagram of Fig. 3-4(*b*) and the fact that pressure force is *always normal* to the area on which it acts.

Remark: The location of this net force is still an open question which we will consider in subsequent sections

EXAMPLE 3-4 FLUID BUOYANCY

A body of some arbitrary shape and volume V is submerged in a fluid of constant specific weight w. What is the lifting force exerted by the fluid on the submerged body?

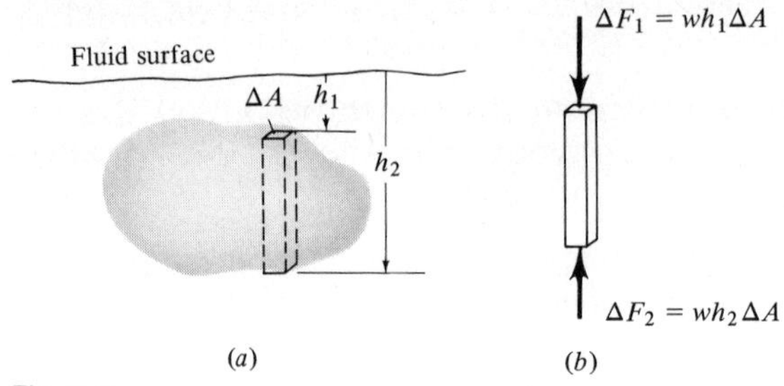

Fig. 3-5

Solution: We envision the submerged body as being composed of small column-like elements of cross section ΔA, as shown in Fig. 3-5(a). Recall from Example 3-1 that the fluid pressure varies according to the depth:

$$p(h) = wh$$

Hence, the downward force due to the fluid pressure on the top of this column is

$$\Delta F_1 = p(h_1)\,\Delta A = wh_1\,\Delta A$$

while the upward force due to the fluid pressure on the bottom of this column is

$$\Delta F_2 = p(h_2)\,\Delta A = wh_2\,\Delta A$$

as indicated in Fig. 3-5(b). Thus the net *upward* force on this column is

$$\begin{aligned}\Delta F = \Delta F_2 - \Delta F_1 &= w(h_2 - h_1)\,\Delta A \\ &= w\Delta V\end{aligned}$$

where ΔV is the volume of this column. Hence, if we sum up all the ΔF's on all the small column-like elements comprising the body, we have formally

$$F = \iiint_V w\,dV = wV$$

where V is the total volume of the submerged body. Thus the magnitude of the *upward buoyant force* due to the fluid pressure is equal to the weight of *fluid displaced* by the submerged body. If the body is totally submerged, V will be the total volume of the body. Otherwise, V will be only the volume of that portion of the body under the surface of the fluid.

EXERCISE PROBLEMS

3-1 A large water tank is partially filled with water. A gage at the bottom indicates a pressure of 20 lb/in^2. If the floor of the tank has 100 ft^2 of area, what is the total force against the floor? How many cubic feet of water are in the tank?

3-2 A pressure chamber has a door measuring $2\frac{1}{2} \times 6$ ft. If the pressure in the chamber is three times the standard atmospheric pressure of 14.7 psi, what is the net force on the door?

3-3 An airplane has a net wing area of 800 ft^2. What must be the average pressure differential between the top side and bottom side of the wing if the plane is to develop a lift of 100,000 lb?

3-4 In Marineland there is a large fish tank 12 ft deep and 20 ft in diameter. At the bottom of this tank there is a viewing window 1 ft high and 18 in wide. Approximate the net force of the water against this window.

3-5 A $20 \times 20 \times 10$-ft bin is partially filled with loosely packed sand weighing 100 lb/ft^3 as indicated. What is the total weight of the sand in the bin? What is the pressure distribution of the sand against the floor of the bin?

PROB. 3-5 PROB. 3-6

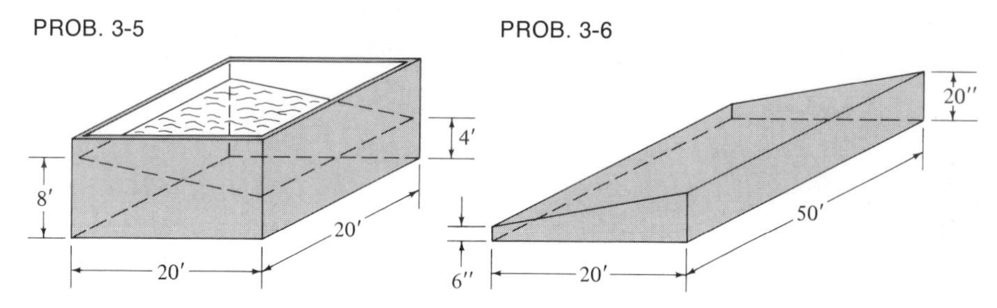

3-6 A snow drift on a flat 20×50-ft roof lies in approximately the shape shown. If the specific gravity of snow is 0.2, what is the total weight of snow on the roof? What is the pressure distribution across the roof?

3-7 A cylindrical air tank has spherical caps on its ends. What is the net force on one of the end caps if the tank is 3 ft long, has an inside diameter of 10 in, and carries a gage pressure of 60 psi?

3-8 Wind blows against the roof of an A-frame cottage as shown. If the average pressure (above atmospheric) on the windward side of the roof is 0.2 psi, what is the total wind force on this side of the roof? What is the net horizontal force on this side of the roof? What is the net vertical force on this side of the roof?

PROB. 3-8

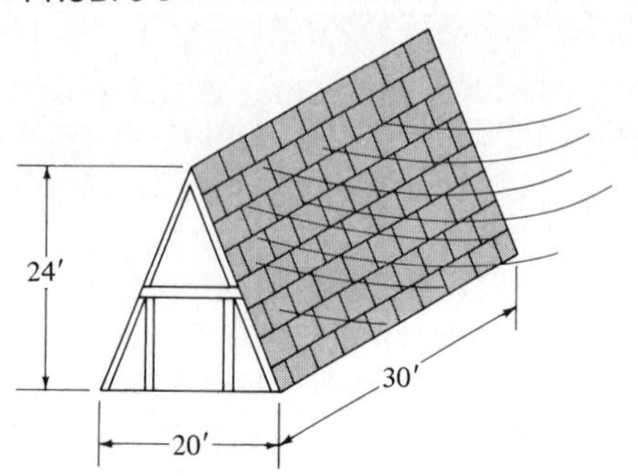

PROB. 3-9

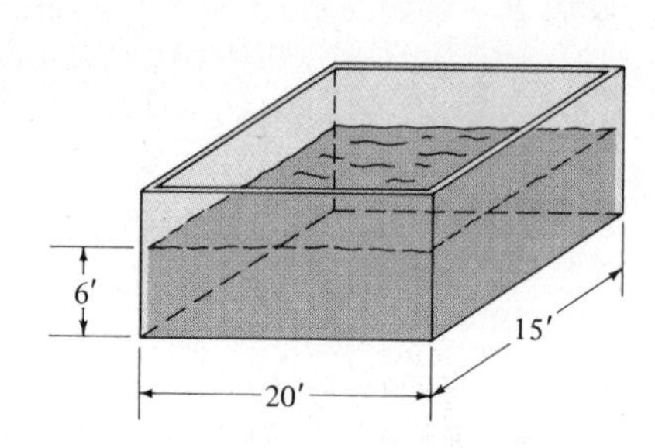

3-9 A tank is partially filled with water to a depth of 6 ft as shown. What is the net force exerted by the water against the front end of the tank? What is the net force exerted by the water against the left side of the tank?

3-10 In Prob. 3-5, what is the net force exerted by the sand against the left side of the bin? What is the net force against the right side of the bin? Treat the sand as if it were a fluid.

3-11 The dam indicated in the illustration holds back water in a 200-ft-wide river having a depth of 40 ft. What is the net horizontal force of the water against the upstream side of the dam?

PROB. 3-11

10′
40′
30′

PROB. 3-12

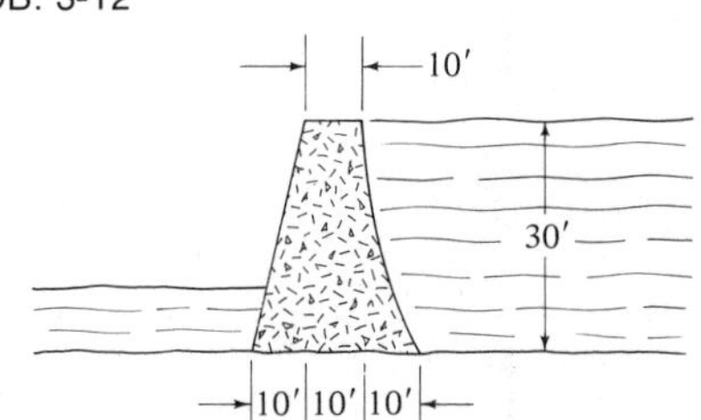

3-12 A dam indicated in the illustration holds back water in a 100-ft-wide river with a depth of 30 ft. What is the net horizontal force of the water against the upstream parabolic-shaped side of the dam?

3-13 In Prob. 3-11, what is the net vertical force of the water against the upstream side of the dam?

3-14 In Prob. 3-12, what is the net vertical force of the water against the upstream side of the dam?

3-15 A rectangular block $6 \times 5 \times 5$ ft weighs 5 tons. Will it float in water? Why?

3-16 If the specific gravity of ice is 0.9, what percentage of an iceberg is visible above the water?

3-17 A coal barge has outside dimensions of 200 ft long, 30 ft wide, and 8 ft deep. The barge itself weighs 50 tons. How many tons of coal can the barge hold before it would submerge?

3-18 A 12-ft-diameter vinyl spherical balloon is filled with helium (0.011 lb/ft^3). What is the buoyant force of the outside air (0.076 lb/ft^3) on the inflated balloon?

3-3 MOMENTS AND RESULTANTS OF DISTRIBUTED FORCES

In the previous section we have seen how to find the net total force or *sum* of a distributed force. We now face the same question that we faced in Chap. 2, namely: What is the simplest force or force system which is *statically equivalent* to the given distributed force? In other words, what is the *statical resultant* of the given force distribution?

To answer this question, we must first define the moment of a distributed force.

Definition 3-3 Let $\mathbf{t}$ be a surface-traction distribution and q be some point in space. Then the moment of this distribution with respect to point q is

$$\mathbf{M}_q(\mathbf{t}) = \iiint_{\text{area}} \mathbf{p} \times \mathbf{t}\, dA \tag{3-3}$$

where $\mathbf{p}$ is the position vector measured from q to the element of surface dA. Both $\mathbf{t}$ and $\mathbf{p}$ usually must be expressed in terms of some surface coordinates in order to perform the indicated integration.

In a similar manner we define the following:

Definition 3-4 Let $\mathbf{b}$ be a body-force density and q be some point in space. Then the moment of this distributed force with respect to point q is

$$\mathbf{M}_q(\mathbf{b}) = \iiint_{\text{vol}} \mathbf{p} \times \mathbf{b}\, dV \tag{3-4}$$

where $\mathbf{p}$ is the position vector measured from q to the element of volume dV. Both $\mathbf{b}$ and $\mathbf{p}$ usually must be expressed in terms of some volume coordinates in order to perform the indicated integration.

Recall from Sec. 2-5 that two force systems $\mathscr{S}$ and $\mathscr{S}'$ are statically equivalent if and only if their sums are equal and their moments with respect to any arbitrary point are equal. Also, recall that the resultant is the *simplest* system which is statically equivalent to the given system. Therefore, if we think of a distributed force as being a *system* made up of a very large number of small forces, we are faced with the problem of finding the statical resultant of this distributed force system.

First of all, the sum of a distributed force (system) is simply the net total force as given by

$$\mathbf{S}(\mathbf{t}) = \iint_{\text{area}} \mathbf{t}\, dA \tag{3-1}$$

or

$$\mathbf{S}(\mathbf{b}) = \iiint_{\text{vol}} \mathbf{b}\, dV \tag{3-2}$$

and the moment of such systems has just been defined by Eqs. (3-3) and (3-4). Let $\mathscr{D}$ represent a distributed force system involving either a surface or volume distribution, and let $\mathscr{R}$ represent its statical resultant. Then, by Sec. 2-7, we must have

$$\mathbf{S}(\mathscr{R}) = \mathbf{S}(\mathscr{D})$$
$$\mathbf{M}_q(\mathscr{R}) = \mathbf{M}_q(\mathscr{D})$$

that is, the resultant will have the same sum and produce the same moment as the distributed force system. Therefore, in general, we are confronted with the same possibilities that were discussed in Sec. 2-7; that is, the resultant could be

1 A single force having a prescribed line of action
2 A couple
3 A wrench (force and couple)
4 The null (zero) system

Fortunately, however, many distributed force systems are made up of parallel forces (recall Examples 3-1, 3-3, and 3-4) which, according to Sec. 2-7, eliminates the wrench possibility and generally makes the problem of finding the resultant much less difficult. As the following sections will illustrate, the idea of a resultant of a distributed force system can lead to some other useful geometric and physical concepts.

3-4 CENTER OF GRAVITY

As we mentioned earlier in this chapter, the most common example of a body-force distribution is the specific weight of a body. Therefore, we ask the question: What is the statical resultant of the distributed weight of a body? In other words, if we represent the weight of a body by a single force vector, where must this vector be located so as to be statically equivalent to the actual distributed weight? We call this location the *center of gravity* of the body. Physically, it is the center of the body insofar as its weight is concerned. Consequently, since all real bodies are three-dimensional, the center of gravity is usually located by giving its location relative to three coordinates. However, there are situations in which one or two coordinates suffice to locate the center of gravity. We will develop these ideas more fully by considering the following examples.

EXAMPLE 3-5

Heavy tubing is laid out in a flat semicircular shape as shown in Fig. 3-6(a). The tubing weighs 1.2 lb/ft of length. Locate the center of gravity for this piece of tubing.

Solution: If the cross section of the tubing is small compared with the dimensions of the semicircle, we visualize the tubing as being made up of small sections of arc length $r\,d\theta$ as shown in Fig. 3-6(b). Each such section will have a weight

$$\begin{aligned} dW &= (1.2 \text{ lb/ft})(r\,d\theta \text{ ft}) \\ &= 1.2r\,d\theta \qquad \text{lb} \end{aligned}$$

The total weight of the tubing is then

$$\begin{aligned} W &= \int_{-\pi/2}^{\pi/2} 1.2r\,d\theta \\ &= \Big[(1.2)(5)\theta\Big]_{-\pi/2}^{\pi/2} = 6\pi \text{ lb} \end{aligned}$$

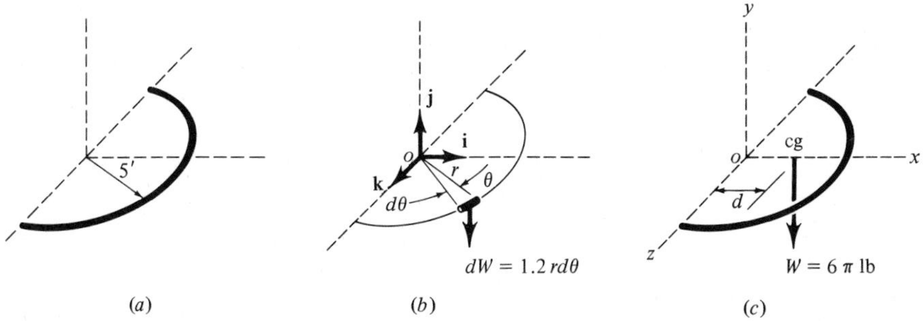

(a) (b) (c)

Fig. 3-6

Thus the resultant must be a force of 6π lb.

To determine the moment of the distributed weight of the tubing, we will use point o as a convenient reference point. Using vector notation, the position vector of the force $\mathbf{dW} = -dW\mathbf{j}$ with respect to point o is

$$\mathbf{p} = r \cos \theta \, \mathbf{i} + r \sin \theta \, \mathbf{k}$$

Thus

$$\begin{aligned}
\mathbf{M}_0(\mathbf{W}) &= \int \mathbf{p} \times \mathbf{dW} = \int_{-\pi/2}^{\pi/2} (r \cos \theta \, \mathbf{i} + r \sin \theta \, \mathbf{k}) \times (-1.2r \, d\theta \, \mathbf{j}) \\
&= -\mathbf{k} \int_{-\pi/2}^{\pi/2} (r \cos \theta)(1.2r \, d\theta) + \mathbf{i} \int_{-\pi/2}^{\pi/2} (r \sin \theta)(1.2r \, d\theta) \\
&= 1.2(5)^2[-\mathbf{k}(2) + \mathbf{i}0] \\
&= -(1.2 \text{ lb/ft})(50 \text{ ft}^2)\mathbf{k} \\
&= -60 \text{ ft-lb } \mathbf{k} = 60 \text{ ft-lb}
\end{aligned}$$

Therefore, the resultant weight of 6π lb must be located so as to produce a clockwise 60-ft-lb **k** moment but no **i** moment with respect to point o. Accordingly, the resultant of the distributed weight of the tubing is a downward vertical force of 6π lb located as shown in Fig. 3-6(c) with

$$d = \frac{60 \text{ ft-lb}}{6\pi \text{ lb}} = \frac{10}{\pi} \text{ ft}$$

By definition then, cg with rectangular coordinates $(10/\pi, 0,0)$ is the center of gravity of the tubing. Before leaving this example, we make some very *important observations.*

1 The result $d = 10/\pi$ is peculiar to this specific problem. However, by reworking the problem for any semicircular arc-shape body of radius r with uniform weight per unit length, we can show that

$$d = \frac{2r}{\pi}$$

2 The configuration of the tubing is symmetric with respect to the x axis, and the center of gravity is located on that axis. This is a general result in that if a body's weight is symmetrically distributed about some axis (or axes), the center of gravity will always lie on that axis (or axes).

3 Just as in this example, the center of gravity need not be a point actually in the body. Usually, it will be within the body, but not always. Quite often you will be able to anticipate the general location of the center of gravity.

4 Since the resultant weight acts through the center of gravity, the moment

$$\mathbf{M}_{cg} = \mathbf{0}$$

Physically, this means that if it were possible to hold the body at its cg, the body would "balance itself" in that its weight would not produce any rotational effect about the cg.

These observations will be quite useful in working subsequent problems.

EXAMPLE 3-6

Find the center of gravity of the flat triangular metal plate shown in Fig. 3-7(*a*). The weight distribution of the plate is given as w lb/ft² of plate area.

Solution: First, the total weight of the plate is

$$W = (w \text{ lb/ft}^2)(\text{area of plate})$$
$$= w(\tfrac{1}{2}\, bh) \text{ lb}$$

If the plate is thin in comparison with its other dimensions h and b, we can visualize the plate as being composed of small strips of length x and width dz, as shown in Fig. 3-7(*b*). The weight of this small strip is

$$dW = (w \text{ lb/ft}^2)(x\ dz \text{ ft}^2) = wx\ dz \text{ lb}$$

and intuitively (from symmetry) the center of gravity of this rectangular strip is at its "middle" $(x/2, z)$, as indicated by the weight dW in Fig. 3-7(*b*).

Now, the moment of dW about the x axis is

$$d\mathbf{M}_x = z\ dW$$

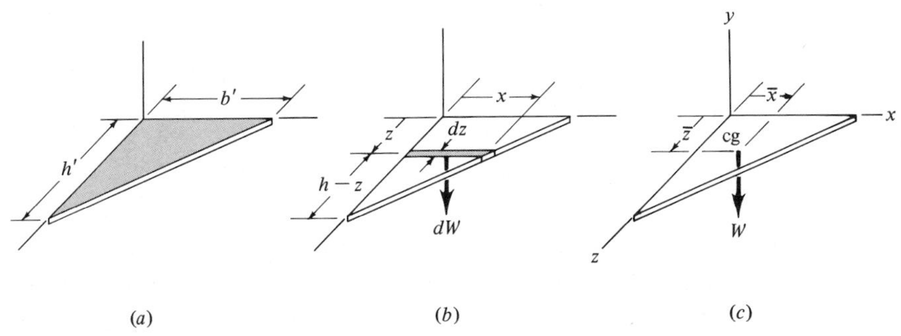

Fig. 3-7

and that about the z axis is

$$d\mathbf{M}_z = \frac{x}{2}\,dW$$

Hence, summing these moments yields, respectively,

$$M_x = \int_o^h zwx\,dz \tag{b}$$

$$M_z = \int_o^h \frac{x}{2} wx\,dz \tag{c}$$

By ratio and proportion we have

$$\frac{x}{h-z} = \frac{b}{h}$$

or

$$x = b\left(1 - \frac{z}{h}\right)$$

Then, integrating (b) and (c), we have

$$M_x = wb\int_o^h \left(1 - \frac{z}{h}\right) z\,dz$$

$$= wb\left[\frac{z^2}{2} - \frac{z^3}{3h}\right]_o^h = \frac{wbh^2}{6}\ \text{ft-lb} \tag{d}$$

and

$$M_z = \frac{wb^2}{2}\int_o^h \left(1 - \frac{z}{h}\right)^2 dz$$

$$= -\frac{wb^2h}{6}\left[\left(1 - \frac{z}{h}\right)^3\right]_o^h = \frac{wb^2h}{6}\ \text{ft-lb} \tag{e}$$

Therefore, referring to Fig. 3-7(c), the center of gravity of the plate is given by

$$\bar{z} = \frac{M_x}{W} = \frac{wbh^2/6}{wbh/2} = \tfrac{1}{3}h\ \text{ft}$$

$$\bar{x} = \frac{M_z}{W} = \frac{wb^2h/6}{wbh/2} = \tfrac{1}{3}b\ \text{ft}$$

Once again we make some *important observations.*

1. Although this plate area was a right triangle, it can be shown that this result is general in that the cg of any triangular-shaped plate would be at one-third the perpendicular distance measured from any side of the triangle.
2. In both this example and Example 3-5, the location of the center of gravity was independent of the specific weight of the body. This is because the specific weight (1.2 lb/ft in Example 3-5 and w lb/ft^2 in Example 3-6) is a constant and ap-

pears in both the expression for the total weight W and the expression for the moment. This will always occur for bodies which have a constant specific weight, i.e., *homogeneous bodies.* For homogeneous bodies the location of the center of gravity is purely a geometrical property of the shape of the body. Thus, if the plate were steel or Fiberglas, while the steel one would be much heavier, both plates would have the same center of gravity.

3. In the light of the previous observation, when dealing with homogeneous bodies, it is often expedient to omit the specific weight from the calculations and deal solely with the volumes (or areas, etc.) of the bodies rather than their weights. When dealing with volumes (or areas), however, we will use the terminology of *centroid of the volume* (or area) rather than the center of gravity of the volume. The reason for distinguishing between the *centroid of a volume* and the *center of gravity of a body* is, of course, that while the two coincide for homogeneous bodies, they will *not* coincide for nonhomogeneous bodies such as loaded dice or a container of a stratified fluid. (See Appendix A for a discussion of centroids.)

EXAMPLE 3-7 CENTER OF BUOYANCY

Determine the center of buoyancy of the submerged body shown in Fig. 3-8(*a*).

Solution: Recalling Example 3-4, the upward buoyant force is equal in magnitude to the weight of fluid displaced by the body. If the body is *totally* submerged in a homogeneous fluid, the weight of displaced fluid will be

$$W = wV$$

where w is the specific weight of the *fluid* and V is the volume of the submerged body. By definition, the center of buoyancy is the center of gravity of this displaced fluid or, since the fluid is homogeneous, the centroid of the *volume* of the submerged body. Physically the center of buoyancy locates the line of action of the buoyant force.

It is convenient to visualize the displaced volume of fluid as being made up of two parts—one a right-triangular prism $6 \times 4 \times 3$ ft and the other a rectangular prism $6 \times 4 \times 4$ ft, as shown in Fig. 3-8(*b*). Then the weight of displaced fluid is

$$\begin{aligned} W_{\text{total}} &= W_1 + W_2 = wV_1 + wV_2 \\ &= w[\tfrac{1}{2}(6)(4)(3)] + w[(6)(4)(4)] \\ &= 36w + 96w = 132w \end{aligned}$$

By symmetry, the centroid of ② is located at the geometric middle

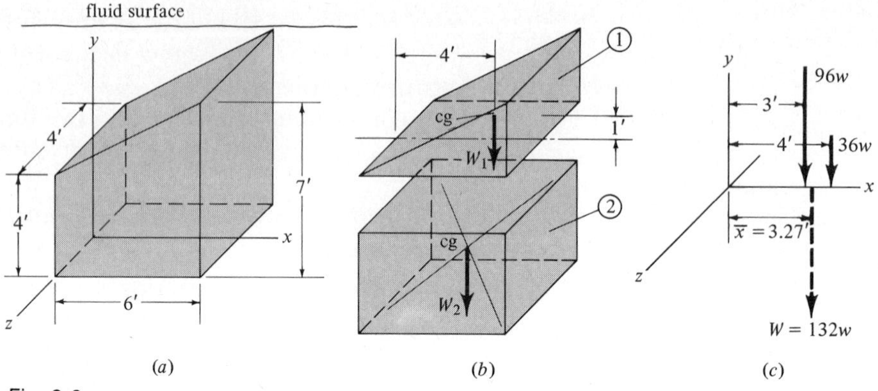

Fig. 3-8

while the centroid of ① is as shown in Fig. 3-8(*b*), where we have utilized the result in Example 3-6 for a triangular-shaped body. Then, treating the weights of the displaced fluid as a parallel force system shown in Fig. 3-8(*c*), we have

$$\mathbf{M}_z = (96w)(3) + (36w)(4)$$
$$= 432w$$

Thus, the resultant weight of 132 w must be located so that

$$W\bar{x} = M_z$$

$$\bar{x} = \frac{M_z}{W} = \frac{432w}{132w} = 3.27 \text{ ft}$$

By symmetry, you should see that

$$\bar{z} = 0$$

While it might appear that the problem is now finished, such is not the case. This submerged body can be oriented in an infinity of positions of which we have considered only one. For example, it

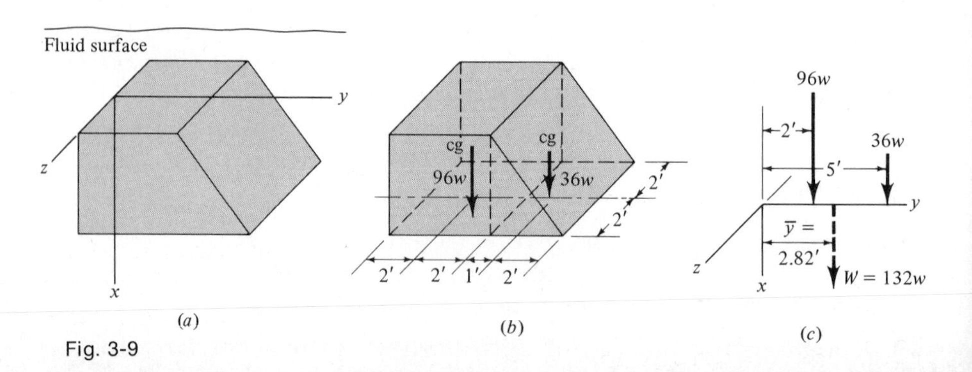

Fig. 3-9

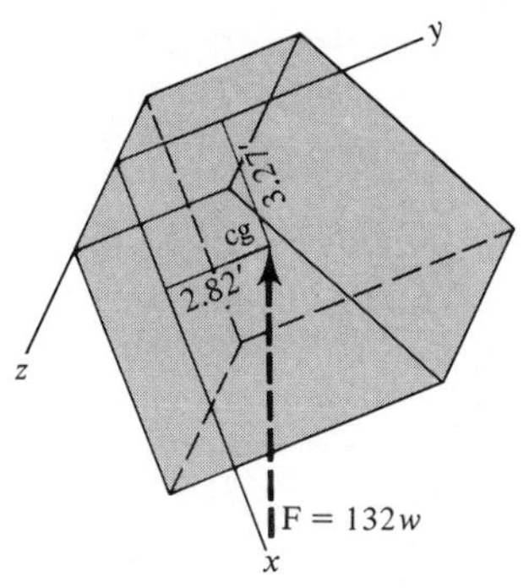

Fig. 3-10

might be oriented in the position indicated in Fig. 3-9(a). Where is the resultant force located for this position? Stated another way, so far we have determined only two coordinates $\bar{x}$ and $\bar{z}$ of the centroid of the volume. To completely locate the centroid, we must determine $\bar{y}$ in the same manner as we did for $\bar{x}$ and $\bar{z}$; that is, from Fig. 3-9(b) and (c),

$$\mathbf{M}_z = 96w(2) + 36w(5)$$
$$= 192w + 180w = 372w$$

and

$$W\bar{y} = M_z$$
$$\bar{y} = \frac{M_z}{W} = \frac{372w}{132w}$$
$$= 2.82 \text{ ft}$$

Hence the center of buoyancy is located at the coordinates (3.27,2.82,0), which we anticipate as being physically reasonable. This means that, regardless of the orientation of this submerged body, the line of action of the resultant upward buoyant force of $132w$ must pass through this point of the body, as illustrated in Fig. 3-10.

EXERCISE PROBLEMS

3-19 to 3-24 A pipe is laid out in the configuration shown. Locate the center of gravity for this configuration.

PROB. 3-19

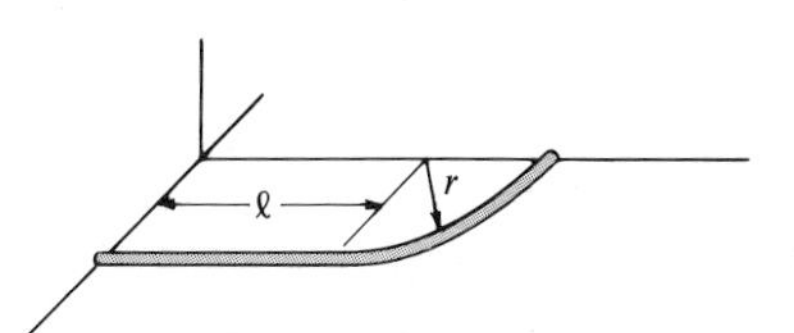

PROB. 3-20

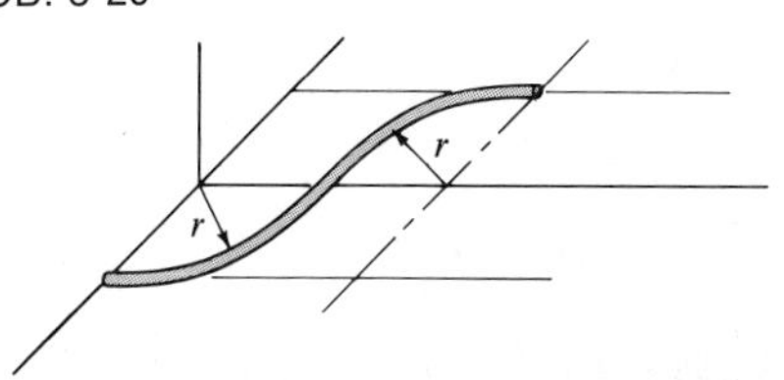

PROB. 3-21

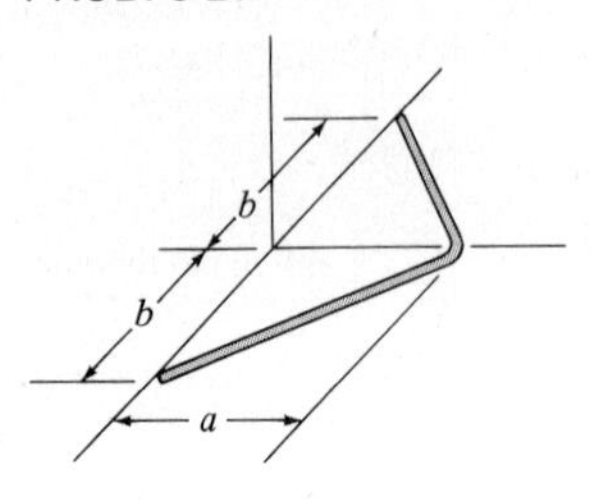

PROB. 3-22

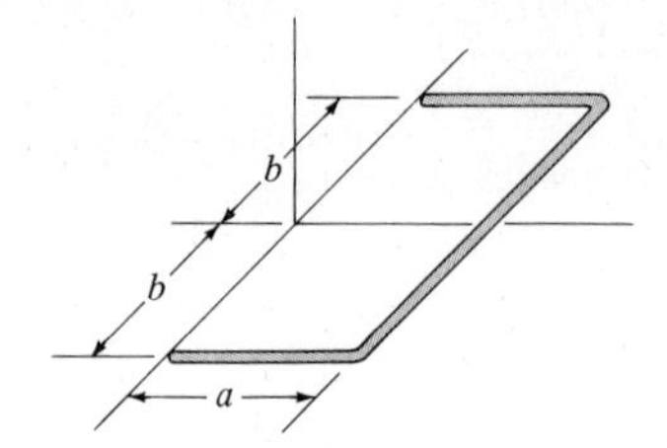

PROB. 3-23

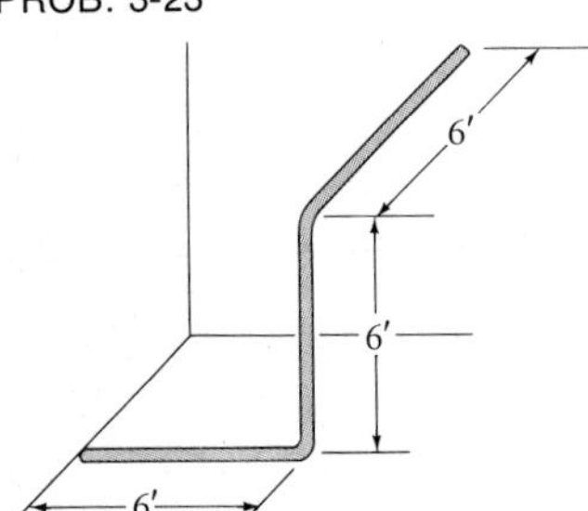

PROB. 3-24

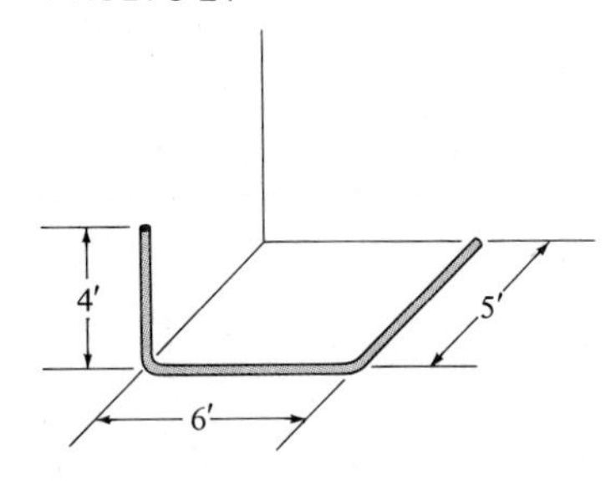

3-25 Locate the center of gravity of the homogeneous flat semicircular plate shown.

PROB. 3-25

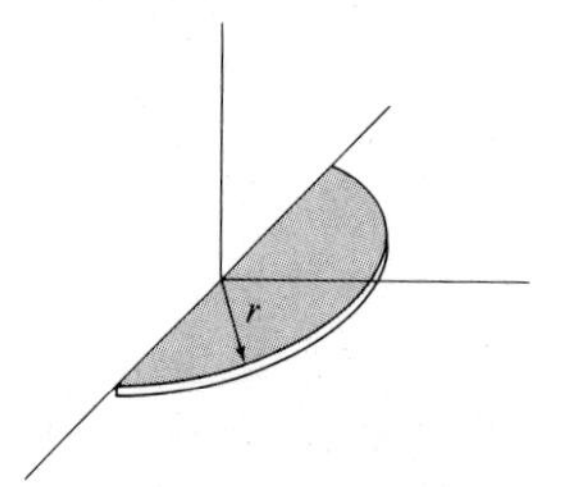

PROB. 3-26

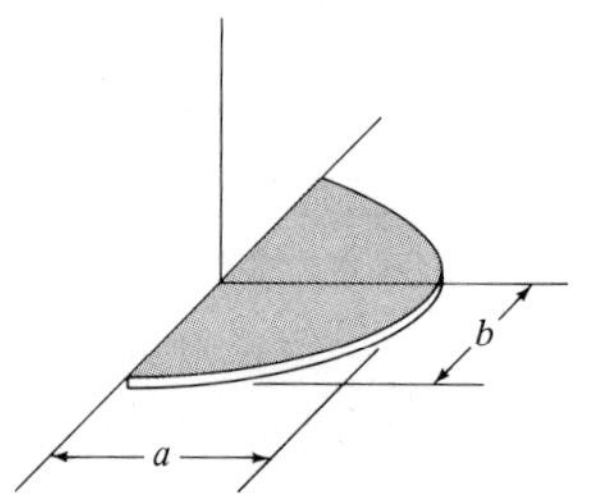

3-26 Locate the center of gravity of the homogeneous flat parabolic-shaped plate shown.

3-27 to 3-30 Locate the center of gravity of the plates shown.

PROB. 3-27

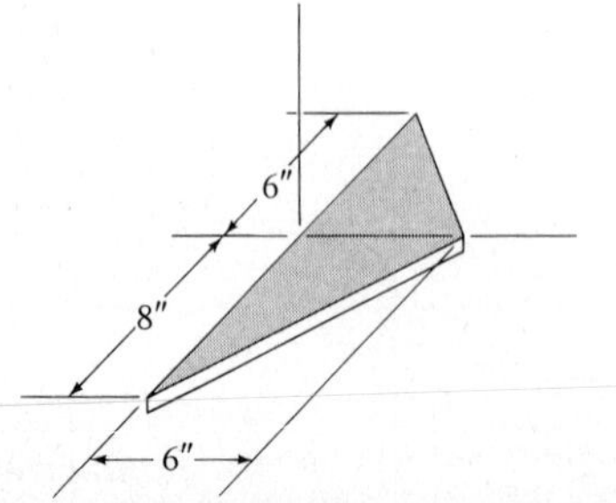

PROB. 3-28

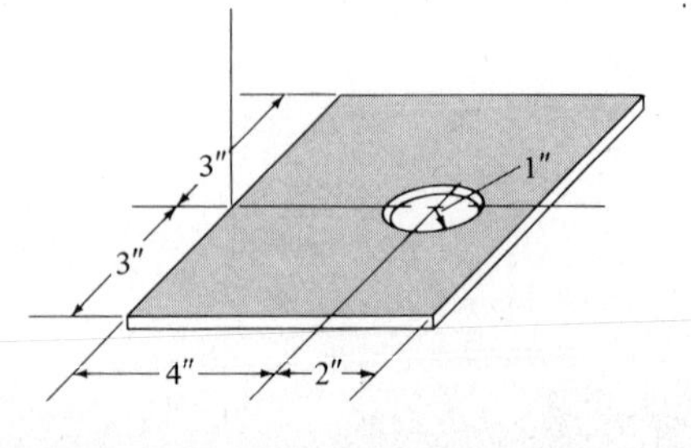

PROB. 3-29

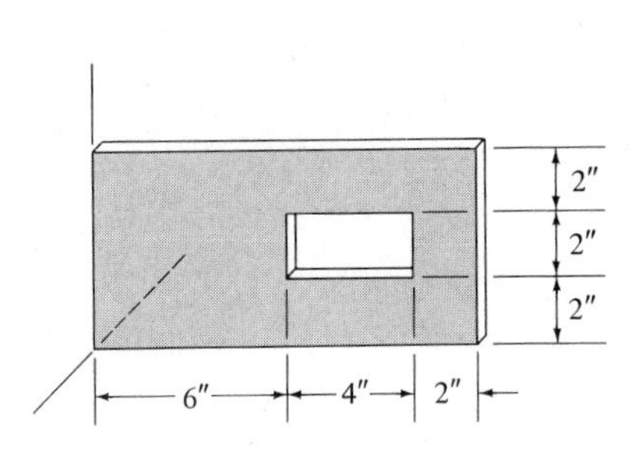

PROB. 3-30

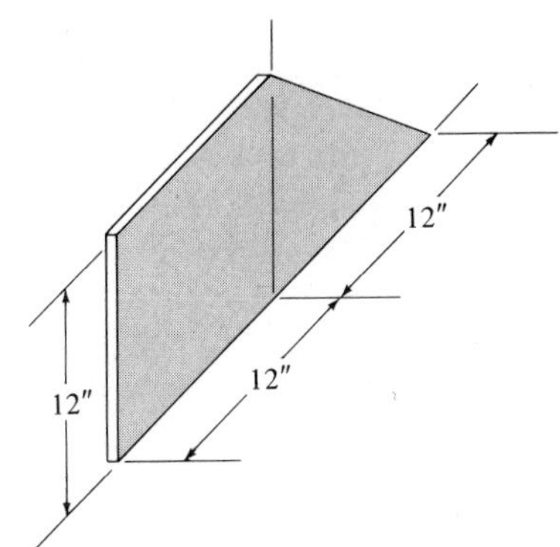

3-31 to 3-38 Locate the center of gravity of the homogeneous bodies shown.

PROB. 3-31

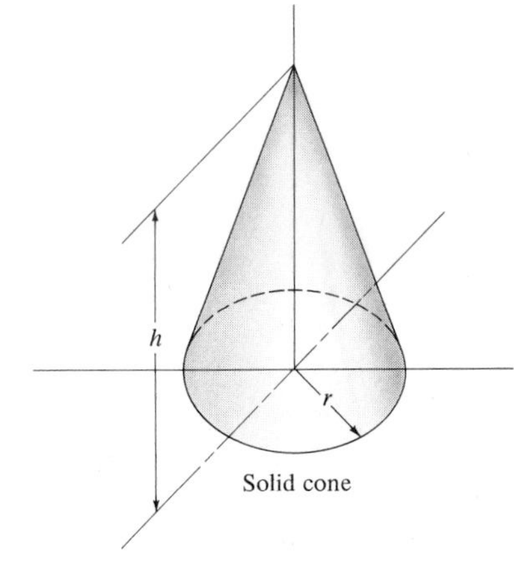

PROB. 3-32

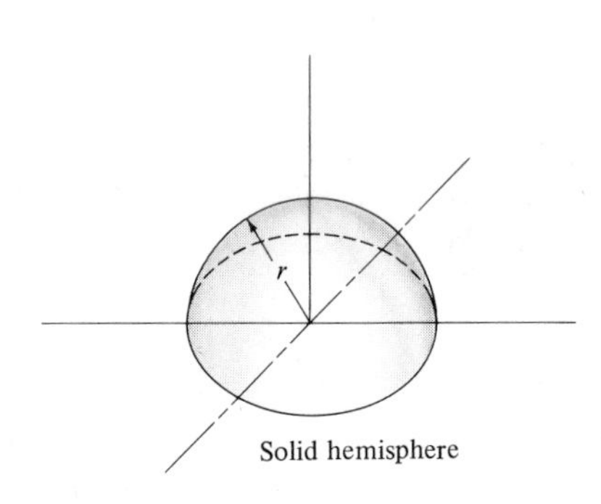

PROB. 3-33

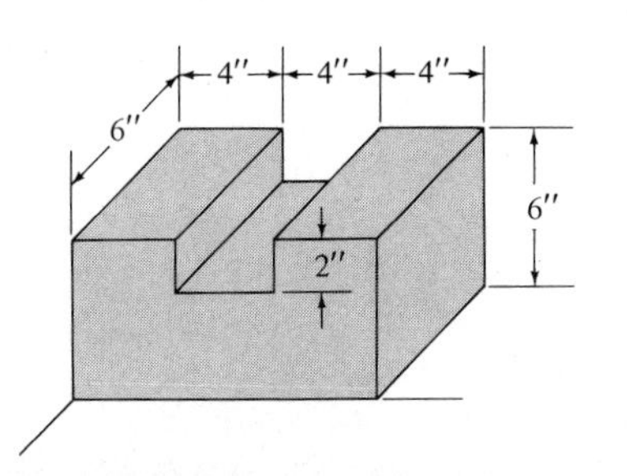

PROB. 3-34

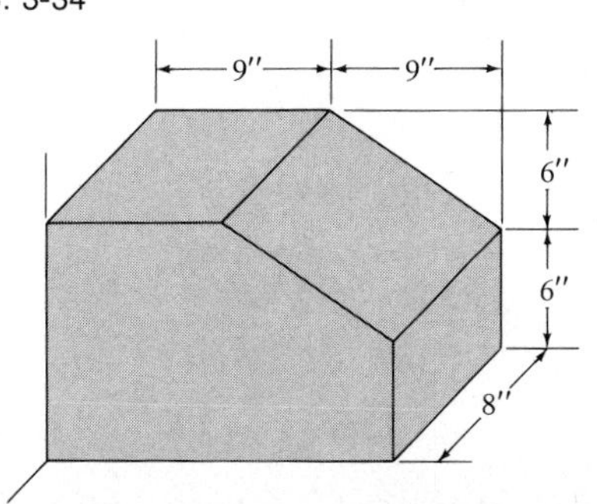

PROB. 3-35

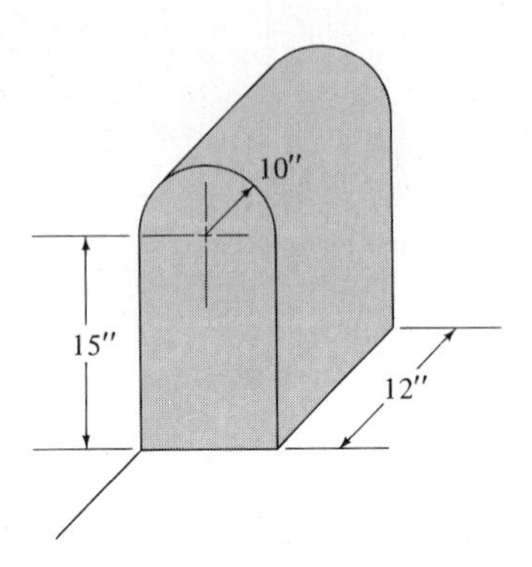

PROB. 3-36

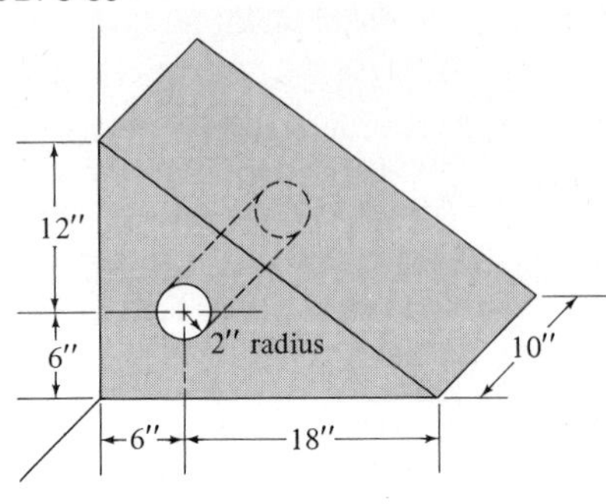

PROB. 3-37

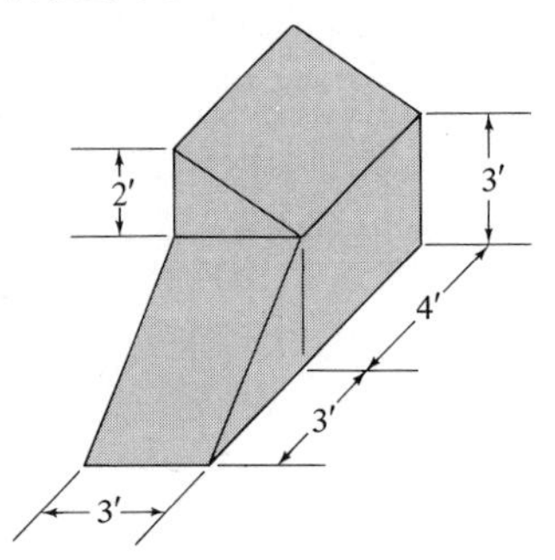

PROB. 3-38

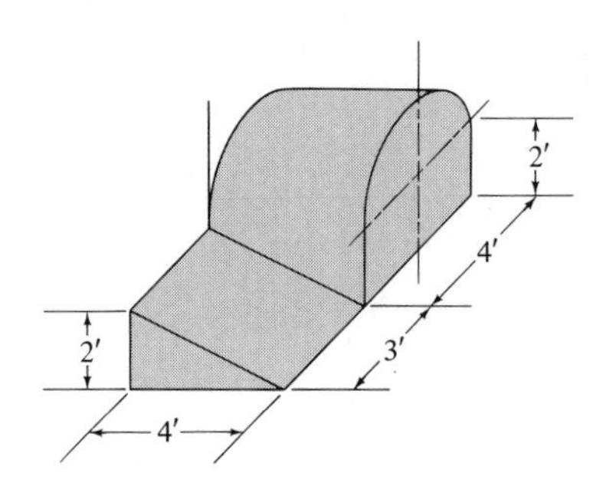

3-39 A composite bar is made by combining a bar of steel weighing 12 lb and a bar of aluminum weighing 6 lb. Locate the cg of the composite bar.

PROB. 3-39

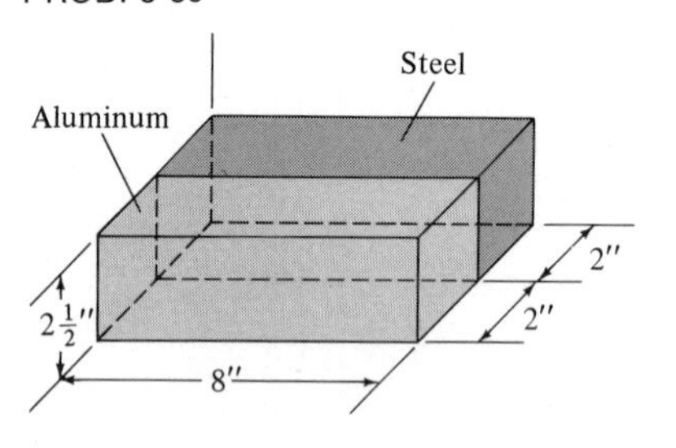

PROB. 3-40

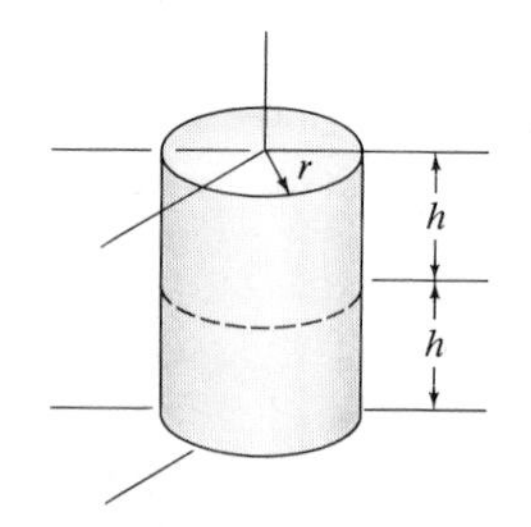

3-40 A cylindrical container contains a stratified fluid, the upper portion having a specific gravity of 0.8 and the lower portion a specific gravity of 1.1. Locate the cg of the fluid.

3-5 CENTER OF PRESSURE

Earlier in this chapter we saw numerous situations in which a load or force was distributed over some surface area, such as the wind load against the A-frame-cabin roof. Visualizing such a situation as a system of distributed forces, once again we ask the question: What is the statical resultant of the given force distribution? In most cases the resultant will be a force equal to the net total force or sum of the surface-traction distribution. (There are occasions in which the resultant of a surface-traction distribution can be a couple or possibly the null system, and some of these situations will be indicated in later chapters.) Our immediate problem, then, is to determine the proper location for the resultant force so that it produces the same moment as the original distributed system of forces. This location is called the *center of pressure* or *center of loading* of the distributed force. The following examples will illustrate some useful ideas and techniques for finding the statical resultants of forces distributed over surfaces.

EXAMPLE 3-8

The dam shown in Fig. 3-11(a) holds back water in a 120-ft-wide river with a depth of 30 ft. Determine the resultant force on the upstream side of this dam.

Solution: Recall from earlier examples that the hydrostatic water pressure varies linearly with depth. Therefore, the pressure at the bottom of the dam is

$$\begin{aligned} p_b &= (62.4 \text{ lb/ft}^3)(30 \text{ ft}) \\ &= 1{,}870 \text{ lb/ft}^2 \end{aligned}$$

and the pressure distribution against the upstream side of the dam is shown in Fig. 3-11(b). Then, by Fig. 3-11(c), if we let y be the distance from the top surface of the water, the net total force is given by

$$\begin{aligned} F &= \iint_{\text{area}} p\, dA \\ &= \int_0^{30} (62.4y)\, 120\, dy = (62.4)(120) \int_0^{30} y\, dy \\ &= (62.4)(120)(\tfrac{1}{2})(30^2 - 0) \\ &= 3{,}370{,}000 \text{ lb} \end{aligned}$$

Thus, the resultant force of the water is 3,370,000 lb, and the question now is its proper location. By symmetry it is apparent that the resultant will act effectively at the middle of the river 60 ft

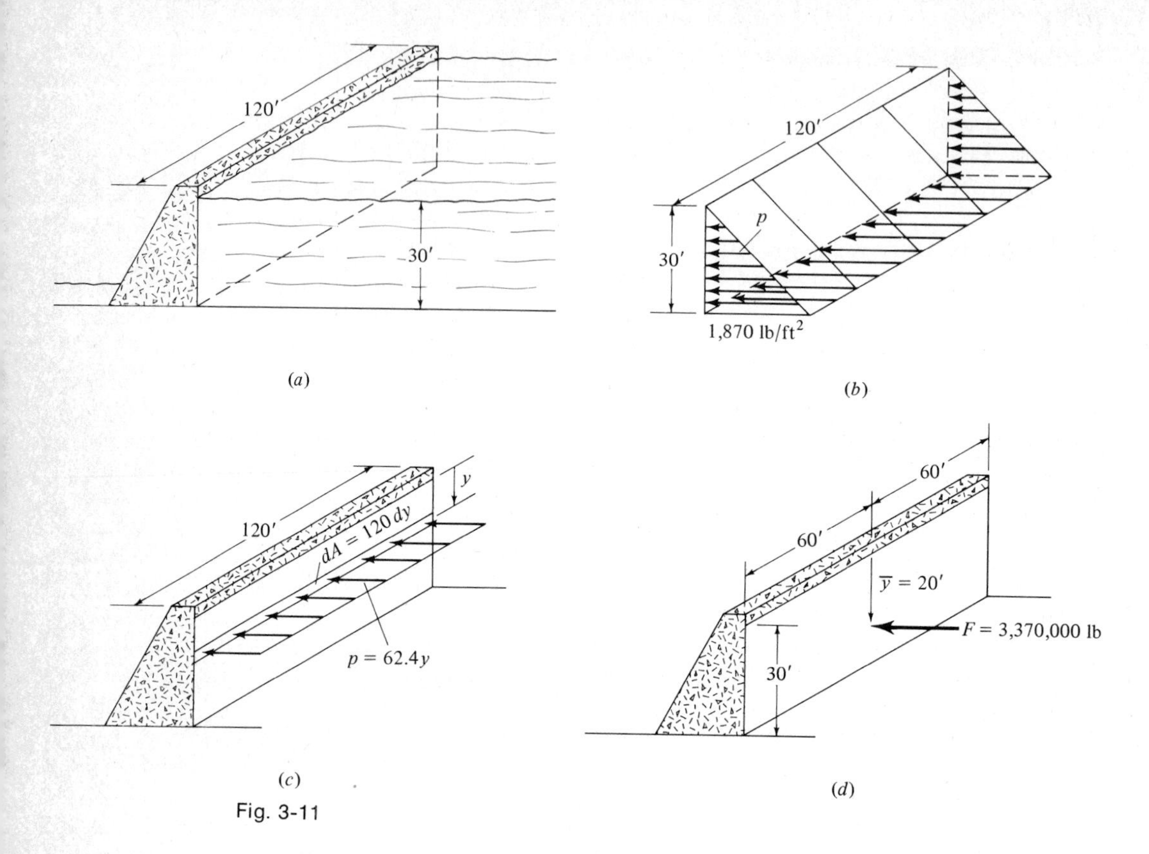

Fig. 3-11

from either side. As for its vertical location, if we consider moments about the top surface of the water, we must have

$$F\bar{y} = \iint_{\text{area}} yp \, dA$$

that is, the resultant force F must produce the same moment produced by the pressure. Thus

$$\begin{aligned} F\bar{y} &= \int_0^{30} y(62.4y)\,120\,dy \\ &= (62.4)(120)\int_0^{30} y^2\,dy \\ &= (62.4)(120)(\tfrac{1}{3})[30^3 - 0] \\ \bar{y} &= \frac{(62.4)(120)(30^3)(1/3)}{(62.4)(120)(30^2)(1/2)} \\ &= \tfrac{2}{3}30 = 20 \text{ ft} \end{aligned}$$

Therefore, the statical resultant of the water pressure against the dam is that shown in Fig. 3-11(d).

Alternate Solution: The integral representing the total net force of the pressure distribution

$$F = \iint_{\text{area}} p \, dA$$

can be interpreted as the *volume* of the prism formed by the pressure-distribution diagram shown in Fig. 3-11(*b*). Since this is a triangular prism, its volume is given by

$$\begin{aligned}\text{Volume of prism} &= \tfrac{1}{2}(30\text{ ft})(120\text{ ft})(1{,}870\text{ lb/ft}^2)\\ &= 3{,}370{,}000\text{ lb}\end{aligned}$$

which is the net resultant force as found previously.

In a similar manner the integral representing the moment of the pressure distribution

$$\iint_{\text{area}} yp \, dA$$

can be interpreted as the *moment of the volume* of the pressure-distribution prism. Hence, the force-moment relation

$$F\bar{y} = \iint_{\text{area}} yp \, dA$$

has the interpretation

$$(\text{Volume of prism})\bar{y} = \text{moment of volume of prism}$$

so that $\bar{y}$ corresponds to the vertical location of the *centroid* of the prism in Fig. 3-11(*b*). Since this is a triangular prism, its centroid is at $\frac{1}{3}$ its height or 10 ft from the bottom of the prism, which is precisely the location of the resultant which was found previously.

> ***Remark:*** The alternate procedure used above is quite advantageous so long as you are careful in properly interpreting the integrals actually involved and provided you know the pressure distribution volumes and the locations of their centroids.

EXAMPLE 3-9

A warehouse floor 16 ft wide and 40 ft long is supported by joists 16 ft long spaced on 16-in centers. Two thousand 100-lb bags of cement are piled on this floor in the shape shown in Fig. 3-12(*a*). Find the resultant force supported by each individual floor joist.

Solution: The floor is 40 ft long and the joists are $1\frac{1}{3}$ ft apart so that each joist effectively supports $1\frac{1}{3}/40$ of the total load. The total

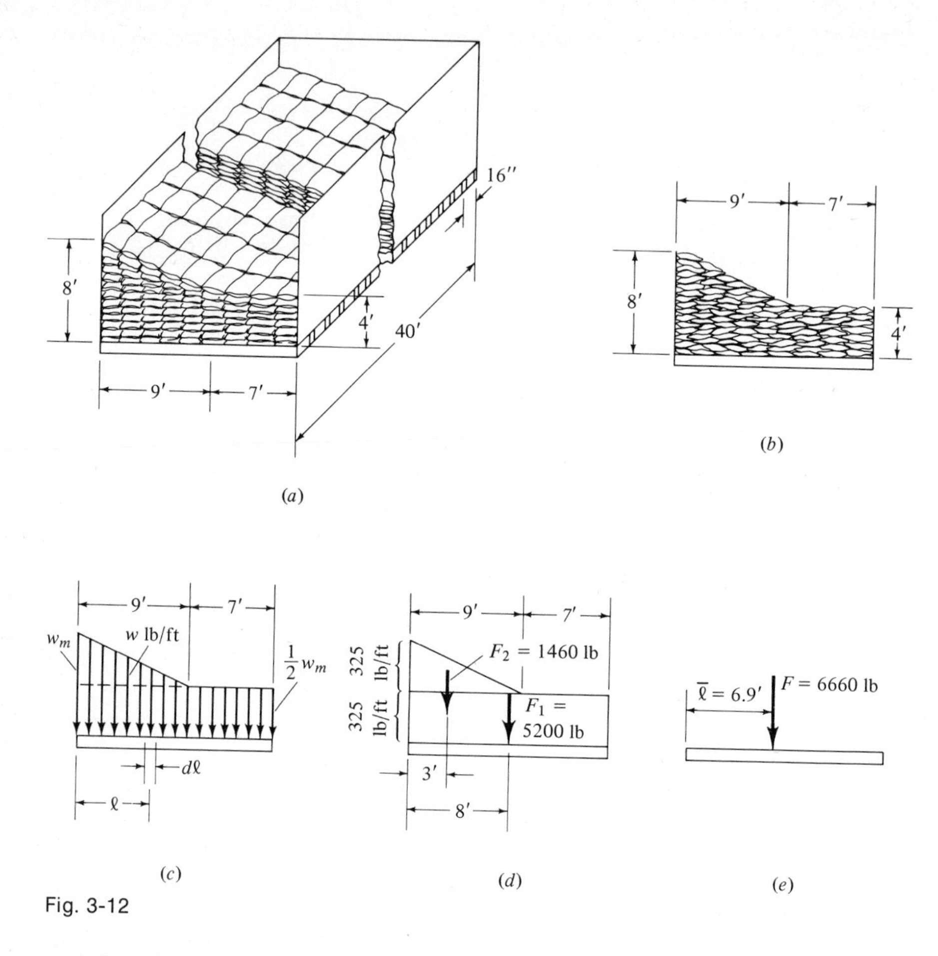

Fig. 3-12

weight of the cement is 2,000 bags × 100 lb/bag = 200,000 lb. Therefore, each joist supports

$$\frac{4/3}{40} \times 200{,}000 \text{ lb} = 6{,}660 \text{ lb/joist}$$

$$\text{or } 66.6 \text{ bags/joist}$$

which are distributed over the 16-ft length of the joist in the manner shown in Fig. 3-12(*b*). However, to determine the location of the resultant load on the joist, we must describe this load distribution much more precisely than just saying the bags are piled to some height. For this purpose, we let w represent the *loading intensity per unit length* of joist—in this case w lb/ft, as shown in Fig. 3-12(*c*). Then, if we can describe how w varies with the length ℓ, that is, $w(\ell)$, the total load on the joist can be written as

$$F = \int_0^{16} w(\ell)\, d\ell$$

which can be interpreted as the *area* of the loading diagram of Fig. 3-12(*c*).

From the way the bags are piled in Fig. 3-12(*b*), we expect the loading to have its maximum value w_m at the left end, decrease linearly for the next 9 ft to half the maximum value, and remain at this value for the last 7 ft. Therefore, by treating the area of the loading diagram in Fig. 3-12(*c*) as a composite area made up of a 16 ft $\times \frac{1}{2}w_m$ rectangle and a 9 ft $\times \frac{1}{2}w_m$ triangle, we have

$$F = \text{area of loading diagram}$$
$$6{,}660 \text{ lb} = (16 \text{ ft})(\tfrac{1}{2}w_m \text{ lb/ft}) + \tfrac{1}{2}(9 \text{ ft})(\tfrac{1}{2}w_m \text{ lb/ft})$$
$$= w_m(8 + 2.25) = 10.25w_m$$

or

$$w_m = 650 \text{ lb/ft}$$

Thus, with this value for w_m we can visualize the total load as being made up of two forces $\mathbf{F}_1$ and $\mathbf{F}_2$ located as shown in Fig. 3-12(*d*) with

$$F_1 = (325 \text{ lb/ft})(16 \text{ ft}) = 5{,}200 \text{ lb}$$
$$F_2 = \tfrac{1}{2}(325 \text{ lb/ft})(9 \text{ ft}) = 1{,}460 \text{ lb}$$

and where we have utilized the fact that the centroid of a rectangle is at its middle while that for a triangle is at its $\frac{1}{3}$ point. The force system composed of $\mathbf{F}_1$ and $\mathbf{F}_2$ is *statically equivalent* to the original distributed loading on the joist. In turn, this two-force system is statically equivalent to the single force $\mathbf{F}$ with

$$F = F_1 + F_2 = 6{,}660$$

located at $\bar{\ell}$ such that

$$F\bar{\ell} = F_1(8) + F_2(3)$$
$$\bar{\ell} = \frac{(5{,}200)(8) + (1{,}460)(3)}{6{,}660}$$
$$= 6.9 \text{ ft}$$

The statical resultant $\mathbf{F}$ is shown in Fig. 3-12(*e*).

Important Remark: Before you complete this chapter by solving some exercise problems, it is important to pause for a minute to put some of our ideas about distributed forces into proper perspective. First of all, when a force is distributed over some surface or volume, there is no other system which is equivalent in all respects to the original system except the system itself (see Sec. 2-1). For example, the effect of your body's weight as you lie on your mattress is not the same as if you were to apply a concentrated force to the mattress equal to your weight. The same observation is true in all the other less obvious situations involving distributed forces. However, the statical resultant which we have been working with in this and the previous chapter is equivalent to the distributed system in a very restricted sense, namely, that it produces the same sum (net force) and the same moment (turning effect) as the original distributed system. Subsequent chapters will show you how this statical equivalence can be utilized to help you solve many types of useful problems involving distributed forces and force systems.

EXERCISE PROBLEMS

3-41 The tank is partially filled with water as shown. Find the resultant force (*a*) on the floor of the tank, (*b*) against the left side of the tank, and (*c*) against the front side of the tank.

PROB. 3-41

18′ 12′ 24′

PROB. 3-42

6′ 30′ 10′ 15′

3-42 and 3-43 A bin is partially filled with dry, loosely packed grain weighing 90 lb/ft^3 as shown. Find the resultant force of the grain (*a*) on the floor of the bin, (*b*) against the left side of the bin, and (*c*) against the right side of the bin. Treat the grain as a fluid.

PROB. 3-43

PROB. 3-44

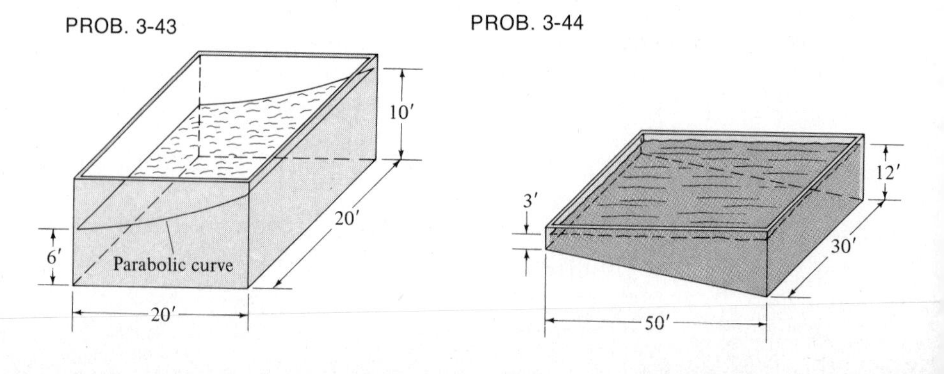

3-44 The swimming pool has a sloped bottom as shown. Find the resultant force of the water (*a*) against the left end of the pool, (*b*) against the right end of the pool, and (*c*) against the floor of the pool.

3-45 The dam shown holds back water in a 140-ft-wide river with a depth of 24 ft. Find the resultant force of the water against the upstream side of the dam.

PROB. 3-45

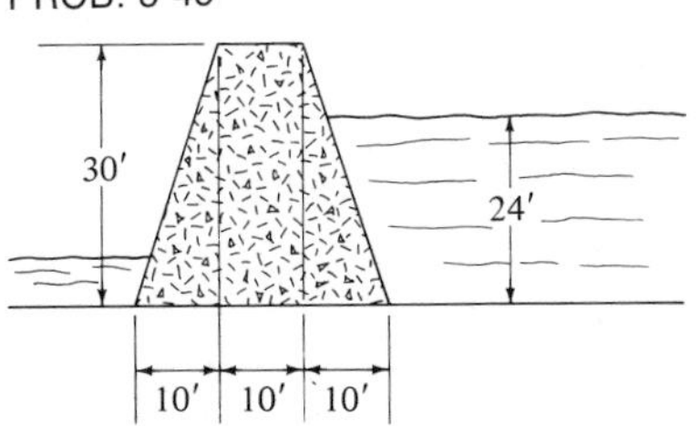

PROB. 3-46

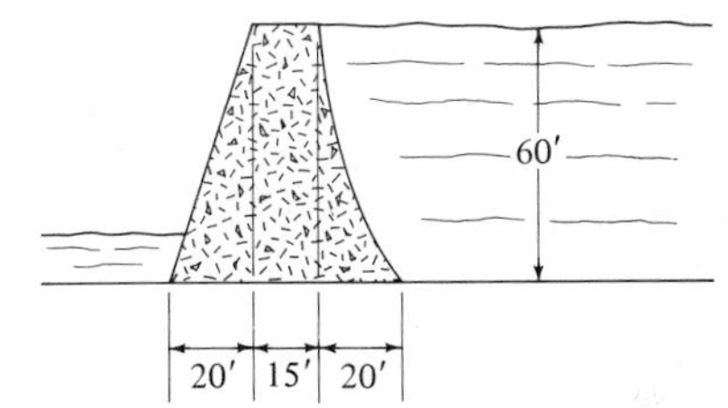

3-46 The dam shown holds back water in a 200-ft-wide river with a depth of 60 ft. Find the resultant force against the parabolic-shaped upstream side of the dam.

3-47 Wind blowing against the A-frame-cabin roof causes an average pressure rise of 0.25 psi (above atmospheric). Find the resultant wind force on the windward side of the roof.

PROB. 3-47

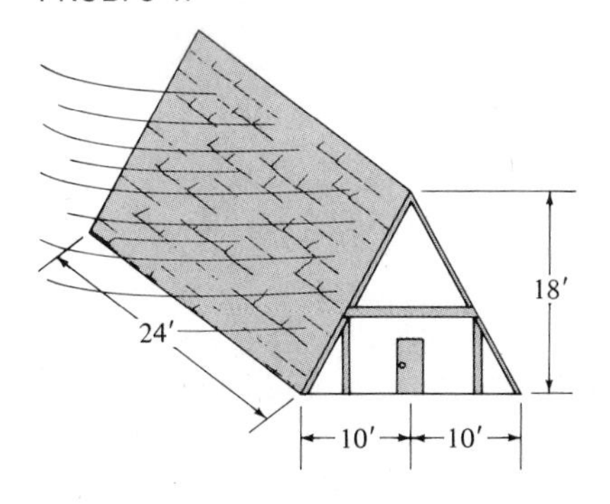

PROB. 3-48

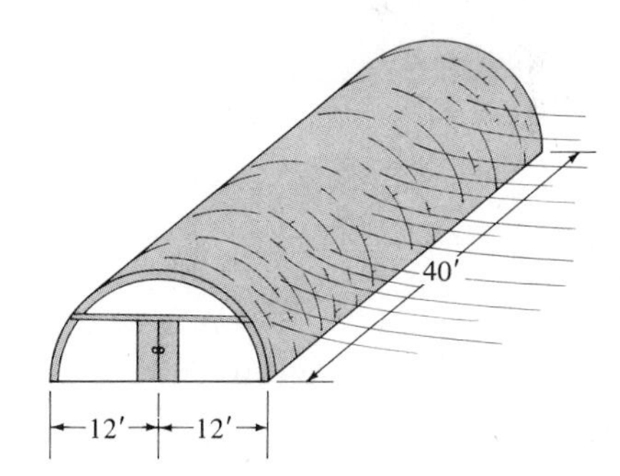

3-48 Wind blowing against the semicircular shaped Quonset hut roof produces an average rise in pressure of 0.3 psi (above atmospheric). Find the resultant wind force on the right half of the roof.

3-49 The closed tank shown is filled with air under a pressure of 40 psig. Find the resultant force on the flat bottom a–b–c–d of the tank. Find the resultant force on the arc-shaped side a–d–e–f.

PROBS. 3-49 AND 3-50

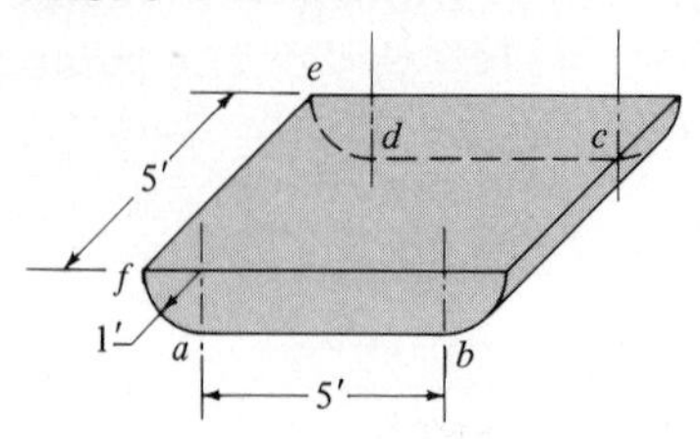

3-50 Same as Prob. 3-49 except the tank is open on top and completely filled with water.

3-51 to 3-54 The floor beam carries the loading indicated in the figure. Determine the resultant load on the beam.

PROB. 3-51

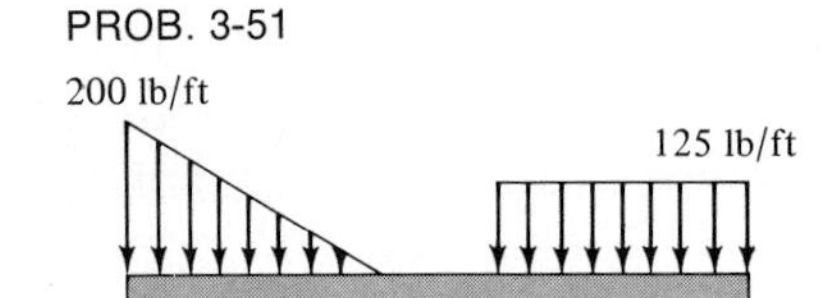

PROB. 3-52

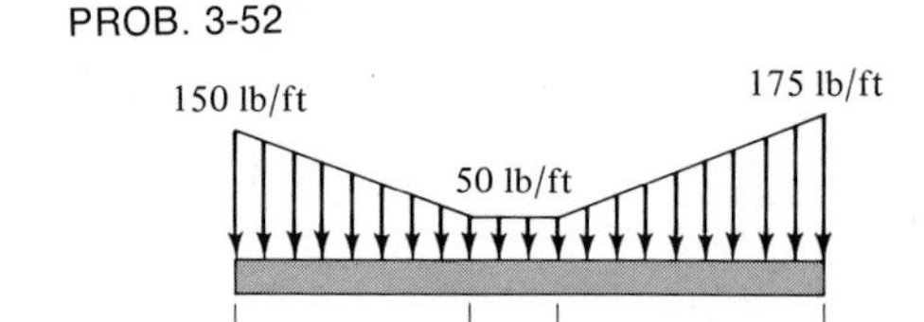

PROB. 3-53

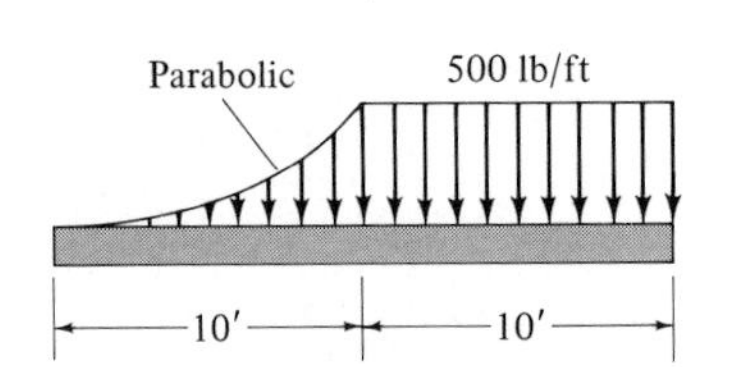

PROB. 3-54

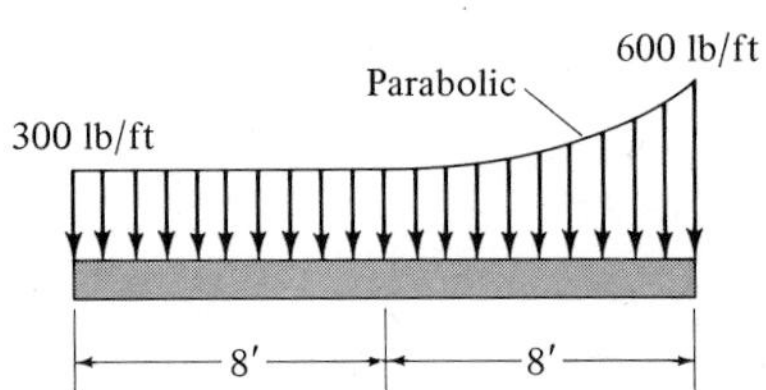

3-55 For the swimming pool in Prob. 3-44, what is the resultant force against the front 50-ft-long side of the pool?

3-56 A cylindrical grain silo 80 ft high and 20 ft in diameter is completely filled with loosely packed grain weighing 75 lb/ft^3. Treating the grain as a fluid, estimate the net resultant force on a semicircular half of the silo wall 80 ft high.

3-57 For the grain in the bin of Prob. 3-42, what is the resultant force against the 15-ft front of the bin?

3-58 A 20 × 30-ft bin is loaded as shown with sand weighing 90 lb/ft^3. The floor is supported by 20-ft joists spaced on 16-in centers. Find the resultant forces on each individual joist.

PROB. 3-58

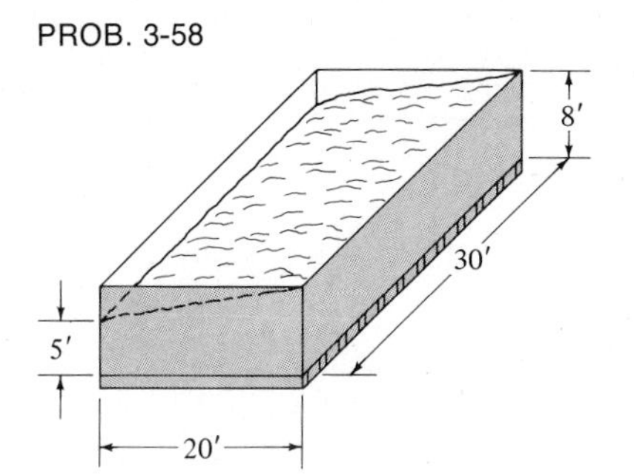

PROB. 3-59

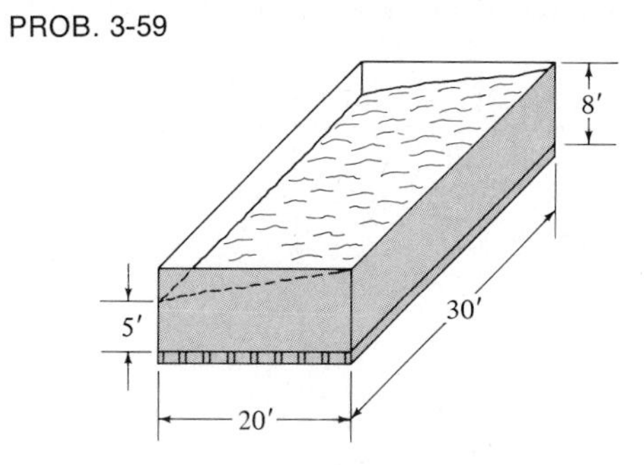

3-59 Same as Prob. 3-58 except the 30-ft joists are spaced on 8-in centers.

chapter 4 STRUCTURAL EQUILIBRIUM

4-1 INTRODUCTION

In the previous chapters we have seen how to represent forces as vectors, find their sums and moments, and find resultants of various types of force systems. At this point you might well ask: Why have we gone through all these analytical gymnastics? What are we trying to do?

Basically, engineers are problem solvers and as such they are generally confronted with two types of situations. In one case, there is some situation which requires a new building, or machine, or instrument for some particular purpose, and the engineer is called upon to design the building, machine, or instrument to fulfill the need in a satisfactory manner. Later design and synthesis courses in your various major areas will delve into this aspect of engineering in considerable depth. The second case deals with a situation in which a given machine, instrument, or system already exists and the engineer is called upon to analyze or evaluate its performance for possible modification, utilization, or some other reason. The remainder of this course is primarily an *analysis* course and one of its main purposes will be to show you how to organize and construct a problem-solving process.

Although we will confine ourselves to mechanical problems involving machines, mechanisms, and structures, the processes of analysis which we use are directly applicable to almost any type of engineering problem involving thermal, electrical, biological, or human factors. As a matter of fact, in the first three chapters of this text we have already taken the first step in the process of engineering analysis. Namely, we have identified the physical phenomena with which we are concerned – in our case, forces and force systems – and we have developed a scheme for representing and working with these phenomena – by the use of vectors and vector operations. Within this framework we are now ready to take the next step in performing engineering analyses of situations involving forces and force systems.

Fig. 4-1

4-2 CONNECTORS AND RESTRAINTS

The primary function of structures and machines, such as those in Fig. 4-1, is to support and/or transmit loads. Generally, the structures and machines are made up of individual pieces, parts, or members which are fastened together in some manner so that the device acts as a single unit. Our immediate objective is to perform a load analysis of a structure or machine to gain a clear picture of how the device is supporting the loads, how the loads are transmitted throughout the device from member to member, and how the device itself is supported. It is only after we have this picture that we are able to understand how the device functions and make rational judgments as to its performance capabilities.

First of all, you should recall that a force is the action of one body on another. Real bodies are usually connected to one another by various methods, such as rivets, bolts, welds, cement, nails, etc. These connectors can transmit certain types of loads from one body to another, the type of load usually being governed by the restraining action of the connector.

There are a multitude of types of connectors each with certain characteristics suitable for a particular application. However, almost all connectors can be categorized into three general classes according to their restraining effects.

1. Those that greatly resist translation in some direction but offer far less resistance to motion in other directions. We call such connectors *rollers, cables, smooth surfaces* and *slots*, etc., and represent them schematically by the various methods shown in Fig. 4-2(*a*). For this type of connector the *direction* of the transmitted force is usually known, as indicated in the diagram directly below the connector.
2. Those that resist translation in several directions but offer little resistance to rotation. Such types are called *pin joints, ball joints,* and *rough* or *friction surfaces* and are represented in Fig. 4-2(*b*). These connectors transmit a force, but, unlike those of type 1, the precise direction of the force is usually unknown. Consequently, these situations are represented by the various schemes shown in the diagrams below the connectors.
3. Those that resist translation *and rotation* in some manner. Such types are called *hinges, sleeve bearings,* and *fixed* or *embedded* supports. Their unique feature is their ability to transmit couples as well as forces. Some examples are shown in Fig. 4-2(*c*).

Classifying real connectors into these three categories is largely a matter of engineering judgment and convenience. For example, the members which make up the A-frame structure shown in Fig. 4-3(*a*) might be riveted together at the various joints. How would we classify such connected joints? Although such riveted joints probably resist translation and to some degree rotation, it would be reasonable to anticipate that if we physically removed member *cde*, the remaining structure would offer little resistance to rotation, as illustrated by the dashed outline in Fig. 4-3(*b*). Hence, without definite information to the contrary, it would seem reasonable to treat all the joints as pin joints which transmit forces but no couples. Thus, we would redraw the frame with pin joints in place of the riveted joints, as in Fig. 4-3(*c*). Whether or not this simplification or idealization of the problem is truly reasonable is, of course, an open question whose answer lies ultimately with the actual performance of the structure.

As another example, whenever two surfaces are in contact, usually there will be a friction effect, with the friction force tending

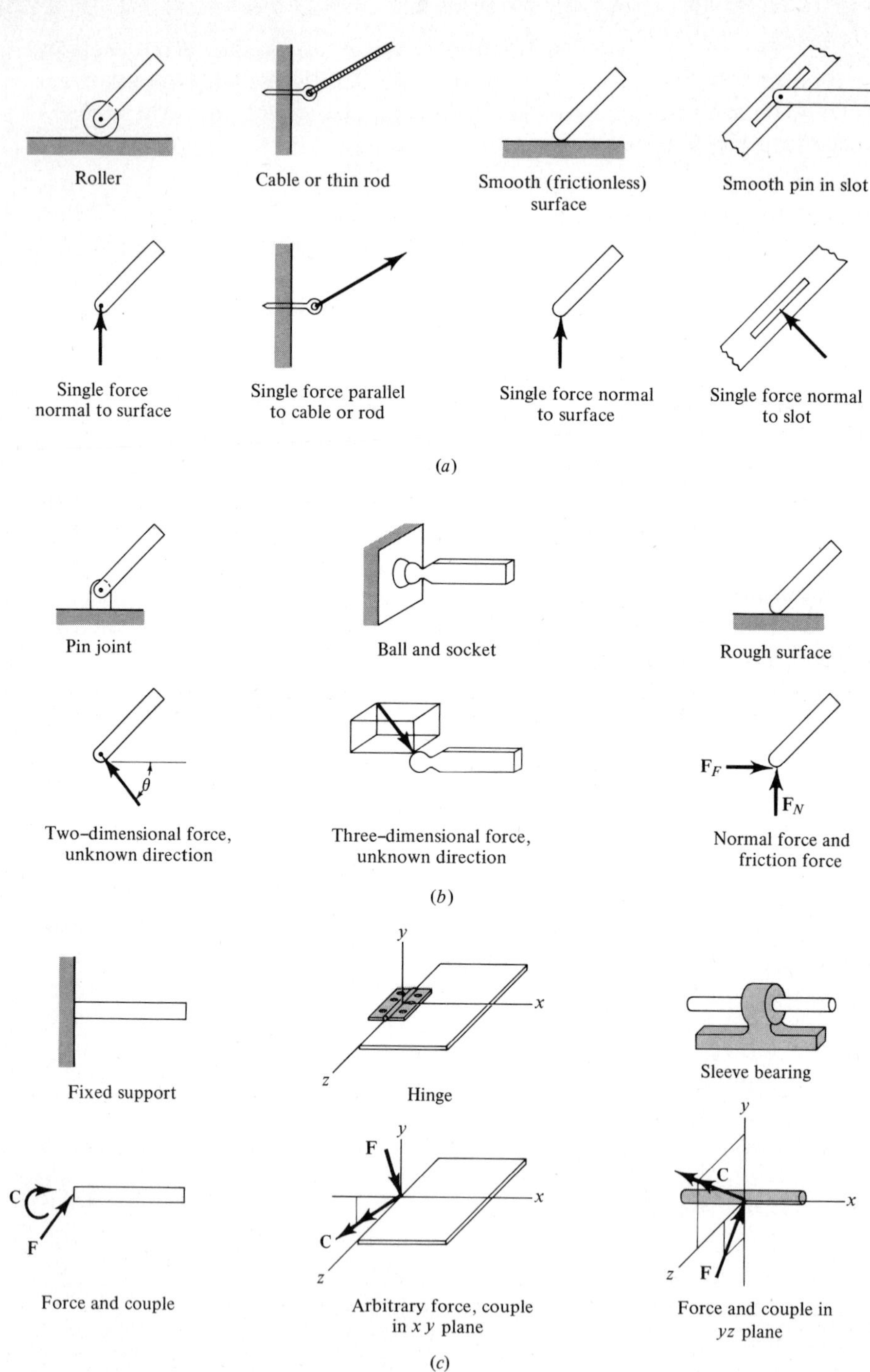
Roller
Cable or thin rod
Smooth (frictionless) surface
Smooth pin in slot
Single force normal to surface
Single force parallel to cable or rod
Single force normal to surface
Single force normal to slot
(a)
Pin joint
Ball and socket
Rough surface
θ
Two-dimensional force, unknown direction
Three-dimensional force, unknown direction
F_F
F_N
Normal force and friction force
(b)
Fixed support
y
x
z
Hinge
Sleeve bearing
C
F
Force and couple
Arbitrary force, couple in xy plane
Force and couple in yz plane
(c)

Fig. 4-2

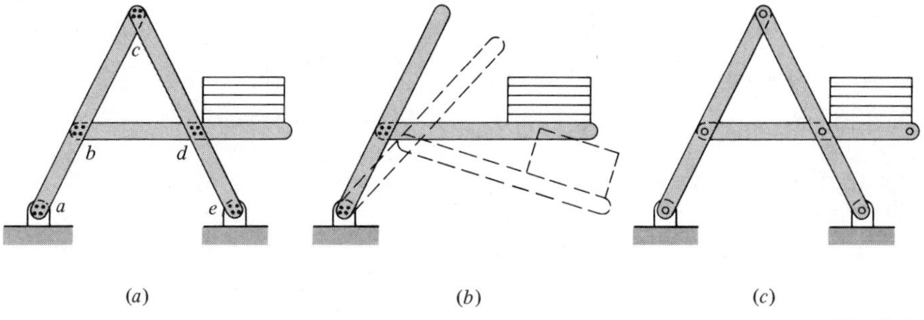

Fig. 4-3

to oppose any intended slipping between the surfaces. Whether or not this friction effect is a significant factor in the problem governs whether we treat the contact surface as rough or smooth.

Thus you can see that classifying connectors involves a great deal of engineering judgment and experience which you presently may not possess. Therefore, this text will temporarily relieve you of this responsibility by classifying most connectors for you; that is, the structures and machines which we will deal with in this chapter will be drawn with pin joints, ball joints, fixed supports, etc., rather than with rivets, bolts, etc. However, it will be up to you to recognize the type of forces and/or couples which these various idealized connectors can transmit, as were shown in Fig. 4-2.

4-3 FREE-BODY DIAGRAMS

The primary tool for the engineer in performing load analysis of structures and machines is the *free-body diagram* or *FBD*. In simple terms, a free-body diagram is a picture of how the engineer visualizes *all* the loads which are acting on a particular device or a particular part of a device. In more general terms, a free-body diagram is a scheme by which the engineer attempts to identify and isolate a relatively small system within a much larger or complex one. From either viewpoint the purpose of the FBD is to identify a force system which is amenable to analysis. By analyzing relatively simple systems, the engineer hopes to obtain answers to relatively complex problems.

It is no simple matter to construct realistic free-body diagrams for a real structure or machine. A great deal of common sense, intuition, and engineering experience is required. To get started, we will lay down a few of the basic ideas needed to draw free-body diagrams and trust that your ability to utilize them will develop as you gain experience.

When we draw an FBD of one particular body we isolate the body by replacing all the connectors attached to that body by vectors representing the types of loads which the connectors transmit to the body. The FBD should also include vectors representing the loads on the body due to gravity or other external effects. In short, the FBD of a particular device or member is a picture of that device or member in which all connectors or other bodies attached to, in contact with, or otherwise influencing that device have been replaced by vectors representing the forces or couples which they exert or transmit to the device. The "free" in free-body diagram means that we have freed the body from its surroundings but have tried to realistically represent the influence of the surroundings on the body, with the key word being "realistically."

To illustrate some of the ideas involved in drawing free-body diagrams, consider the frame shown in Fig. 4-4(*a*). You should realize from the previous discussions that this figure with its pin joints and cable probably represents a first degree of idealization of some real structure. Also, no information is given about the weights of the individual members, so that in all probability the members are relatively light in comparison with the 1,000-lb weight shown. Proceeding from this figure, we isolate the frame from its surroundings by replacing the connector at *a* and the cable by the forces which they transmit to the frame. The cable pulls horizontally, and its effect is shown by the force **T**. On the other hand, the pin joint at *a* transmits a force **A** of unknown direction [recall Fig. 4-2(*b*)]. Such a force will necessarily have two vector components $\mathbf{A}_x$ and $\mathbf{A}_y$. Thus, we can represent the effect of the pin joint *a* on

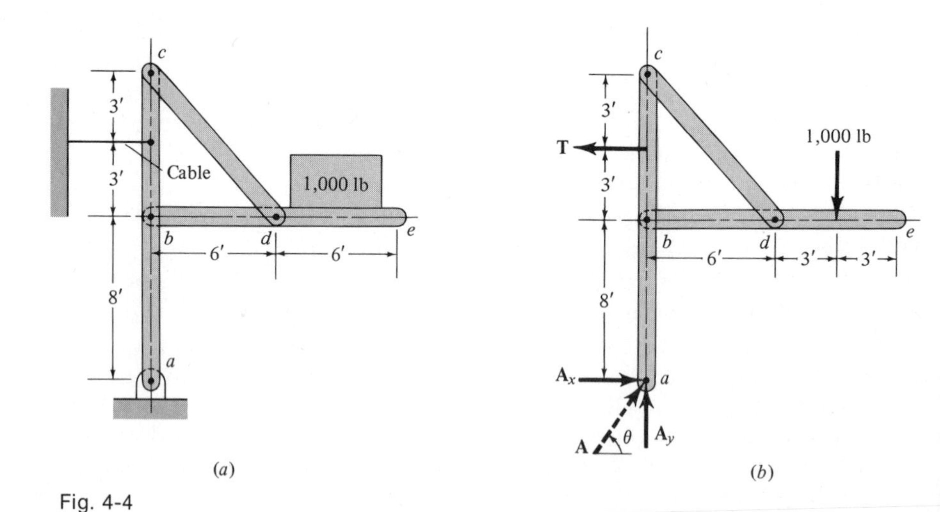

Fig. 4-4

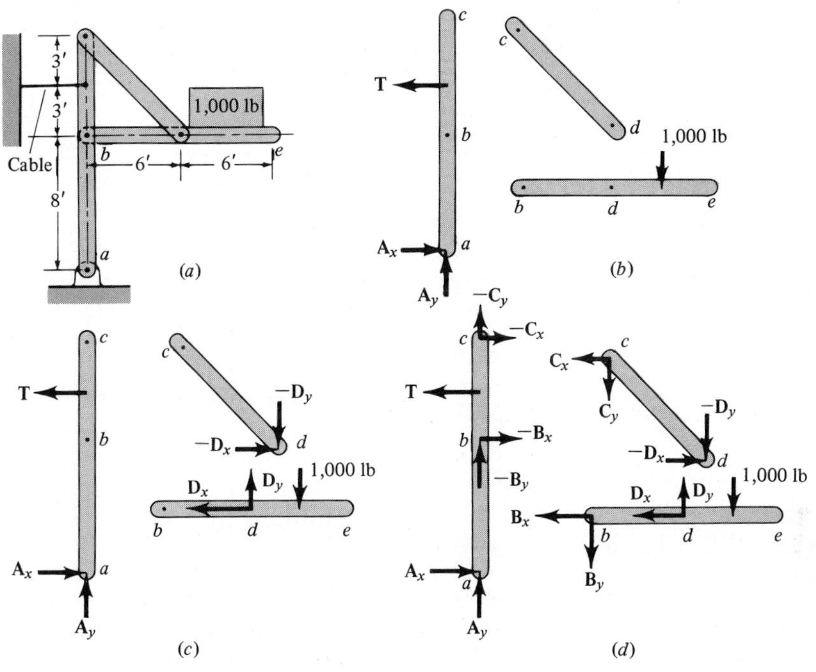

Fig. 4-5

the frame either by a single force **A** of unknown orientation θ or by two component forces $\mathbf{A}_x$ and $\mathbf{A}_y$. Either scheme is permissible, although the two-component method is usually preferred. Finally, the 1,000-lb weight is removed from the frame and replaced by its statical resultant of 1,000 lb located at the center of gravity of the weight (recall Chap. 3). The first FBD of the frame is shown in Fig. 4-4(*b*).

Continuing with this example, suppose we are interested in the forces exerted on the individual members making up the frame. The first step, of course, is to draw a picture of each individual member and to indicate those forces which we already know to be acting on the members, namely, $\mathbf{A}_x$, $\mathbf{A}_y$, and **T** on member *abc* and the 1,000-lb force on member *bde*, as shown in Fig. 4-5(*b*). We must now indicate the forces which the members exert on each other. The fundamental principle used in doing this is the following:

> ***Newton's Third Law:*** Given two bodies which exert forces on each other, the force which the first body exerts on the second is equal in magnitude but opposite in direction to that which the second body exerts on the first; i.e., for every action there is an equal and opposite reaction.

This "law," like all other physical laws, is simply a statement of our physical experiences and observations. To apply it to the problem at hand, suppose we assume that the member *cd* is pulling up and to the left on member *bde* at the pin joint *d*, as shown by the force vectors $\mathbf{D}_y$ and $\mathbf{D}_x$ on member *bde* in Fig. 4-5(*c*). Then, by Newton's third law, member *bde* exerts the opposite forces $-\mathbf{D}_x$ and $-\mathbf{D}_y$ on member *cd*, as also shown in this figure. Now, carrying out this same procedure for pin joints *c* and *b*, we eventually arrive at the free-body diagrams shown in Fig. 4-5(*d*).

Before leaving this example, a few comments might be useful.

Comment 1 The labeling of forces such as $\mathbf{B}_x$ and $-\mathbf{B}_x$ is quite arbitrary; that is, although we have shown $\mathbf{B}_x$ and $\mathbf{B}_y$ as the forces on member *bde*, and, therefore, we have $-\mathbf{B}_x$ and $-\mathbf{B}_y$ on member *abc*, we could just as easily reverse this labeling. The main point is to be sure that we have applied Newton's third law correctly so that whatever force is shown on one member, its *opposite* must be shown and labeled on the other.

Comment 2 In drawing these free-body diagrams, we have done considerable speculating as to how these members are affecting each other; that is, we have assumed member *cd* to be pulling up on member *bde*, member *abc* to be pulling down and to the left on member *bde*, etc. These are merely reasonable guesses which might or might not be correct. This will be looked into later, but for now our main concern is to be sure that we have handled the connectors properly. That is, since each of these connectors were assumed to be pin joints capable of transmitting both horizontal and vertical forces, we must be sure that we have not overlooked or omitted any possible forces transmitted to the various members.

Comment 3 In drawing the vectors in Fig. 4-5 representing the forces at the various joints, sometimes we have shown the tips (arrows) of the vectors at the joint, such as joint *a* on member *abc*

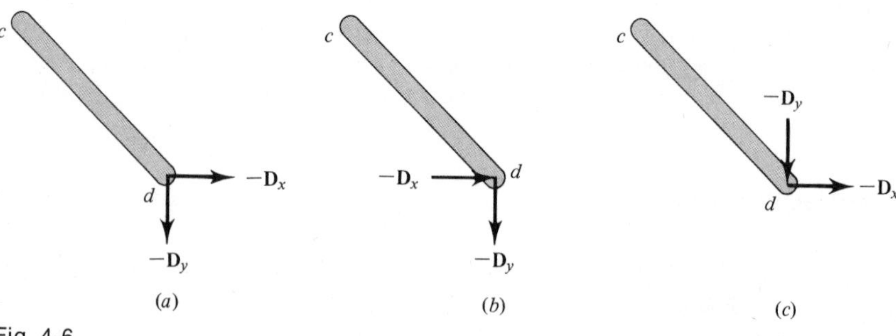

Fig. 4-6

and joint d on member cd. In other cases, we have shown the tails of the vectors at the joints, such as joints b and d on member bde and joint c. At joint b on member abc, the vector $-\mathbf{B}_y$ has its tip at b, while the vector $-\mathbf{B}_x$ has its tail at b. The point of these observations is that whether the tip or tail of the vector is drawn at the joint is purely a matter of convenience or choice. The essential thing is that the vector has the proper *direction*. Thus, forces at joint d on member cd in Fig. 4-5(c) could just as well have been drawn as shown in Fig. 4-6(a), (b), or (c), since all three represent the same effect as that shown in Fig. 4-5(c). *Carefully restudy Fig. 4-5(d)!*

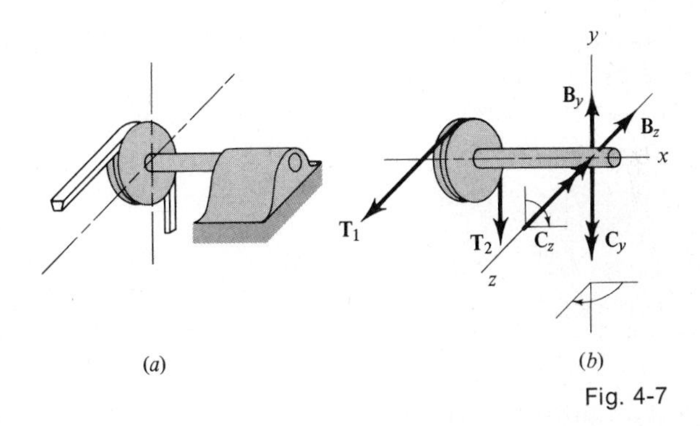

Fig. 4-7

As another illustration, consider the shaft–bearing–V-belt–pulley assembly shown in Fig. 4-7(a). The key feature about this example is the effect of the bearing on the shaft. First of all, the effect of the V belt is shown by the forces $\mathbf{T}_1$ and $\mathbf{T}_2$ tangent to the pulley. It is reasonable to expect that the bearing exerts a vertical force $\mathbf{B}_y$ and a horizontal force $\mathbf{B}_z$, as shown in Fig. 4-7(b). However, to keep the shaft horizontal, the bearing must also prevent rotation about the z axis, and thus it exerts a couple $\mathbf{C}_z$. Similarly, the bearing must prevent the shaft from pivoting about the y axis, and, therefore, it exerts a couple $\mathbf{C}_y$. Finally, we do not expect the bearing to exert any significant force parallel to the x axis, and, of course, the purpose of the bearing is to easily permit rotation about the x axis; thus, no x couple would be expected.

EXERCISE PROBLEMS

4-1 to 4-9 Draw a free-body diagram indicating the forces acting on the device or body.

PROB. 4-1 A frame.

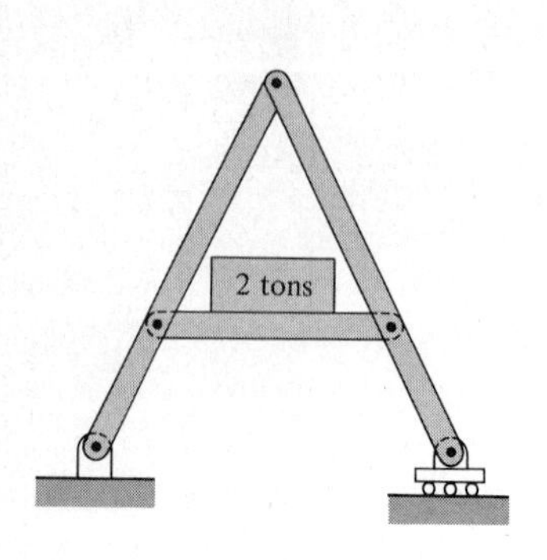

PROB. 4-2 Pin-connected truss.

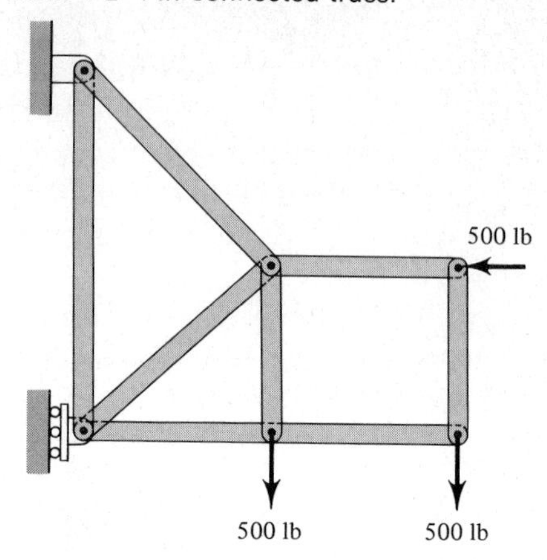

PROB. 4-3 Heavy door.

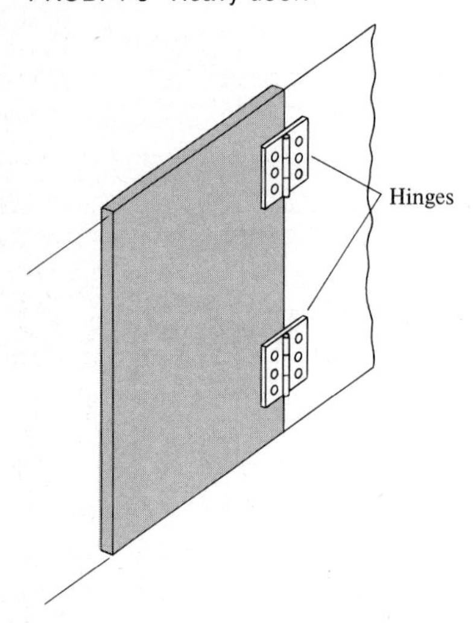

PROB. 4-4 Heavy sign.

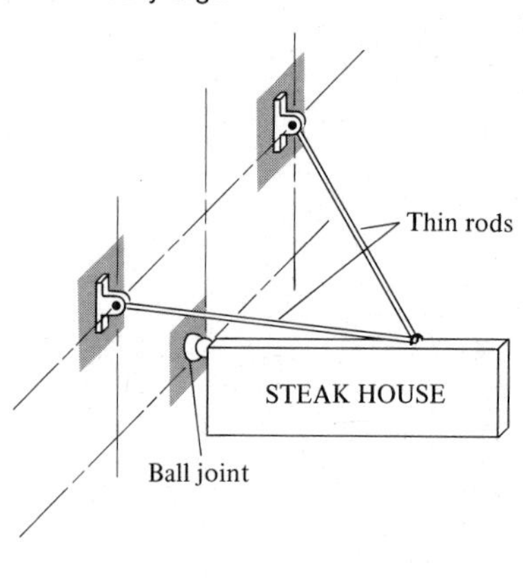

PROB. 4-5 Load frame.

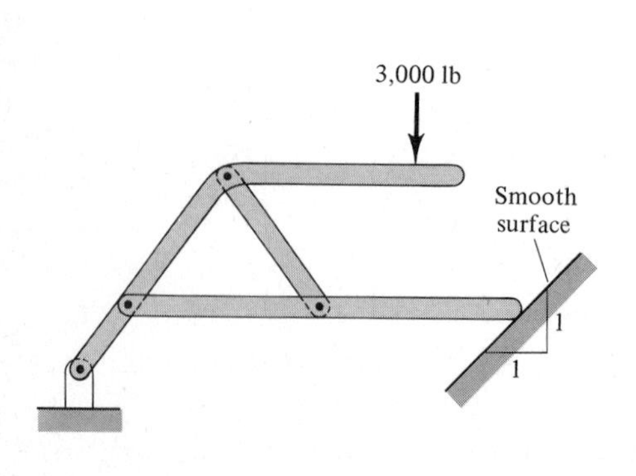

PROB. 4-6 Fixed hoist frame.

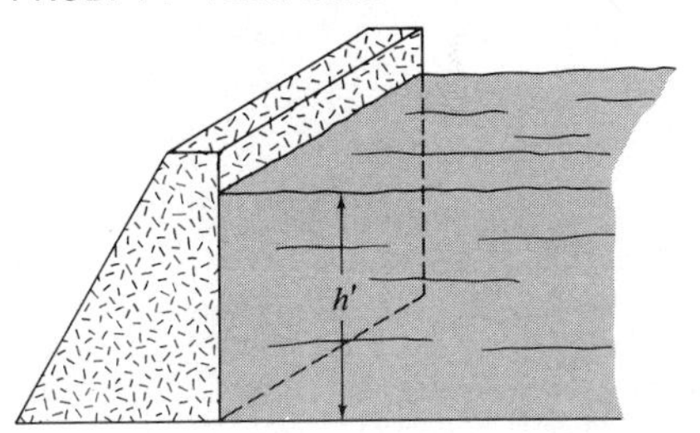

PROB. 4-7 River dam.

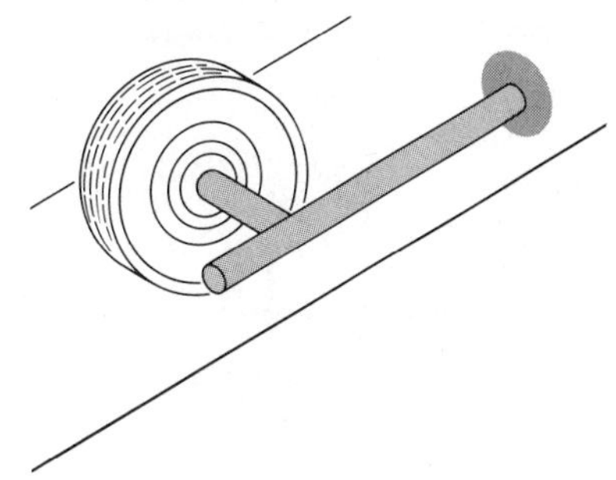

PROB. 4-8 Automobile torsion bar.

PROB. 4-9 Man pushing crate up incline.

4-10 Draw an FBD of each of the following for conditions under normal use: (*a*) a pencil; (*b*) connecting rod in an auto engine; (*c*) a half-opened dresser drawer.

4-11 Draw an FBD of each of the following for conditions under normal use: (*a*) a ladder leaning against a wall; (*b*) a frying pan held over a camp fire; (*c*) an adjustable crescent wrench.

4-12 Draw an FBD of each of the following for conditions under normal use: (*a*) thigh bone (femur) for a standing man; (*b*) hydraulic car hoist in a gasoline station; (*c*) a pair of tweezers.

4-13 to 4-20 Draw an FBD of piece *A* and piece *B* in each of the devices shown.

PROB. 4-13

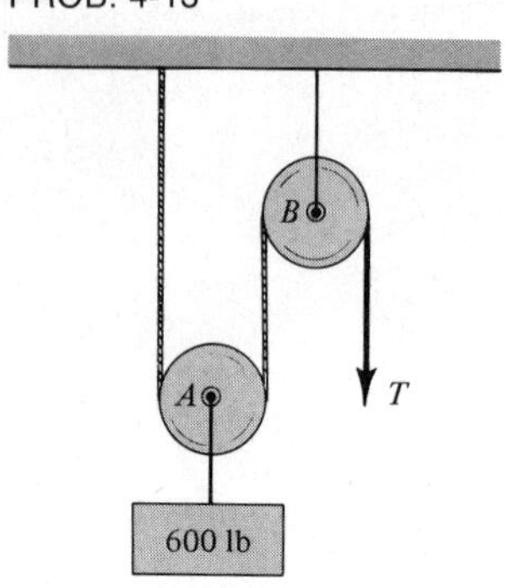

PROB. 4-14

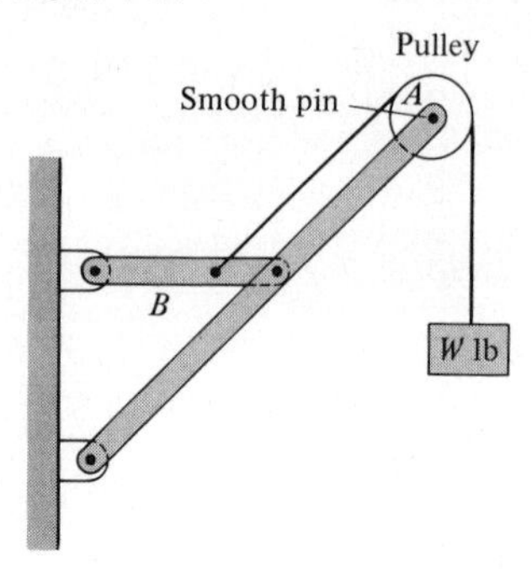

PROB. 4-15

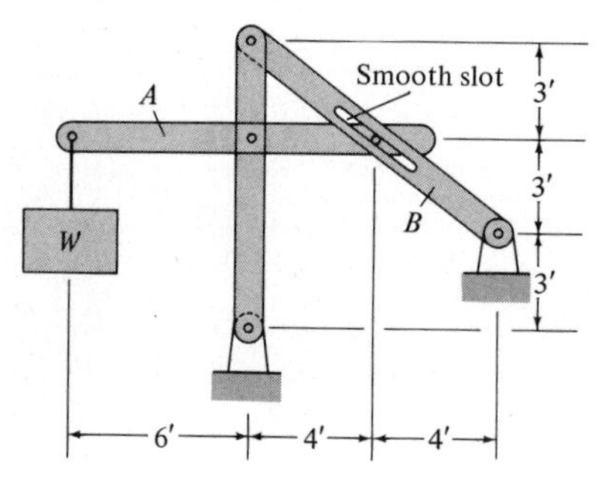

PROB. 4-16

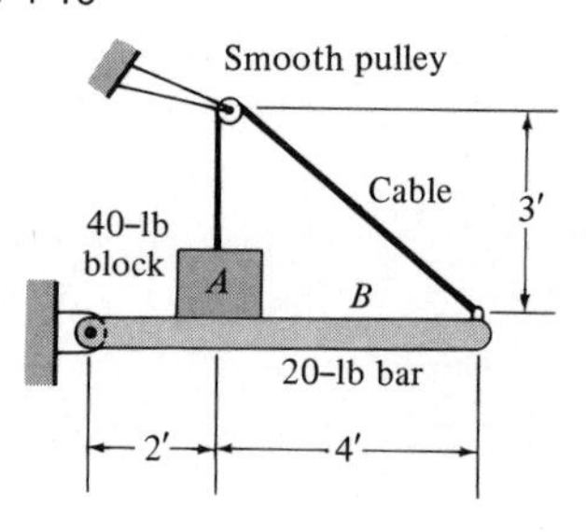

PROB. 4-17 Piston–connecting rod–crankshaft.

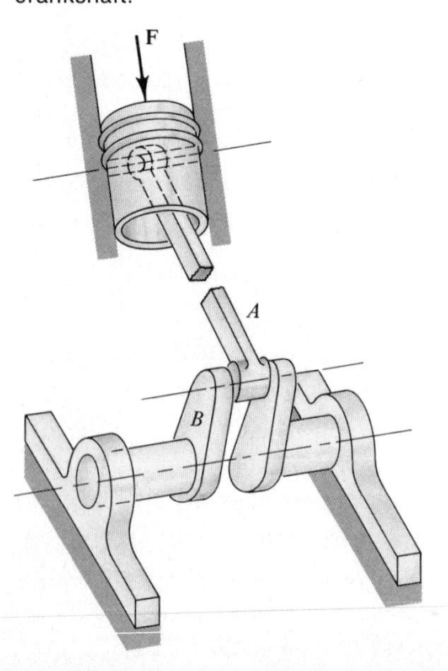

PROB. 4-18

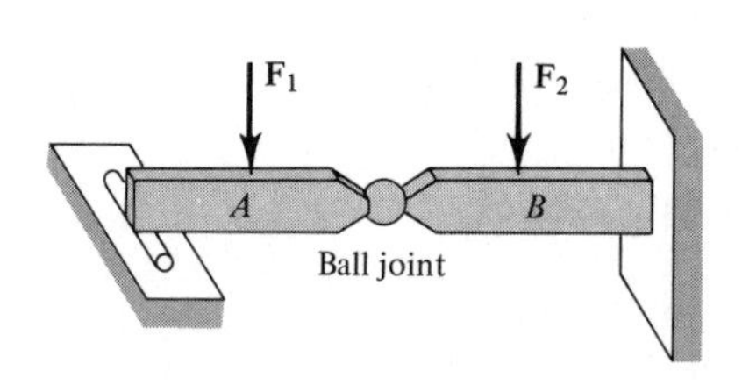

PROB. 4-19 Tripod support.

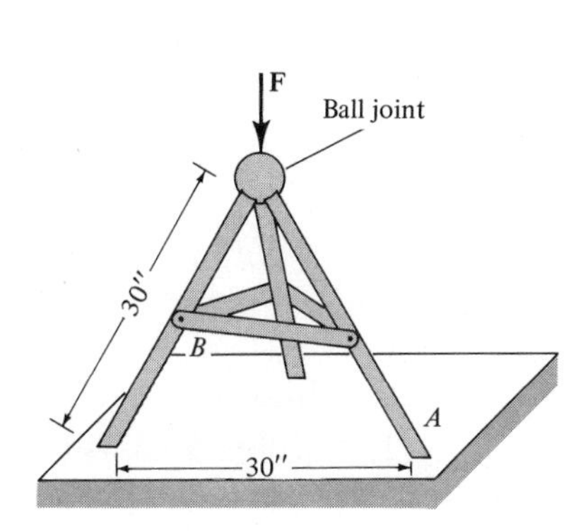

PROB. 4-20

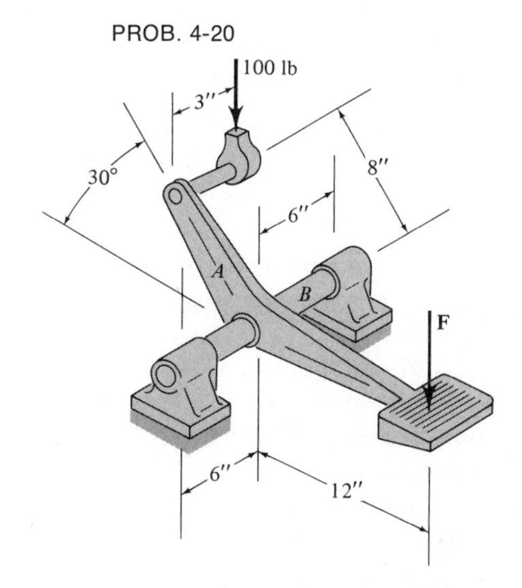

4-4 STRUCTURAL EQUILIBRIUM

When a force is applied to a free, unattached body, our experience tells us that the body will tend to move. This is a manifestation of Newton's second law of motion which, in general terms, states that a force acting on a body tends to produce a proportional change in the momentum of the body.

But what happens when a force is applied to a body that is attached to some sort of restraining device or connector? If this device is strong enough it will prevent the body from moving in some manner; that is, it will restrain the motion of the body. If this restraining force is the right type and size, it will "balance out" the original applied force so that the net force on the body is zero and the body as a whole will not move. This is precisely the situation that occurs in most structures and machines whose main function is to support or transmit loads. This situation is called *structural equilibrium* and will be our main concern in the remainder of this chapter. Of course, many machines are designed to produce some sort of motion rather than completely restraining it, and in many of these cases the forces acting on the machine or its various parts will not be "balanced out." These nonequilibrium situations also are of great importance to the engineer and will be studied in a later dynamics course.

Let us now put these notions of structural equilibrium into a more precise form. First of all, the idea that the forces acting on the structure or machine must "balance out" is made precise by stating that the resultant of all the forces acting on the structure or machine must be the null resultant. Thus, by Sec. 2-7,

$$\text{Sum of all forces} = \mathbf{0} \tag{4-1}$$

and, since rotational motion is also totally restrained, the net moment of all the forces and couples acting on the structure or machine must also be zero, so that by Sec. 2-7,

$$\text{Total moment } \mathbf{M}_q = \mathbf{0} \tag{4-2}$$

where q is any arbitrary point. Equations (4-1) and)4-2), then, are the necessary conditions for the entire structure or machine to be in structural equilibrium.

Of course, whether or not the structure or machine is actually in equilibrium depends upon whether or not it is capable of supporting the applied loads without breaking. This question depends upon whether or not the individual pieces or parts of the structure or machine are capable of supporting their required loads. Therefore, the same criteria for structural equilibrium must be applied to all the individual members making up the structure or machine; that is, the individual members must also be acted upon by balanced or equilibrated force systems, so that Eqs. (4-1) and (4-2) are necessary conditions for the force systems acting on the individual members as well. We summarize these ideas with the following statement:

> A structure or machine is in structural equilibrium if and only if all parts or members which make up the structure or machine are also in structural equilibrium. Moreover, for any body in structural equilibrium, the force system $\mathscr{S}$ acting on that body must satisfy
>
> $$\mathbf{S}(\mathscr{S}) = \mathbf{0}$$
> $$\mathbf{M}_q(\mathscr{S}) = \mathbf{0}$$

Applying these criteria to the free-body diagrams shown in Fig. 4-8, the force system in Fig. 4-8(a) acting on the entire structure must be an equilibrated one, the force system on member abc in Fig. 4-8(b) must be an equilibrated one, the force system on bde must be an equilibrated one, and the force system on member cd must be an equilibrated one. Thus each of these force systems must individually and collectively have the null resultant and, therefore, must satisfy Eqs. (4-1) and (4-2).

Before delving into a detailed discussion of equilibrated force systems, it is important to make the following observation. When a load is applied to a real body made of a real material, the body will always deform in some manner; that is, the body will stretch, compress, bend, twist, or deform by some combination of these, depending upon the type of load that is applied. In some cases, this

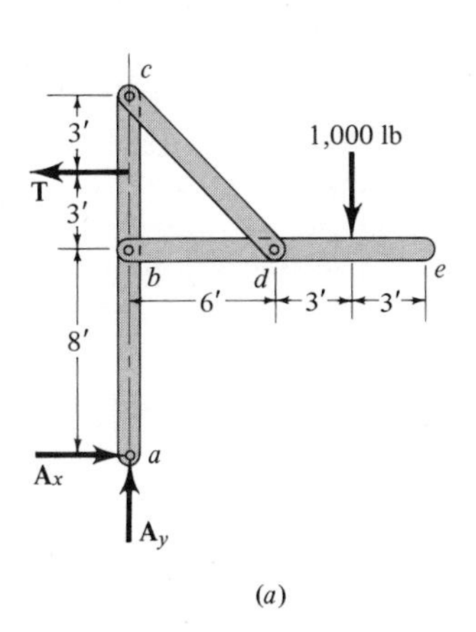

(a)

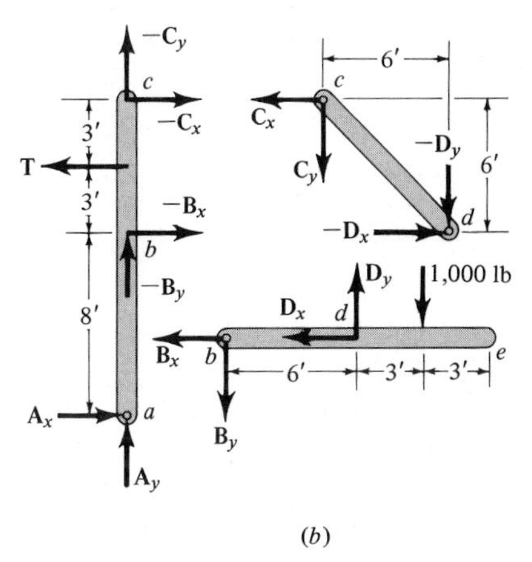

(b)

Fig. 4-8

deformation will be easily visible to the human eye, such as the compression of an innerspring mattress or the bending of a piece of electrical cable during installation. On the other hand, when you walk across the classroom floor, the amount of deformation that your weight causes is not discernible to the naked eye. Likewise, in most machines and structures whose primary function is to transmit or support loads, the deformations which are produced are extremely small relative to the original undeformed dimensions of the members, perhaps in the order of 1 unit of deformation to 1,000 units of original dimension. Later chapters will be devoted to determining the actual deformations which the loads produce.

Strictly speaking, in drawing free-body diagrams and writing the corresponding equations of equilibrium, you should use the stabilized deformed dimensions of the body or members since it is the *deformed* body which is in equilibrium under the applied loads. However, as is evident from the previous paragraph, in most structures and machines the deformed dimensions will differ from the undeformed dimensions by an amount several orders of magnitude smaller than the dimensions themselves. Therefore, from a practical viewpoint, we usually make no distinction between deformed and undeformed dimensions and use whichever are available to us. These situations are usually described by saying that we are considering the bodies to be *rigid* and that we are performing a *rigid-body analysis.*

4-5 EQUILIBRATED FORCE SYSTEMS

Let us suppose that you, the engineer, have been called upon to perform a load analysis of some actual load-carrying device. Accordingly, you have carefully drawn all the appropriate free-body diagrams indicating, to the best of your knowledge and judgment, how the loads are transmitted throughout the device and its various members. You are now ready to apply the required criteria for the device to be in structural equilibrium, namely, Eqs. (4-1) and (4-2). You are now down to the essential question: Is the problem solvable? That is, do Eqs. (4-1) and (4-2) provide enough information to actually determine values for the various forces shown on the free-body diagrams? Stated more formally, are the requirements of structural equilibrium *sufficient* to perform the load analysis? As you might anticipate, the answer is sometimes yes and sometimes no.

Definition 4-1 When the requirements of structural equilibrium, Eqs. (4-1) and (4-2), are sufficient to determine all the unknown forces, the problem is called *statically determinate.*

Definition 4-2 When the requirements of structural equilibrium, Eqs. (4-1) and (4-2), are *not* sufficient to determine all the unknown forces, the problem is called *statically indeterminate.*

If the problem which you have constructed via your free-body diagrams is statically determinate, you can proceed to determine the forces transmitted throughout the device. If, on the other hand, the problem you have constructed is statically indeterminate, you have several alternatives. You could go back to the original actual device and make some additional simplifying assumption(s) regarding the restraining characteristics of some connectors or joints. The problem as then constructed might prove to be statically determinate and you can proceed to solve it. A second (and often more realistic but much more difficult) alternative in the case of a statically indeterminate problem is to adjoin to the requirements of structural equilibrium [Eqs. (4-1) and (4-2)] the further requirements of *structural compatibility.* In simple terms, structural compatibility means that the deformations of the various pieces or members making up the structure or machine must be geometrically compatible with one another so that the entire assembly can fit together. Hopefully, these additional requirements will provide enough information, so that the problem is now solvable.

In the progress of this course we will encounter both statically determinate and statically indeterminate situations, and, by the time you are finished, you should be able to readily distinguish one situation from the other. Generally, a structure will be statically indeterminate when it is restrained by its supports and connectors more than minimally necessary to maintain equilibrium. Our immediate concern, however, is to learn how to solve problems involved with statically determinate situations. To this end we will look at some special but rather important force systems and examine them in regard to their statical determinacy.

We begin by making the observation that any force system composed of a single force, a single couple, or a single force and a single couple cannot possibly be an equilibrated one since its sum and/or total moment could *not* be null.

Two-force System

Suppose the total force system acting on some body is composed of two forces only, such as shown in Fig. 4-9(*a*). If this system is to be an equilibrated one, its sum must be a null, so that

$$\mathbf{F}_1 + \mathbf{F}_2 = \mathbf{0}$$

Thus, $\mathbf{F}_1$ and $\mathbf{F}_2$ must be opposite, or

$$\mathbf{F}_1 = -\mathbf{F}_2$$

and we can redraw the system like that shown in Fig. 4-9(*b*). But this system forms a couple and cannot satisfy the second criterion of equilibrium, namely, its total moment must be null. In fact, the only way this criterion can be satisfied is if the two forces are opposite *and collinear*, as shown in Fig. 4-9(*c*). Therefore, we arrive at the following conclusion:

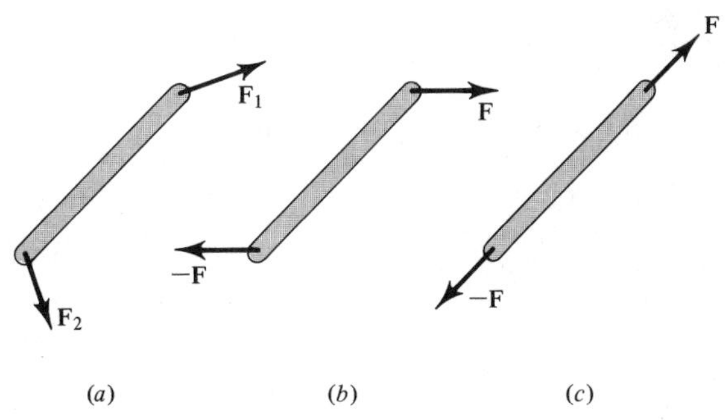

Fig. 4-9

> If a body in structural equilibrium is acted upon by only two forces, these forces *must* be equal in magnitude, opposite in direction, and have the same line of action.

Recognizing this situation when it occurs in various problems can often be a great aid in eventually solving the problem.

Concurrent Force Systems

Recall from Sec. 2-7 that a concurrent force system $\mathscr{S}$ is one for which the lines of action of all the forces intersect at one common point, called the point of concurrency. Letting q be the point of concurrency, it is clear that

$$\mathbf{M}_q(\mathscr{S}) = \mathbf{0} \tag{4-2}$$

which satisfies the moment requirement for equilibrium. The remaining requirement is that the sum must be null,

$$\mathbf{S}(\mathscr{S}) = \mathbf{0} \tag{4-1}$$

If we write each force in the system in terms of its components as, for example,

$$\mathbf{F}_1 = F_{1x}\mathbf{i} + F_{1y}\mathbf{j} + F_{1z}\mathbf{k}$$

then Eq. (4-1) is simply a shorthand way of saying that

$$\sum_{r=1}^{n} \mathbf{F}_r = \mathbf{i}\sum_{r=1}^{n} F_{rx} + \mathbf{j}\sum_{r=1}^{n} F_{ry} + \mathbf{k}\sum_{r=1}^{n} F_{rz} = \mathbf{0} \tag{4-3}$$

where n is the number of forces in the system and F_{rx} is the x-scalar component of the rth force. Equation (4-3) implies that

$$\sum_{r=1}^{n} F_{rx} = 0 \qquad \sum_{r=1}^{n} F_{ry} = 0 \qquad \sum_{r=1}^{n} F_{rz} = 0 \tag{4-4}$$

which represents three independent algebraic equations. Thus, we conclude:

> In general, a problem involving an equilibrated concurrent force system is statically determinate (solvable) provided the problem contains *no more than three unknowns.*

Before looking at an example problem involving a concurrent force system, we remark that any equilibrated force system made up of only three nonparallel forces must be a concurrent force system. This observation can sometimes be of use in solving certain problems, and its partial proof is left as an exercise problem (Prob. 4-21).

EXAMPLE 4-1

The crate is held in place by two cables as shown in Fig. 4-10(a). If the tension in cable B is 450 lb, what is the weight of the crate? ***Solution:*** A free-body diagram of the crate is shown in Fig. 4-10(b), and by the remark in the previous paragraph, the three forces $\mathbf{T}_a$, $\mathbf{T}_b$, and $\mathbf{W}$ form a concurrent force system with o being the point of concurrency. With the crate in equilibrium, we have

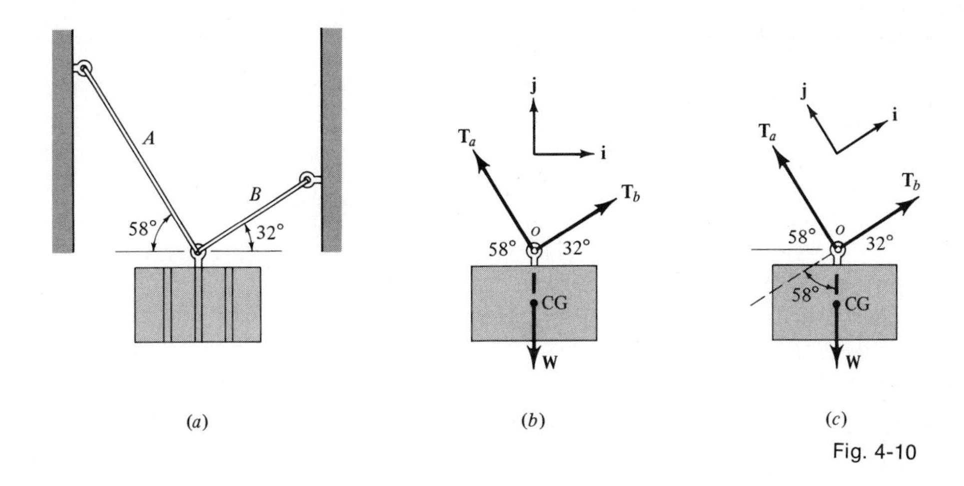

Fig. 4-10

$$\Sigma \mathbf{F} = \mathbf{0}$$

$$\mathbf{T}_a + \mathbf{T}_b + \mathbf{W} = \mathbf{0} \qquad (a)$$

Letting **i** and **j** be horizontal and vertical unit vectors, respectively,

$$\mathbf{T}_a = -T_a \cos 58^\circ\, \mathbf{i} + T_a \sin 58^\circ\, \mathbf{j}$$

$$\mathbf{T}_b = 450 \cos 32^\circ\, \mathbf{i} + 450 \sin 32^\circ\, \mathbf{j}$$

$$\mathbf{W} = -W\mathbf{j}$$

Then, by Eq. (a)

$$\mathbf{i}(-T_a \cos 58^\circ + 450 \cos 32^\circ) = 0$$

$$T_a = 720 \text{ lb}$$

and

$$\mathbf{j}(T_a \sin 58^\circ + 450 \sin 32^\circ - W) = 0$$

$$W = 850 \text{ lb}$$

Alternate Solution: Instead of using horizontal and vertical components, let us choose unit vectors as shown in Fig. 4-10(c). Then

$$\mathbf{T}_a = T_a\, \mathbf{j}$$

$$\mathbf{T}_b = 450\, \mathbf{i}$$

$$\mathbf{W} = -W \cos 58^\circ\, \mathbf{i} - W \sin 58^\circ\, \mathbf{j}$$

Then, by Eq. (a)

$$\mathbf{i}(450 - W \cos 58^\circ) = 0$$

$$W = 850 \text{ lb}$$

Thus, with this choice of components, the solution becomes much more direct. Such a scheme can often save time and errors.

Coplanar Force Systems

Although real bodies and structures are necessarily three dimensional, quite often the force system acting on a structure or member can be idealized (simplified) into a two-dimensional (coplanar) one. The structure depicted earlier in this chapter in Fig. 4-4 was such a situation, as were several exercise problems. The ability to analyze such coplanar problems is of paramount importance to much of what you will do later in this course and in other engineering courses.

Consider the coplanar force system $\mathscr{S}$ shown in Fig. 4-11(*a*). Each of the forces can be expressed in the form

$$\mathbf{F}_r = F_{rx}\,\mathbf{i} + F_{ry}\,\mathbf{j}$$

where **i** and **j** are the usual unit vectors. Equilibrium requires

$$\text{Sum}\ (\mathscr{S}) = \sum_{r=1}^{n} \mathbf{F}_r = \mathbf{0}$$

or

$$\mathbf{i} \sum_{r=1}^{n} F_{rx} + \mathbf{j} \sum_{r=1}^{n} F_{ry} = \mathbf{0}$$

which implies the two independent algebraic conditions

$$\sum_{r=1}^{n} F_{rx} = 0 \qquad \text{and} \qquad \sum_{r=1}^{n} F_{ry} = 0 \tag{4-5}$$

where n is the number of forces in the system. Equilibrium also requires

$$\text{Sum of moments}\ \mathbf{M}_q(\mathscr{S}) = \sum_{r=1}^{n} \mathbf{qp}_r \times \mathbf{F}_r = \mathbf{0} \tag{4-6}$$

where $\mathbf{qp}_r$ is the position vector of the rth force relative to point q in the plane. However, because the system is coplanar, $\mathbf{M}_q$ will always have only one component, namely, the **k** component. Hence Eq. (4-6) represents only one additional requirement for equilibrium. Thus, the two conditions from Eq. (4-5) along with the one

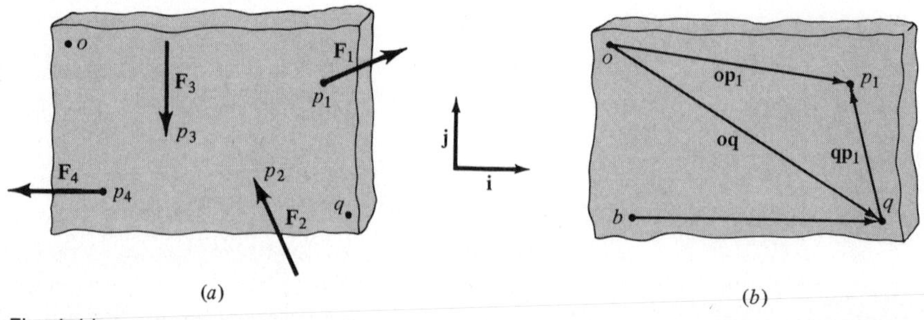

Fig. 4-11

from Eq. (4-6) means that there are three algebraic conditions which the system must satisfy to be an equilibrated one. Therefore, in general, we have the conclusion:

> For an equilibrated coplanar force system there are three, and only three, independent algebraic conditions which must be fulfilled. Hence, in general, a coplanar force system is statically determinate provided it contains no more than three unknowns.

We remark that this conclusion is the same whether or not the system contains couples, so long as the couples are vectors perpendicular to the plane of the forces.

Although Eqs. (4-5) and (4-6) represent two force equations and one moment equation, in many problems we often find it convenient to replace one or even both force conditions by other moment conditions. To see this, consider moments with respect to point o in Fig. 4-11(a):

$$\mathbf{M}_o(\mathscr{S}) = \sum_{r=1}^{n} \mathbf{op}_r \times \mathbf{F}_r$$

where $\mathbf{op}_r$ is the position vector of $\mathbf{F}_r$ relative to point o. But

$$\mathbf{op}_r = \mathbf{oq} + \mathbf{qp}_r$$

when $\mathbf{oq}$ is the position vector from o to q, as illustrated in Fig. 4-11(b). Thus

$$\begin{aligned} \mathbf{M}_o(\mathscr{S}) &= \sum_{r=1}^{n} (\mathbf{oq} + \mathbf{qp}_r) \times \mathbf{F}_r \\ &= \sum_{r=1}^{n} \mathbf{oq} \times \mathbf{F}_r + \sum_{r=1}^{n} \mathbf{qp}_r \times \mathbf{F}_r \\ &= \mathbf{oq} \times \sum_{r=1}^{n} \mathbf{F}_r + \sum_{r=1}^{n} \mathbf{qp}_r \times \mathbf{F}_r \\ &= \mathbf{oq} \times \mathbf{S}(\mathscr{S}) + \mathbf{M}_q(\mathscr{S}) \end{aligned} \tag{4-7}$$

Equilibrium requires

$$\mathbf{M}_o(\mathscr{S}) = \mathbf{0}$$

But the last term in Eq. (4-7) vanishes by virtue of Eq. (4-6). Hence

$$\mathbf{M}_o(\mathscr{S}) = \mathbf{0} \Longrightarrow \mathbf{oq} \times \mathbf{S}(\mathscr{S}) = \mathbf{0} \tag{4-8}$$

This last result leads to the conclusion that the sum of the forces $\mathbf{S}(\mathscr{S})$ must either be a force parallel to $\mathbf{oq}$ or zero. Thus the condition $\mathbf{M}_o(\mathscr{S}) = \mathbf{0}$ implies a condition on the sum of the forces, namely, that the sum of components perpendicular to $\mathbf{oq}$ must be zero.

Similarly, by selecting a third point not on the line **oq**, such as b in Fig. 4-11(b), we obtain

$$\mathbf{M}_b(\mathscr{S}) = \mathbf{bq} \times \mathbf{S}(\mathscr{S}) = \mathbf{0} \tag{4-9}$$

This implies that $\mathbf{S}(\mathscr{S})$ must be zero or parallel to **bq**. But $\mathbf{S}(\mathscr{S})$ cannot be parallel to **bq** and **oq** simultaneously. Hence Eqs. (4-8) and (4-9) together imply

$$\mathbf{S}(\mathscr{S}) = \mathbf{0}$$

The judicious selection of points for taking moments can greatly simplify the solving of many problems. However, remember that regardless of how many moment and force equations you write, for a coplanar problem there are only *three independent* conditions.

EXAMPLE 4-2

The cantilever beam shown in Fig. 4-12(a) carries a uniformly distributed load of 100 lb per foot of length. Determine the reactions on the fixed left end of the beam.

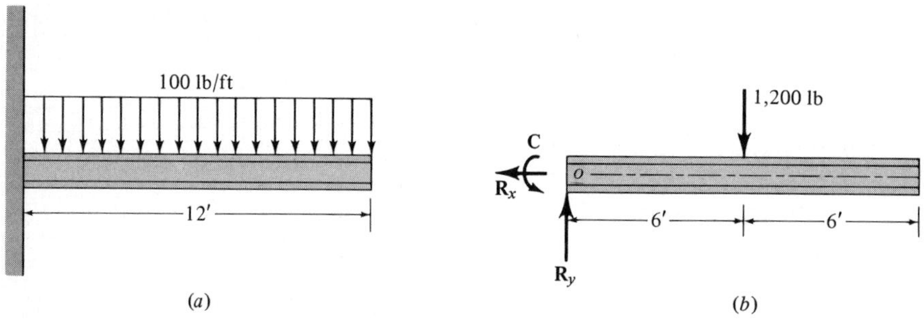

Fig. 4-12

Solution: An FBD of the beam is shown in Fig. 4-12(b). In this diagram, we have replaced the distributed load by its statically equivalent resultant of 1,200 lb at the midpoint of the beam. Also, the fixed support at the wall resists translation and rotation, so that we have indicated the three unknowns as forces $\mathbf{R}_x$ and $\mathbf{R}_y$ and a couple **C**.

The force conditions (4-5) for equilibrium yield

$$\text{Sum of horizontal components} = -R_x + 0 = 0$$

$$\therefore R_x = 0$$

$$\text{Sum of vertical components} = R_y - 1{,}200 = 0$$

$$\therefore R_y = 1{,}200 \text{ lb}$$

We now apply the moment requirement (4-6) by taking moments with respect to the left end of the beam:

$$\Sigma \mathbf{M}_o = (\overset{\curvearrowright}{6})(1{,}200) + \overset{\curvearrowleft}{C} = \mathbf{0} \tag{a}$$

or, equivalently,

$$6(1{,}200)(-\mathbf{k}) + C\mathbf{k} = \mathbf{0} \tag{b}$$

where $\mathbf{k}$ is the usual unit vector perpendicular to the plane. From either Eq. (*a*) or (*b*), we arrive at the algebraic condition

$$-6(1{,}200) + C = 0$$
$$C = 7{,}200 \text{ in-lb}$$

Hence the wall reactions are

$$\mathbf{R}_x = \mathbf{0} \qquad \mathbf{R}_y = 1{,}200 \text{ lb } \mathbf{j} = 1{,}200 \text{ lb} \uparrow \qquad \mathbf{C} = 7{,}200 \text{ in-lb } \mathbf{k} = 7{,}200 \text{ in-lb} \curvearrowleft$$

Thus far we have looked at equilibrium of concurrent and coplanar force systems. There are, of course, many other special types of force systems which can be equilibrated, and we could spend considerable time examining the various conditions for equilibrium. However, from what we have done thus far, it is apparent that the equilibrium conditions (4-1) and (4-2) yield at most six independent equations (three force conditions and three moment conditions) for the solution of problems involving an equilibrated force system. Thus the most general problem for an equilibrated force system cannot contain more than six unknowns if it is to be statically determinate.

We close this section with an example involving a nonconcurrent, noncoplanar force system to illustrate some of the techniques which can be useful in solving such problems.

EXAMPLE 4-3

The thick triangular plate shown in Fig. 4-13(*a*) weighs 30 lb and is loaded and supported horizontally as indicated. Determine all the forces exerted on the plate by these various connectors.

Solution: An FBD of the plate is shown in Fig. 4-13(*b*), in which the force of 30 lb represents the weight of the plate acting at its center of gravity; three component forces are shown at point *a* in place of the ball and socket, a single pushing force is shown at *c* in place of the ball, and a single tension is shown at *b* and *d* in place of the cables.

To apply the requirements of equilibrium, namely, the sums of forces and moments must be zero, there is no set procedure to

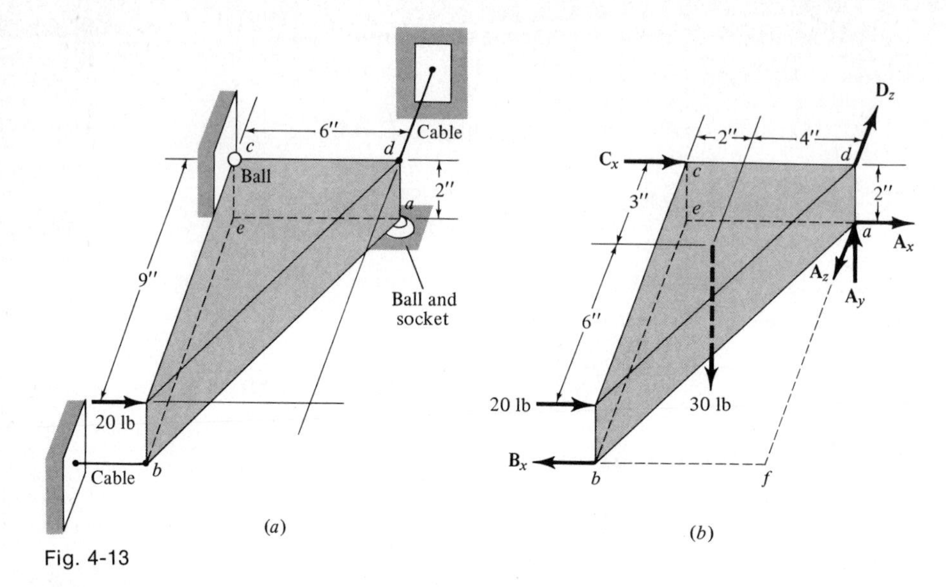

Fig. 4-13

follow, and what we do is mostly a matter of judgment and observation.

First of all, consider moments about the vertical line ad. Why this? Well, none of the unknowns $\mathbf{D}_z$, $\mathbf{A}_x$, $\mathbf{A}_y$, $\mathbf{A}_z$, or $\mathbf{C}_x$ will produce moments about this line, since they intersect this line. We have

$$\Sigma \mathbf{M}_{ad} = 20(9) + B_x(9)$$

Hence, the moment requirement for equilibrium means that

$$+20(9)\mathbf{j} - B_x(9)\mathbf{j} = \mathbf{0} \qquad (a)$$

$$B_x = 20 \text{ lb} \qquad \text{(as shown)}$$

$$\mathbf{B}_x = -20\mathbf{i} \text{ lb}$$

Similarly, considering moments about line ae yields

$$\Sigma \mathbf{M}_{ae} = 30(3) + D_z(2)$$

or

$$30(3)\mathbf{i} - D_z(2)\mathbf{i} = \mathbf{0} \qquad (b)$$

$$D_z = 45 \text{ lb} \qquad \text{(as shown)}$$

$$\mathbf{D}_z = -45\mathbf{k} \text{ lb}$$

Similarly, considering moments about line af yields

$$\Sigma \mathbf{M}_{af} = 20(2) + 30(4) + C_x(2)$$

or

$$-20(2)\mathbf{k} + 30(4)\mathbf{k} - C_x(2)\mathbf{k} = \mathbf{0} \qquad (c)$$

$$C_x = 40 \text{ lb} \qquad \text{(as shown)}$$

$$\mathbf{C}_x = 40\mathbf{i} \text{ lb}$$

Three unknowns still remain, namely, $\mathbf{A}_x$, $\mathbf{A}_y$, and $\mathbf{A}_z$. Summing forces in the **i** direction yields

$$\mathbf{i}(C_x + A_x + 20 - B_x) = \mathbf{0}$$
$$A_x = B_x - 20 - C_x$$
$$= 20 - 20 - 40 = -40 \text{ lb}$$

The minus sign indicates that the arrow on the vector A_x should be reversed, so that

$$\mathbf{A}_x = 40 \text{ lb} \quad \leftarrow$$
$$= -40\mathbf{i} \text{ lb}$$

Similarly, summing **j** and **k** forces yields

$$\mathbf{j}(-30 + A_y) = \mathbf{0}$$
$$A_y = 30 \text{ lb} \qquad \text{(as shown)}$$
$$\mathbf{A}_y = 30\mathbf{j} \text{ lb}$$
$$\mathbf{k}(-D_z + A_z) = \mathbf{0}$$
$$A_z = 45 \text{ lb} \qquad \text{(as shown)}$$
$$\mathbf{A}_z = 45\mathbf{k} \text{ lb}$$

Alternate Solution: Instead of considering moments about lines, as we did in the above solution, we can consider moments with respect to, say, *point a*. Thus

$$\Sigma \mathbf{M}_a = (-6\mathbf{i} + 2\mathbf{j}) \times C_x\mathbf{i} + 2\mathbf{j} \times (-D_z\mathbf{k})$$
$$+ (-4\mathbf{i} + 1\mathbf{j} + 3\mathbf{k}) \times (-30\mathbf{j}) + (-6\mathbf{i} + 2\mathbf{j} + 9\mathbf{k}) \times 20\mathbf{i}$$
$$+ (-6\mathbf{i} + 9\mathbf{k}) \times (-B_x\mathbf{i}) = \mathbf{0}$$
$$= -2C_x\mathbf{k} - 2D_z\mathbf{i} + 120\mathbf{k} + 90\mathbf{i} - 40\mathbf{k} + 180\mathbf{j}$$
$$-9B_x\mathbf{j} = \mathbf{0}$$
$$= \mathbf{i}(90 - 2D_z) + \mathbf{j}(180 - 9B_x) + \mathbf{k}(120 - 40 - 2C_x) = \mathbf{0}$$

which requires

$$90 - 2D_z = 0$$
$$180 - 9B_x = 0$$
$$80 - 2C_x = 0$$

which are precisely the conditions (*b*), (*a*), and (*c*), respectively, found previously. We will not pursue this method further since the remainder of this solution would be the same as before.

EXERCISE PROBLEMS

4-21 Suppose we have an equilibrated force system composed of: (*a*) Three nonparallel coplanar forces. Show that they must be concurrent. (*b*) Three concurrent forces. Show that they must be coplanar. (*c*) Three parallel forces. Show that they must be coplanar.

4-22 The 40-lb cylinder rests in the rack as shown. Determine the forces exerted on the cylinder by the rack.

PROB. 4-22

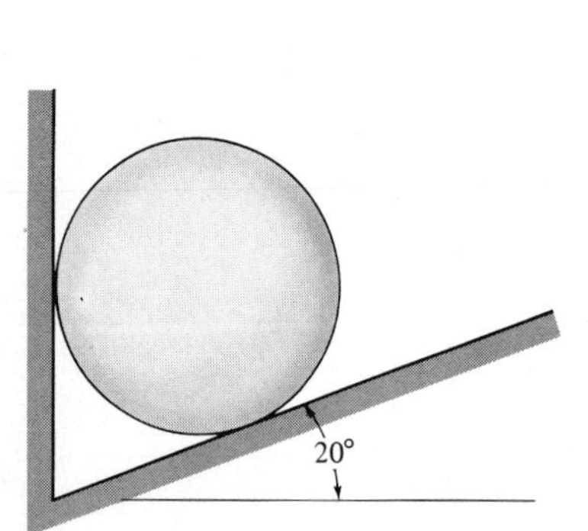

PROB. 4-23

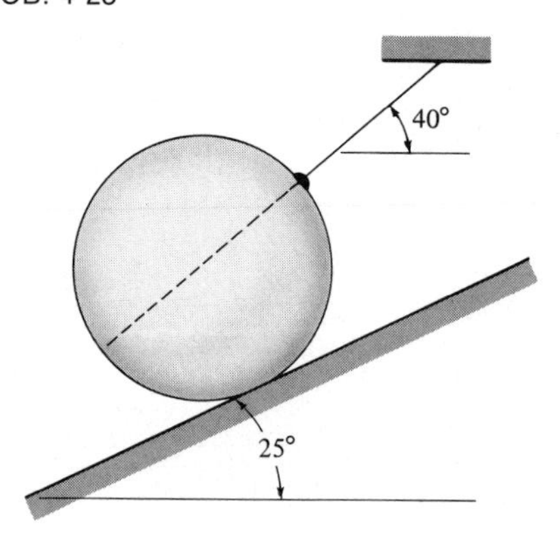

4-23 The 80-lb sphere rests against the inclined plane and is held in position by a cord as shown. Determine the forces exerted on the sphere.

4-24 For the boom crane shown, determine the compressive force in the boom and the tensile force in the cable.

PROB. 4-24

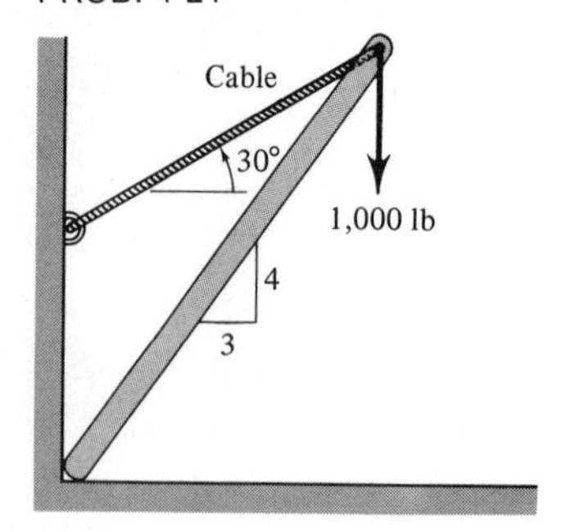

4-25 to 4-27 Determine the forces in cables *A*, *B*, and *C*.

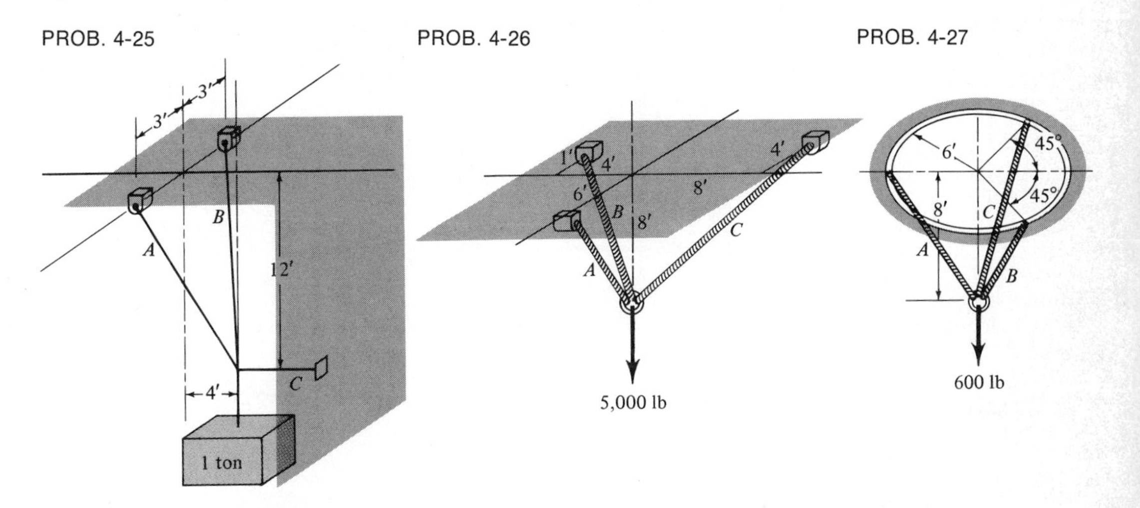

4-28 The derrick supports a load of 3,000 lb. Determine the forces in legs c–d and b–d and the guy wire a–d.

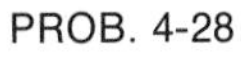

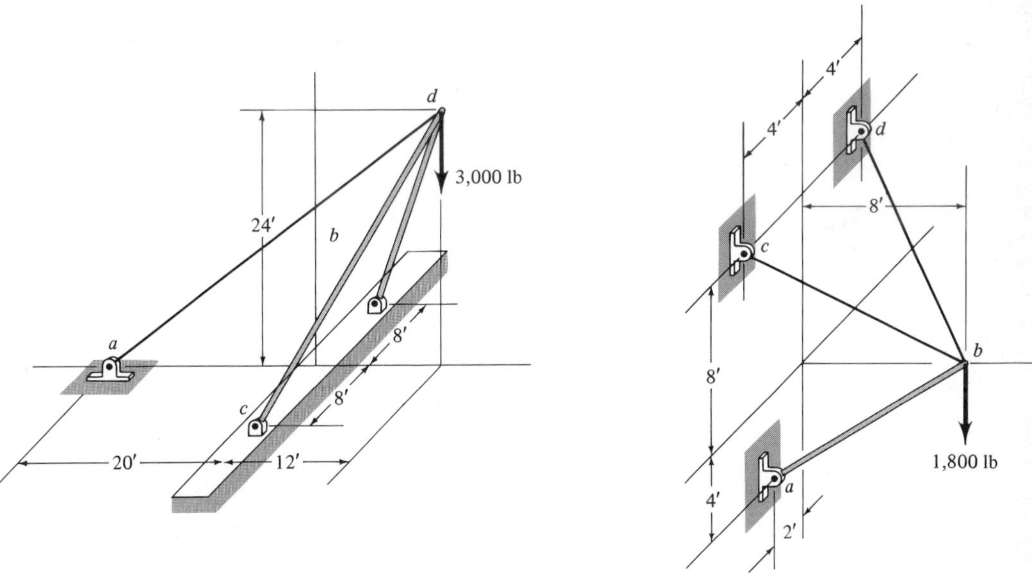

4-29 The boom a–b supports a 1,800-lb load and is held by a ball joint at a and cables b–c and b–d. Determine the force in the boom and the tension in each cable.

4-30 The weight of the stiff bar abc can be neglected and all contact surfaces considered smooth. Determine the reactions at b and a.

PROB. 4-30

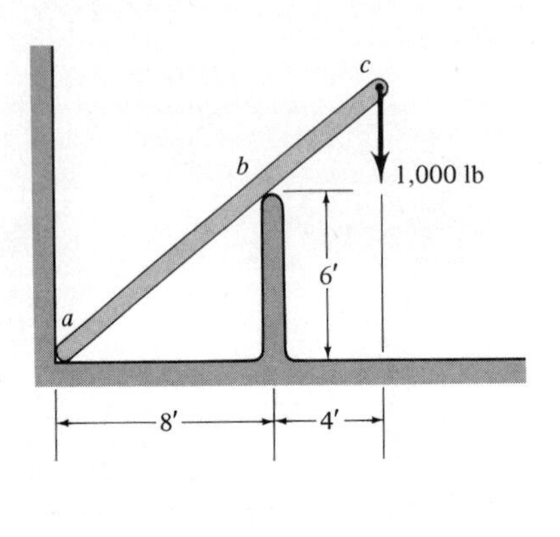

PROB. 4-31

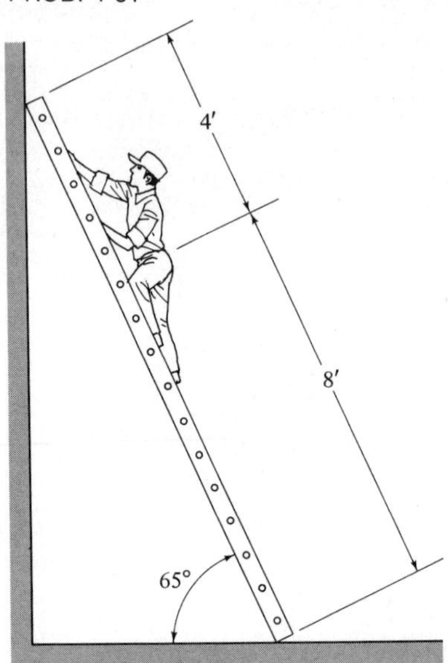

4-31 The man on the ladder weighs 160 lb. If the vertical wall is considered smooth, determine the force which the ground must exert to hold the ladder in place.

4-32 and 4-33 Find the reactions of a and b on the structure shown.

PROB. 4-32 A frame.

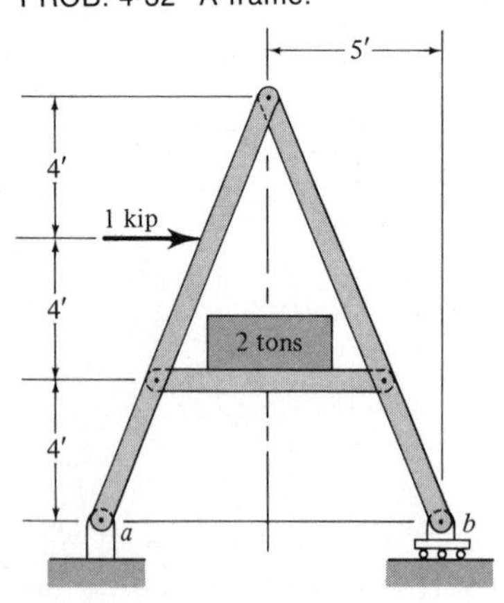

PROB. 4-33 Pin-connected truss.

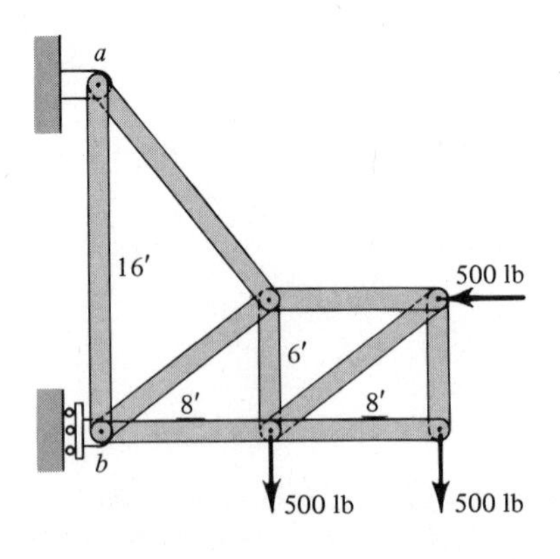

4-34 For the 30-lb door shown, determine the reactions at the hinges a and b. Assume each hinge supports half of the vertical weight.

PROB. 4-34 Heavy door.

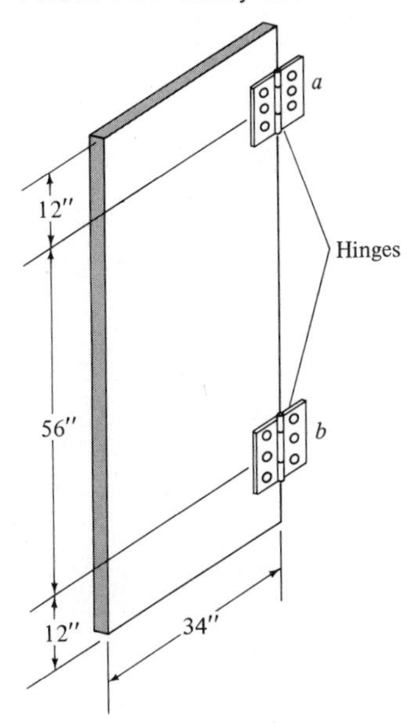

4-35 The 50-lb sign is supported as shown. Determine the forces in the rods and the reaction at the ball joint.

PROB. 4-35 Heavy sign.

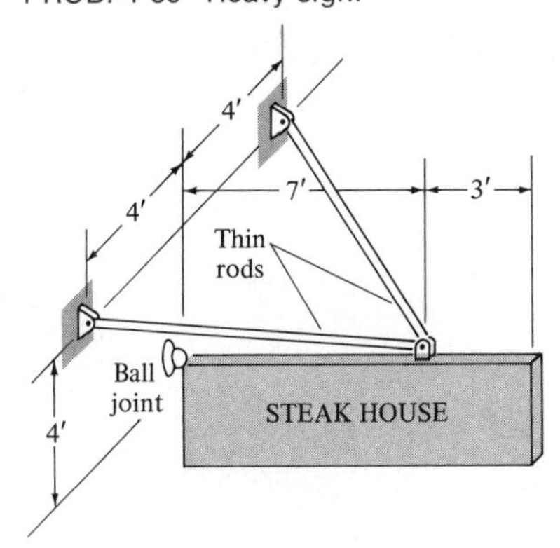

PROB. 4-36

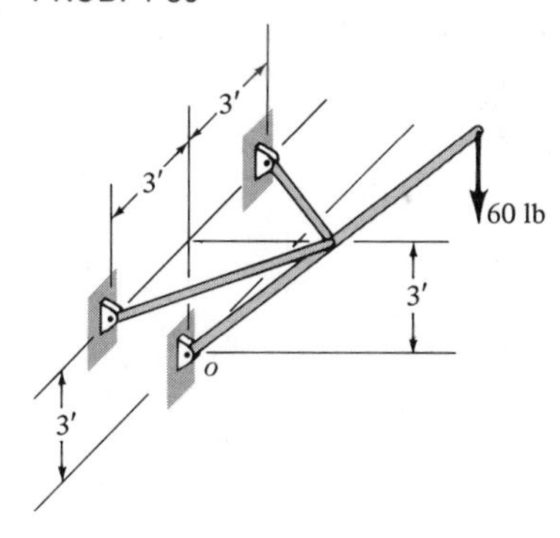

4-36 The 10-ft boom has two horizontal braces attached at the middle of the boom as shown. Determine the forces in the braces and the reactions at the ball and socket at *o*.

4-37 Using the hoist frame shown, the man is holding a 140-lb weight suspended off the ground. Determine the reaction of the ground on the frame at point *o*.

PROB. 4-37 Fixed-hoist frame.

PROB. 4-38

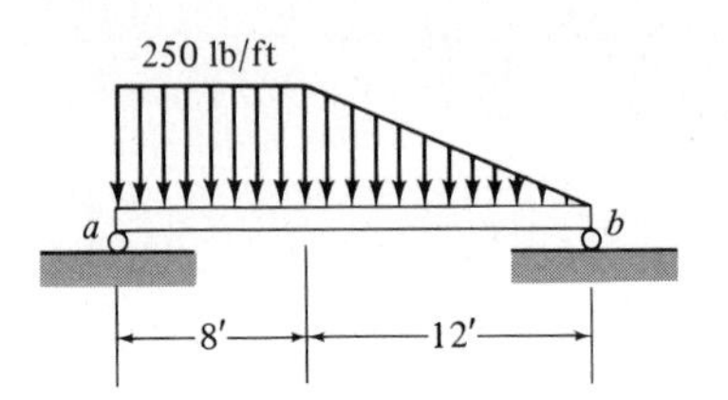

4-38 The simply supported beam carries a distributed load as shown. Find the reactions at points a and b.

4-39 If the resultant downward force on the wing is 250 lb, determine the required forces exerted by the pin and strut.

PROB. 4-39

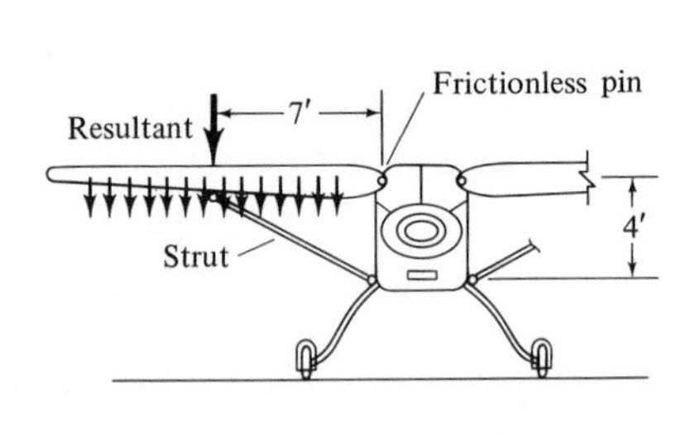

PROB. 4-40 Automobile torsion bar.

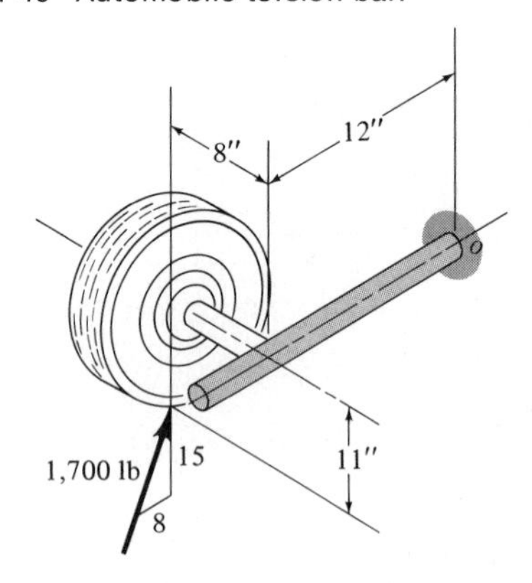

4-40 Determine the reactions at o on the automobile torsion bar shown.

4-41 The crate shown weighs 200 lb and is at rest on the incline. Assume its center of gravity is at the geometric center. Determine the friction force exerted on the crate by the rough incline. Find the magnitude and location of the net normal force against the crate.

PROB. 4-41

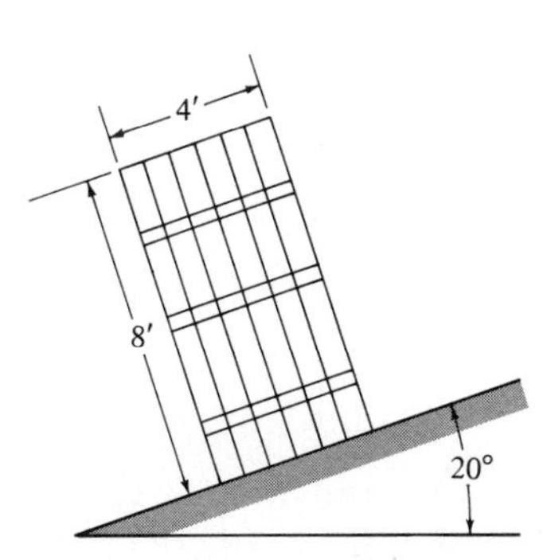

PROB. 4-42

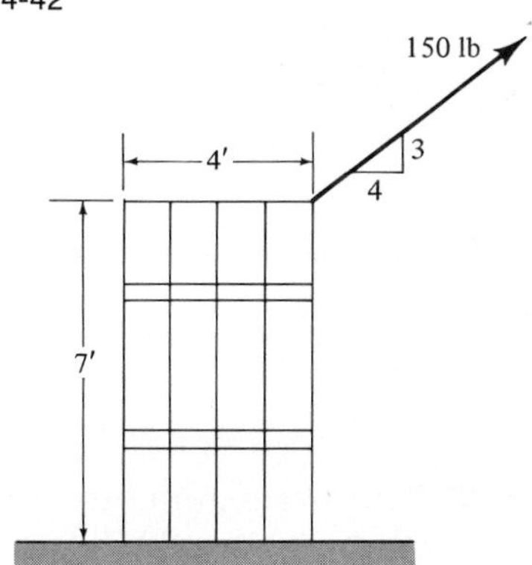

4-42 and 4-43 The 600-lb crate is in equilibrium. Determine the friction force exerted on the crate by the ground. Find the magnitude and location of the net normal force against the crate.

PROB. 4-43

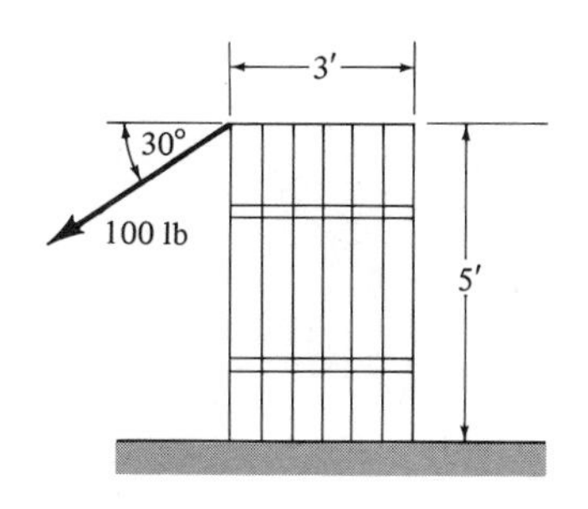

PROB. 4-44

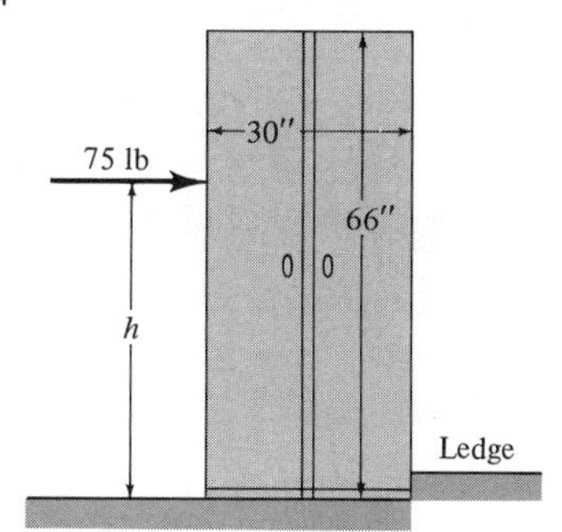

4-44 At what height h should the 75-lb horizontal force be applied to tip the 250-lb cabinet?

4-45 The man is pushing on the 300-lb crate parallel to the incline. What minimum force must the man exert to prevent the crate from tipping over on top of him?

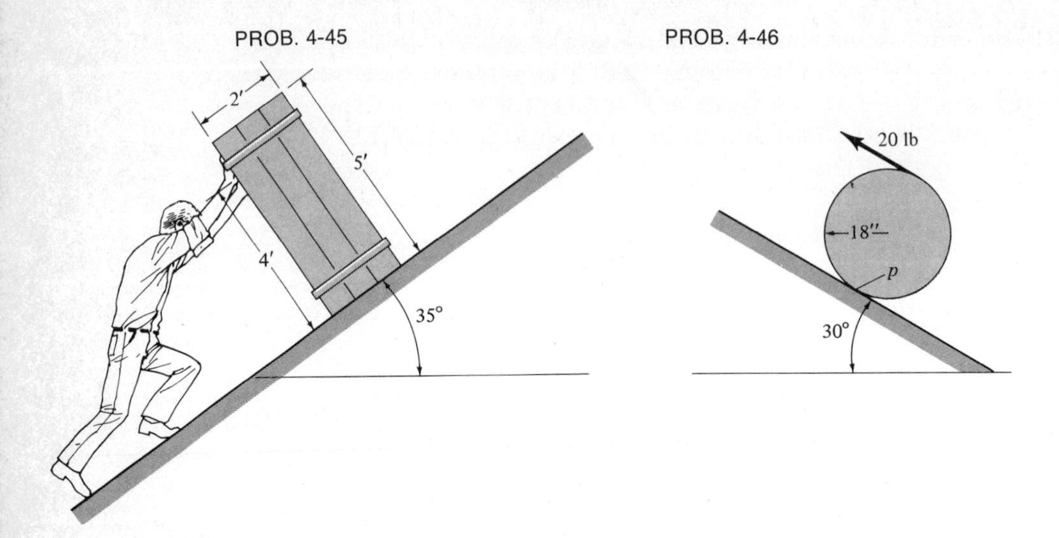

4-46 The homogeneous disk is held in equilibrium on the rough plane by the 20-lb force. What is the weight of the disk? Find the normal and friction forces at the contact point p.

4-47 The homogeneous 30-lb disk is at rest. Determine the force F. Find the normal and friction force at the contact point p.

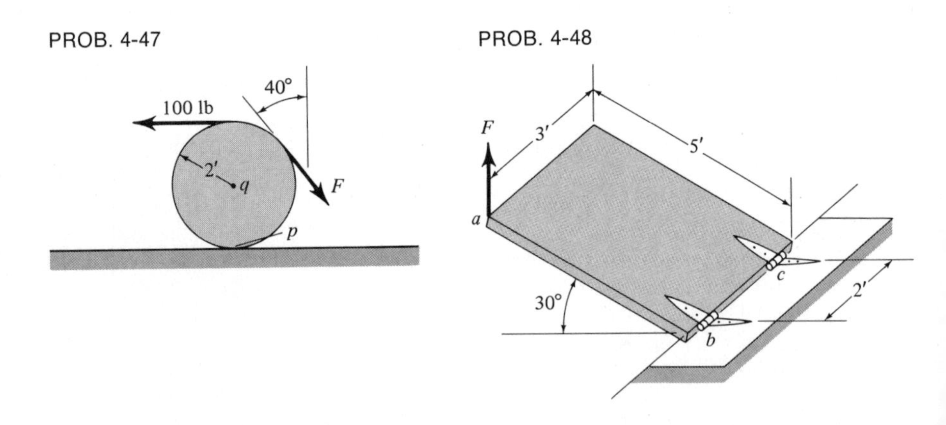

4-48 The heavy trapdoor weighs 40 lb and is held partially opened by a vertical force at a. What are the significant reactions at the narrow hinges at b and c?

4-49 and 4-50 For the cantilevered fixture shown, determine the reactions at the fixed support p.

PROB. 4-49

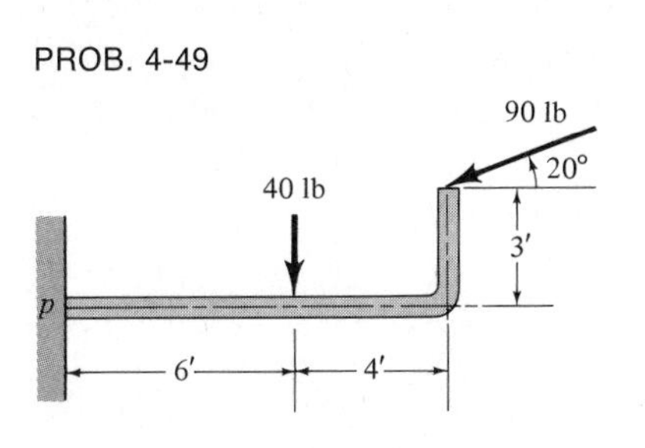

PROB. 4-50

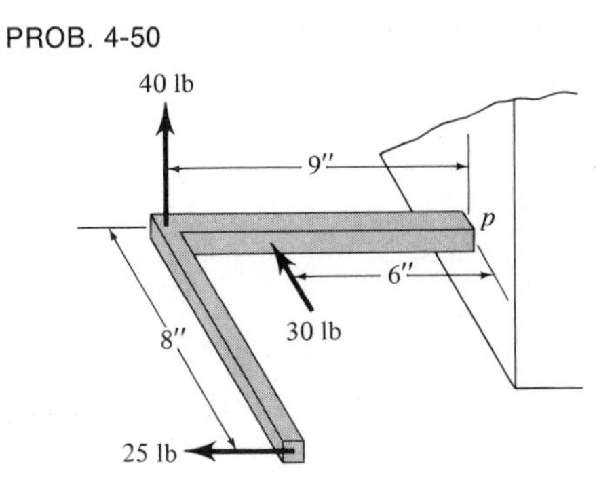

4-51 If the shaft is in equilibrium, find the reactions at the ball **and** bearings at a and b.
4-52

PROB. 4-51

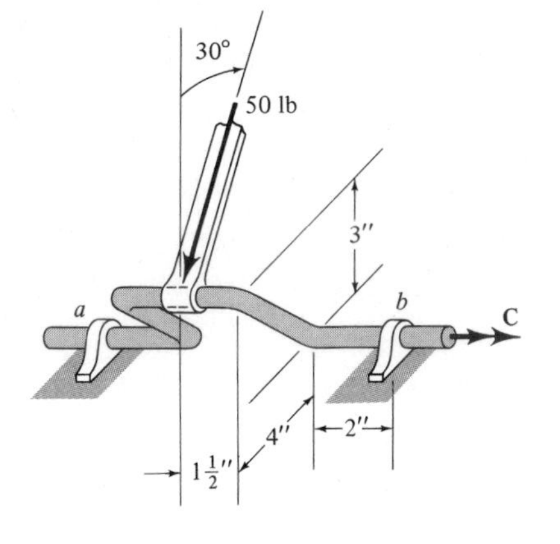

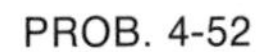
PROB. 4-52

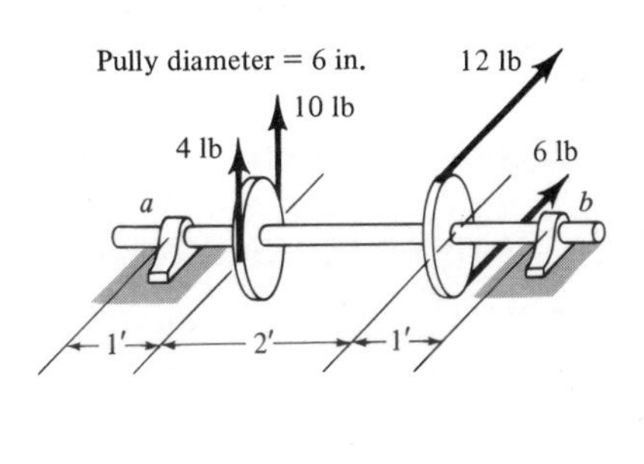

4-53 The 30-lb trapdoor is held in place by the force F directed as shown. Find F. Also, what are the reactions at the hinges at a and b?

PROB. 4-53

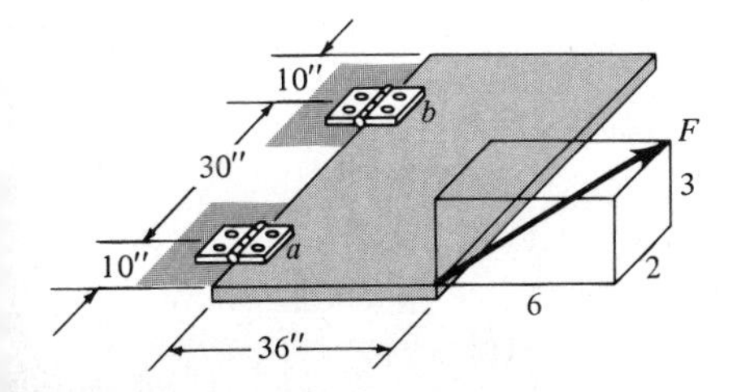

4-54 The flat semicircular homogeneous 60-lb plate is supported horizontally by three cables as shown. Find the tension in each cable.

PROB. 4-54 PROB. 4-57

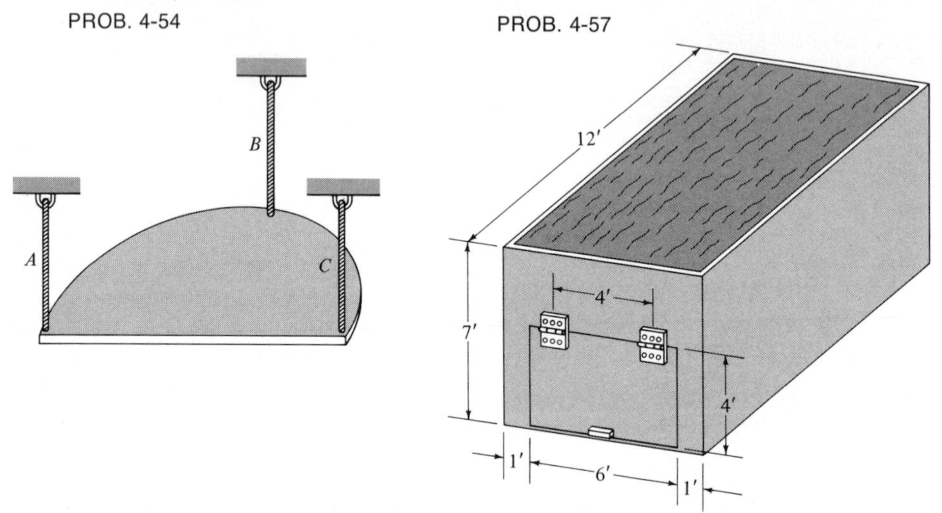

4-55 A 2 × 2 × 2 ft cube weighs 600 lb and is resting on the bottom of a tank of water 10 ft deep. What is the force between the block and the floor of the tank?

4-56 A 2-ft-diameter steel ball rests at the bottom of a tank of water 6 ft deep. What is the force between the ball and the floor of the tank? The specific weight of steel is 0.28 lb/in^3.

4-57 The grain bin shown has a side door. The door has two hinges at the top and a locking device at the bottom. If the bin is filled with grain weighing 80 lb/ft^3, what are the forces on the hinges and lock?

4-58 The concrete dam holds back water to a depth of 24 ft. What is the net horizontal reaction exerted on the bottom of the dam by the riverbed? What is the net vertical reaction, and where is its line of action located? Concrete weighs 150 lb/ft^3 and water, 62.4 lb/ft^3.

PROB. 4-58 River dam.

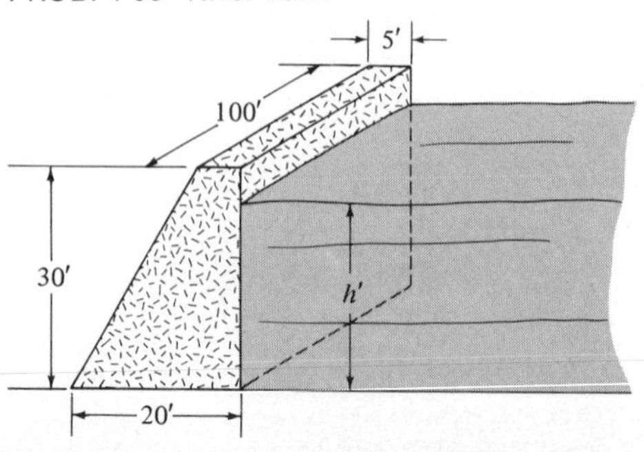

4-59 The trapezoidal-shaped door weighs 25 lb. Find the reactions at the hinges assuming each hinge supports half the vertical weight.

PROB. 4-59

PROB. 4-60

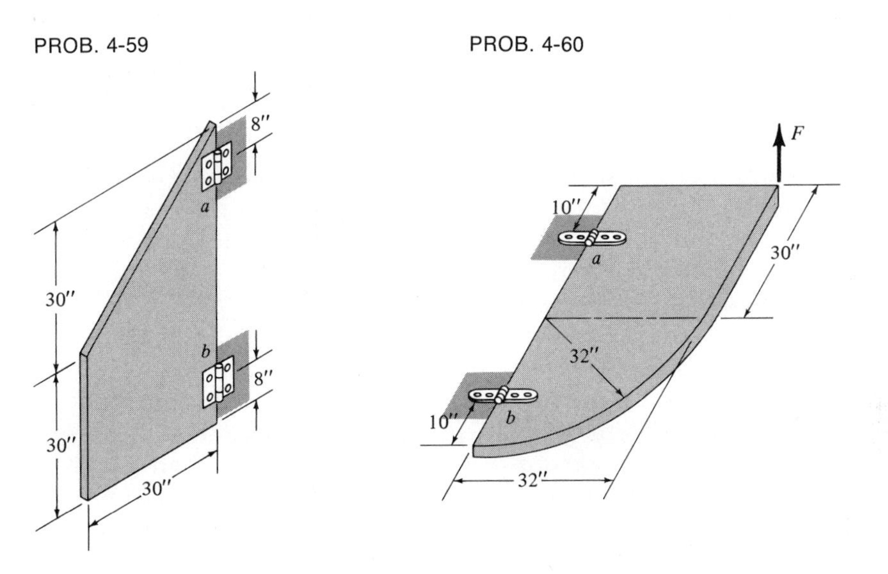

4-60 The 35-lb horizontal plate is supported by a vertical force F and two hinges as shown. Determine F and the reactions at the hinges.

4-6 LOAD ANALYSIS OF STRUCTURES AND MACHINES

Recall that the primary objective of this chapter was to develop methods and techniques to perform a load analysis of force-carrying machines and structures. In the preceding sections, we have developed the basic tools and requirements for doing this job. The basic tool is the free-body diagram, and the basic requirements are those for structural equilibrium. We are now ready to achieve our objective.

Although there is no set procedure for performing a load analysis of a machine or structure, in many cases the procedure will proceed along the following lines:

1. *First!* Carefully draw a free-body diagram of the entire structure or machine. Do not omit any possible forces unless there is some definite reason for doing so.
2. Using the requirements for equilibrated force systems, determine, if possible, all the unknown forces acting on the entire device.

3 Now, draw free-body diagrams for each individual member making up the machine or structure. Take care to be sure that the connectors are replaced by the proper type and numbers of forces and couples. But, most importantly, be sure you have correctly applied Newton's third law (equal and opposite reactions)! Label the forces clearly and unambiguously.

4 Apply the requirements for equilibrium to each individual member, and determine all the unknown forces. Quite often this will involve the solution of simultaneous algebraic equations, but, in many situations, the computational work can be greatly simplified by exercising some good judgment in writing the equations of equilibrium. For example, in summing forces you need not always consider vertical and horizontal components but rather choose components more convenient for the problem. Also, taking moments with respect to carefully selected points can often simplify the algebra involved.

5 Finally, you should perform some type of check to verify your solution, and we do *not* mean by simply looking at the answer in the back of the book. Out in the field you will not have the "back of the book," and you will have to be ready to defend your answer by your ability to verify your results. Usually a fairly reliable check can be made by reworking part of the problem by using a different set of equilibrium equations and comparing answers. This is not a certain check since the entire formulation of the problem might be incorrect, but it quite often will suffice.

There will, of course, be exceptions to this procedure, but, for the most part, this procedure and good common sense will get the job done. We close this section with several illustrative problems which demonstrate the above procedure.

EXAMPLE 4-4

For the structure shown in Fig. 4-14(a), determine the forces acting on each individual member of the structure.

Solution: Recall that this is the structure discussed earlier in Figs. 4-4 and 4-8, and its FBD is shown in Fig. 4-14(b). The forces are obviously coplanar and constitute a statically determinate system. Taking moments with respect to point a yields

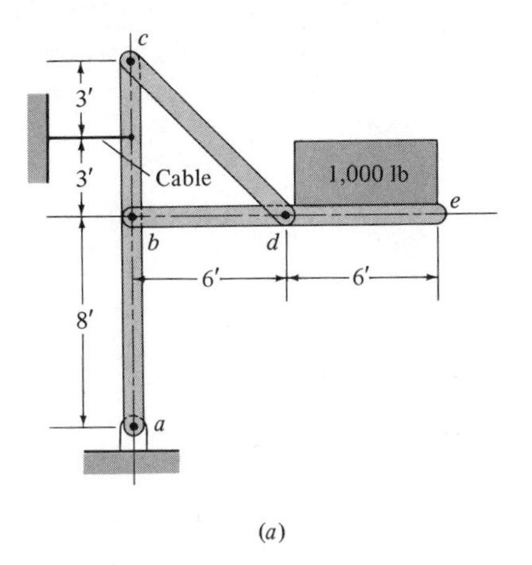

(a)

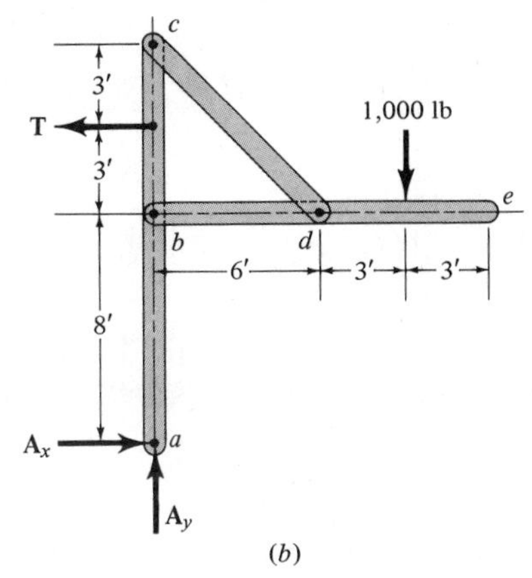

(b)

Fig. 4-14

$$\Sigma \mathbf{M}_a = \mathbf{0}$$
$$= (1{,}000)(9) + T(11)$$
$$= -9{,}000\mathbf{k} + T(11)\mathbf{k} = \mathbf{0}$$
$$\therefore\ 11T - 9{,}000 = 0$$
$$T = 818 \text{ lb} \qquad \text{(as shown)}$$

Summing forces yields

$$\mathbf{i}(A_x - T) = \mathbf{0}$$
$$\mathbf{j}(A_y - 1{,}000) = \mathbf{0}$$
$$\therefore\ A_x = T = 818 \text{ lb} \qquad \text{(as shown)}$$
$$A_y = 1{,}000 \text{ lb} \qquad \text{(as shown)}$$

Our next step is to draw an FBD of each individual member. This was done previously in Fig. 4-8 and is shown in Fig. 4-15(*a*). Before plunging headlong into a morass of simultaneous equations of equilibrium, let us pause to study this figure a little more carefully. While at first it might seem that there is an overabundance of unknowns in the problem, things are not quite as complicated as they appear. We have already determined $\mathbf{T}$, $\mathbf{A}_x$, and $\mathbf{A}_y$. Also, member *cd* is actually a two-force member since $\mathbf{C}_x$ and $\mathbf{C}_y$ are merely components of a single force at *c*, and similarly for $-\mathbf{D}_x$ and $-\mathbf{D}_y$ at *d*. Thus, by the remarks in Sec. 4-5 on two-force systems, we can redraw the free-body diagrams as shown in Fig. 4-15(*b*), in which the forces at *c* and *d* are drawn parallel to member *cd*. Also, notice in this figure that we have labeled the forces with their magnitudes only, letting the arrow indicate the direction of the vector. Thus, for example, the force $\mathbf{B}_x$ on member *bde* and the force $-\mathbf{B}_x$ on member *abc* are both labeled with a magnitude of B_x, and similarly

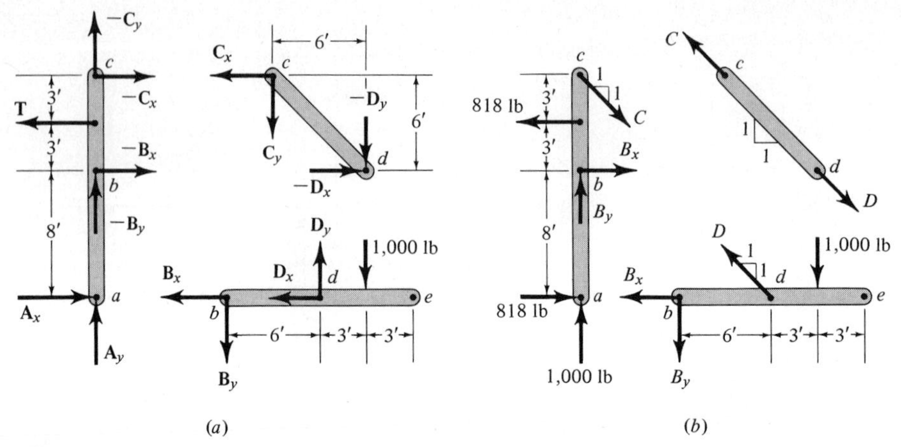

Fig. 4-15

for all the other equal but opposite forces. This scheme makes our computation somewhat more simple.

Taking moments with respect to point b on member bde yields

$$\Sigma \mathbf{M}_b = \mathbf{0}$$

$$= \frac{1}{\sqrt{2}} \overset{\curvearrowleft}{D}(6) + 1{,}000\overset{\curvearrowright}{(9)}$$

$$= \mathbf{k}\left(\frac{6}{\sqrt{2}} D - 9{,}000\right) = \mathbf{0}$$

$$\therefore D = 1{,}500\sqrt{2} \text{ lb} \qquad \text{(as shown)}$$

Hence, both vectors labeled D are shown correctly on Fig. 4-15(*b*). Now, by inspection of member *cd*, we have

$$C = D = 1{,}500\sqrt{2} \text{ lb} \qquad \text{(as shown)}$$

since the forces must be equal but opposite for equilibrium. Finally, by summing forces on member *bde* we obtain

$$\mathbf{i}\left(-B_x - \frac{D}{\sqrt{2}}\right) = \mathbf{0}$$

$$\mathbf{j}\left(-B_y + \frac{D}{\sqrt{2}} - 1{,}000\right) = \mathbf{0}$$

$$\therefore B_x = -\frac{D}{\sqrt{2}} = -1{,}500 \text{ lb}$$

$$B_y = \frac{D}{\sqrt{2}} - 1{,}000 = 500 \text{ lb} \qquad \text{(as shown)}$$

The minus sign on B_x means that our original guess as to the direction of $\mathbf{B}_x$ (and also $-\mathbf{B}_x$) was incorrect and that the arrows of both should be reversed, so that

$$\mathbf{B}_x = 1{,}500 \text{ lb} \rightarrow \text{ on member } bde$$
$$-\mathbf{B}_x = 1{,}500 \text{ lb} \leftarrow \text{ on member } abc$$

We *will not* actually change the arrows on the figure since the figure would then not agree with our previous work. As long as we correctly interpret our answer, it is not necessary to actually change the arrow.

We have now determined all the unknown forces transmitted at b, c, and d. As a check on our solution, let us sum forces on member abc, since we have not used this FBD in our work thus far:

$$\mathbf{i}\left(B_x + \frac{C}{\sqrt{2}} - 818 + 818\right) = \mathbf{i}\left(-1{,}500 + \frac{1{,}500\sqrt{2}}{\sqrt{2}} - 818 + 818\right)$$
$$= \mathbf{0} \qquad \text{(as required)}$$
$$\mathbf{j}\left(B_y - \frac{C}{\sqrt{2}} + 1{,}000\right) = \mathbf{j}\left(500 - \frac{1{,}500\sqrt{2}}{\sqrt{2}} + 1{,}000\right)$$
$$= \mathbf{0} \qquad \text{(as required)}$$

In substituting into the first equation above, we have utilized the fact that, for our FBD, B_x came out to be a negative 1,500. From this check it appears that our results are correct, and we stop here.

EXAMPLE 4-5

For the mechanism shown in Fig. 4-16(a), determine all the forces on member bc for a given weight W.

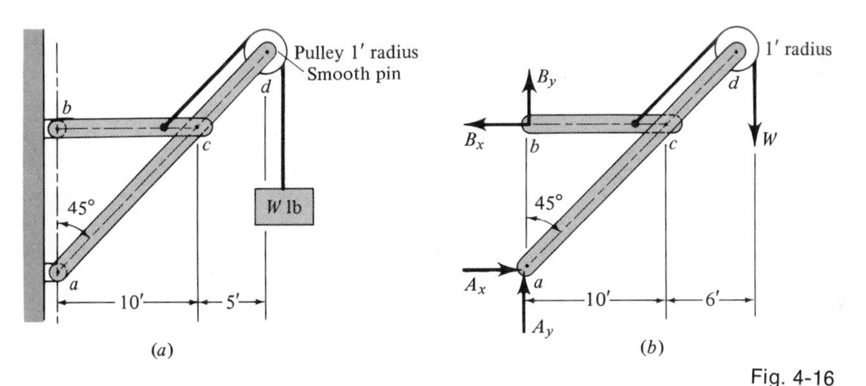

Fig. 4-16

Solution: As before, we begin by drawing an FBD of the entire mechanism, as shown in Fig. 4-16(b). At first glance, it might appear that our problem is statically indeterminate, since this coplanar force system contains four unknown forces. However, such a judgment would be premature until we have examined the situation in more detail. Actually, the mechanism is *not* overrestrained, and both pin joints a and b are required to maintain equilibrium.

Taking moments with respect to point a yields

$$\begin{aligned}\Sigma \mathbf{M}_a &= \mathbf{0}\\ &= \mathbf{k}[B_x(10) - W(16)] = \mathbf{0}\\ &\therefore 10B_x - 16W = 0\\ B_x &= \tfrac{8}{5}W \qquad \text{(as shown)}\end{aligned}$$

and summing horizontal forces yields

$$\mathbf{i}(A_x - B_x) = \mathbf{0}$$

$$A_x = B_x = \tfrac{8}{5}W \qquad \text{(as shown)}$$

Summing vertical forces yields

$$\mathbf{j}(A_y + B_y - W) = \mathbf{0}$$

from which we get the result

$$A_y + B_y = W$$

This is as much information as we can obtain from the FBD in Fig. 4-16(b).

Our next step is to draw an FBD of each individual member of the mechanism, as in Fig. 4-17(a). We first turn our attention to the pulley and consider moments about its center.

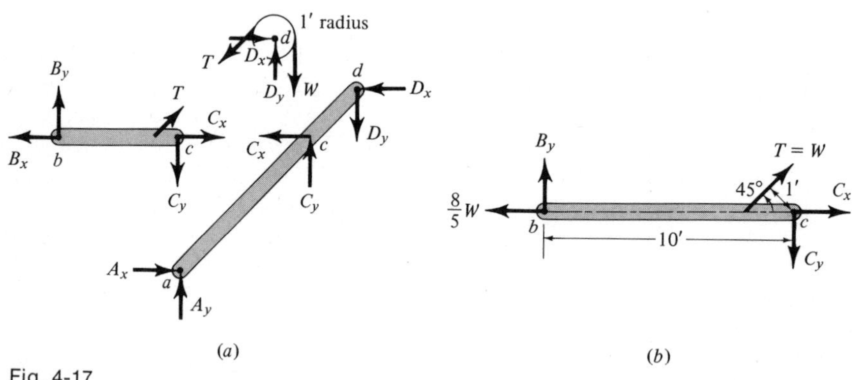

Fig. 4-17

$$\begin{aligned}\Sigma \mathbf{M}_d &= \mathbf{0}\\ &= \mathbf{k}[T(1) - W(1)] = \mathbf{0}\\ \therefore T &= W \qquad \text{(as you probably anticipated)}\end{aligned}$$

We now redraw the FBD of member bc in Fig. 4-17(b), indicating the forces we have already determined. Taking moments with respect to point c gives

$$\begin{aligned}\Sigma \mathbf{M}_c &= \mathbf{0}\\ &= \mathbf{k}[-B_y(10) - T(1)] = \mathbf{0}\\ &\therefore B_y(10) + T = 0\\ B_y &= -\frac{W}{10}\end{aligned}$$

Thus the vector $\mathbf{B}_y$ should be reversed, so that

$$\mathbf{B}_y = \frac{W}{10}\downarrow$$

Continuing with our FBD of member *bc*, summing forces yields

$$\mathbf{i}\left(-\tfrac{8}{5}W + \frac{1}{\sqrt{2}}\,T + C_x\right) = \mathbf{0}$$

$$\therefore C_x = \tfrac{8}{5}W - \frac{1}{\sqrt{2}}\,W = 0.89W \qquad \text{(as shown)}$$

$$\mathbf{j}\left(B_y + \frac{1}{\sqrt{2}}\,T - C_y\right) = \mathbf{0}$$

$$\begin{aligned}\therefore C_y = B_y + \frac{1}{\sqrt{2}}\,T &= \left(-\frac{W}{10} + \frac{1}{\sqrt{2}}\,W\right)\\ &= 0.61W \qquad \text{(as shown)}\end{aligned}$$

Thus, we have now determined all the forces acting on member *bc*. As a check, let us take another moment equation, say, with respect to point *b*.

$$\begin{aligned}\Sigma \mathbf{M}_b &= \mathbf{k}\left[\frac{1}{\sqrt{2}}\,T(10 - \sqrt{2}) - C_y(10)\right]\\ &= \mathbf{k}\left[\frac{W}{\sqrt{2}}\,(10 - \sqrt{2}) - 6.1W\right]\\ &= \mathbf{k}W(7.1 - 1 - 6.1) = \mathbf{k}W(0) = \mathbf{0}\end{aligned}$$

as required, and things appear to be okay.

EXERCISE PROBLEMS

4-61 Determine the tensions in cables A, B, C, and D.

PROB. 4-61

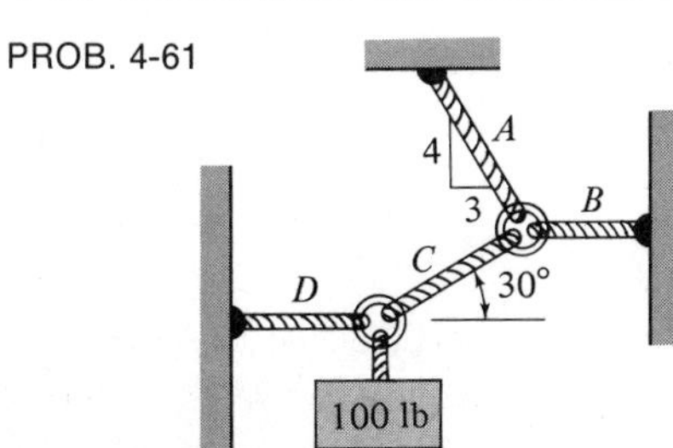

4-62 Find the tension T if the pulley system is equilibrated. Find the tension in cable C.

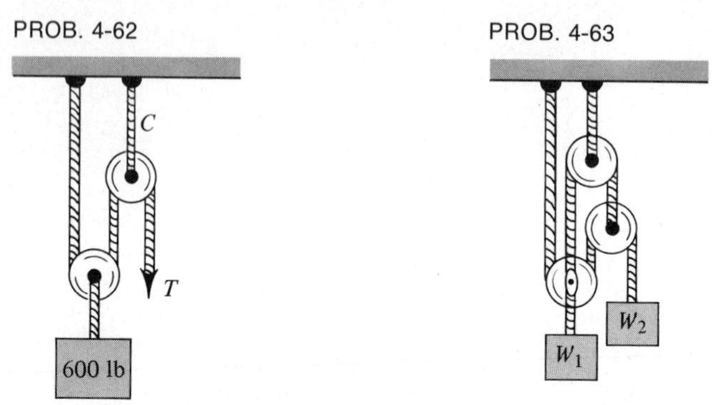

PROB. 4-62 PROB. 4-63

4-63 If $W_1 = 120$ lb, find W_2 which will equilibrate the pulley system.

4-64 The two cylinders have the same diameter and each weighs 100 lb. Consider the surfaces at a, b, c, and d to be smooth. Find the reactions at a and b.

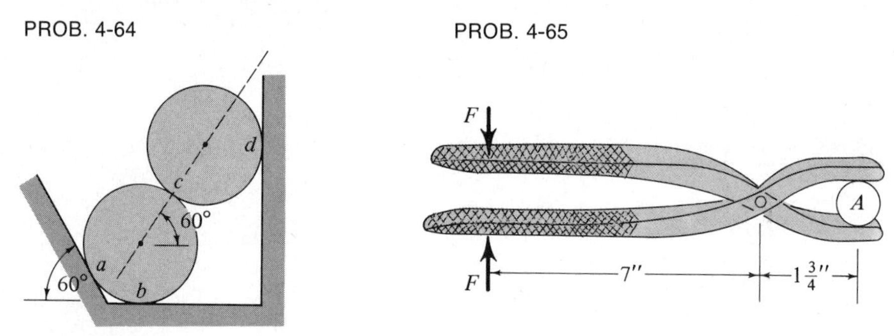

PROB. 4-64 PROB. 4-65

4-65 The force F exerted on each of the tongs is 16.5 lb. Determine the gripping force on the body A.

4-66 Body A weighs 1,000 lb, and the bar bcd weighs 200 lb. Determine the reactions at b and c.

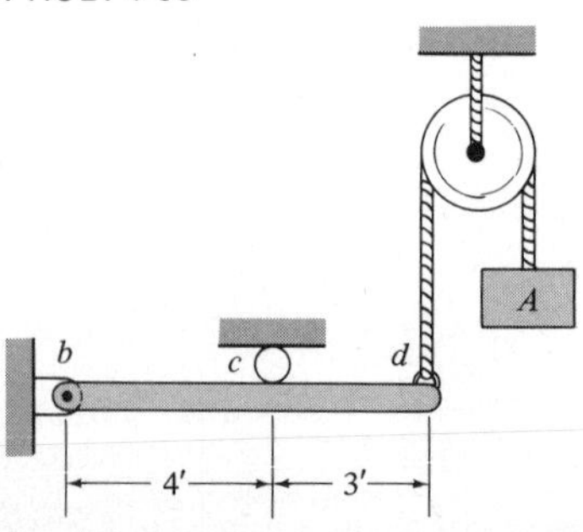

PROB. 4-66

4-67 Determine the pin reactions at b and c on member bc.

PROB. 4-67

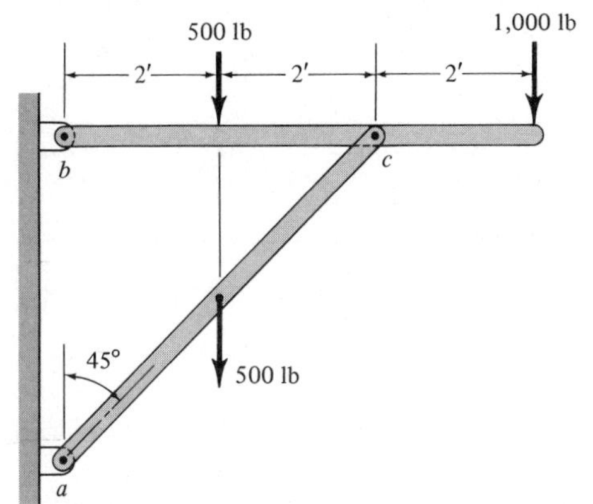

4-68 Determine all the forces exerted on the curved member $abcd$.

PROB. 4-68

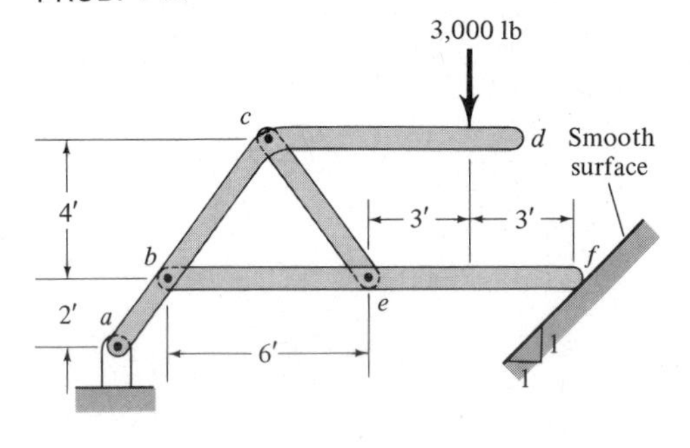

PROB. 4-69

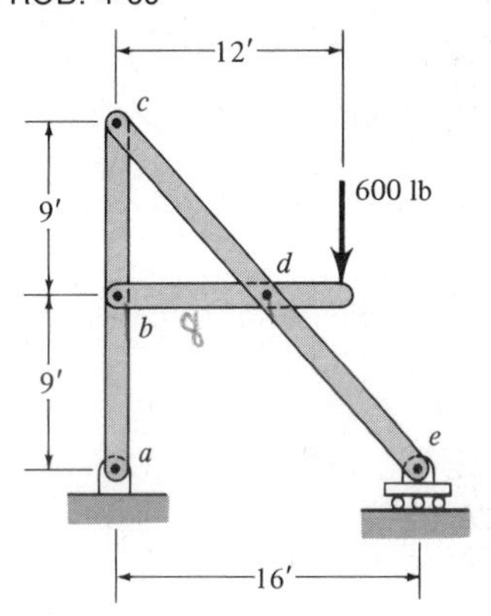

4-69 Determine the forces transmitted at all of the pin joints in the structure.

4-70 In the machine shown, the drum diameter is $1\frac{1}{2}$ ft and the crank arm radius is 3 ft. If P is 150 lb and the weight of the rock is negligible, determine the forces on the long member A.

PROB. 4-70

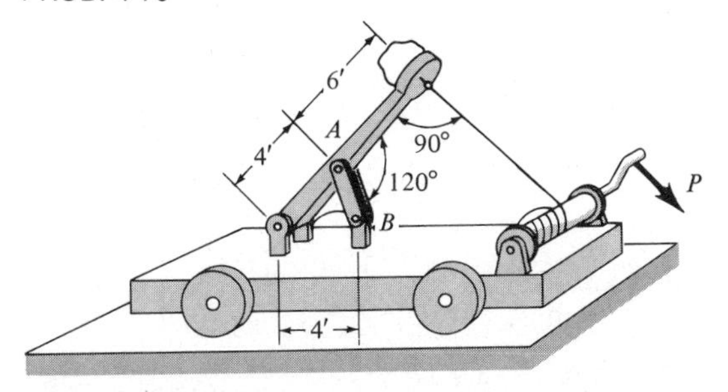

4-71 For the pin-connected device shown, find the reactions at b and d on member bde.

PROB. 4-71

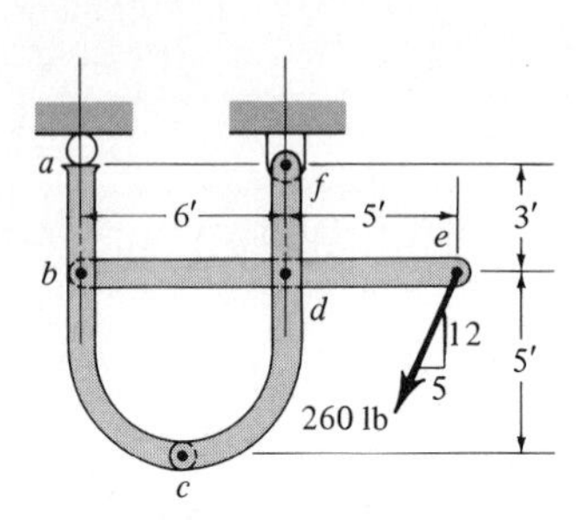

PROB. 4-72

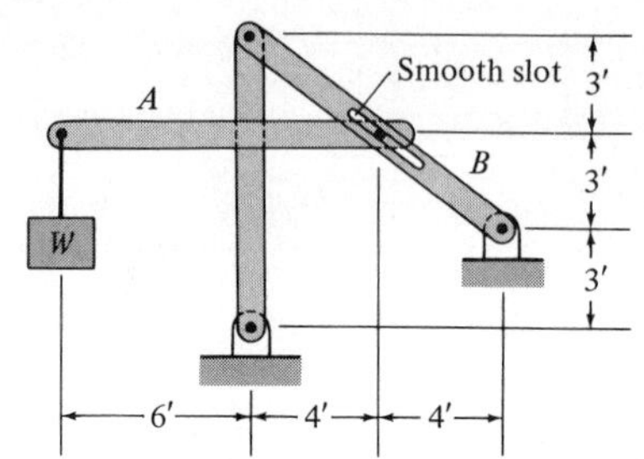

4-72 In the structure shown, the pin is free to slide in the smooth slot. If $W = 1{,}600$ lb, find the forces exerted on member A and member B.

4-73 Find all the forces acting on the 40-lb block A and the uniform 20-lb bar B.

PROB. 4-73

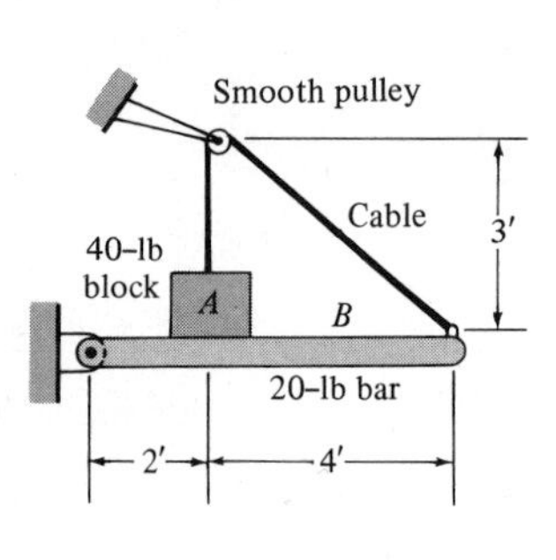

PROB. 4-74

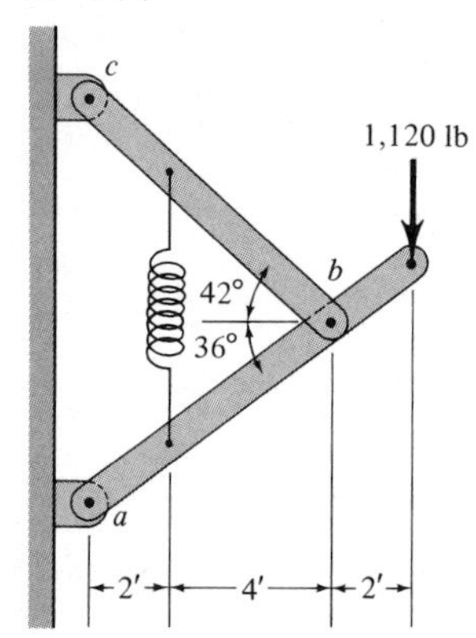

4-74 The tension in the spring is 540 lb. Find the reactions at b on member cb.

4-75 Determine the reactions at d on member adg.

PROB. 4-75

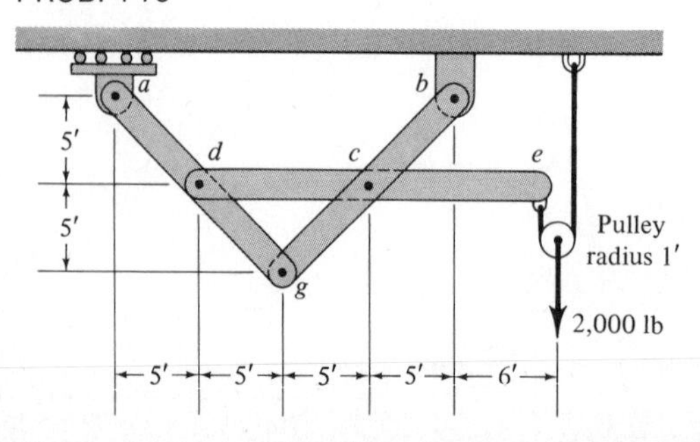

4-76 Find the reactions at c and d on member cde.

PROB. 4-76

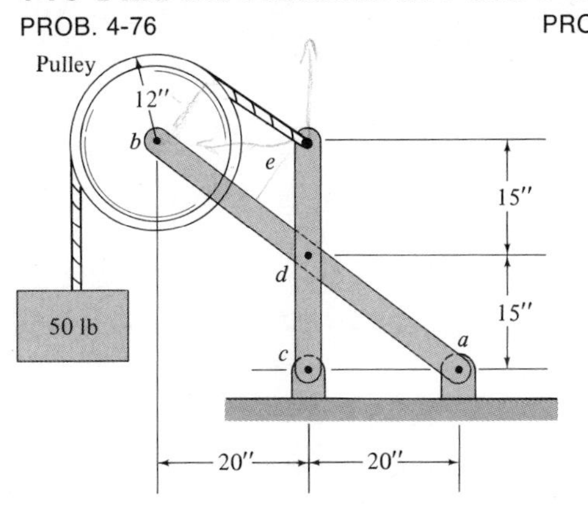

PROB. 4-77

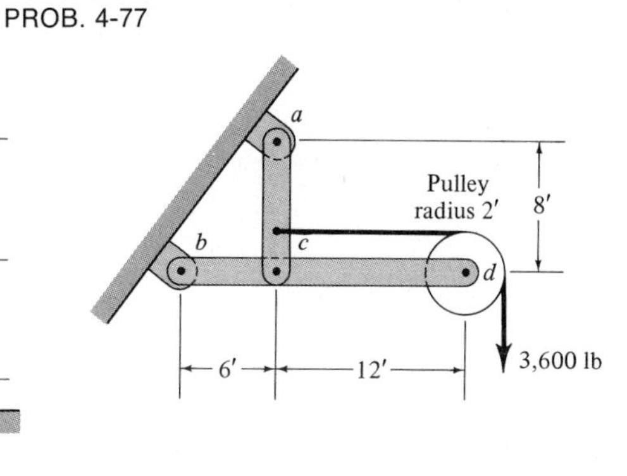

4-77 Find the reactions at a, b, and c for the device shown.

4-78 For the structure shown, find all the forces exerted on member A and on member B.

PROB. 4-78

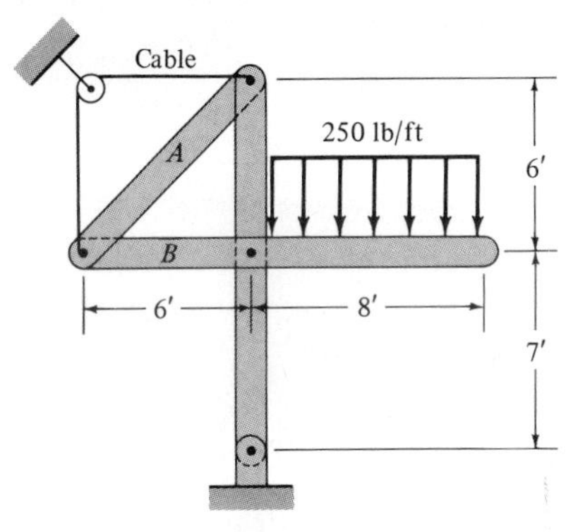

PROB. 4-79

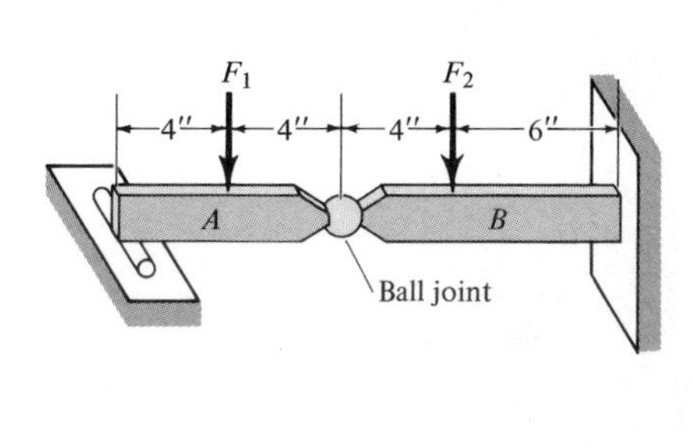

4-79 For the device shown, find the reactions at the right fixed support and at the left roller. $F_1 = 80$ lb and $F_2 = 100$ lb.

PROB. 4-80

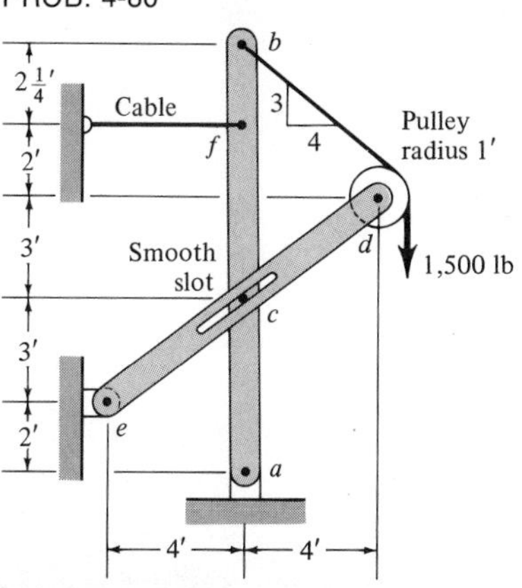

PROB. 4-81

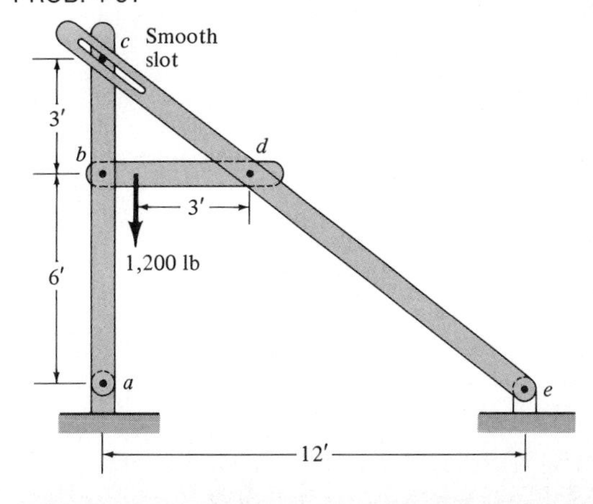

4-80 and 4-81 In the structure shown, the pin at c is free to slide in the smooth slot. Determine the reactions at a and e.

4-82 If the piston rod–shaft system shown is in equilibrium, find the forces exerted on the rod A, and find the couple **C**. $F = 600$ lb.

PROB. 4-82 Piston–connecting rod–crankshaft.

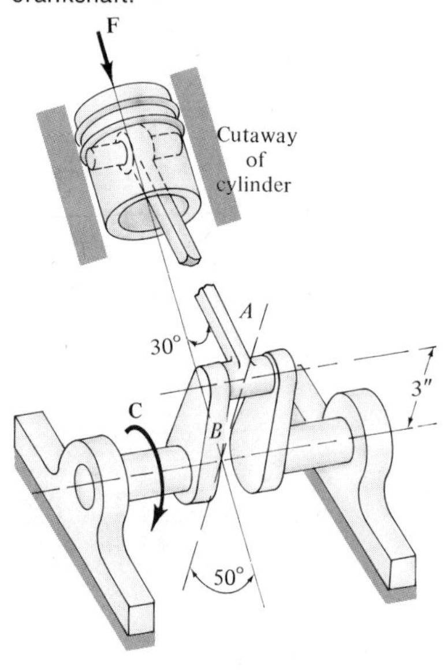

Piston–connecting rod–crankshaft

PROB. 4-83

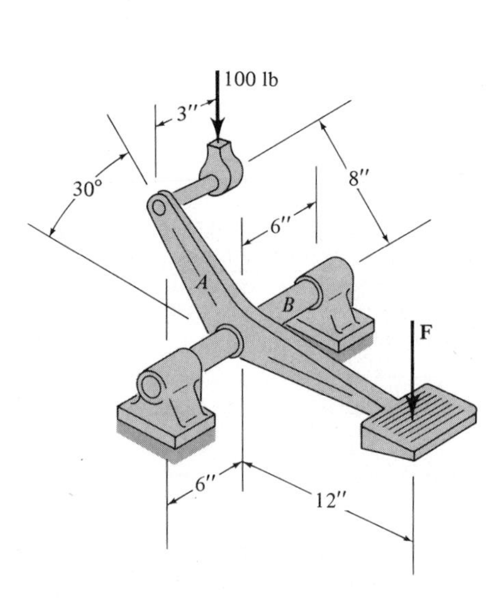

4-83 Find the force **F** on the pedal. Also, find the reaction between the pedal A and shaft B.

4-84 The surface on which the tripod rests can be considered smooth. If $F = 30$ lb, find the forces on one of the legs A and the forces on one of the horizontal cross braces B attached at the midpoint of the legs.

PROB. 4-84 Tripod support.

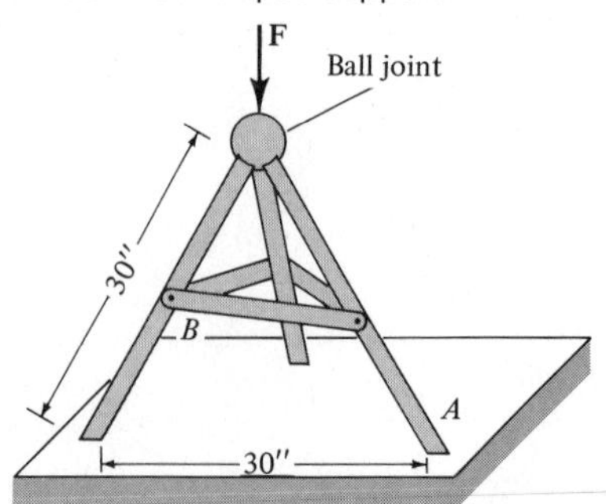

4-7 TRUSSES

A truss is a particular type of load-carrying structure which is constructed in such a way that certain simplifying assumptions can be made in regard to its load-carrying characteristics. Truss-like structures are often found in bridges, scaffolding, frames, and various other devices in which a lightweight structure but large load-carrying capacity is desired.

The basic truss element is composed of three straight lightweight members as shown in Fig. 4-18(*a*). The members may or may not be the same length. A truss structure is made up of a series of such basic truss elements as, for example, the bridge truss shown in Fig. 4-18(*b*). Although the members are usually fastened together with rivets or welds, the slenderness of the members is such that they cannot transmit structurally significant couples. Consequently, the joints can be considered to be pin joints (or ball joints in three-dimensional trusses).

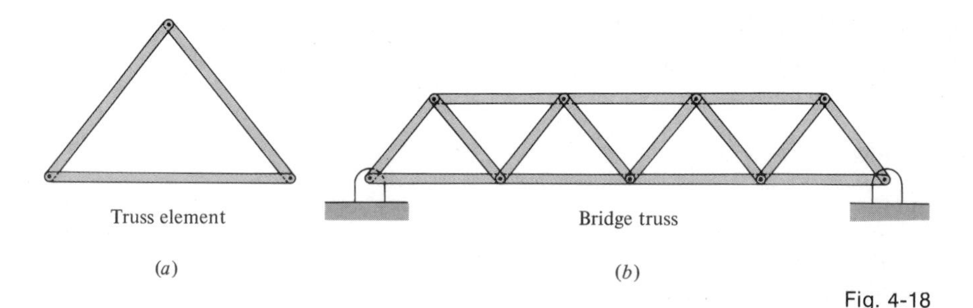

Fig. 4-18

In addition to the structural characteristics of a truss described above, further assumptions usually are made in regard to the manner in which the structure is loaded.

1. The weight of each member is negligible in comparison with the other loads on the structures.
2. All significant loads acting on the truss structure are considered to act effectively at the joints and *not* on the members directly.

These structural characteristics and loading assumptions lead to the basic feature that distinguishes a truss structure from the other types of load-carrying structures we have considered thus far, namely:

In a truss, all members can be considered to be pin-connected two-force members.

This means that each member in a truss structure transmits a force *parallel* to the member. The force can be a pulling force (tension) or a pushing force (compression) depending upon the load-carrying function of the member. Figure 4-19(*a*) illustrates the assumptions for a truss structure, while Fig. 4-19(*b*) and (*c*) illustrate situations which fail to satisfy the usual assumptions for a truss.

The fundamental objective in the load analysis of a truss structure is to determine the tensile or compressive force in the individual members of the truss. There are several well-established

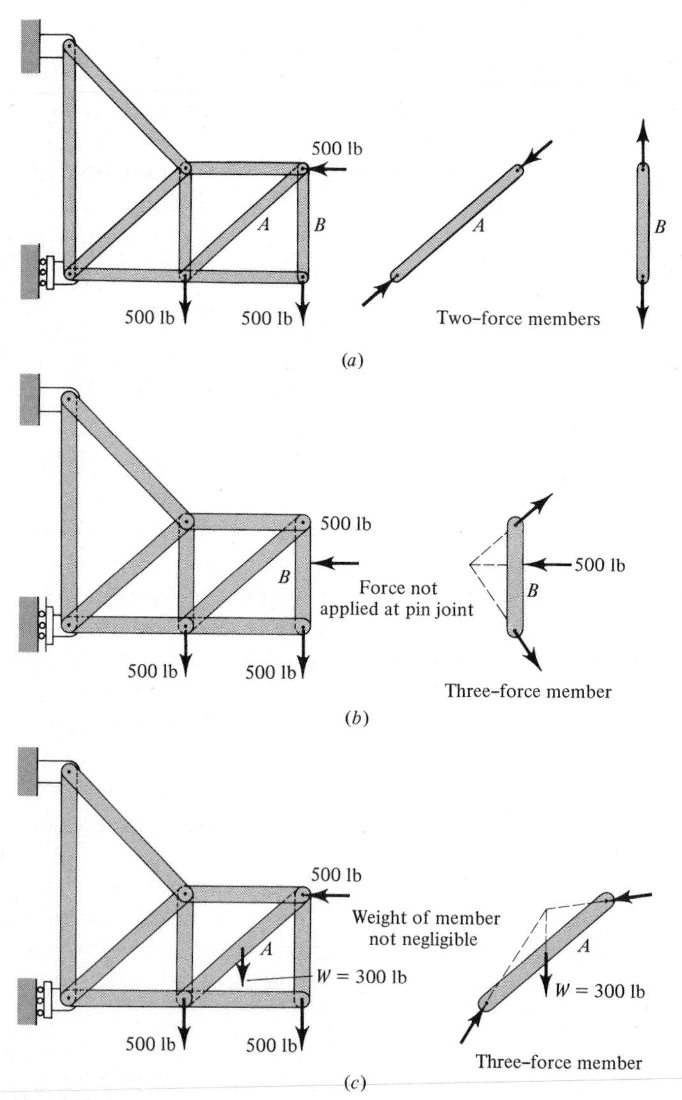

Fig. 4-19

techniques, all of which are simply specialized applications of the free-body diagram and the equilibrium requirements we have already used extensively.

Method of Joints

The main idea in this method of analysis is the free-body diagram of an individual pin joint of the truss. The forces on such a pin joint, such as joints *a* and *b* in Fig. 4-20(*a*), will necessarily be concurrent, with the pin being the point of concurrency. Moreover, since the force exerted by each member *must be parallel* to that member, the line of action of each force acting on the joint is known beforehand, although whether it is pulling on the joint (tension) or pushing on the joint (compression) is yet to be determined.

Some typical FBD's of joints for the truss in Fig. 4-20(*a*) are shown in Fig. 4-20(*b*). In these FBD's, the forces are labeled with subscripts which name the member which actually transmits the force. Thus, F_{ab} represents the magnitude of the force transmitted by member *ab*, etc. The order of the subscripts is immaterial: $F_{ab} = F_{ba}$, etc.

In all probability, some of the members are in tension while others are in compression, and perhaps with a little study we could probably make a pretty good guess as to which ones are in tension and which ones in compression. However, there is little need to worry about this since a minus sign in the solution will tell us if we guessed wrong. In the FBD of Fig. 4-20(*b*), we have guessed all the forces to be tensile (they are pulling on the joints), so that a minus sign in our final solution would indicate compression.

In drawing FBD's of joints you must be very careful! Do not leave out any forces acting on the joint. Do not forget Newton's third law! If a member is pulling (or being pulled) on one end, it must also pull (or be pulled) on its other end. Thus, if a force such as F_{ab} is pulling (or pushing) on joint *a*, it must also pull (or push) on joint *b*, and vice versa, as indicated by the force F_{ab} and F_{ba} in Fig. 4-20(*b*).

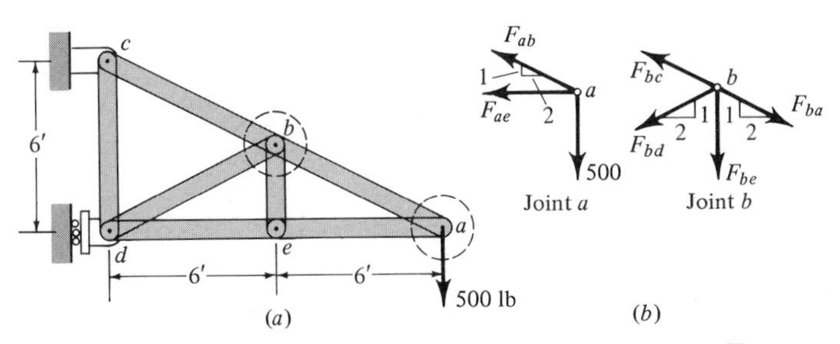

Fig. 4-20

After we have drawn an FBD of a joint, our problem is to determine the values of the forces. Since each FBD of a joint constitutes a concurrent force system, equilibrium requires that

$$\text{Sum of all forces on joint} = \mathbf{0} \tag{4-10}$$

In a two-dimensional problem, Eq. (4-10) represents two independent algebraic equations, and in a three-dimensional problem it represents three independent equations. From our examples in Fig. 4-20(*b*), it is obvious that quite often an FBD of a joint will contain more than two (or three) unknown forces. Consequently, in solving problems using the method of joints, it is often necessary to solve several joints simultaneously, and the solution can become quite tedious. Algebraic errors can easily occur and they tend to propagate themselves throughout the entire problem. Thus, care and accuracy are of paramount importance. Some of these ideas are illustrated in the following example.

EXAMPLE 4-6

For the truss structure shown in Fig. 4-21(*a*), determine the force in each member.

Solution: Generally, we begin problems of this type in the same manner as we have done in other types of structures, namely, by first determining the forces acting on the structure as a whole,

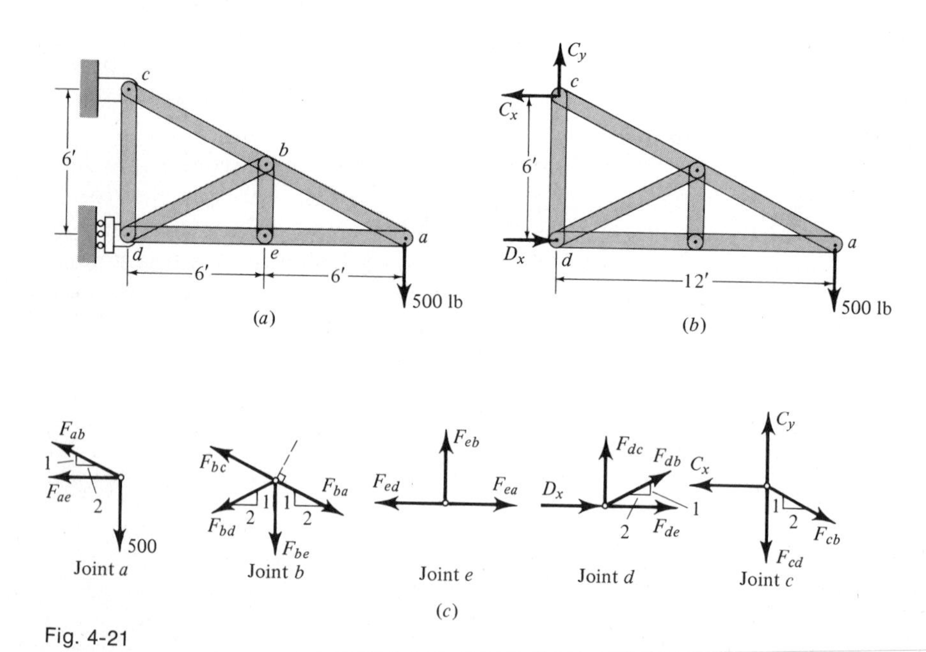

Fig. 4-21

although this is sometimes unnecessary. An FBD of the structure is shown in Fig. 4-21(*b*), and by summing forces and summing moments with respect to point d, we obtain

$$C_x = 1{,}000 \text{ lb}$$
$$C_y = 500 \text{ lb}$$
$$D_x = 1{,}000 \text{ lb}$$

We now begin the load analysis of the members by first considering joint a, whose FBD is shown in Fig. 4-21(*c*). Summing forces we obtain

$$\mathbf{i}\left(-F_{ae} - \frac{2}{\sqrt{5}} F_{ab}\right) = \mathbf{0}$$
$$\mathbf{j}\left(-500 + \frac{1}{\sqrt{5}} F_{ab}\right) = \mathbf{0}$$
$$F_{ab} = 500\sqrt{5} \text{ lb} \qquad \text{(tension)}$$
$$F_{ae} = -2(500) = -1{,}000 \text{ lb} \qquad \text{(compression)}$$

Proceeding to joint b, we see that even though we now know F_{ab}, the FBD of joint b still contains three unknowns. Therefore, we will first examine joint e. Summing forces on joint e yields

$$\mathbf{i}(F_{ea} - F_{ed}) = \mathbf{0}$$
$$\mathbf{j}(F_{eb}) = \mathbf{0}$$
$$\therefore F_{eb} = 0$$
$$F_{ed} = F_{ea} = -1{,}000 \text{ lb} \qquad \text{(compression)}$$

Returning to joint b, and summing forces,

$$\mathbf{i}\left(\frac{2}{\sqrt{5}} F_{ba} - \frac{2}{\sqrt{5}} F_{bc} - \frac{2}{\sqrt{5}} F_{bd}\right) = \mathbf{0}$$
$$\mathbf{j}\left(-F_{be} - \frac{1}{\sqrt{5}} F_{ba} + \frac{1}{\sqrt{5}} F_{bc} - \frac{1}{\sqrt{5}} F_{bd}\right) = \mathbf{0}$$
$$\therefore F_{bd} = 0$$
$$F_{bc} = F_{ab} = 500\sqrt{5} \text{ lb} \qquad \text{(tension)}$$

Remark: These last two results could have been obtained more or less by inspection, thus avoiding some messy algebra. To see this, recall that $F_{be} = 0$ and then sum force components perpendicular and parallel to the direction of F_{bc} and F_{ba}.

Finally, for joint c we obtain

$$\mathbf{i}\left(\frac{2}{\sqrt{5}} F_{cb} - C_x\right) = \mathbf{0}$$

$$\mathbf{j}\left(C_y - F_{cd} - \frac{1}{\sqrt{5}} F_{cb}\right) = \mathbf{0}$$

$$\therefore C_x = \frac{2}{\sqrt{5}} F_{cb} = 1{,}000 \text{ lb} \qquad \text{(as found earlier)}$$

$$F_{cd} = C_y - \frac{1}{\sqrt{5}} F_{cb}$$

$$= 500 - \frac{1}{\sqrt{5}} 500\sqrt{5} = 0$$

As a check, considering joint d and substituting our previously obtained values, we see that

$$\mathbf{i}\left(D_x + F_{de} + \frac{2}{\sqrt{5}} F_{db}\right)$$

$$= \mathbf{i}(1{,}000 - 1{,}000 + 0) = \mathbf{0} \qquad \text{(as required)}$$

and $$\mathbf{j}\left(F_{dc} + \frac{1}{\sqrt{5}} F_{db}\right) = \mathbf{j}(0 + 0) = \mathbf{0} \qquad \text{(as required)}$$

Things appear to be okay.

Remark: It is not uncommon for the force in a member to come out to be zero, as F_{be}, F_{bd}, and F_{cd} did in this example. This might lead you to believe that these members *eb*, *bd*, and *cd* serve no function and could be removed from the truss. This is usually *not* true if for no other reason than the fact that we have idealized the actual structure and loading in obtaining our solution. Also, suppose we actually removed member *eb* of Fig. 4-21(a). Recalling that members *ea* and *ed* are in compression, what would prevent joint e from rotating and causing collapse of the structure? Thus, *eb* actually serves a very important function, namely, stabilizing joint e. Similarly, members *db* and *dc* are needed to stabilize the structure.

Method of Sections

This technique, like the method of joints, is simply a specialized application of the FBD. In this method of analysis of trusses, instead of drawing an FBD of a single joint of the truss, we draw an FBD of a "section" of the truss made up of several truss elements. In drawing such an FBD, we utilize the basic principle of truss analysis, namely, that every member of the truss can transmit only a force parallel to the member.

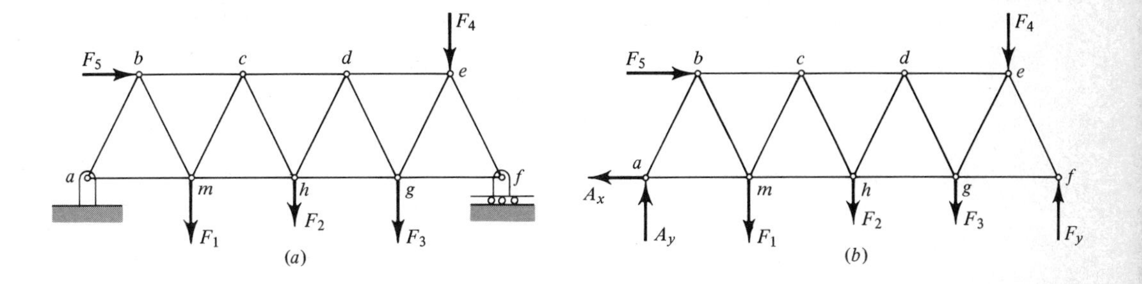

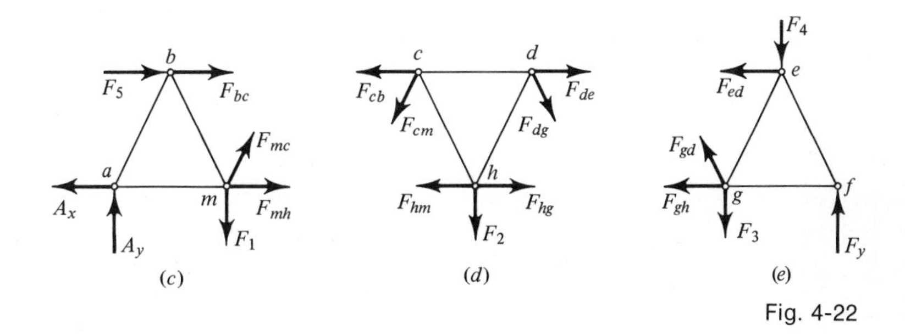

Fig. 4-22

Referring to the truss in Fig. 4-22(*a*), an FBD of the entire truss structure is shown in Fig. 4-22(*b*). However, we can, if we wish, draw an FBD of a section of the structure, as, for example, the section composed of the truss element *abm* shown in Fig. 4-22(*c*), the section *cdh* in Fig. 4-22(*d*), or the section *efg* in Fig. 4-22(*e*). The important thing to note in this series of free-body diagrams is the consistent application of Newton's third law of equal but opposite reactions. Just as in the method of joints, F_{bc} represents the force transmitted by member *bc*, etc. Although the FBD's in Fig. 4-22(*c*) to (*e*) are of individual basic truss elements, by employing exactly the same principles we can draw an FBD of as large or as small a section of the entire truss structure as we please.

How do we use an FBD of a section of a truss in performing a load analysis of the truss? Basically, we use the same idea that we used throughout this chapter, namely, if the entire truss structure is in equilibrium, any section of the structure must also be in equilibrium. This means that the equilibrium requirements

$$\mathbf{S}(\mathscr{S}) = \mathbf{0} \tag{4-11}$$

$$\mathbf{M}_q(\mathscr{S}) = \mathbf{0} \tag{4-12}$$

are applicable to each of the FBD's in Fig. 4-22(*b*) to (*e*). Notice that these FBD's do not, in general, comprise a concurrent force system, so that the moment requirement is nontrivial and must be utilized. Thus, in two-dimensional problems, Eqs. (4-11) and (4-12) represent three independent conditions, while in a three-dimensional problem they comprise six independent conditions.

Rather than develop a general theory for the method of sections, we will illustrate the essential ideas by means of an example problem. No new principles are involved, but the method does require good judgment and perception, which you should be developing by this stage of the course.

EXAMPLE 4-7

For the bridge truss shown in Fig. 4-23(*a*), find the forces transmitted by members *dm*, *nm*, and *em*. The letter *k* stands for kip.

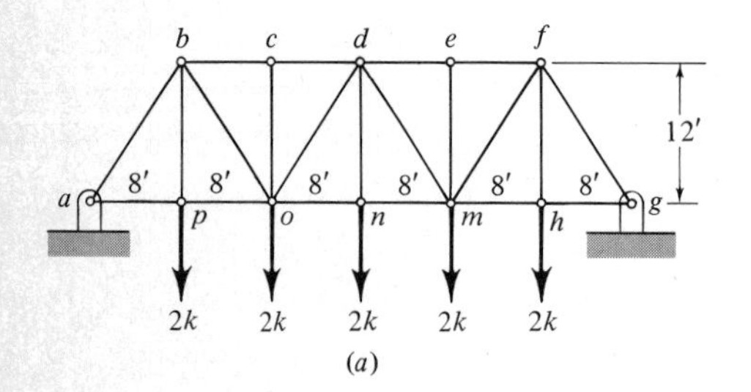

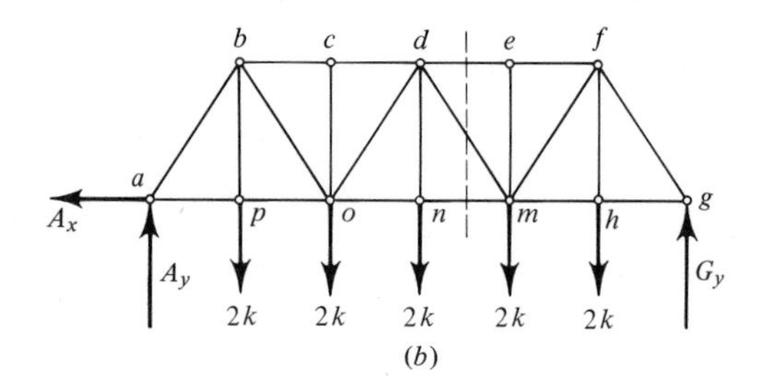

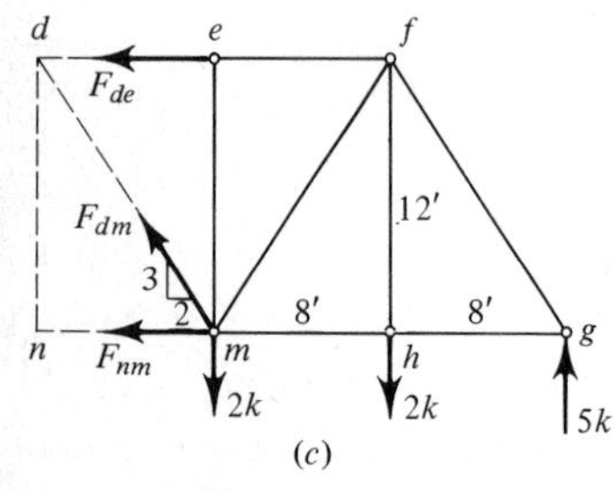

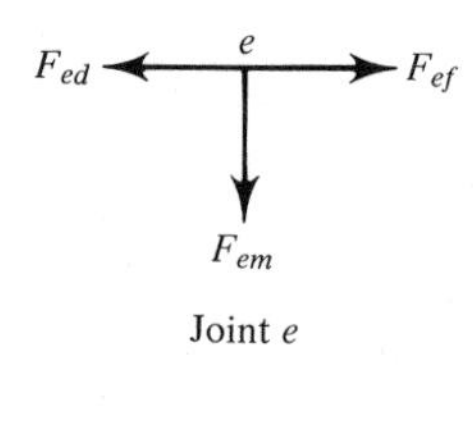

Fig. 4-23

Solution: As usual, we begin by drawing an FBD of the entire structure and finding the reactions at the supports. In this case, by symmetry we can see that

$$A_y = G_y = \tfrac{1}{2}(2 + 2 + 2 + 2 + 2) = 5 \text{ kips}$$

$$A_x = 0$$

Since we are asked to determine the force in members *dm*, *nm*, and *em*, we might think that an FBD of joint *m* would do the job. But, as we can see, such an FBD would show five (*hm*,*fm*,*em*,*dm*,*nm*) unknowns. On the other hand, if we were to "cut" the truss structure along the dashed line in Fig. 4-23(*b*) and draw an FBD of the section to the right of this cut, we would have the FBD of Fig. 4-23(c). This FBD contains two of the forces we wish to find, which is primarily the reason why we chose this particular section. Now, summing forces in the vertical direction yields

$$\mathbf{j}\left(\frac{3}{\sqrt{13}}F_{dm} - 2 - 2 + 5\right) = \mathbf{0}$$

$$F_{dm} = -\frac{\sqrt{13}}{3}\ \text{kips} \qquad \text{(compression)}$$

which is one of the desired forces. By summing moments with respect to point d, we can obtain F_{nm} directly. Thus

$$\begin{aligned}\mathbf{M}_d &= F_{nm}\overset{\curvearrowright}{(12)} + 2\overset{\curvearrowright}{(8)} + 2\overset{\curvearrowright}{(16)} + 5\overset{\curvearrowleft}{(24)}\\ &= \mathbf{k}(-12F_{nm} - 16 - 32 + 120) = \mathbf{0}\\ &\therefore F_{nm} = 6\ \text{kips} \qquad \text{(tension)}\end{aligned}$$

Although we were not asked to find F_{de}, by summing horizontal forces we obtain

$$\mathbf{i}\left(-F_{de} - \frac{2}{\sqrt{13}}F_{dm} - F_{nm}\right) = \mathbf{0}$$

$$\begin{aligned}F_{de} &= -\frac{2}{\sqrt{13}}F_{dm} - F_{nm}\\ &= +\tfrac{2}{3} - 6 = -5\tfrac{1}{3}\ \text{kip} \qquad \text{(compression)}\end{aligned}$$

As a check, taking moments with respect to point m yields

$$\begin{aligned}\mathbf{M}_m &= F_{de}\overset{\curvearrowleft}{(12)} + 2\overset{\curvearrowright}{(8)} + 5\overset{\curvearrowleft}{(16)}\\ &= \mathbf{k}[-5\tfrac{1}{3}(12) - 16 + 80]\\ &= \mathbf{k}(-64 - 16 + 80) = \mathbf{0} \qquad \text{(as required)}\end{aligned}$$

To finish the problem, consider an FBD of joint e, as shown in Fig. 4-23(d). By summing vertical forces, it is immediately apparent that

$$F_{em} = 0$$

Remark: As this illustration demonstrates, both the method of joints and the method of sections can be used advantageously in a given problem. Generally speaking, the method of sections is more sophisticated and, when used intelligently, can save you a lot of algebraic computations. However, this advantage can be negated if you are careless in writing correct moment equations. In this regard experience will be your best teacher.

4-8 SUMMARY

This concludes the chapter on structural equilibrium but, in fact, only begins our analysis of load-carrying machines and structures. At this stage you should, with appropriate simplifying assumptions, be able to perform a load analysis of various types of real and idealized structures. Beginning with the next chapter, we will turn to the equally important question: Can the structure under consideration actually support the applied loads without breaking or failing in some manner? This question involves various factors which heretofore have not been considered, such as the material of which the structure is made, the deformations which occur due to the loads, and other important considerations.

EXERCISE PROBLEMS

4-85 to 4-88 Determine the force in each of the members of the pin-connected truss shown.

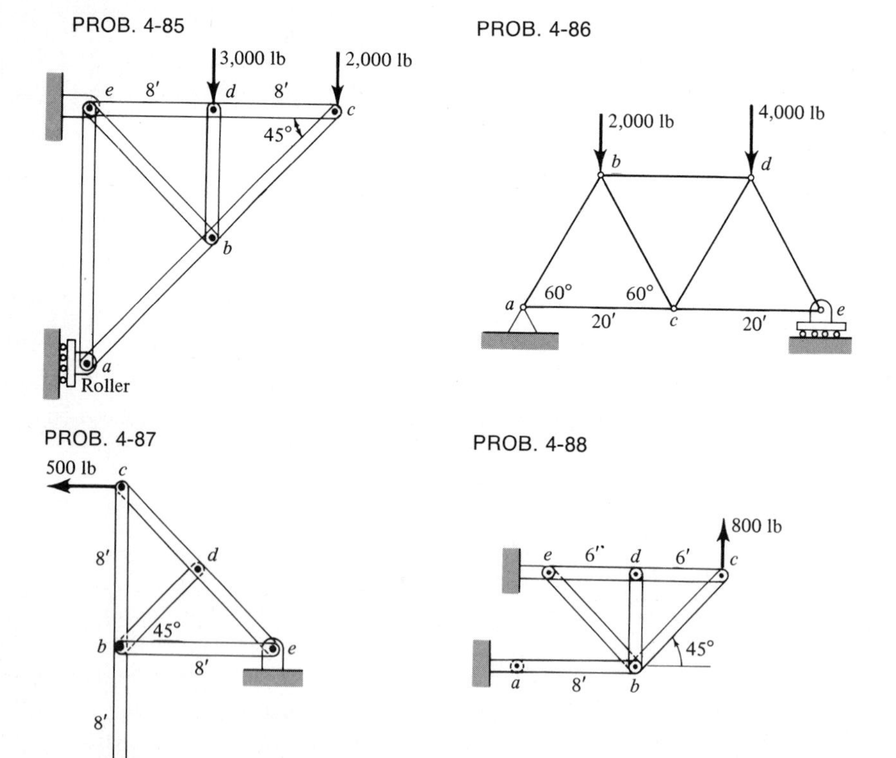

4-89 Determine the force in members *ec*, *ed*, and *ef*.

PROB. 4-89

PROB. 4-90

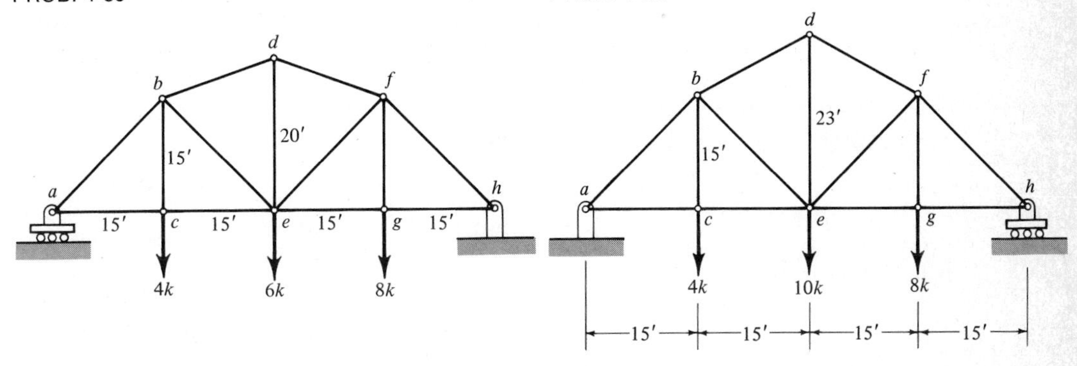

4-90 Determine the force in members *bc*, *bd*, *be*, and *ef*.

4-91 Determine the force in members *cf*, *ef*, and *bf*.

PROB. 4-91

PROB. 4-92

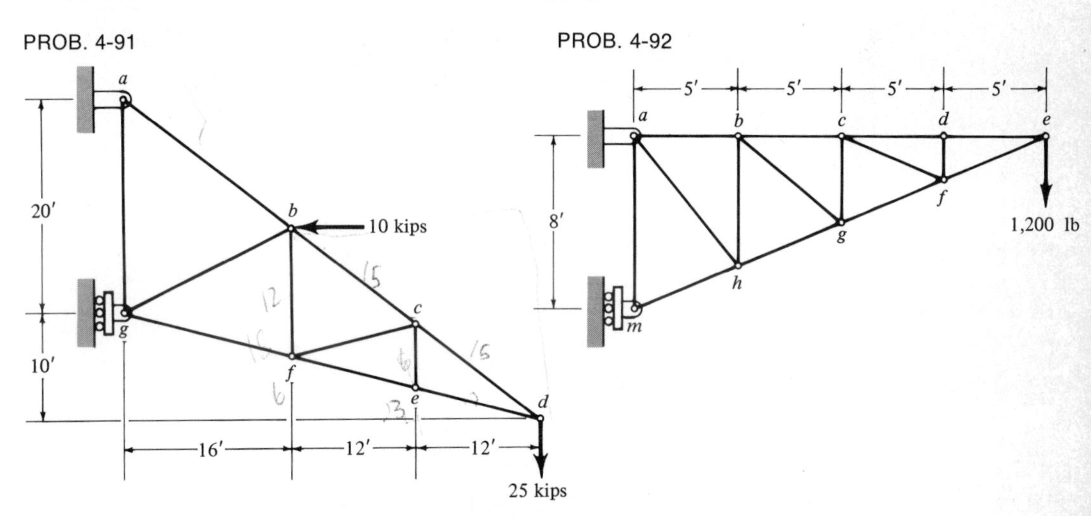

4-92 Determine the force in members *bc*, *cg*, and *bg*.

4-93 Calculate the force in members *cj*, *cd*, and *dj*.

PROB. 4-93

PROB. 4-94

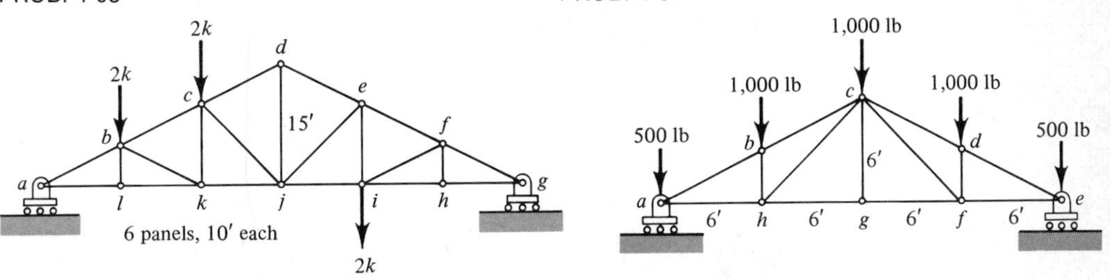

4-94 Calculate the force in members *bh*, *bc*, and *ch*.

4-95 Calculate the force in members bd, cf, ce, and eb.

PROB. 4-95

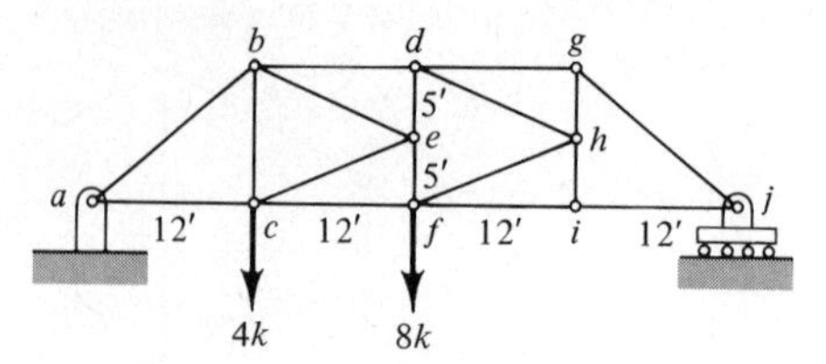

4-96 Find the force in members hm, hg, and cd.

PROB. 4-96

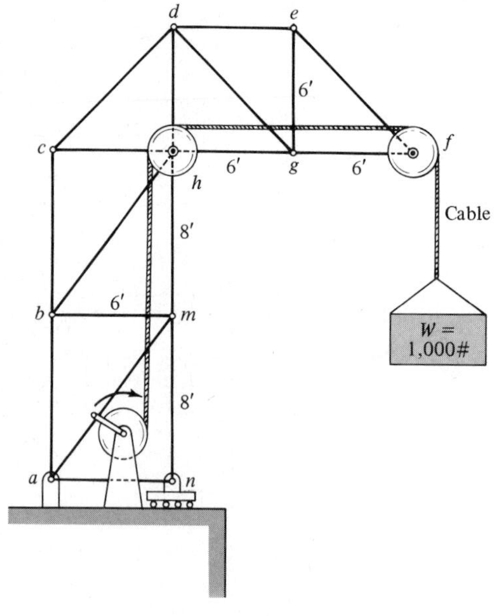

PROB. 4-97

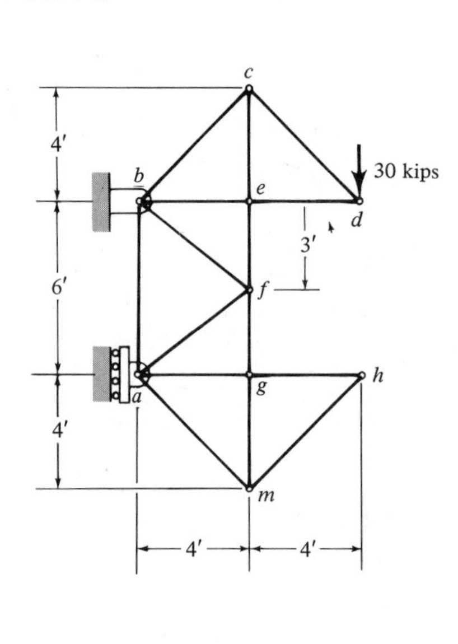

4-97 Determine the force in members gf, af, bf, and eb.

4-98 Find the force in members hg, df, and fc.

PROB. 4-98

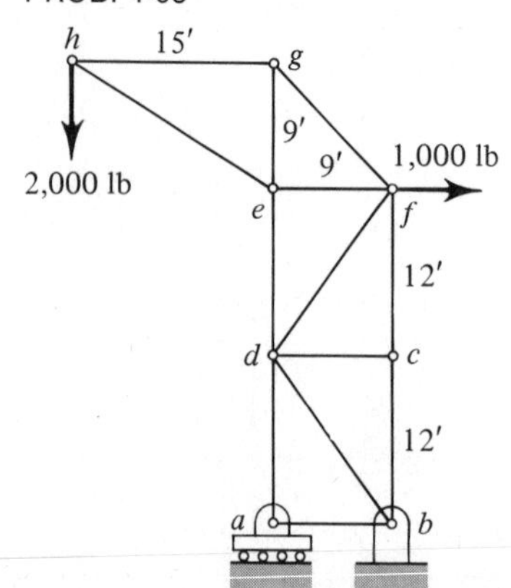

PROB. 4-99

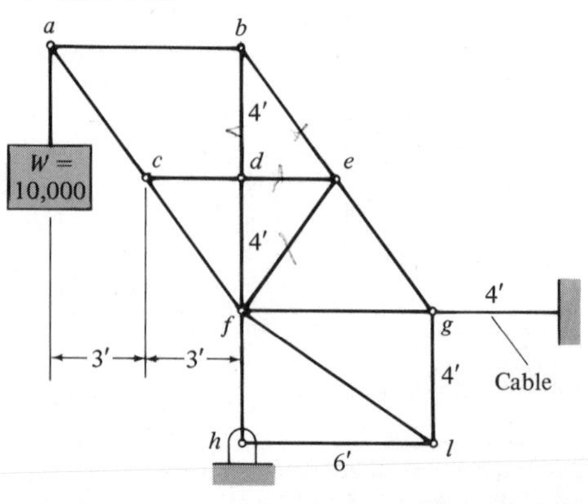

4-99 Calculate the force in members *eb*, *ed*, *bd*, and *ef*.

4-100 Find the force in members *df*, *dg*, and *fg*.

PROB. 4-100 PROB. 4-101

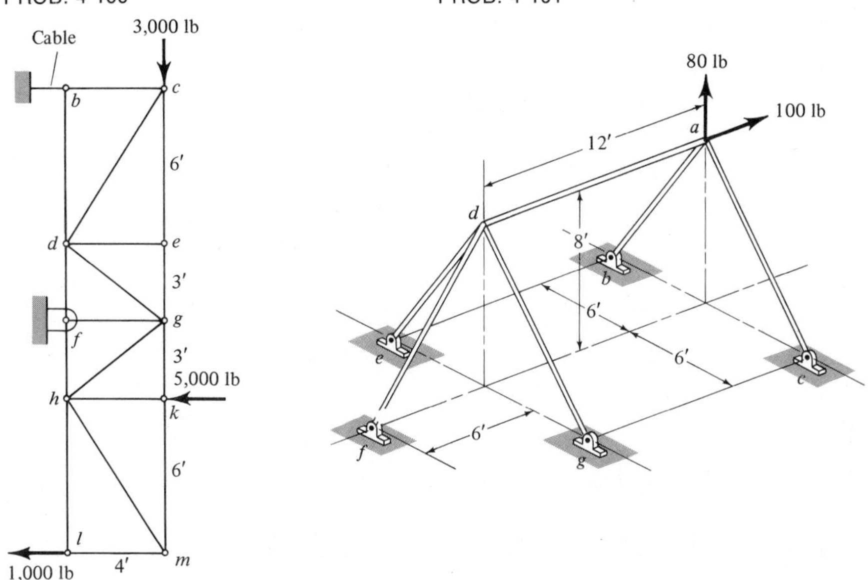

4-101 For the structure shown, determine the force in members *ab* and *df*.

4-102 The turnbuckle is tightened so that the center diagonal tensile rod has a force of 4,000 lb. Estimate the forces in members *ab* and *df*.

PROB. 4-102 PROB. 4-103

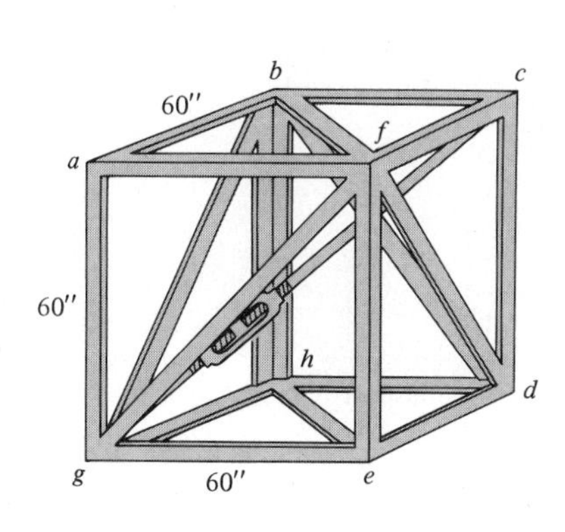

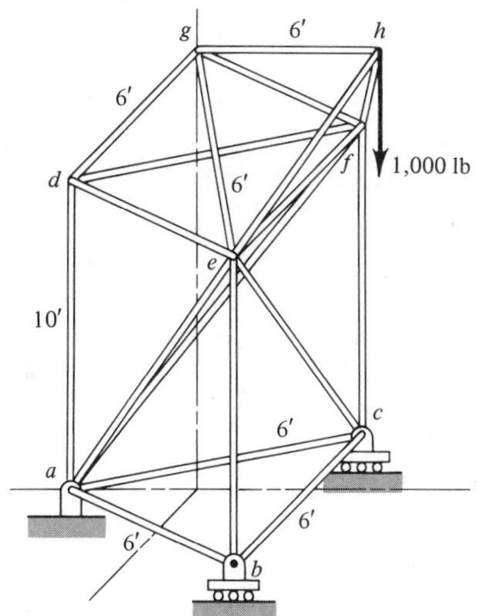

4-103 The portable crane is anchored at *a* and on rollers at *b* and *c*. Find the forces in members *eh* and *eb*.

4-104 The three turnbuckles on the guy wires are tightened to a tension of 800 lb each. All members are the same length. Find the force in each member.

PROB. 4-104

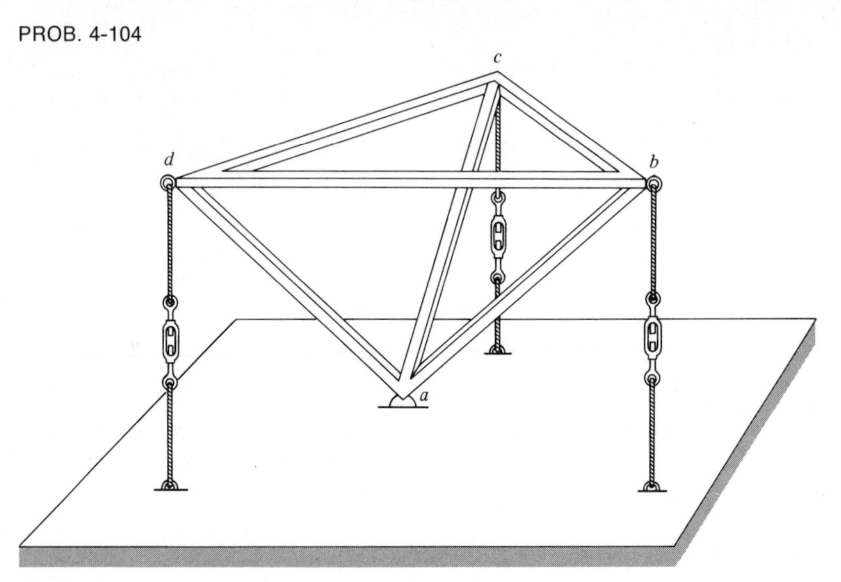

4-105 Find the forces in members *ed*, *ad*, and *bf*.

PROB. 4-105

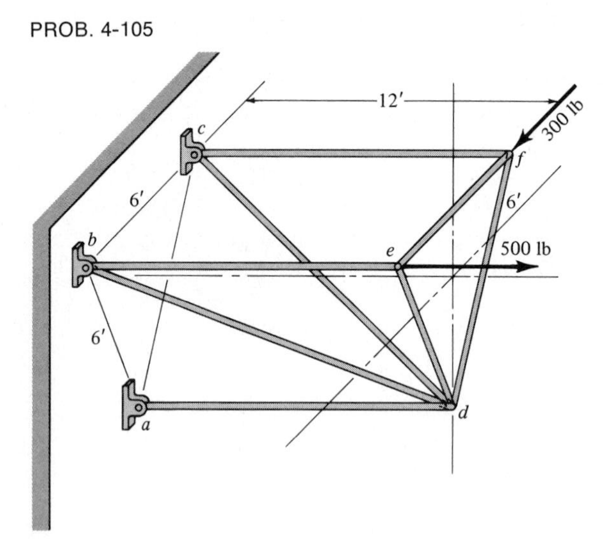

chapter 5 INTERNAL REACTIONS—THE STRESS CONCEPT

5-1 INTRODUCTION

In the previous chapter we learned how to perform a load analysis of various structures and machines in which the primary objective was to determine the loads transmitted by and to the individual members making up the structure or machine. In all of our analyses it was presumed that all of the members were capable of supporting the applied loads without breaking or failing in some other manner such as bending, stretching, or twisting excessively. However, undoubtedly the most important question in the design or analysis of any load-carrying device is whether or not the device is actually strong enough to properly perform its load-carrying function.

The primary purpose of this chapter is to introduce some concepts and ideas which will subsequently enable us to answer the question of whether or not a member is "strong enough." Some of these ideas will be apparent to you from your physical experience, while others will be less apparent and thereby more difficult to accept and understand. However, you should keep in mind that these concepts and ideas are not an end in themselves but rather are simple but ingenious schemes by which engineers are able to answer difficult practical questions regarding the load-carrying capabilities of structures and machines.

5-2 INTERNAL FORCES AND COUPLES

We have defined a force as being the action of one body on another which tends to produce some mechanical effect, such as motion or deformation. Therefore, by this definition, the forces exerted on a given body are produced by other bodies which are *external* to the given body, and the purpose of the free-body diagram is to indicate these *external* forces acting on the given body.

But what about the body itself? By the above definition a body cannot exert a force on itself. However, one *part* of a body can exert a force on another *part* of the body, as, for example, your hand can squeeze your leg, although both are *parts* of your one body. This situation can be made consistent with our definition of a force by considering the hand and leg as two distinct bodies rather than as parts of the same body. This idea of subdividing a body into pieces or parts has already been used extensively in the previous chapter whereby we isolated individual parts or sections of various frames, machines, and trusses.

A very similar idea can be used for an individual structural member. To illustrate this point, consider the equilibrated beam shown in Fig. 5-1(*a*). A free-body diagram with all the external forces on the entire beam is shown in Fig. 5-1(*b*). Suppose we now mentally visualize the beam as being made up of two parts, a "left half" and a "right half," as indicated by the dashed line in Fig. 5-1(*b*). In Fig. 5-1(*c*), we have drawn each of these "halves" with the external forces acting on the appropriate half. In the case of the $\mathbf{R}_3$ force, its point of application is directly on the line separating the "halves," but we have considered it to be acting on the left

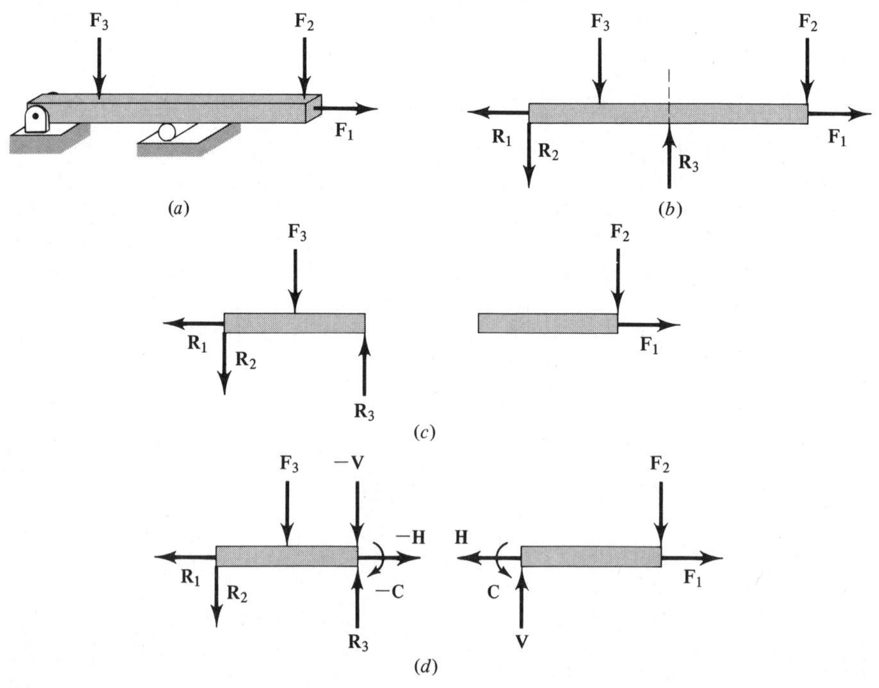

Fig. 5-1

"half" rather than the right half, or, rather than splitting it into two forces, one acting on each half of the beam.

Let us now look at these halves more closely. As it is drawn in Fig. 5-1(*c*), the right half cannot possibly be in equilibrium—forcewise or momentwise. Intuitively, we expect that the left half of the beam must exert some forces on the right half, otherwise the right half would "drop off." More precisely, as far as the right half of the beam is concerned, the left half of the beam is an *external* body which exerts on it some forces and/or couples necessary to keep the right half in place, that is, to keep the halves together as a whole beam. In Fig. 5-1(*d*), we have redrawn these halves, indicating the *net effective* forces and couple which each half exerts upon the other. Note carefully the application of Newton's third law (equal and opposite reactions). The quantity **V** represents a net vertical force, **H** a net horizontal force, and **C** a net couple necessary to maintain the equilibrium of the right half.

As we have said, in treating the halves as separate bodies, **V**, **H**, and **C** (−**V**, −**H**, and −**C**) can be thought of as external reactions, just as we have always done. On the other hand, in reality, since the beam is actually one solid piece, **V**, **H**, and **C** (−**V**, −**H**, and −**C**) are actually *internal* reactions keeping the halves together as a unit. Notice that if we put the two halves in Fig. 5-1(*d*) back together, these internal forces **V**, **H** and −**V**, −**H**, and the couples **C** and −**C**, will "cancel out," and we recover the FBD of Fig. 5-1(*b*).

At this time you might well ask: Why are we interested in internal forces or couples? The answer to this question is simply that the material making up the body must possess the structural ability to exert the required internal reactions in order to keep the body together and maintain its structural integrity. If the material is unable to do this, the structure will fail to fulfill its load-carrying function. From the engineer's viewpoint, he must be aware of what is being demanded of the material, and, therefore, he must know what internal reactions will be required of the material; that is, he must know how strong the material must be in terms of the internal forces it must withstand.

Experience would lead us to believe that the internal reactions will vary from location to location throughout the body and that a body will break at whatever location the internal reactions become too large. This is partially true, and the engineer must be able to determine the internal reactions at any desired location or to determine the location at which the reactions are most critical. Returning to the example of Fig. 5-1, an engineer should be able to determine the value and location of the largest horizontal reaction **H**, or the vertical reaction **V**, or the bending couple **C** throughout the entire beam. For the time being, however, we will concentrate our attention on determining the types and values of the internal reactions at a *given* specific location.

EXAMPLE 5-1

Find the internal reactions at locations a and b for the fixture shown in Fig. 5-2(a).

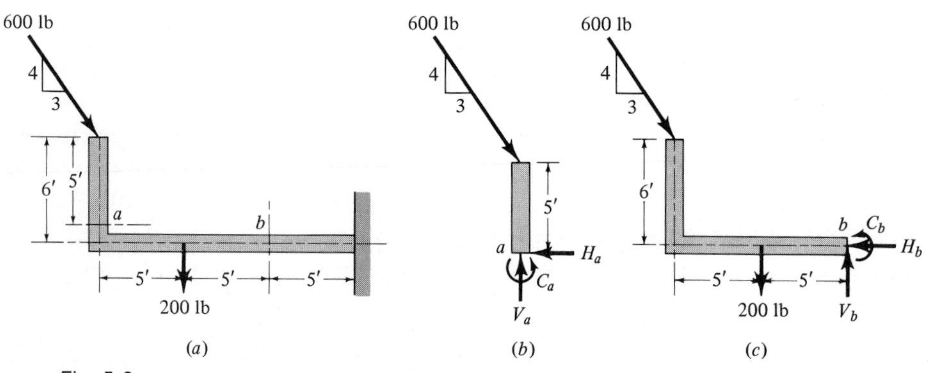

Fig. 5-2

Solution: We first cut the fixture at a and draw a free-body diagram of the upper section, as shown in Fig. 5-2(b). Applying the requirements of statical equilibrium we have

$$\text{Sum of forces} = \mathbf{0}$$

$$\mathbf{i}(\tfrac{3}{5}600 - H_a) + \mathbf{j}(V_a - \tfrac{4}{5}600) = \mathbf{0}$$

$$\therefore H_a = 360 \text{ lb} \qquad \text{(as shown)}$$

$$V_a = 480 \text{ lb} \qquad \text{(as shown)}$$

$$\text{Moments with respect to } a = \mathbf{0}$$

$$\mathbf{M}_a = \mathbf{k}[C_a - \tfrac{3}{5}600(5)] = \mathbf{0}$$

$$\therefore C_a = 1{,}800 \text{ ft-lb} \qquad \text{(as shown)}$$

For b we draw the free-body diagram shown in Fig. 5-2(c). Once again we apply the requirements of equilibrium.

$$\mathbf{S} = \mathbf{0}$$

$$\mathbf{i}(\tfrac{3}{5}600 - H_b) + \mathbf{j}(V_b - 200 - \tfrac{4}{5}600) = \mathbf{0}$$

$$\therefore H_b = 360 \text{ lb} \qquad \text{(as shown)}$$

$$V_b = 680 \text{ lb} \qquad \text{(as shown)}$$

and

$$\mathbf{M}_b = \mathbf{0}$$

$$\mathbf{k}[200(5) + \tfrac{4}{5}600(10) - \tfrac{3}{5}600(6) + C_b] = \mathbf{0}$$

$$C_b = -1{,}000 - 4{,}800 + 2{,}160 = -3{,}640 \text{ ft-lb}$$

$$\therefore \mathbf{C}_b = -3{,}640\mathbf{k} \text{ ft-lb}$$

Thus we guessed right in drawing all the internal reactions except for the couple at b, which should be opposite to that shown.

5-3 NORMAL AND SHEAR FORCES; BENDING AND TWISTING COUPLES

The examples given in the previous section dealt with specific internal reactions in the form of a horizontal and vertical force and a **k** couple. We are now ready to examine these ideas of internal forces and couples in more physically meaningful terms.

Consider a body of rather arbitrary size and shape loaded by some arbitrary types and number of forces, as shown in Fig. 5-3(a). As before, let us cut this body along the indicated plane into two sections and draw a free-body diagram of the, say, left section. In reality, the forces which the two sections of the body exert on each other are distributed over the entire surface of the cut, as indicated in Fig. 5-3(b) by the numerous small force vectors. Physically, we might think of these small forces as being the internal cohesive forces holding the material of the body together as a solid. In any case, we know that this distribution $\mathscr{D}$ of internal forces will have a resultant and, in general, will be statically equivalent to a single force-couple system such as the one shown in Fig. 5-3(c) or the one shown in Fig. 5-3(d). In Fig. 5-3(c), the net force $\mathbf{R}$ is the sum of the distributed internal forces in Fig. 5-3(b) and, if located at o, must be

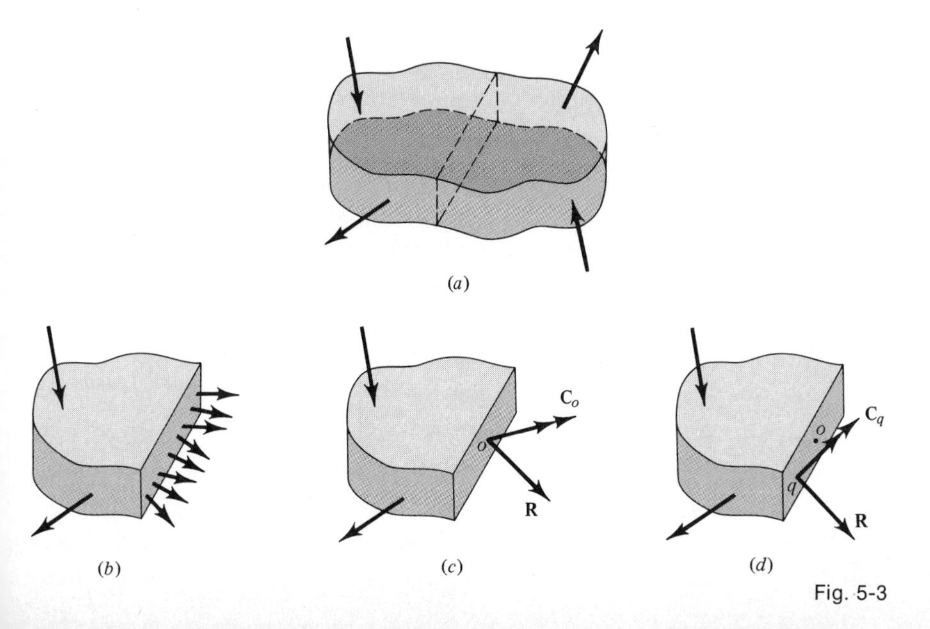

Fig. 5-3

accompanied by a couple $\mathbf{C}_o$ equal to the moment with respect to point o of the distributed forces.

$$\begin{aligned} \mathbf{R} &= \mathbf{S}(\mathscr{D}) \\ \mathbf{C}_o &= \mathbf{M}_o(\mathscr{D}) \end{aligned} \tag{5-1}$$

Similarly, in Fig. 5-3(d)

$$\begin{aligned} \mathbf{R} &= \mathbf{S}(\mathscr{D}) \\ \mathbf{C}_q &= \mathbf{M}_q(\mathscr{D}) \end{aligned}$$

with the $\mathbf{R}$ in Fig. 5-3(c) and (d) being equal but

$$\mathbf{C}_o = \mathbf{C}_q + \mathbf{oq} \times \mathbf{R} \tag{5-2}$$

Thus we arrive at the following conclusion:

In general, the internal reaction is statically equivalent to a force-couple system, with the couple being dependent upon the choice of location of the force.

In most cases, and unless stated otherwise, we will choose the location o to be the *centroid* of the cross-sectional area of the cut. As we saw in the example of the previous article, we can usually determine the net internal force $\mathbf{R}$ and couple $\mathbf{C}_o$ by applying the requirements of equilibrium to the individual sections.

We know that any force or couple can be resolved into components, such as the vertical and horizontal components used extensively in our work up to now. However, "horizontal" and "vertical" often are not very meaningful terms in describing the effect which a force has on a structural body. For example, the horizontal force $\mathbf{H}$ in Fig. 5-4 has two very different effects on the structural member shown. In the one case it tends to bend the member, while in the other case it tends to stretch the member, as indicated by the dashed lines.

These same remarks are even more appropriate insofar as internal reactions are concerned. "Horizontal" and "vertical" do not truly describe the type of internal effect which the material of the body must withstand. For example, in Fig. 5-2(b) the vertical reaction V_a is pushing on the cut cross section at a, whereas in Fig. 5-2(c) it is the horizontal force H_b which is pushing on the cut cross section at b. In dealing with problems involving the strength of structural materials, we must introduce terminology which is more indicative of the physical situation existing within the body.

Although there are many adjectives that one might use in describing the physical effects which forces and/or couples have on real structural bodies, there are a few fundamental ones which are commonly accepted in engineering terminology.

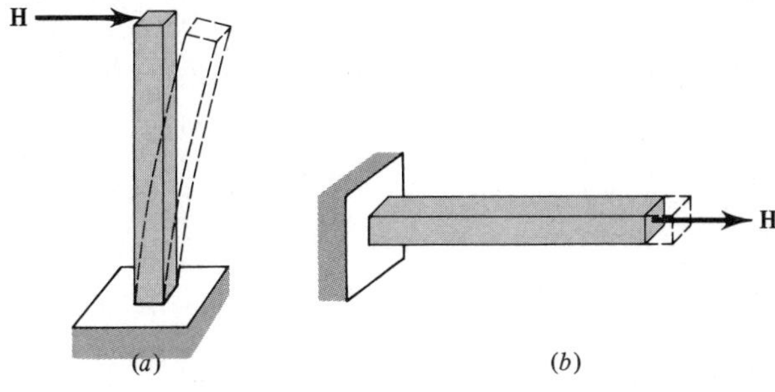

Fig. 5-4

1 A body can be *stretched* or *compressed,* and the corresponding forces are usually called *tensile* or *compressive.*
2 A force can tend to *shear* a body.
3 A body can be *twisted* by a *twisting couple* or *torque.*
4 A body can be *bent* by a *bending couple.*

Each of these situations is depicted in Fig. 5-5. Any one individual force might very well produce a combination of these effects, such as illustrated in Fig. 5-5(f), where the single force **F** tends to bend and shear the cantilever beam.

Returning now to our internal reactions of Fig. 5-3(c), which we redraw in Fig. 5-6(a), we will resolve this net internal force **R** and couple **C** into components of the type describing their physical effects. The force **R** is resolved into two components, $\mathbf{R}_N$ normal (perpendicular) to the cut cross section and $\mathbf{R}_S$ tangent (parallel) to the cut cross section. The normal force exerts a push or pull on the material, whereas the tangent force exerts a shearing effect on the material, and hence the subscript S.

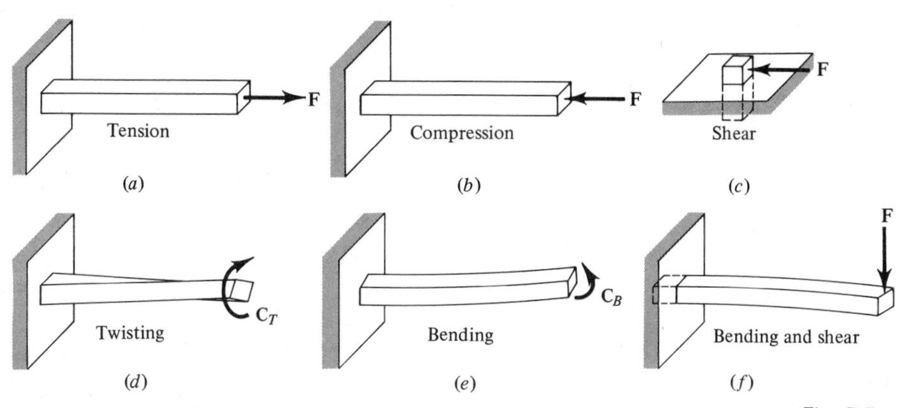

Fig. 5-5

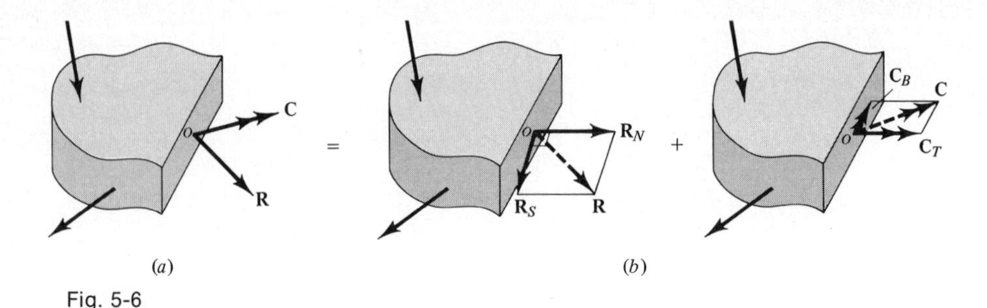

Fig. 5-6

Similarly, the couple **C** is resolved into two components, $\mathbf{C}_T$ normal to the cross section and $\mathbf{C}_B$ tangent to the cross section. Recalling the physical interpretation of a vector couple (right-hand rule), we see that $\mathbf{C}_T$ tends to twist the body and $\mathbf{C}_B$ tends to bend the body, and hence the subscripts T and B.

With this type of decomposition of the internal reactions into normal and shear forces and twisting and bending couples, we are able to physically interpret their effects on the structural member. The following examples illustrate these ideas.

EXAMPLE 5-2

Find the physical-component internal reactions at the indicated plane for the imbedded fixture shown in Fig. 5-7(*a*).

Solution: A free-body diagram of the left portion of the fixture is shown in Fig. 5-7(*b*), with **R** and **C** representing the net internal reactions on the cut cross section. Note the location of these reactions at point *o*, the centroid of the area. Also, we have chosen the **i, j, k** unit vectors in the usual manner.

Applying the requirements of equilibrium we have

$$\text{Sum of forces} = \mathbf{0}$$
$$-300\mathbf{j} + \mathbf{R} = \mathbf{0}$$
$$\therefore \mathbf{R} = 300\mathbf{j} \text{ lb}$$
$$\text{Moments with respect to point } o = \mathbf{0}$$
$$(-5\mathbf{i} + 1\tfrac{1}{2}\mathbf{j} + 3\tfrac{1}{2}\mathbf{k}) \times (-300\mathbf{j}) + \mathbf{C} = \mathbf{0}$$
$$\therefore \mathbf{C} = (-1{,}050\mathbf{i} - 1{,}500\mathbf{k}) \text{ in-lb}$$

Now, recalling the choice of **i, j, k** unit vectors and the definitions of the physical components, we have

$$\mathbf{R}_S = \mathbf{R} = 300\mathbf{j} \text{ lb} \qquad \text{and} \qquad \mathbf{R}_N = \mathbf{0}$$
$$\mathbf{C}_T = -1{,}050\mathbf{i} \text{ in-lb} \qquad \text{and} \qquad \mathbf{C}_B = -1{,}500\mathbf{k} \text{ in-lb}$$

These physical components are shown in Fig. 5-7(*c*), with the circular arrows indicating the rotational tendencies of the couples.

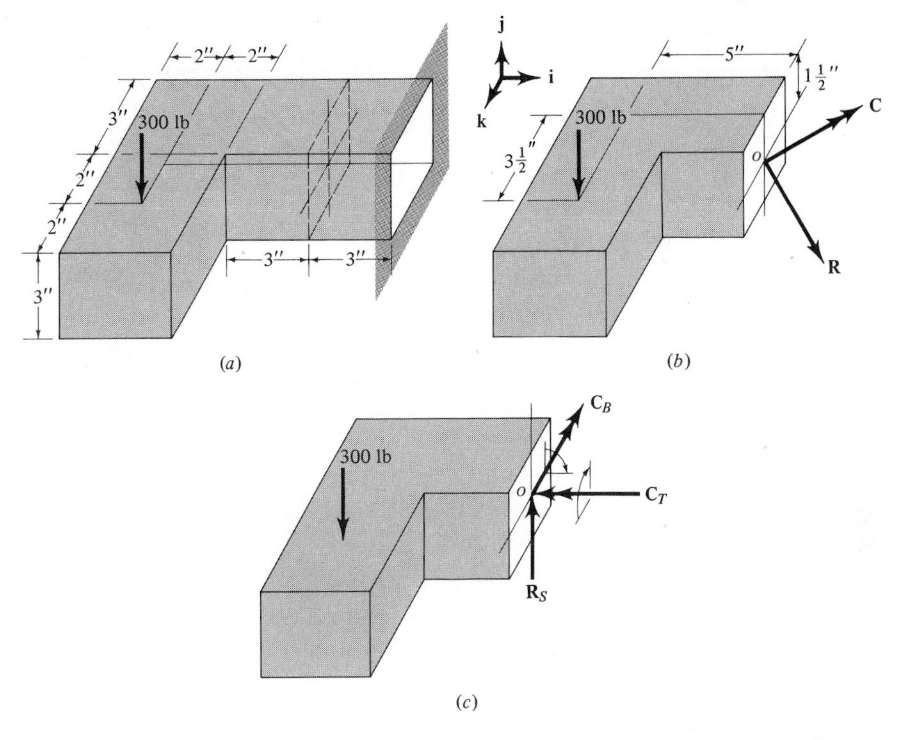

Fig. 5-7

Thus, this fixture is being sheared, twisted, and bent simultaneously at the given cross section.

EXAMPLE 5-3

Find the internal reactions at sections ① and ② in the structure shown in Fig. 5-8(a).

Solution: We first draw an FBD of the entire structure, as in Fig. 5-8(b). Applying the requirements of equilibrium

$$\mathbf{M}_d = \mathbf{0}$$

$$\mathbf{k}[A_x(6) - 1{,}200(10)] = \mathbf{0}$$

$$\therefore A_x = 2{,}000 \text{ lb} \qquad \text{(as shown)}$$

$$\text{Sum of forces} = \mathbf{0}$$

$$\mathbf{i}(D_x - A_x) + \mathbf{j}(A_y + D_y - 1{,}200) = \mathbf{0}$$

$$\therefore D_x = A_x = 2{,}000 \text{ lb} \qquad \text{(as shown)}$$

$$A_y + D_y = 1{,}200$$

In Fig. 5-8(c) are shown the FBD's of the individual members. We observe that member db is actually a two-force member, and, therefore,

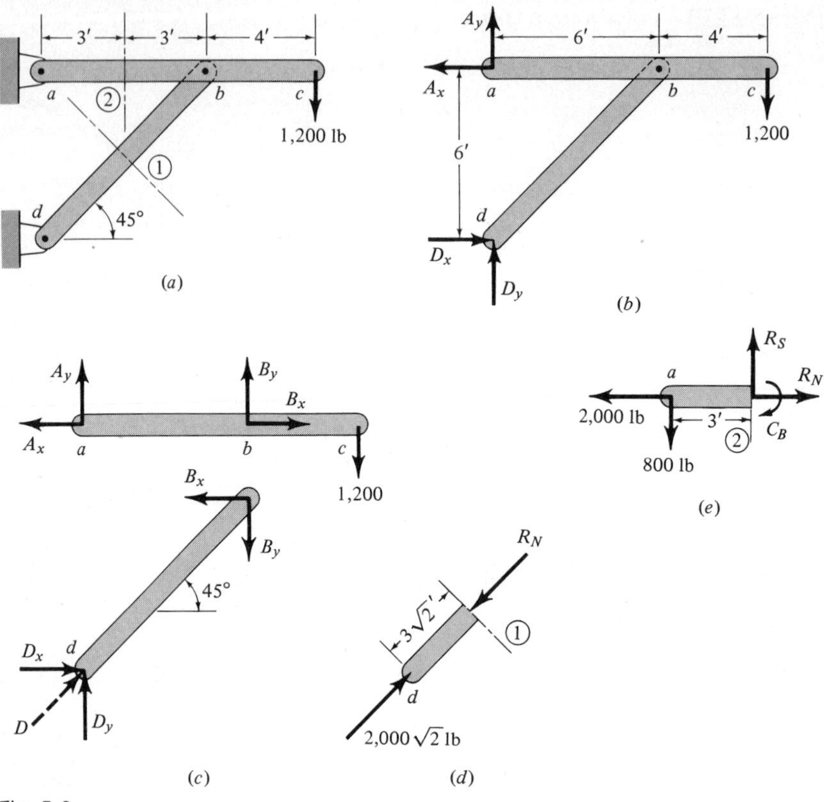

Fig. 5-8

$$D_y = D_x = 2{,}000 \text{ lb} \qquad \text{(as shown)}$$

or

$$D = 2{,}000\sqrt{2} \text{ lb} \qquad \text{(as shown)}$$

and

$$B_x = B_y = 2{,}000 \text{ lb} \qquad \text{(as shown)}$$

Also, taking moments with respect to point b on member abc yields

$$\mathbf{M}_b = \mathbf{0}$$

$$\mathbf{k}[-A_y(6) - 1{,}200(4)] = \mathbf{0}$$

$$A_y = -800 \text{ lb} \qquad \text{(opposite to that shown)}$$

Now, for the internal reactions at ①, we use the FBD in Fig. 5-8(*d*). Obviously, the only internal reaction is a compressive force of

$$R_N = 2{,}000\sqrt{2} \text{ lb} \qquad \text{(compression)}$$

For the reactions of section ② we have, from Fig. 5-8(*e*), by summing forces and moments,

$$\mathbf{j}(R_S - 800) = \mathbf{0}$$
$$R_S = 800 \text{ lb} \qquad \text{(as shown)}$$
$$\mathbf{i}(R_N - 2{,}000) = \mathbf{0}$$
$$R_N = 2{,}000 \text{ lb} \qquad \text{(tension, as shown)}$$
$$\mathbf{k}[800(3) - C_B] = \mathbf{0}$$
$$C_B = 2{,}400 \text{ ft-lb} \qquad \text{(as shown)}$$

Hence, member *abc* is being stretched, bent, and sheared, while member *bd* is being compressed only.

> ***Remark:*** In this problem the dimensions of the cross-sectional areas were not given and probably can be considered small in comparison with the overall dimensions of the structure.

EXERCISE PROBLEMS

5-1 to 5-4 Find the internal reactions at sections ① and ② for the beams shown in terms of physical components.

PROB. 5-1

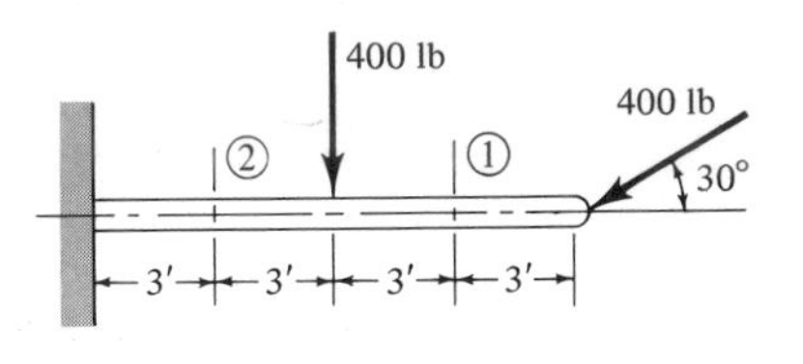

PROB. 5-2

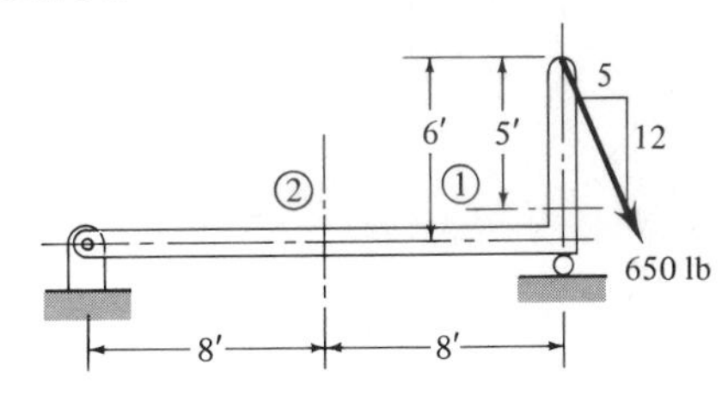

PROB. 5-3

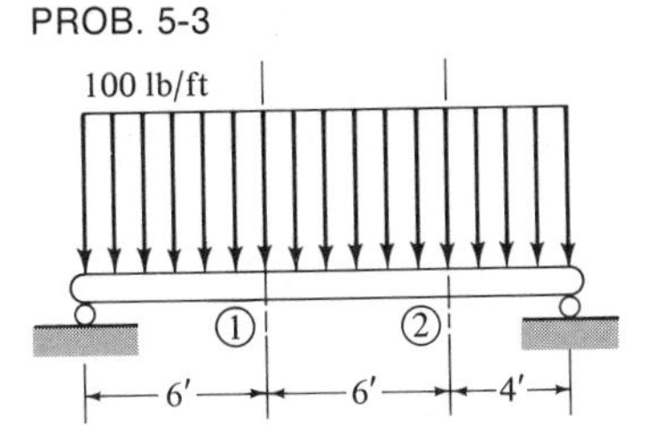

PROB. 5-4

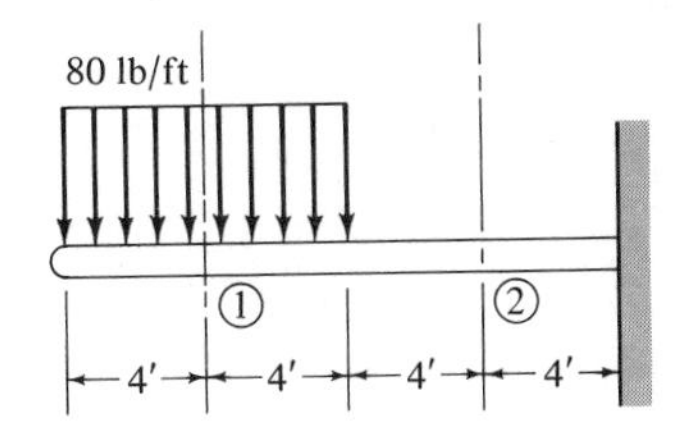

5-5 to 5-8 For the fixture shown, determine the component internal reactions at the indicated locations *a* and *b*.

PROB. 5-5

PROB. 5-7

PROB. 5-6

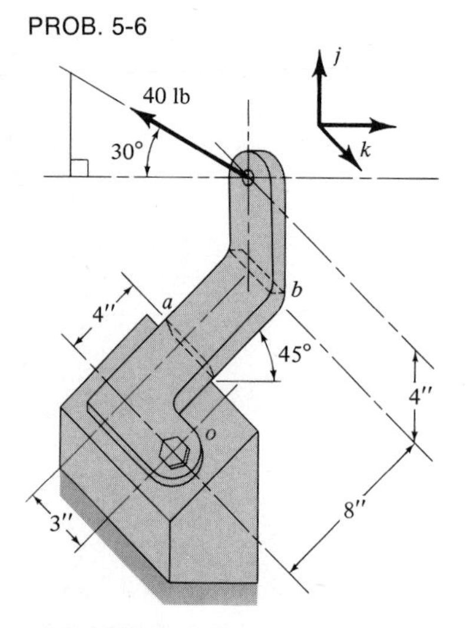

PROB. 5-8

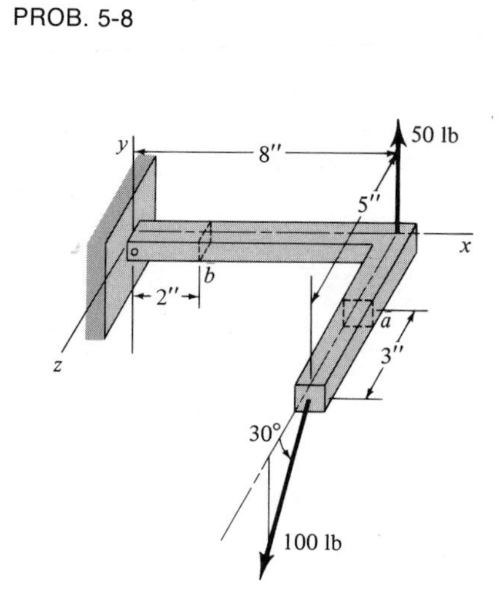

5-9 and 5-10 If the shaft is in equilibrium, find the internal reactions at sections ① and ②.

PROB. 5-9

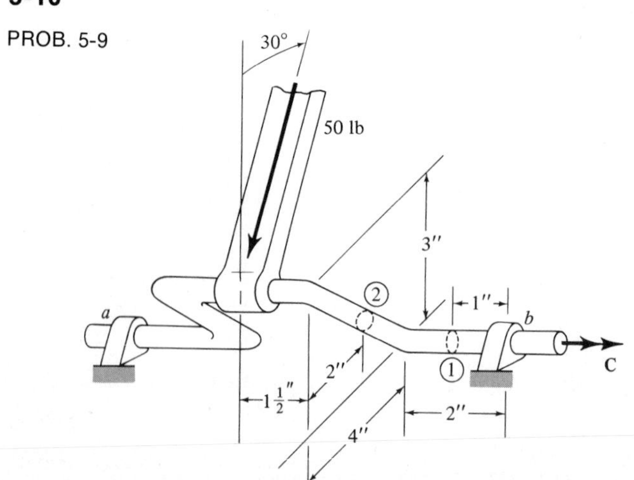

PROB. 5-10

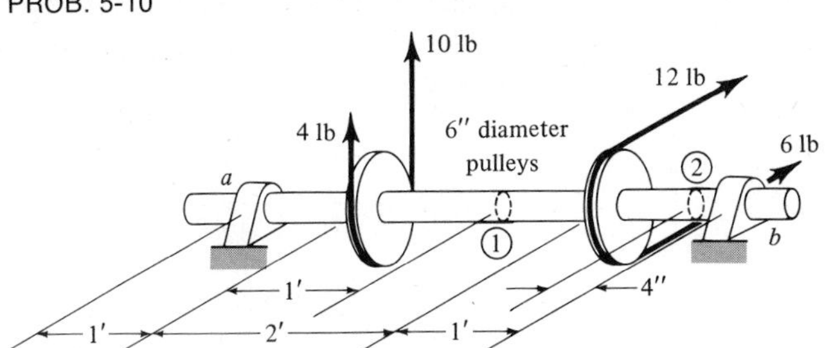

PROB. 5-11

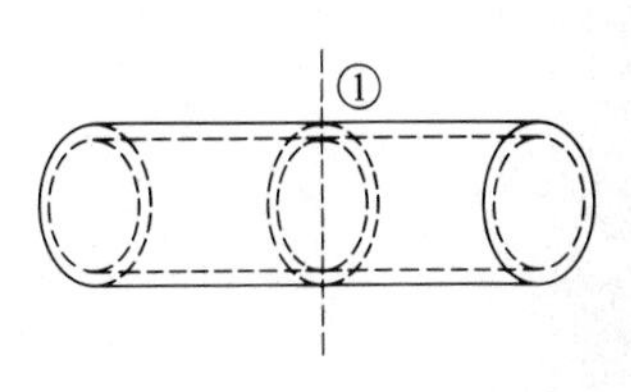

5-11 A cylindrical tank of diameter 12 in and 30 in long holds gas under a pressure of 40 psig. What is the net internal reaction on the indicated cylindrical cross section of the wall of the tank?

5-12 to 5-18 Find the internal reactions at the indicated sections for the structures shown.

PROB. 5-12 Load frame.

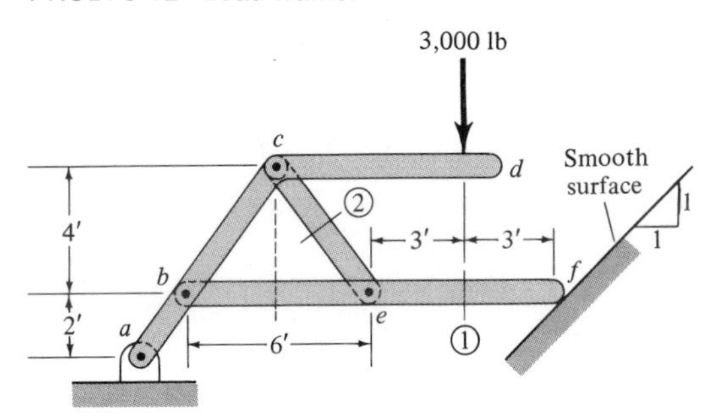

PROB. 5-13

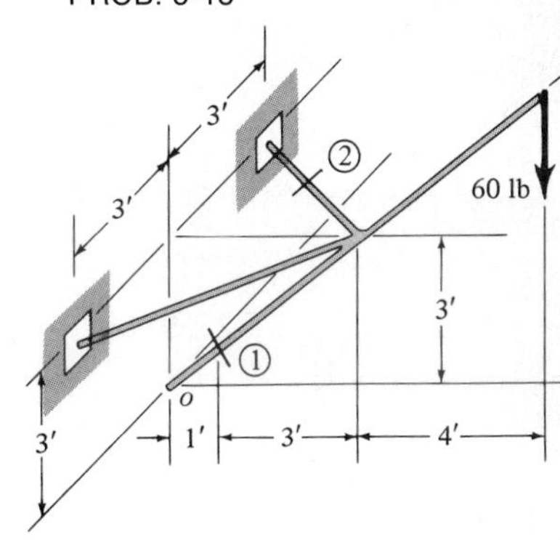

PROB. 5-14 Pin-connected truss.

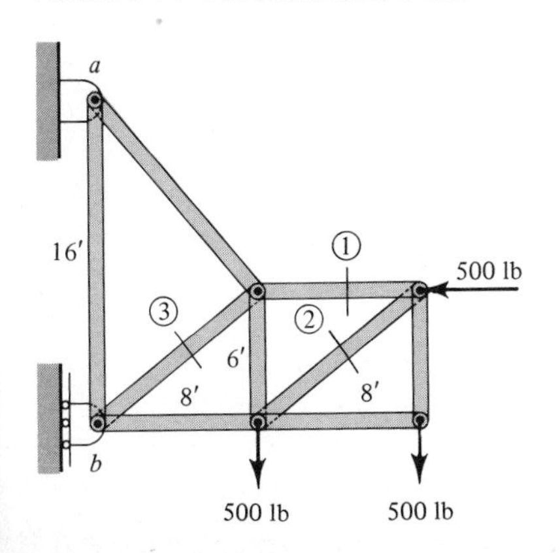

PROB. 5-15

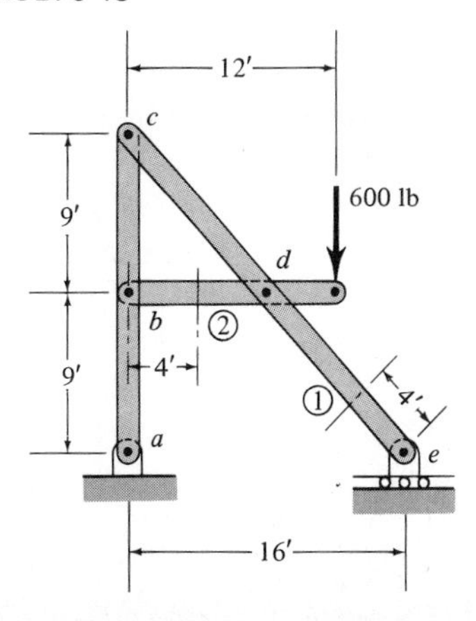

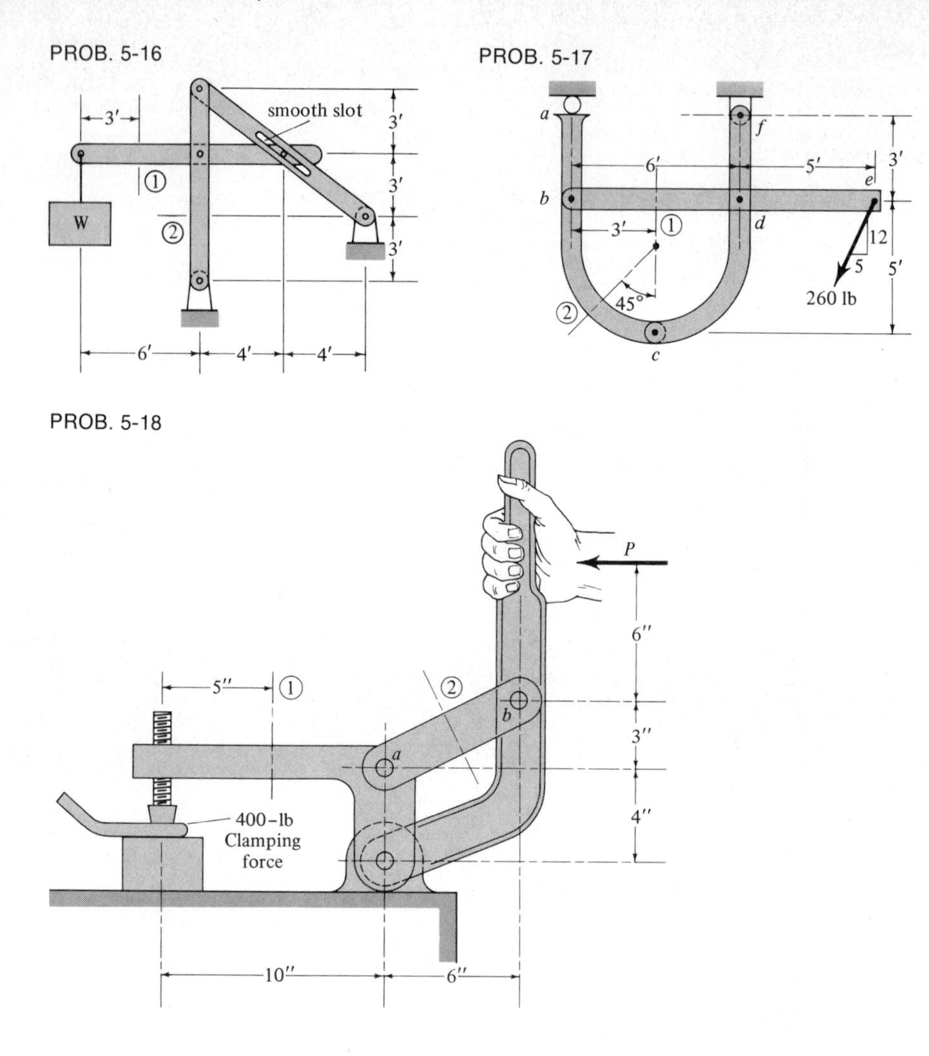

5-4 THE STRESS CONCEPT

The idea of internal reactions is a useful scheme for determining how strong a structural member must be in terms of the magnitude and type of forces or couples the member must transmit. But the question as to the actual strength of a structural member is inherently tied to such factors as the size and shape of the member and the material of which it is made. For example, our experience leads us to surmise that a $\frac{1}{2}$-in-diameter bolt is considerably stronger than a similar $\frac{1}{4}$-in-diameter bolt, and we attribute this simply to the fact that the $\frac{1}{2}$-in bolt is bigger and has more material than the $\frac{1}{4}$-in bolt. Hence, we anticipate that the size and shape will enter into the evaluation of the strength of a structural member.

As a second example, our experience leads us to believe that steel is stronger than, say, wood. Probably, what we mean by stronger is that, given a piece of steel and a piece of wood of the *same* size and shape, the steel one can withstand a correspondingly greater load of any type than the wood before breaking. Intuitively, then, when comparing the relative strength of two bodies of material, the comparison should be made on the basis of the same geometric and loading configurations. However, it is quite apparent that all structural members do not have the same geometry and loading conditions, and, therefore, to compare the relative strengths of two different structural members, we must devise some common denominator which incorporates the size, shape, and loading. Accordingly, we are led to introduce the concept of stress.

In its simplest terms, stress is the *internal force per unit area*. Recall from the preceding sections that the *net* internal reaction was the accumulative effect of a distribution of internal cohesive forces across the entire cut cross section. [See Fig. 5-3(*b*).] Now, if we think of this distribution in terms of a force per unit area, we can formally define:

Definition 5-1 The stress vector distribution $\boldsymbol{\sigma}$ (or simply stress vector) is the internal force per unit area, such that

$$\mathbf{R} = \iint_{\substack{\text{cut} \\ \text{cross section}}} \boldsymbol{\sigma}\, dA \tag{5-3}$$

and

$$\mathbf{C}_o = \iint_{\substack{\text{cut} \\ \text{cross section}}} \mathbf{p} \times \boldsymbol{\sigma}\, dA \tag{5-4}$$

where $\mathbf{p}$ is the position vector of dA relative to o, as shown in Fig. 5-9(*a*). The integrals are to be interpreted as summing operations over the entire cut cross section.

Although this definition of the stress vector is formal rather than physical, we can infer some physically significant ideas about

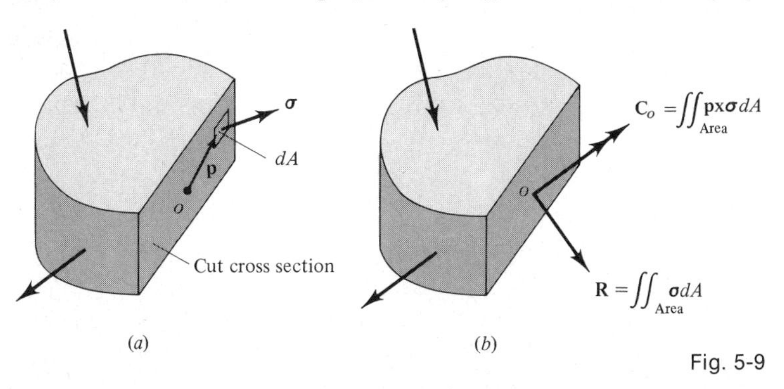

Fig. 5-9

stress. First of all, since the stress vector is a force per unit area, this vector can be resolved into a normal stress vector $\boldsymbol{\sigma}_N$ and a shear stress vector $\boldsymbol{\sigma}_S$, so that

$$\boldsymbol{\sigma} = \boldsymbol{\sigma}_N + \boldsymbol{\sigma}_S \tag{5-5}$$

Normal stresses are either pushing or pulling effects and, accordingly, are called compressive and tensile stresses, respectively. Shear stress represents a tearing effect tangent to the surface area. Secondly, by the definition of normal and shear forces and bending and twisting couples

$$\iint_{\text{area}} \boldsymbol{\sigma}\, dA = \mathbf{R} = \mathbf{R}_N + \mathbf{R}_S \tag{5-6}$$

$$\iint_{\text{area}} \mathbf{p} \times \boldsymbol{\sigma}\, dA = \mathbf{C}_o = \mathbf{C}_B + \mathbf{C}_T \tag{5-7}$$

we see that the stress vector is related to all four component internal reactions $\mathbf{R}_N$, $\mathbf{R}_S$, $\mathbf{C}_B$, and $\mathbf{C}_T$. Consequently, the stress distribution in a member is related to the net physical effects of pushing, pulling, shearing, bending, and twisting.

A structural member subjected to all of these physical effects will usually have a very complicated stress distribution. To simplify our development of the stress concept, in the remainder of this chapter we will restrict our study to members subjected primarily to pushing, pulling, and shearing effects. Chapter 7 will take up members subjected to twisting, and Chap. 9 will consider members subjected to bending.

Before progressing on this idea of stress, let us pause briefly. Our objective in introducing the concept of stress was to develop some scheme which would reduce the manner of loading and the size and shape of a member to some common denominator. As we have defined it, the stress vector is a common effect produced by all four types of physical internal reactions. Also, since stress is force *per unit area*, it is defined in the same manner for a large area as for a small area. Hence, stress appears to be a common denominator for all types of load-carrying members. Whether or not this denominator is physically significant is still an open question, which we will pursue in subsequent discussions.

5-5 STRESSES IN TWO-FORCE MEMBERS

Let us assume for the time being that stress is a physically significant quantity as a measure of the strength of a load-carrying member. As engineers, we now face the question: Given a structural member of some size, shape, and material loaded in some manner, how do we determine the stress in the member? First of all, you should realize that stress as defined in the previous section is an *internal effect* and is *not* physically measurable in the manner that we measure an external force (such as weight) or a surface traction (such as pressure). An attempt to implant a "stress-measuring device" in a solid body would involve actually cutting the body, implanting the device, and then patching the body back together, all of which would unquestionably alter and damage the body, making our measurements virtually worthless insofar as the original unaltered body is concerned. Also, to determine the stress *everywhere* in the body would require an innumerable number of such experimental measurements throughout the body. Faced with this dilemma, the engineer tries to develop some rational theory relating the external applied loads and external geometry of the member to the internal stresses. Sometimes these theories are quite sophisticated, other times rather crude, with the choice often dictated by the peculiar requirements, economics, or complexities of a given problem. The engineer's job is to do the best he can within the imposed limitations and difficulties.

What do we know about stress thus far? Recall Eqs. (5-6) and (5-7) relating the net internal reactions to the stress vector $\boldsymbol{\sigma}$

$$\iint_{\text{area}} \boldsymbol{\sigma}\, dA = \mathbf{R} = \mathbf{R}_N + \mathbf{R}_S \tag{5-6}$$

$$\iint_{\text{area}} \mathbf{p} \times \boldsymbol{\sigma}\, dA = \mathbf{C}_o = \mathbf{C}_B + \mathbf{C}_T \tag{5-7}$$

In Secs. 5-2 and 5-3 we learned how to determine the internal reactions from the externally applied loads by invoking the requirements of equilibrium on a section of the body. The question now before us is how to determine the stress vector $\boldsymbol{\sigma}$ from the net internal forces $\mathbf{R}_S$ and $\mathbf{R}_N$.

> ***Remark:*** For the present, we are considering only internal-force reactions, $\mathbf{R}_N$ and $\mathbf{R}_S$. We will consider the stress related to the internal-couple reactions $\mathbf{C}_B$ and $\mathbf{C}_T$ in subsequent chapters.

Let us consider a straight two-force member of uniform cross section as shown in Fig. 5-10(*a*). The line of action of the loads is assumed to pass through the centroid of the cross-sectional area. If we cut the member across a cross section perpendicular to the longitudinal axis of the member, we see in Fig. 5-10(*b*) that the net internal reaction is the normal force $\mathbf{R}_N$. Now, by the stress concept, this normal force can be visualized as the accumulative effect

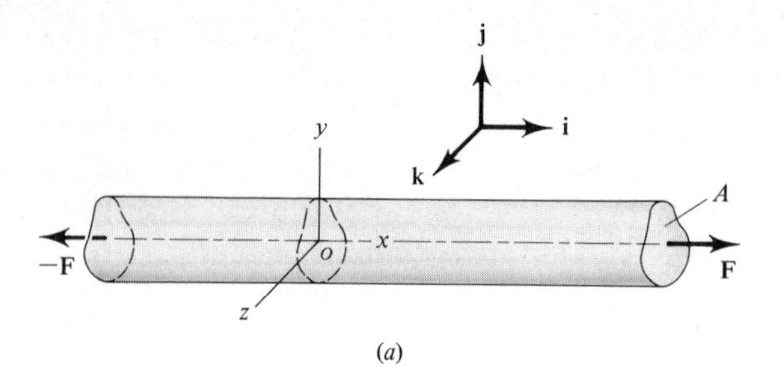

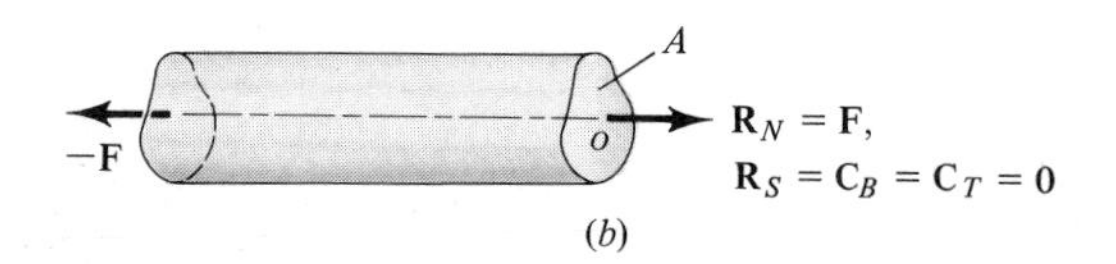

−F

$\iint_A \sigma_N \, dA = \mathbf{R}_N, \iint_A \mathbf{p} \times \sigma_N \, dA = 0$

(c)

Fig. 5-10

of the stress vectors over the entire cut cross section. According to Eq. (5-5), the stress vector at any point is generally composed of a normal (pushing-pulling) stress and a shearing stress. Although we cannot prove it, it seems physically reasonable for the situation under consideration in Fig. 5-10 that the stress vector will be entirely a normal stress, which in this case can be written as

$$\boldsymbol{\sigma} = \boldsymbol{\sigma}_N = \sigma_N \mathbf{i}$$

and is indicated in Fig. 5-10(c). With this assumption, Eq. (5-6) reduces to

$$\iint_A \sigma_N \mathbf{i} \, dA = R_N \mathbf{i}$$

or

$$\iint_A \sigma_N \, dA = R_N = F \tag{5-8}$$

while Eq. (5-7) becomes

$$\iint_A (y\mathbf{j} + z\mathbf{k}) \times (\sigma_N \mathbf{i}) \, dA = \mathbf{0}$$

or

$$\iint_A y\sigma_N \, dA = 0 \qquad \text{and} \qquad \iint_A z\sigma_N \, dA = 0 \tag{5-9}$$

where we have used $\mathbf{p} = y\mathbf{j} + z\mathbf{k}$, as shown in Fig. 5-11. Hence, Eq. (5-8) relates the normal stress σ_N to the net internal normal force

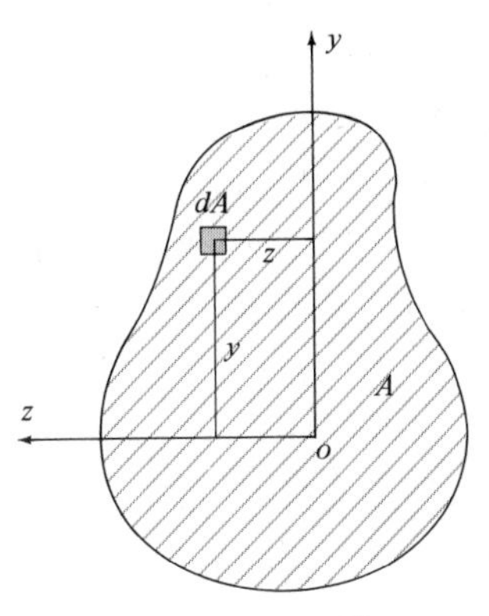

Fig. 5-11

$R_N = F$, subject to the requirements of Eqs. (5-9). In general, σ_N might well vary in magnitude from point to point over the cut cross section, and thus we seek to find σ_N as a function of the cross-section variables y and z:

$$\sigma_N = \sigma_N(y,z)$$

From a purely mathematical viewpoint, for any given cross section, there might be a multitude of functions $\sigma_N(y,z)$ which will satisfy Eqs. (5-8) and (5-9). However, we will look at the situation from an engineer's viewpoint.

Referring to Fig. 5-10, recall that the line of action of F and R_N is through the centroid o of the cross-sectional area. Hence, since the net load acts at the "middle" of the area, let us assume that it is *evenly* distributed over the entire cross-sectional area; that is, we will assume

$$\sigma_N(y,z) = \text{constant } \sigma_N$$

For this assumption, Eq. (5-8) yields

$$\iint_A \sigma_N \, dA = \sigma_N \iint_A dA = \sigma_N A$$

$$\therefore R_N = \sigma_N A$$

or

$$\sigma_N = \frac{R_N}{A} = \frac{F}{A} \tag{5-10}$$

That is, the stress is simply the net internal force divided by the cross-sectional area.

> ***Remark:*** At first glance, it might seem that this result should have been apparent from the definition of stress as force per unit area and that we have made a very elaborate presentation of a very simple idea. However, just because we have arrived

at a rather simple and logical answer does *not* guarantee that it is correct, and, indeed, the history of mechanics is full of instances in which incorrect answers were obtained by very logical methods.

Thus far in our development here, we know for sure that the stress given by Eq. (5-10) will satisfy Eq. (5-8). What about Eqs. (5-9), which must also be satisfied? For σ_N a constant, we see that

$$\iint_A y\sigma_N \, dA = \sigma_N \iint_A y \, dA \tag{5-11a}$$

$$\iint_A z\sigma_N \, dA = \sigma_N \iint_A z \, dA \tag{5-11b}$$

Referring to the cross-sectional area shown in Fig. 5-11, the integrals on the right of the equalities represent the first moment of the area with respect to the z and y axes, respectively. (Refer to Appendix A.) But, recalling that o is the centroid of the area, by the definition of centroidal axes we have

$$\iint_A y \, dA = 0 \qquad \text{and} \qquad \iint_A z \, dA = 0$$

Hence, Eqs. (5-9) are automatically satisfied, and thus we arrive at a very important observation.

Observation 1: For the normal stress to be uniform over the cross section of a two-force member, it is *necessary* that the line of action of the forces pass through the centroid of the transverse cross-sectional area of the member; i.e., the member must be *centroidally* (*centrally*) *loaded*. This follows from the fact that if o were not the centroid, the integrals in Eq. (5-11) would not vanish, and, therefore, Eqs. (5-9) would not be satisfied.

We wish to emphasize that centroidal loading is a *necessary* condition for uniform stress, but it is *not* sufficient. For the engineer this means that if the member is centroidally loaded, Eq. (5-10) has a reasonable chance of being correct, but is surely incorrect if the member is not centroidally loaded. In the next chapter, when we look at the behavior of materials, we will further discuss the correctness of Eq. (5-10), but for now let us accept it as being a reasonable result giving the *average* value of the normal stress.

Let us return to the original centroidally loaded two-force member of Fig. 5-10(*a*), which we redraw in Fig. 5-12(*a*). Suppose we cut this member across a cross section whose normal **N** is

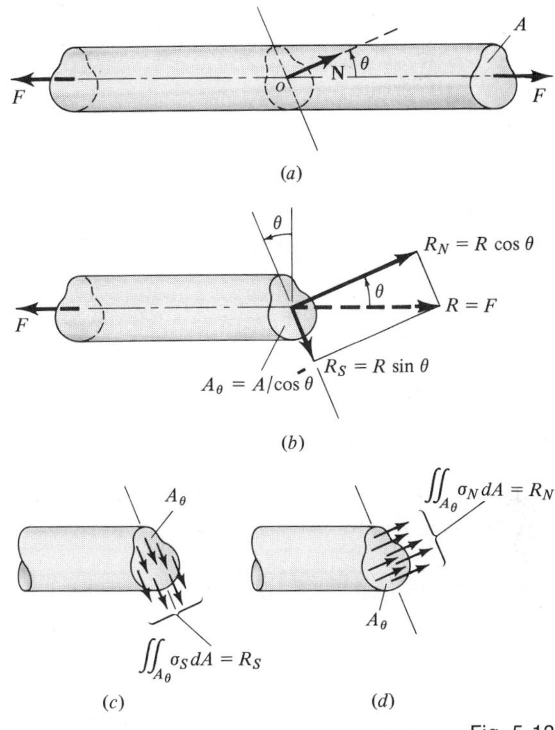

Fig. 5-12

directed at an angle θ with, say, the longitudinal axis. Then the cut cross section will have an area A_θ of

$$A_\theta = \frac{A}{\cos\theta}$$

and net internal forces of

$$R_N = R\cos\theta$$
$$R_S = R\sin\theta$$

as shown in Fig. 5-12(*b*). In this case, we have reason to expect both a normal stress σ_N and a shear stress σ_S to exist on this area A_θ, such that

$$\iint_{A_\theta} \sigma_N \, dA = R_N$$

$$\iint_{A_\theta} \sigma_S \, dA = R_S$$

as indicated in Fig. 5-12(*c*) and (*d*). Using the same arguments as before, we assume these stresses to be uniform over the entire area, and, therefore,

$$\sigma_N = \frac{R_N}{A_\theta} = \frac{R}{A} \cos^2 \theta = \sigma \cos^2 \theta \tag{5-12a}$$

$$\sigma_S = \frac{R_S}{A_\theta} = \frac{R}{A} \sin \theta \cos \theta = \sigma \sin \theta \cos \theta \tag{5-12b}$$

where

$$\sigma = \frac{R}{A} = \frac{F}{A}$$

is the normal stress on the area A perpendicular to the longitudinal axis of the member, as given previously by Eq. (5-10). Equations (5-12) point out a very important observation, namely:

> ***Observation 2:*** The normal and shear stresses in a centroidally loaded two-force member depend directly upon the orientation of the cut cross-sectional area.

In most instances, the engineer is interested in knowing the *maximum* normal and shear stresses. To find the angle θ which will maximize the normal stress, we differentiate Eq. (5-12*a*) and set the derivative equal to zero; similarly for the shear stress—by differentiating Eq. (5-12*b*) and setting the derivative to zero. In this way we arrive at the following:

Theorem 5-1 In a centroidally loaded two-force member, the largest average normal stress is

$$\sigma_N(\max) = \frac{F}{A} \tag{5-13}$$

on the transverse area perpendicular to the line of action of the load F, and the largest average shear stress is

$$\sigma_S(\max) = \left(\frac{F}{A}\right)\left(\frac{1}{2}\right) \tag{5-14}$$

on any area inclined at 45° with the transverse area.

The main purpose in obtaining this theorem was to point out the seemingly unexpected result that even though the external forces on a member may be tensile or compressive longitudinal forces, the material in the member will be subjected to *both* normal and *shear* stresses. This fact can be very important in ultimately determining the load-carrying capability of the member.

EXAMPLE 5-4

The eye bar in Fig. 5-13(*a*) is fusion welded across a 30° seam as shown. Find the average normal and shear stresses along the

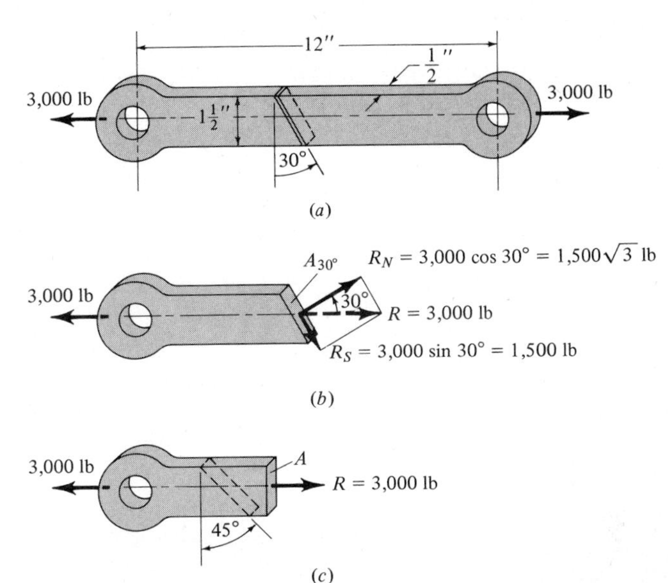

Fig. 5-13

seam. What are the maximum tensile and shear stresses in the main body of the eye bar?

Solution: A free-body diagram of the left portion of the eye-bar cut along the weld is shown in Fig. 5-13(*b*), in which R has been resolved into normal and shear components. The transverse area of the bar is

$$A = (\tfrac{1}{2})(1\tfrac{1}{2}) = \tfrac{3}{4}\ \text{in}^2$$

so that the area of the welded seam is

$$A_{30°} = \frac{A}{\cos 30°} = \left(\frac{3}{4}\right)\left(\frac{2}{\sqrt{3}}\right) = \frac{\sqrt{3}}{2}\ \text{in}^2$$

Assuming the lines of action of the applied 3,000-lb loads pass through the centroid of the cross section of the bar, the normal and shear stresses along the seam are given by

$$\sigma_N = \frac{R_N}{A_\theta} = \frac{1{,}500\sqrt{3}}{\sqrt{3}/2} = 3{,}000\ \text{lb/in}^2 \qquad \text{(tensile)}$$

$$\sigma_S = \frac{R_S}{A_\theta} = \frac{1{,}500}{\sqrt{3}/2} = 1{,}732\ \text{lb/in}^2$$

According to Theorem 5-1, the maximum tensile stress in the bar will occur on the cross section perpendicular to the axis of the bar. Therefore, from Fig. 5-13(*c*)

$$\sigma_N(\text{max}) = \frac{R}{A} = \frac{3{,}000}{3/4} = 4{,}000\ \text{lb/in}^2 \qquad \text{(tensile)}$$

Also, by Theorem 5-1, the maximum shear stress will occur on a plane at 45° with the transverse area, and is given by

$$\sigma_S(\max) = \frac{1}{2}\frac{R}{A} = 2{,}000 \text{ lb/in}^2$$

Remark: In a solid fusion weld, the maximum normal and shear stresses are usually taken to be those corresponding to the maximums in the solid bar of virgin material, regardless of the weld angle—in this case 4,000 psi and 2,000 psi, respectively.

Remark: Engineering stresses are most commonly given in pounds per square inch rather than pounds per square foot or dynes per square centimeter or some other combination. Also, stresses of the order of thousands of pounds per square inch are quite reasonable in many structural members.

5-6 NOMINAL NORMAL AND SHEAR STRESSES IN CONNECTORS

In Chap. 4 we discussed various types of connectors and connecting devices used to fasten together two or more load-carrying members. For purposes of external load analysis, the connected joints were idealized as pin joints, ball joints, fixed joints, and several others shown in Fig. 4-2. However, quite often the load-carrying capability of a structure or machine is governed by the strength of the actual connecting device, which may include bolts, rivets, welds, keys, etc. This relative weakness of a connection might be due to purposeful design, e.g., a cotter pin holding a gear on a shaft is cheaper to replace than the more expensive gear or shaft. On the other hand, a connector might fail due to careless or faulty design. The intention of this section is to look at some basic ideas used to determine the strength of connectors and connected joints.

Generally, a connected joint is composed of three elements: two members which we wish to fasten together and a connector such as a bolt, rivet, key, etc. To function properly, all three elements must be strong enough to withstand the loads which are to be transmitted by the joint. Our objective here is to relate the applied loads and the geometry of the connection to some physically significant quantities indicative of the strength of the joint. To this end we give the following definition:

Definition 5-2 An average or nominal stress in a connected joint is the force per unit area evaluated by dividing the net transmitted force by the net *effective* area over which the force is transmitted. Such a stress can be tensile, compressive, or shear, depending upon the physical effect of the force.

> ***Remark:*** In most cases the actual stresses in a connected joint are difficult if not impossible to determine in a completely rigorous manner due to the geometry and the complex force distribution throughout the joint. The nominal stresses as defined above serve as a useful and practical starting point for evaluating the load-carrying capacity of a connected joint. Some further refinements will be discussed in the next chapter.

To illustrate the idea of nominal stresses in a connected joint, consider the two flat plates held together by a single rivet, as shown in Fig. 5-14(a). The purpose of this connection is to transmit the load F from the left plate to the right plate. *Simplified* free-body diagrams of each of the elements of the joint are shown in Fig. 5-14(b) to (d). By "simplified," we mean that we have shown the net effective forces on each element, but their locations are somewhat imprecise. In particular, note that the forces on the rivet in Fig. 5-14(c) form a couple, and, therefore, the rivet cannot be in equilibrium. As we mentioned earlier, the actual force distribution in a connecting device is usually quite complex, and some liberties are usually taken in simplifying the situation, as we have done. In this particular case we are ignoring bending effects in the joint, both in the plates and the rivet, so that whatever couples are missing for equilibrium requirements are assumed to be relatively unimportant physically insofar as the stresses are concerned.

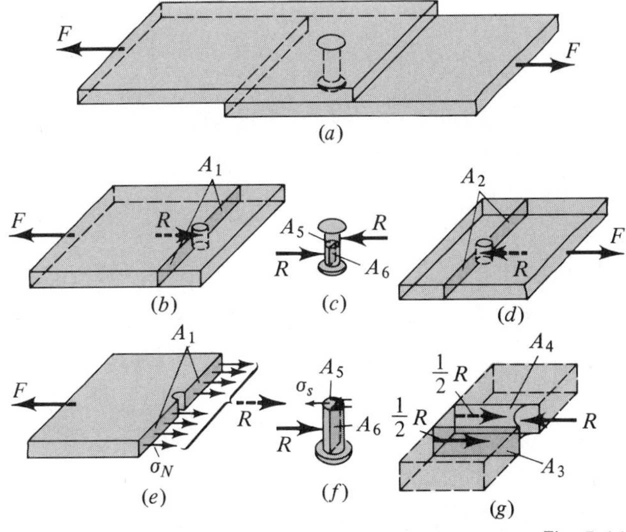

Fig. 5-14

Looking first at the plates in Fig. 5-14(*b*) and (*d*), the force R exerts a pull on the net area A_1 (or A_2), which is the area of the solid plate minus the portion cut away by the hole, as indicated in Fig. 5-14(*e*). Hence, the corresponding *nominal* tensile stress in, for example, the left plate would be

$$\sigma_N = \frac{R}{A_1}$$

As for the rivet in Fig. 5-14(*c*), the forces are tending to shear the rivet across its cross-sectional area A_5. The corresponding *nominal* shear stress, indicated in Fig. 5-14(*f*), is given by

$$\sigma_S = \frac{R}{A_5}$$

Although the tensile stresses in the plates and the shear stress in the rivet are probably the most important stresses in the joint, two other stresses can sometimes be of significant importance. The contact force between the inside surface of the hole in the plate and the outer-cylindrical surface of the rivet is a pushing or *bearing* force, so that the rivet is tending to be *crushed* by the plate. The corresponding nominal compressive *bearing stress* is evaluated by dividing the bearing force by the projected *plane* bearing area. In this case, the projected plane bearing area is

$$A_6 = \text{diameter of rivet} \times \text{thickness of plate}$$

so that the *nominal* bearing stress between the right plate and the rivet is

$$\sigma_B = \frac{R}{A_6}$$

Finally, the force exerted on the plate by the rivet tends to *shear* a plug out of the plate, as indicated in Fig. 5-14(*g*). This plug is being sheared on two sides, A_3 and A_4, and the corresponding *nominal* shear stress is given by

$$\sigma_S = \frac{1/2\ R}{A_3} \quad \text{or} \quad \frac{1/2\ R}{A_4}$$

We wish to emphasize that there is no need to memorize any of the above equations for evaluating nominal stresses. Nominal stresses are simply reasonable guesses and estimates of the significant tensile, compressive, and shear stresses in a connecting device. Each type of device has its own characteristics as to how it transmits the loads and the effective areas over which these loads are transmitted. Therefore, in determining nominal stresses, each device requires its own careful analysis of its load-transmitting characteristics.

EXAMPLE 5-5

The connecting device shown in Fig. 5-15(*a*) is composed of a sleeve A, a key B, and a solid rod C. For a net load P of 10,000 lb, find (*a*) the nominal tensile stress in the sleeve, (*b*) the nominal shear stress in the key, and (*c*) the bearing stress between the key and the rod.

Solution: The essential free-body diagram is that of the key shown in Fig. 5-15(*b*). The two 5,000-lb forces represent the contact forces between the key and the outer shell of the sleeve, while the 10,000-lb force is the contact force between the key and the rod.

(1) The net effective tensile area in the sleeve is the cross section of the outer shell minus the cutout for the key. Thus

$$A_{\text{net}} = \frac{\pi}{4}\left[(1\tfrac{1}{2})^2 - (1)^2\right] - (\tfrac{1}{8})(\tfrac{1}{2})$$
$$= 0.93 \text{ in}^2$$

then
$$\sigma_N = \frac{P}{A_{\text{net}}} = \frac{10{,}000}{0.93} = 10{,}700 \text{ lb/in}^2 \qquad \text{(tensile)}$$

(2) The loading situation on the key depicted in Fig. 5-15(*b*) is commonly referred to as "double shear" in that each

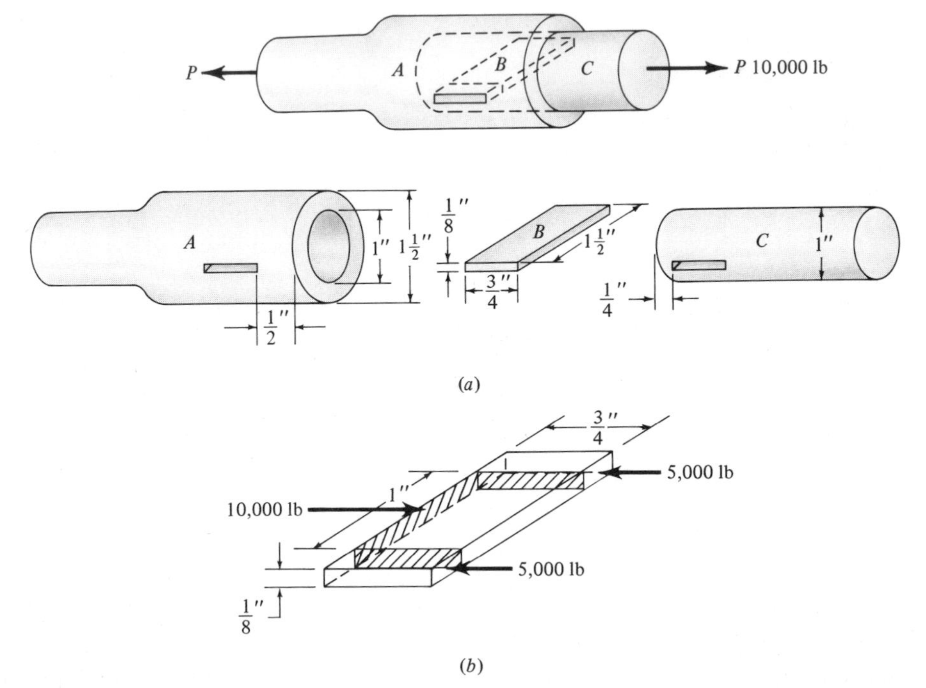

Fig. 5-15

5,000-lb load is shearing half of the key. Thus, each 5,000-lb force is shearing an area of $\frac{1}{8} \times \frac{3}{4}$ and the nominal shear stress is given by

$$\sigma_S = \frac{5{,}000}{(3/4)(1/8)} = 53{,}300 \text{ lb/in}^2$$

(3) As mentioned previously, the contact force between the key and the rod is 10,000 lb, and the contact area is given by the thickness of the key and the diameter of the rod. Hence the bearing stress is

$$\sigma_B = \frac{10{,}000}{(1/8)(1)} = 80{,}000 \text{ lb/in}^2$$

This closes our introduction to the stress concept, in which we have tried to give a physical interpretation of stress as an internal force per unit area. Subsequent chapters will have a great deal more to say about stress and its relation to the load-carrying capability of various materials and structural members. For now, it is important that you learn to visualize the internal forces and the corresponding areas on which these forces act. In this regard, there is no substitute for a well-drawn free-body diagram to help you gain a physical feel for stress.

EXERCISE PROBLEMS

5-19 A $\frac{1}{2}$-in-diameter Lucite plastic rod must carry an axial tensile load of 68 lb. To increase the length of the rod, a splice was necessary. If the splice surface makes an angle of 30° with the axis of the rod, what are the average normal and shear stresses on the adhesive of the splice?

5-20 A broken spruce spar in a light aircraft is to be repaired with an adhesive splice. The cross section of the spar is a $\frac{3}{4} \times 4$-in rectangle. If the spar must transmit an axial load of 2,400 lb and the shear and normal stresses in the adhesive cannot exceed 500 psi and 300 psi, respectively, what is the maximum angle that the splice surface should make with the axis of the spar?

5-21 Same as Prob. 5-20 except that neither the shear stress nor normal stress can exceed 300 psi.

5-22 Two pieces of brass are fused together along a V-shaped surface as shown. What are the normal and shear stresses on the fused surface when a tensile load of 6 tons is applied?

PROB. 5-22

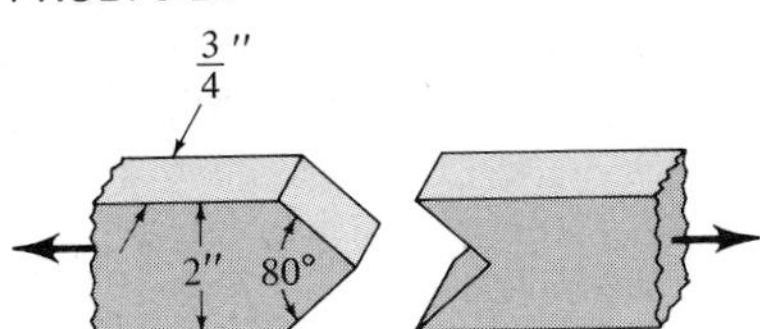

5-23 Refer to Prob. 5-13. The horizontal braces have an elliptical-shaped cross section with major diameter 0.5 in and minor diameter 0.2 in. What are the largest normal and shearing stresses in the braces?

5-24 Refer to Prob. 5-14. All the members of the truss have a standard I-shaped cross section S4 × 7.7 (see Table 3 of Appendix B). Determine the maximum normal and shear stresses at the sections indicated.

5-25 Refer to Prob. 5-18. Member *ab* is $\frac{5}{8}$ in thick and 2 in wide. Find the maximum normal and shear stress at section ②.

5-26 Refer to Prob. 5-25. The pins at *a* and *b* have a diameter of $\frac{3}{4}$ in. Find (*a*) the nominal shear stress in the pin at *b*, and (*b*) the nominal bearing stress between the pin at *a* and member *ab*.

5-27 The left rear wheel of a 4,200-lb automobile has been removed, and the brake housing is resting on a 4 × 4-in block sitting as shown. What is the shear stress parallel to the wood grain? What is the normal stress perpendicular to the wood grain? Assume the front wheels support 60 percent of the weight of the automobile.

PROB. 5-27

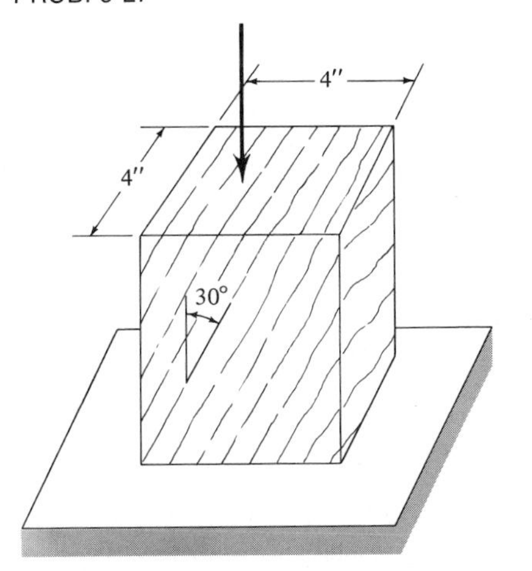

PROB. 5-28

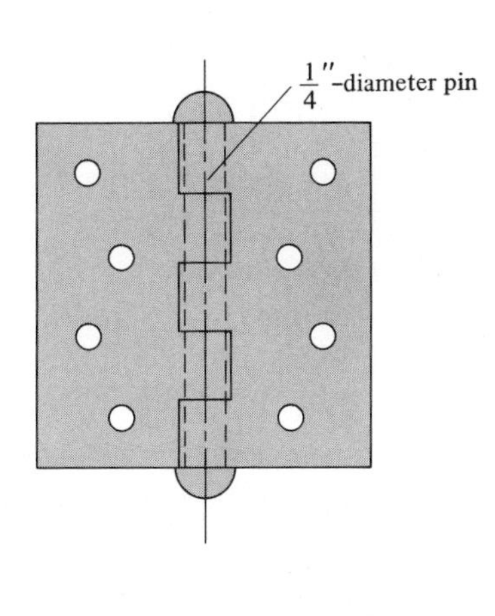

5-28 A 6-ft 8-in by 3-ft 0-in metal door weighs 234 lb and is hung with two hinges, one being 1 ft from the bottom and the other 1 ft from the top. The hinge detail is shown in the figure. Find the nominal shear stress in the top hinge pin.

5-29 A reasonably flat $4\frac{1}{2}$-lb picture is to be mounted on a vertical wall using $\frac{3}{4}$-in-wide two-sided sticky tape. What minimum length of the sticky tape must be applied to the back of the picture if the shear stress on the stickum cannot exceed 0.5 psi?

5-30 The T bolt in the clamping device shown is tightened until the tensile stress in the unthreaded portion is 9,400 psi. If the slot is just wide enough to accommodate the bolt, what is the nominal bearing stress between the slot and bolt head? What is the nominal shear stress in the bolt head?

PROB. 5-30

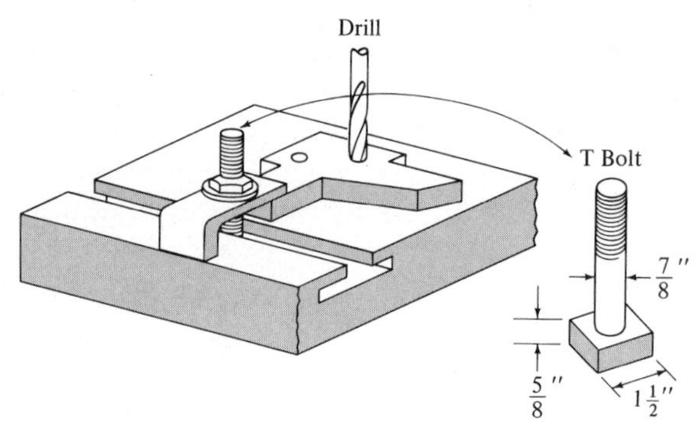

5-31 The solid $\frac{3}{4}$-in-diameter shaft is held in the hollow sleeve by a $\frac{1}{8}$-in-diameter steel pin. Estimate the nominal shear stress in the pin due to the combined effects of the 600-lb tensile force and the 350-in-lb twisting couple. What is the tensile stress in the sleeve due to the 600-lb force?

PROB. 5-31

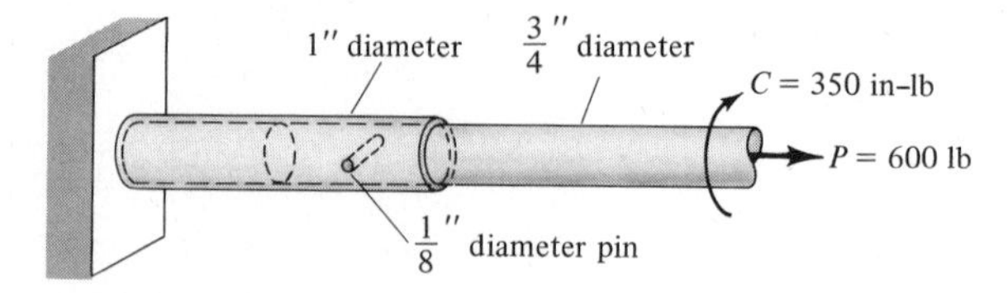

5-32 The $\frac{1}{4}$-in-diameter stud bolt has a tensile stress of 28,000 psi. If the nominal bearing stress between the washer B and the wood block C is not to exceed 500 psi, what minimum size washer is needed?

PROB. 5-32

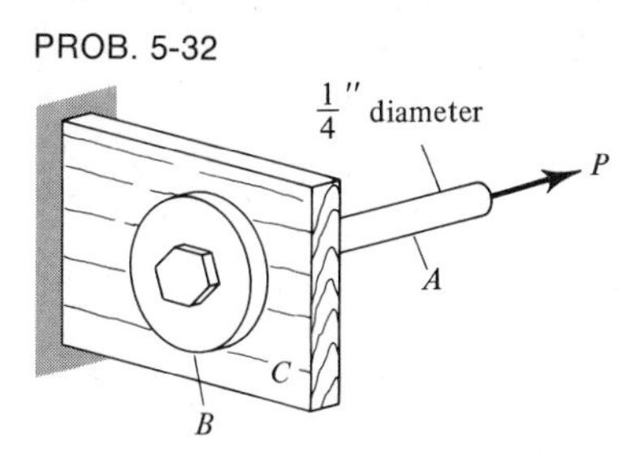

PROB. 5-34

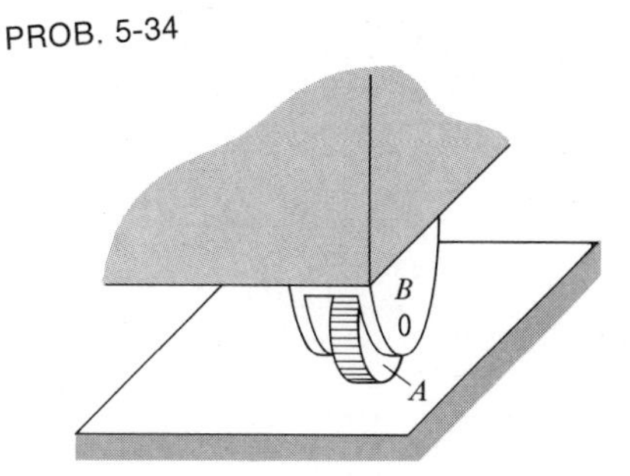

5-33 Refer to Prob. 5-32. If the nominal shear stress in the steel-bolt head is not to exceed 10,000 psi, what minimum bolt-head thickness is required?

5-34 The caster shown supports a load of 800 lb. How thick should the support plates B be if the nominal bearing stress between them and the $\frac{3}{8}$-in-diameter axle is not to exceed 4,000 psi? Find the shear stress in the axle.

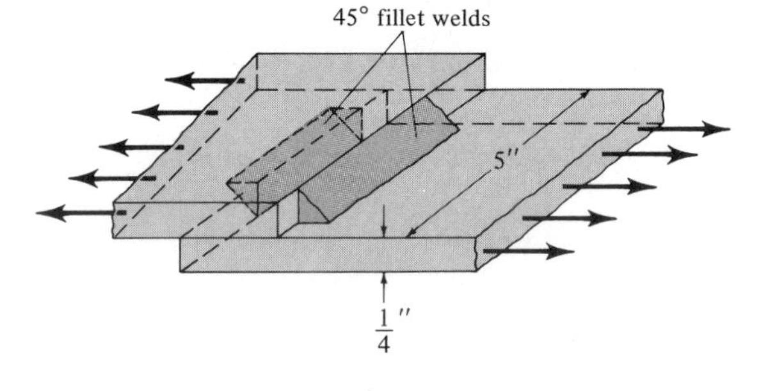

PROB. 5-36

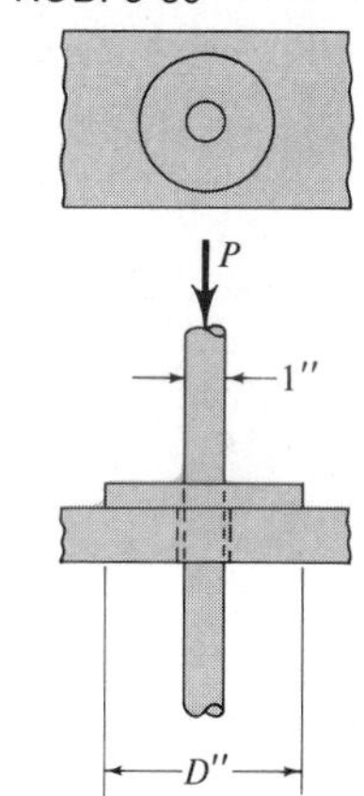

5-35 The two overlapped $\frac{1}{4}$-in plates are 45° fillet welded on top and bottom as shown. If the tensile stress in the plate is 16,000 psi and the nominal shear stress in the welds cannot exceed 10,000 psi, how many total inches of weld are required?

5-36 The vertical shaft shown is supported by a collar bearing and is loaded in compression so that the compressive stress in the shaft is 24,000 psi and the nominal bearing stress between the collar and the bearing is 3,000 psi. Determine the diameter D of the collar.

5-37 For the fixture shown, find the nominal shear stress in the 45° fillet weld. Find the shear stress in the two 1-in-diameter rivets.

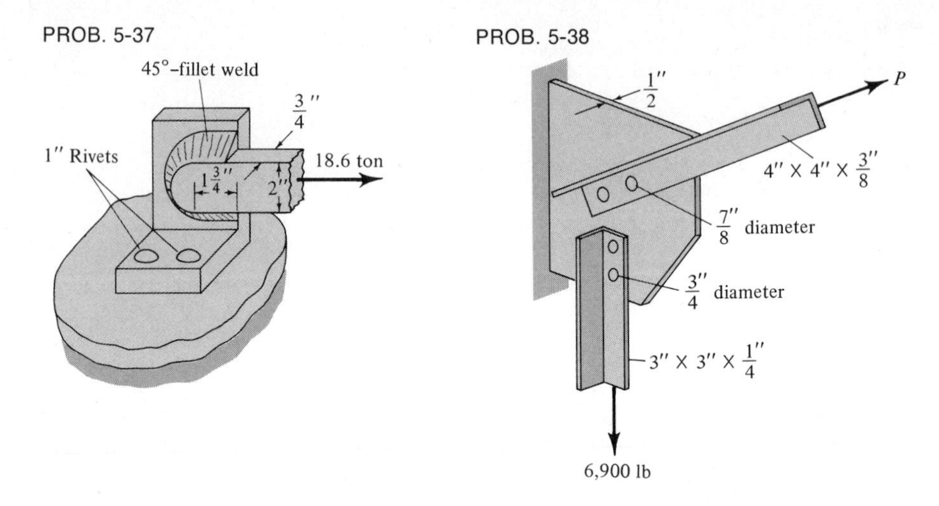

5-38 Find the largest load P if the maximum allowable nominal shear stress in the two $\frac{7}{8}$-in-diameter rivets is 15,000 psi. Assume each rivet supports half the load. Estimate the largest nominal tensile stress in the $4 \times 4 \times \frac{3}{8}$ in angle piece.

5-39 Refer to the figure for Prob. 5-38. Find the bearing stress between the lower angle piece and the $\frac{3}{4}$-in-diameter rivets. Assume each rivet supports half the load. Estimate the largest nominal tensile stress in the angle piece.

5-40 Details of the clevis-type connector at a and b are shown in the figure. Find the nominal shear stress in the pin at a. Find the nominal tensile stress in the clevis at b.

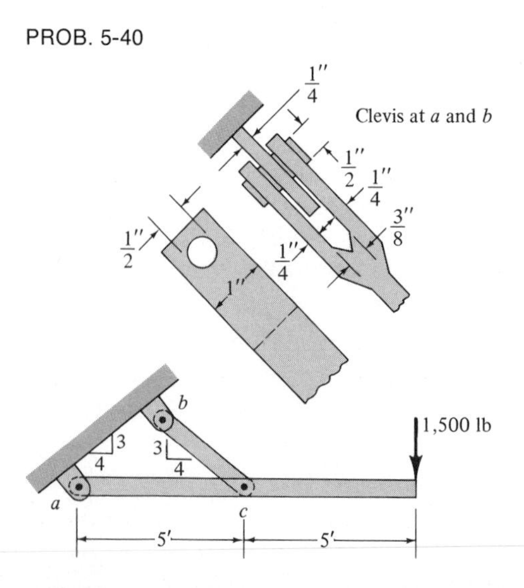

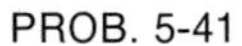

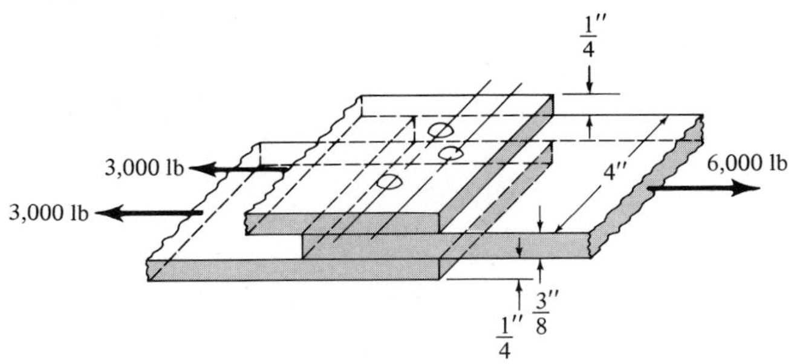

5-41 The three plates are held together by three $\frac{1}{2}$-in-diameter rivets. Find the nominal shear stress in the rivets. What is the largest nominal tensile stress in the $\frac{1}{4}$-in plates? What is the largest nominal tensile stress in the $\frac{3}{8}$-in plate?

chapter 6 AXIAL STRESS AND STRAIN: MECHANICAL PROPERTIES

6-1 INTRODUCTION

We know from experience that when a force acts on a real body which is restrained in some manner, quite often the most pronounced physical effect is that the body deforms in some manner such as stretching, bending, or twisting (refer to Fig. 5-5). How much the body deforms depends upon how big the force is, the size and shape of the body, and, last but not least, the material of which the body is made. In this chapter we will address ourselves primarily to this third factor.

What do you mean when you say one material is "stronger" than another material, or that this material is "soft" while that material is "hard"? Certainly such adjectives as "strong," "soft," "hard," etc., are quite acceptable as everyday terminology, but an engineer must be more precise and definite in comparing the characteristics of two materials. He must define acceptable, quantitative measures of such terms as strength, hardness, stiffness, etc.

In the previous chapter, we introduced the concept of stress as an internal force per unit area and indicated that it was in some way a measure of the strength of a structural member. We will now pursue more fully this idea of using stress and other indices as quantitative measures of the structural characteristics of materials. Before we do this, however, the following brief philosophical statement might help to explain the engineering approach to the study of materials.

What gives a *solid* body its structural ability to transmit various types of forces as distinguished, say, from a body of *fluid* which does not possess this structural ability? The answer to this question is, of course, tied to the molecular and atomic structure of the material, and such questions are properly the domain of chemists, metallurgists, etc. However, as engineering mechanicians, we are primarily interested in the *fact* that a material possesses certain structural characteristics rather than *why* it has them. Putting it another way, in engineering mechanics we deal with the overall phenomenological behavior of materials rather than with the molecular and atomic mechanisms which produce this behavior. Our objective, then, is to develop a scheme to describe and analyze this overall phenomenological behavior of materials.

6-2 THE STRESS-STRAIN CURVE

The structural characteristics, or mechanical properties, of a material are usually measured by means of controlled laboratory experiments on representative specimens of the material, in which loads are applied and the resulting deformations are measured. In analyzing such experiments, the first thing we must recognize is that if some quantity is to be a measure of a certain structural characteristic of a material, this quantity must be reproducible in that its value should be the same for *different specimens* of the *same material*, provided, of course, that the experimental procedure and material sample are essentially the same. By "essentially the same" we mean that the experimental environment (temperature, humidity, etc.), the manner of loading (rapid, slow, etc.), the type of loading (tensile, shear, bending, twisting, etc.), and the geometry of the experiment (specimen shape, grain orientation, etc.) are within reasonable limits of being the same, while such things as the machines or devices used to apply the loads, the actual size of the loads and specimens, and the person performing the experiment are of no significant influence on the mechanical property being measured. (Idealistically, a measured property of a material should be totally independent of all experimental conditions, but, as a practical matter, it is usually necessary to control certain factors in experiments which measure properties of materials.)

The simplest and most common experiment for measuring some mechanical properties of solid materials is the uniaxial stress-strain test. In this test, a centroidally loaded two-force member is subjected to a slowly increasing tensile (or compressive) load, and the corresponding elongation (or contraction) parallel to the load is measured for the corresponding load, either continu-

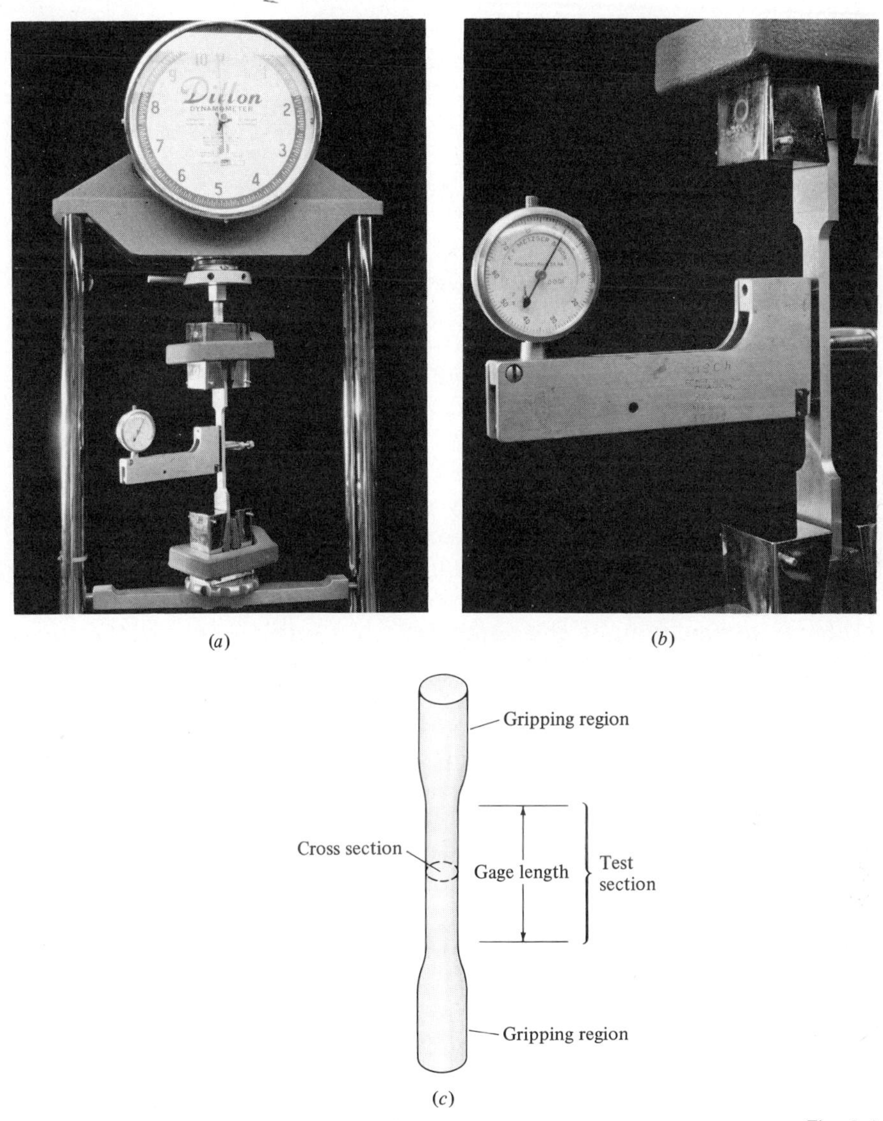

Fig. 6-1

ously or at some prescribed intervals. In practice, the tests are conducted in specially designed machines in which one end of the test specimen is fastened to a hydraulic or screw-activated load ram and the other end to a load-sensing fixture, such as shown in the photograph in Fig. 6-1(*a*). The specimen is usually shaped similarly to that shown in Fig. 6-1(*b*) or (*c*), with the ends of the specimen significantly thicker than the main test section. This is done to aid in gripping the specimen and to minimize the possibility of premature breaking of the specimen due to damage in the gripping region. Usually, the test section is of some prescribed length, such as 2 in, or 8 in, called the *gage length*, and its cross section may be circular or rectangular. The gripping ends may or may not be threaded, depending upon the type of gripping device being used.

In the basic test two things are recorded during the test:

1. The load on the specimen
2. The elongation (or contraction) of the gage length of the test section

In the more modern machines these data are recorded automatically and continuously in the form of a load vs. deformation curve, whose general shape will appear similar to that shown in Fig. 6-2(*a*). No actual numbers are indicated in this figure for a very important reason. For any given material being tested, the load and the deformation data would greatly depend upon the size of the test specimen. For example, intuitively we would expect a specimen of 8-in gage length to stretch in the order of four times as much as a specimen of the same area but having only a 2-in gage length. Similarly, we would expect that a thick specimen would require more load than a thin specimen to produce the same amount of deformation. Hence, while a load-deformation curve might give a qualitative indication of the behavior of the material, such a curve does not give a quantitative measure of the characteristics of the material. In other words, while we might conclude from Fig. 6-2(*a*) that the tested material seemed to behave linearly during the early portion of the test from 0 to ① and that it broke at ②, the available data in this form do not enable us to quantify this behavior in a precise manner since points ① and ② will depend upon the size of the test specimen. For example, Fig. 6-2(*b*) might well be the data from another test on the *same* material with a specimen which was thinner and longer than the specimen used in Fig. 6-2(*a*).

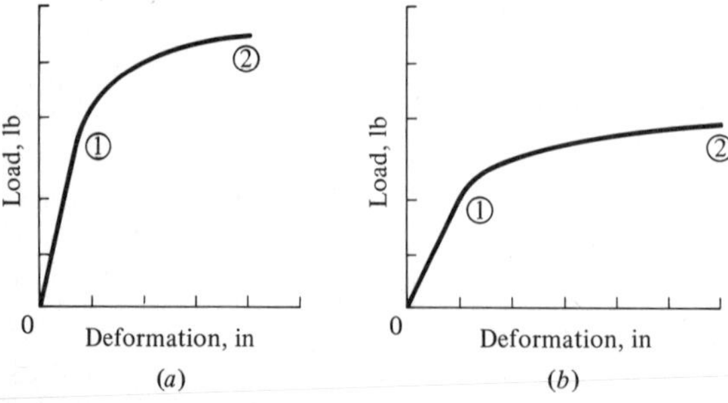

Fig. 6-2

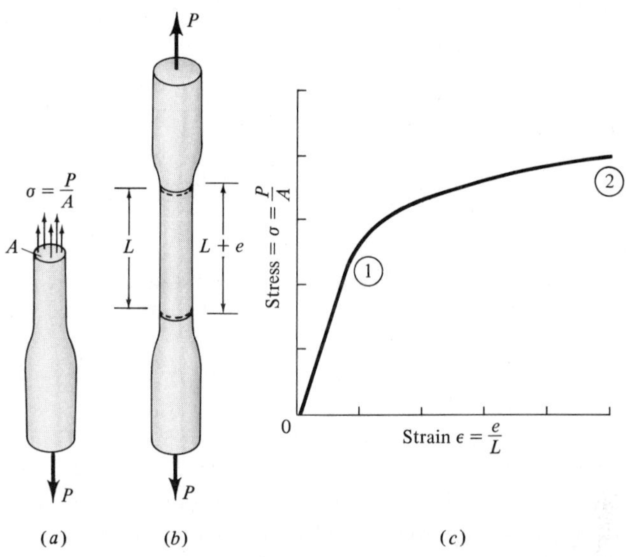

Fig. 6-3

From the foregoing discussion, it is apparent that to quantify the results of a load-deformation test, we must eliminate the dependence upon the dimensions of the test specimen; that is, we must reduce the data to some common denominators independent of the specimen size. To this end, we reduce the load scale to a *load per unit area* by dividing the values of the load by the cross-sectional area of the test section. In view of the discussion of stress in the previous chapter and the fact that the specimen is assumed to be centroidally loaded, we are rationally justified in calling this load per unit area the *axial stress* (normal stress) in the test section [recall Eq. (5-10) and Theorem 5-1]. Thus we write

$$\text{Stress} = \sigma = \frac{\text{load}}{\text{cross-sectional area}} = \frac{P}{A} \tag{6-1}$$

as illustrated in Fig. 6-3(a). (*Note:* Traditionally, P rather than F has been used to denote axial load.)

In a similar manner, we reduce the deformation scale to a *deformation per unit length* by dividing the value of the deformation by the gage length of the test section. This deformation per unit length is called the *axial strain*, and we write

$$\text{Strain} = \epsilon = \frac{\text{deformation}}{\text{gage length}} = \frac{e}{L} \tag{6-2}$$

as illustrated (though exaggerated) in Fig. 6-3(b).

Therefore, if our reasoning is correct and we were to reduce the original load-deformation data in Fig. 6-2(a) and (b) to corresponding stress-strain data, *both* load-deformation curves would produce the *same*[1] stress-strain curve indicated in Fig. 6-3(*c*). Within reasonable limitations, this is indeed the case, and a stress-strain curve for a structural material is a very reproducible quantitative indication of its mechanical behavior.

Some of the limitations can be anticipated. The test specimen cannot be so small that its behavior would not be indicative of the bulk behavior of the material, as, for example, a single crystal of graphite would not indicate the behavior of a large block of crystalline graphite. Also, as mentioned earlier, the test environment and rate of loading must be reasonably controlled within some prescribed bounds. These factors are particularly significant in the testing of biological materials and other nonmetallic materials. Furthermore, two specimens of a nonhomogeneous material might logically be expected to behave quite differently, as, for example, two specimens of pine wood with distinctly different grain size and orientation. However, despite these limitations, the uniaxial stress-strain curve is the most widely accepted indication of the quantitative mechanical behavior of materials.

6-3 MECHANICAL PROPERTIES

We begin this discussion of the mechanical properties of materials obtained via a uniaxial load-deformation test by observing some typical stress-strain curves. Figure 6-4 shows a typical tensile stress-strain curve for (1) hot-rolled steel and (2) an aluminum alloy; and a compressive stress-strain curve for (3) birchwood and (4) concrete. Carefully observe the stress and strain scales for each of these curves to gain some feel for the orders of magnitude involved.

Even though these curves are distinctly different, most of them have common qualitative features, although there are exceptions. First, each of the curves except the one for concrete appears to have an initial linear portion in which the stress and strain are directly proportional to each other. Second, after the initial linearity, the curves tend to become concave downward, indicating that the strain increases in a greater proportion than the corresponding increase in stress; that is, the material tends to deform more easily at the higher stresses. Finally, each of the curves reaches some maximum value of stress and/or strain, at which time the test data were discontinued either purposely or because of some type of failure, such as rupture of the test specimen. Assuming the latter case, the rupture point is indicative of the behavior of the

[1] Except near the rupture point, depending upon whether or not rupture of the specimen occurs within the test gage length.

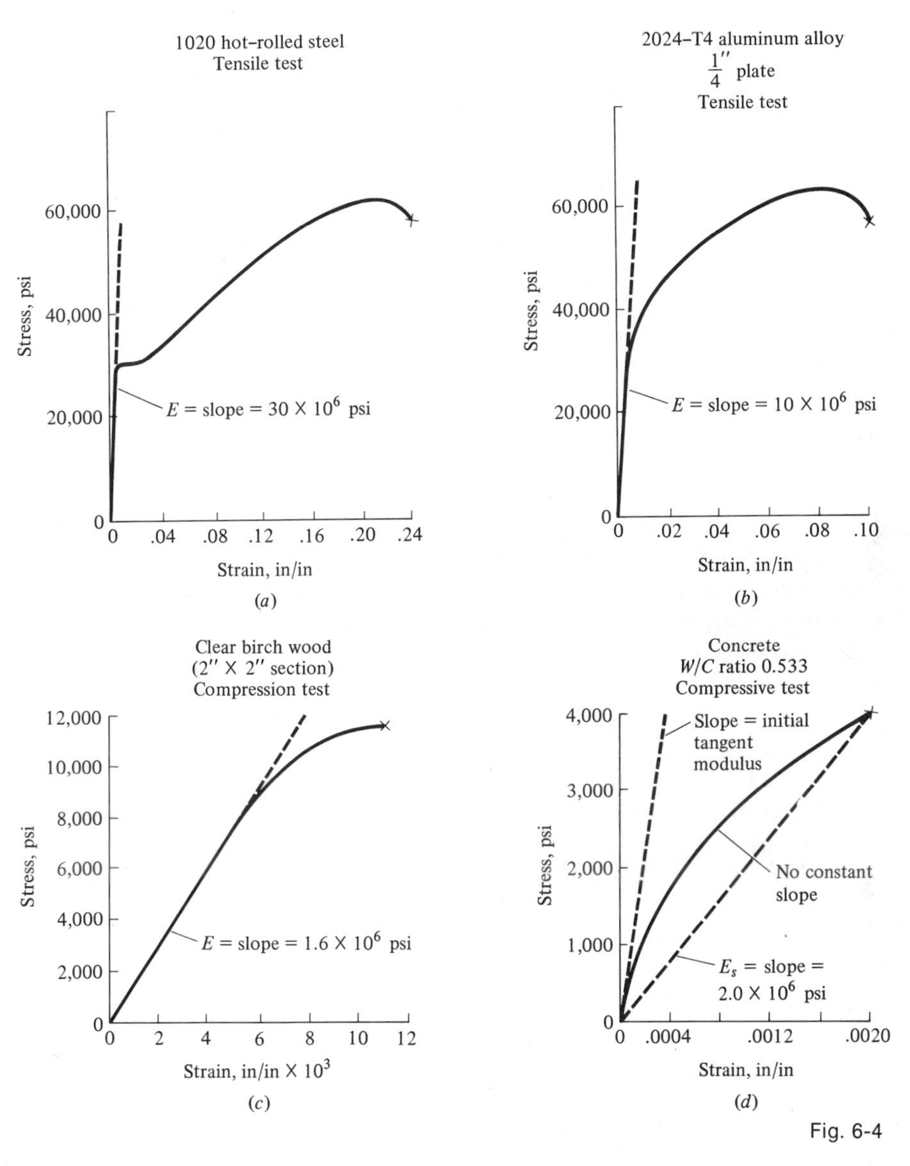

Fig. 6-4

material. We will now attempt to quantify each of these qualitative observations by defining some measurable quantities, which are called mechanical properties of the materials.

Modulus of Elasticity (Young's Modulus)

The slope of the initial linear portion of the stress-strain curve is

$$E = \frac{\sigma}{\epsilon} \qquad \frac{\text{lb/in}^2}{\text{in/in}} = \text{lb/in}^2 \tag{6-3}$$

The elastic modulus E is considered to be a reproducible measure of the initial *stiffness* of the material. Some typical values are

$$
\begin{aligned}
E &= 30 \times 10^6 \text{ psi for steel (tensile)} \\
&= 10 \times 10^6 \text{ psi for aluminum (tensile)} \\
&= 15 \times 10^6 \text{ psi for brass (tensile)} \\
&= 1 \text{ to } 3 \times 10^6 \text{ psi for wood (compressive)} \\
&= 2 \text{ to } 3 \times 10^6 \text{ psi for concrete (compressive)} \\
&= 2.0 \text{ to } 4.0 \times 10^6 \text{ psi for human bone (compressive)}
\end{aligned}
$$

Hence, it is appropriate to say that steel is "stiffer" than wood or concrete, etc.

Proportional Limit

The proportional limit is the point at which linearity ceases. The corresponding values of stress and strain are actually somewhat difficult to ascertain for some materials as they involve some judgment as to exactly where the linearity ends and nonlinearity begins. However, despite this questionable reproducibility, values for the proportional-limit stress are often listed as mechanical properties for many materials.

> ***Remark:*** In the case of hot-rolled steel [Fig. 6-4(*a*)], the end of linearity is marked by a pronounced flattening of the stress-strain curve, which we call the *yield phenomenon.* This sudden loss of stiffness is associated with an instability in the crystalline structure of the material and is peculiar to hot-rolled steel. The stress level at which this phenomenon occurs is called the *yield-point stress.*

Tangent Modulus

The slope of the stress-strain curve beyond the proportional limit is

$$E_t = \frac{\Delta\sigma}{\Delta\epsilon} \qquad \frac{\text{lb/in}^2}{\text{in/in}} = \text{lb/in}^2 \tag{6-4}$$

The tangent modulus is indicative of the incremental stress-strain relation and is, therefore, also a measure of the stiffness of the material. Its value relative to the elastic modulus indicates how rapidly the stiffness of the material decreases as the stress level increases. Accordingly, there is no one value for the tangent modulus of a material, and, therefore, it is usually not listed in a table of material properties (Table 1 of Appendix B), although sometimes for nonlinear materials, such as in Fig. 6-4(*d*), the *initial tangent modulus* is listed as the value for the *modulus of elasticity.* The tangent modulus often plays an important role in structural design and is a very useful measure of a material's mechanical response.

Secant Modulus

Another useful measure of the decreasing stiffness of a material is the secant modulus, which is the ratio of stress to strain for all levels of stress and strain:

$$E_s = \frac{\sigma}{\epsilon} \qquad \frac{\text{lb/in}^2}{\text{in/in}} = \text{lb/in}^2 \tag{6-5}$$

The difference between the three measures of stiffness E, E_t, and E_s is illustrated in Fig. 6-5.

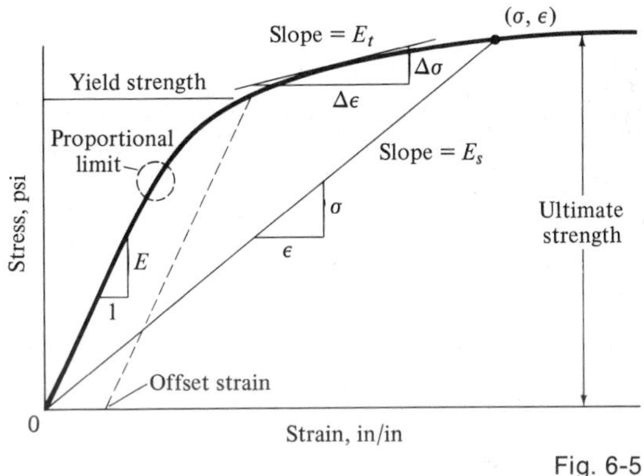

Fig. 6-5

Yield Strength

The yield strength is defined as the stress level corresponding to a prescribed offset strain, usually 0.2 percent (0.002 in/in). Referring to Fig. 6-5, at the prescribed offset strain of 0.2 percent, a projection line is drawn parallel to the initial linear portion, and its intersection with the actual stress-strain curve identifies the value of the yield strength of the material. This quantity, being very reproducible, is usually listed in tables of mechanical properties in lieu of the less reproducible proportional-limit stress. The yield strength is very useful as a design parameter since it represents a liberal upper bound on the proportional limit while at the same time is usually a good measure of the usable strength of the material for design purposes.

Ultimate Strength

The ultimate strength is defined as the highest stress level attained before rupture (or other failure of the test specimen) occurs.

This value is fairly reproducible so long as the failure occurs within the test section of the specimen [refer to Fig. 6-1(*c*)]. However, variations in this value can have many causes, such as a local imperfection or damage in the material. Also, this value is sensitive to the rate of loading and the geometry of the test specimen. Nevertheless, its value is statistically meaningful and is, therefore, an important design parameter and mechanical property. Some typical values are

$$
\begin{aligned}
\sigma_{\text{ult}} &= 65{,}000 \text{ psi (tensile) for hot-rolled low-carbon steel} \\
&= 250{,}000 \text{ psi (tensile) for heat-treated alloy steel} \\
&= 4{,}000 \text{ to } 10{,}000 \text{ psi (compression) for wood} \\
&= 2{,}000 \text{ to } 5{,}000 \text{ psi (compression) for concrete} \\
&= 8{,}000 \text{ to } 16{,}000 \text{ psi (compression) for human skull bone}
\end{aligned}
$$

Summarizing, most uniaxial stress-strain curves can be characterized by an initial linear portion with a well-defined slope, a nonlinear portion representing a steadily decreasing stiffness, and an upper bound on the stress at or shortly before failure. These qualitative and quantitative features are illustrated in Fig. 6-5.

Before closing this discussion of basic mechanical properties, we make an important comment. Our discussion of stress, strain, and mechanical properties has been based on what is commonly referred to as *engineering stress and strain*. In Eqs. (6-1) and (6-2) defining axial stress and strain, respectively, the cross-sectional area A and gage length L were both based on the *undeformed* dimensions of the test specimen. During a tensile test, say, the gage length of the test section will increase while the cross section of the test section will decrease. In many structural materials, these changes in dimensions will be a very small percentage of the original dimensions throughout most of the duration of the test, as is evident from the strain values on the curves in Fig. 6-4. On the other hand, rubbery-like materials and some soft metals can deform a significant amount during such a test. For these and most fluid-like materials that tend to "flow" when subjected to forces, it is often more reasonable to base the definitions of stress and strain on the *instantaneous* deformed dimensions of the test specimen. In these cases, we use the terminology of *true stress and strain*. There are also several other ways in which stress and strain can be defined which have useful application to particular situations. However, engineering stress and strain based on the original undeformed dimensions of the test specimen are undoubtedly the most widely used in discussing structural materials.

EXERCISE PROBLEMS

6-1 The following are data from a tensile test at room temperature of a specimen of Plexiglas. Plot the stress-strain curve. Use a straight edge for the linear portion and an irregular curve for the nonlinear portion. Take great care in plotting the transition near the proportional-limit stress.

Stress, psi	Strain, in/in	Stress, psi	Strain, in/in
0	0	6,250	0.020
1,000	0.002	6,650	0.024
1,900	0.004	7,000	0.028
2,750	0.006	7,350	0.032
3,650	0.008	7,600	0.036
4,500	0.010	7,900	0.040
5,100	0.012	8,150	0.044
5,450	0.014	8,400	0.048
5,750	0.016	8,500*	0.050

* Rupture.

6-2 Determine the proportional-limit stress, the yield-strength stress (0.2 percent offset), and the modulus of elasticity for the material of Prob. 6-1.

6-3 For the material of Prob. 6-1, determine the tangent modulus and the secant modulus for a stress of 7,000 psi.

6-4 Following are data from a tensile test of a specimen of heat-treated steel (0.62 percent carbon). Plot the stress-strain curve. Use a straight edge for the linear portion and an irregular curve for the nonlinear portion. Take great care in the transition near the proportional-limit stress.

Stress, psi	Strain, in/in
0	0
20,000	0.0007
40,000	0.0014
60,000	0.0020
80,000	0.0027
90,000	0.0030
100,000	0.0035
105,000	0.0040
110,000	0.0045
118,000	0.0060
123,000	0.0070
128,000	0.0080
132,000	0.0090
134,000	0.0100
136,000	0.0110
138,000*	0.0120

* Rupture stress.

6-5 Find the proportional-limit stress, the elastic modulus, and the yield strength for an offset of 0.2 percent for the material of Prob. 6-4.

6-6 For the material of Prob. 6-4, determine the tangent modulus and the secant modulus for a stress of 130,000 psi.

6-7 The following are data from a tensile test of a specimen of Class 20 gray cast iron. The specimen was of 0.505 in diameter and the deformations were measured over a 2.00-in gage length. Plot the stress-strain curve.

Load, lb	Deformation, in	Load, lb	Deformation, in
0	0	1,760	0.00240
320	0.00020	2,080	0.00360
640	0.00040	2,400	0.00500
800	0.00046	2,680	0.00680
960	0.00080	3,200	0.01040
1,120	0.00086	3,520	0.01360
1,280	0.00110	3,840	0.01760
1,440	0.00150	4,000*	0.02000
1,600	0.00200		

* Rupture load.

6-8 For the material of Prob. 6-7, determine the proportional-limit stress, the yield-strength stress (for an offset of 0.2 percent), and the modulus of elasticity.

6-9 For the material of Prob. 6-7, determine the tangent modulus and the secant modulus for a stress of 16,000 psi.

6-10 The following are data from a compressive test of a 0.25 × 1.25-in cross section of a fiber glass laminate made with 181 glass fabric and epoxy resin (MIL-R-9300 spec.). Test load was in the direction (0°) of the fabric weave. The deformations were measured over a 4-in gage length. Plot the stress-strain curve and find the modulus of elasticity and ultimate strength.

Load, lb	Deformation, in	Load, lb	Deformation, in
0	0	10,934	0.0372
1,562	0.0052	12,496	0.0436
3,124	0.0104	14,058	0.0500
4,683	0.0156	15,620	0.0568
6,248	0.0208	17,182	0.0644
7,810	0.0260	18,744	0.0728
9,372	0.0316	19,950*	0.0788

* Rupture.

6.4 FURTHER TERMINOLOGY AND MECHANICAL PROPERTIES

In addition to the basic mechanical properties defined in the previous section, there are numerous other terms and parameters which have significant use in describing the behavior of materials.

According to the dictionary, the word *elastic* implies the ability to recover, while the word *plastic* means capable of being shaped or permanently deformed. If a material test specimen is loaded and then unloaded, the specimen might or might not recover its original size and shape. Of course, such an observation would depend upon the instruments used to measure the deformations. If, for all practical purposes, there is no permanent deformation after unloading, the material is said to have behaved *elastically*, otherwise the material has behaved *plastically*. In terms of the corresponding stress-strain data, elastic behavior is depicted as in Fig. 6-6(*a*) to (*c*), while plastic behavior is depicted as in Fig. 6-6(*d*) to (*f*). The basic difference between these sets of curves is that, for elastic behavior, the plastic, permanent, or residual strain ϵ_p remaining in the specimen after unloading is zero.

Quite often, a material will behave elastically for low-level stresses, while plastic behavior will occur at the higher-level stresses. For example, perhaps Fig. 6-6(*a*) and (*d*) give data for the same material except that Fig. 6-6(*d*) was loaded to a substantially higher stress level than Fig. 6-6(*a*); similarly for Fig. 6-6(*b*) and (*e*), and for Fig. 6-6(*c*) and (*f*). That is, most materials have a stress level below which they behave elastically but above

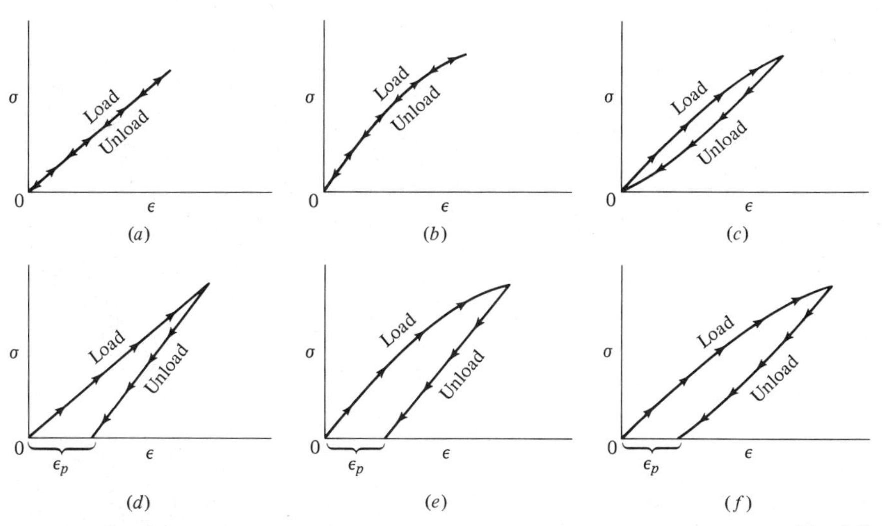

Fig. 6-6

which they begin to exhibit pronounced plastic effects. This bound for elastic behavior is called the *elastic limit.* As a practical matter, this bound is not easily determined or very reproducible in laboratory experiments, although it is a useful concept. Generally, an engineer associates elastic behavior with linear response of a material and plastic behavior with nonlinear response, although such an association does not always hold, as is evident from the curves in Fig. 6-6(*b*) and (*c*) which exhibit nonlinear elastic behavior, and Fig. 6-6(*d*) which exhibits linear plastic behavior. Since values for the elastic limit are often not available, the proportional limit or sometimes the yield strength is used as the limit for elastic behavior.

Some materials such as mild steel, soft aluminum, soft plastics, etc., can withstand a relatively large amount of plastic strain before rupture, whereas others such as cast iron, hard plastics, concrete, glass, etc., withstand relatively little plastic strain before rupture. The former are called *ductile* materials while the latter are called *brittle.* Figure 6-7(*a*) and (*b*) shows a stress-strain curve of a brittle fiber glass composite and a ductile aluminum alloy, respectively. The key word here is *relative*, as there is no specific amount of plastic strain that distinguishes a brittle material from a ductile one.

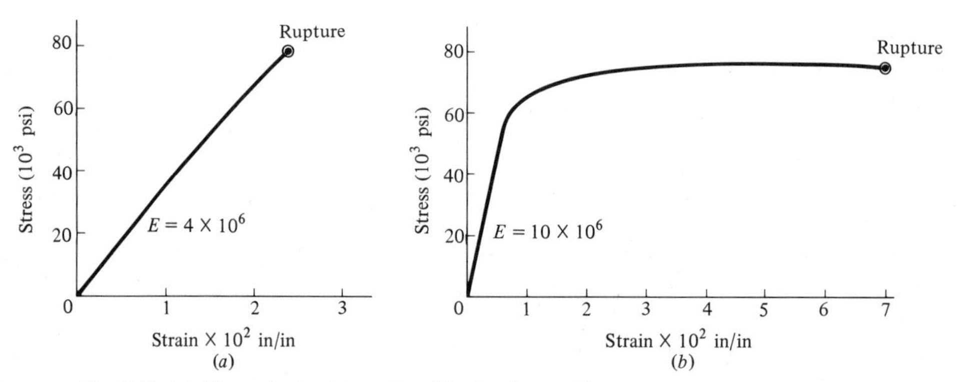

Fig. 6-7 (*a*) Fiber glass composite; (*b*) aluminum alloy.

Another structurally significant feature of a ductile material is its mode of failure compared with that of a brittle material. Generally, a brittle material ruptures very abruptly when the normal *tensile* stresses reach the ultimate stress level of the material. On the other hand, a ductile material, because of its ability to "flow" while undergoing large plastic deformations, is much more sensitive to *shear* stresses which promote this "flowing" of the material; and, in fact, the failure of a ductile material can usually be attributed to shear stresses rather than normal stresses, with final rupture being preceded by a significantly visible amount of plastic deformation. The photographs in Fig. 6-8(*a*) and (*b*) show the fracture of a test specimen of a brittle Lucite plastic and a ductile aluminum alloy, respectively. Recalling Theorem 5-1, in a tensile test the maximum tensile stresses will be on a plane cross section perpendicular to the axis of the test specimen, while the maximum shear stresses occur on a plane at 45° to the axis of the member—hence the cup-and-cone rupture surface for the ductile material.

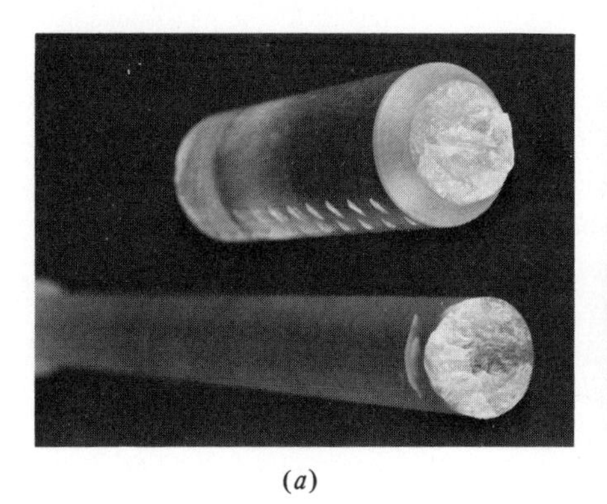

(*a*)

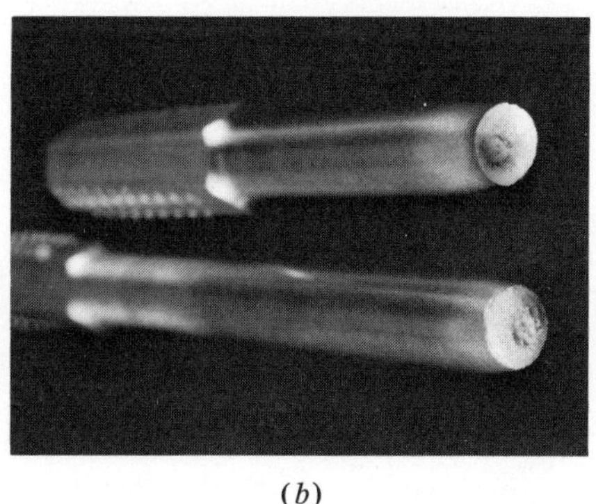

(*b*)

Fig. 6-8

The properties and parameters discussed thus far have been obtained from the axial stress-strain data obtained by measuring the axial load and deformation during a tensile or compressive test. As we mentioned at the end of the previous article, during such a test the cross-sectional area will decrease (contract) in a tensile test and increase (expand) in a compressive test. This lateral change in dimension is due to the so-called "Poisson effect," which you can readily observe by stretching a rubber band. To quantify this effect, we can measure the original lateral or transverse dimensions of the test specimen and, during the test, record the changes in these dimensions. Accordingly, we can define the transverse strain by

$$\epsilon_{\text{transverse}} = \frac{\text{transverse deformation}}{\text{original transverse length}} \tag{6-6}$$

We should observe that for any given axial direction, there are an infinity of directions which are transverse (perpendicular) to the axial one. Thus there might be an infinity of such transverse strains, as defined by Eq. (6-6). Fortunately, most structural materials exhibit the same response in all directions, so that Eq. (6-6) yields the same value regardless of which transverse direction is considered. Such materials are called *isotropic*, and for them we can talk about *the* transverse strain for a uniaxial stress-strain test. For a tensile test in which the axial strain is taken to be positive, the transverse strain will be contractive, which we take to be a negative strain. Conversely, the opposite signs would prevail in a compressive test. With this sign convention we define the *Poisson ratio* μ by

$$\mu = -\frac{\epsilon_{\text{transverse}}}{\epsilon_{\text{axial}}} \tag{6-7}$$

for a uniaxial tensile or compressive test. Carefully note that this is a ratio of *strains*, not deformations, and that with the minus sign in Eq. (6-7) μ should be a positive number. Indeed, for most materials, μ will have a value somewhere between 0.2 and 0.5, with 0.3 being rather common for elastic behavior and approaching 0.5 for extensive plastic behavior. Poisson's ratio plays a very important role in problems dealing with combined states of stress, which we will consider in some detail in Chap. 8. Our reason for introducing it now is that it is a mechanical parameter which can be obtained experimentally from a uniaxial stress-strain test.

Before closing this section, it should be recalled that this discussion of mechanical response and properties has been within the context of so-called "standard stress-strain tests" in which a slowly increasing (quasi-static) load is applied to a reasonably stable solid-like material. For other rheological materials that exhibit fluid-like characteristics, other types of tests must be designed and conducted to measure the significant response parameters, whatever they may be, such as viscosity, strain rate, etc. However, even for solid-like materials, the rate of loading can greatly influence the mechanical response, as exemplified by Silly Putty, which stretches like taffy when pulled slowly but shatters like glass when hit with a hammer. Although most structural materials are not this extreme in their range of response, an engineer must be cognizant of possible differences between the dynamical response of a material as compared with its statical response.

6-5 STRAIN ENERGY

The last paragraph of the preceding section raised the possibility that factors other than stress or strain might be significant in measuring the mechanical response of materials, particularly insofar as dynamical applications are concerned. One such basic and very useful concept is that of strain energy.

We begin by giving the following definition:

Definition 6-1 *Work* is done when a force $\mathbf{F}$ produces a displacement $\mathbf{u}$, with the amount of work given by

$$W = \mathbf{F} \cdot \mathbf{u} = \|\mathbf{F}\| \, \|\mathbf{u}\| \cos(\mathbf{F},\mathbf{u}) \tag{6-8}$$

Accordingly, work is a scalar quantity (positive or negative) resulting from a combination of a vector force acting through some vector displacement.

Maximum work occurs when the force $\mathbf{F}$ and displacement $\mathbf{u}$ are parallel, while no work occurs when the force and displacement are perpendicular. In many situations, the force varies with the displacement; for example, when you stretch a coil spring, you must increase the force as you increase the elongation. For these situations we must formally modify our definition as follows:

Definition 6-1*a* The work done by a varying force $\mathbf{F}_u$ as it produces the displacement $\mathbf{u}$ is given by

$$W = \int_{\mathbf{o}}^{\mathbf{u}} \mathbf{F}_u \cdot d\mathbf{u} \tag{6-9}$$

with the scalar (dot) product having its usual interpretation (see Chap. 1).

Let us apply this definition to the force-deformation situation that exists in the test region of a specimen of material during a uniaxial tensile (or compressive) test. Referring to Fig. 6-9, for convenience assume that the lower end of the specimen does not move, so that the force $-\mathbf{P}_e$ on that end does not move through any

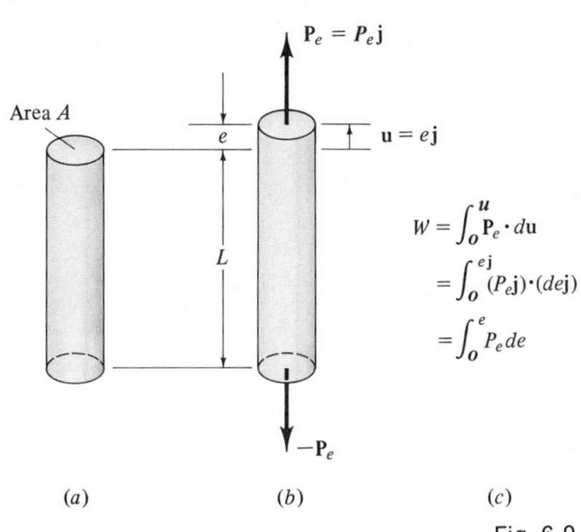

Fig. 6-9

displacement and hence does no work. On the other hand, applying Eq. (6-9) to the force $\mathbf{P}_e$ as it increases with the deformation e, and recalling the definition of the unit vector $\mathbf{j}$, the work done on the test section is given by the integral expression in Fig. 6-9(c). If the load-deformation data were continuously recorded during the test, this integral would represent the *area* under the load-deformation curve up to whatever final deformation (or load) is reached, as indicated in Fig. 6-10.

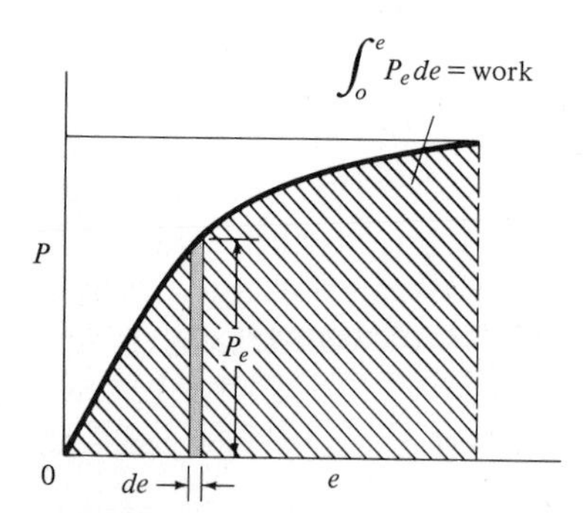

Fig. 6-10

Just as in our earlier discussions in this chapter, the load-deformation curve and, therefore, its area would greatly depend upon the size of the test specimen. Hence, the work, as given by this area, would be rather meaningless as a material parameter. To overcome this dilemma, we give the following definition:

Definition 6-2 The *stress work per unit volume* or *strain-energy density*, denoted by U, is the mechanical work per unit volume done by the applied forces in causing a body to deform.

Applying this definition to the case of the uniaxial load-deformation test, referring to Fig. 6-9, we have

$$U = \frac{\text{work}}{\text{volume}} = \frac{\int_o^e P_e\, de}{(A)(L)}$$
$$= \int_o^e \frac{P_e\, de}{AL}$$

But, by the definitions of axial stress [Eq. (6-1)] and axial strain [Eq. (6-2)], we see that

$$U = \int_o^\epsilon \sigma_\epsilon\, d\epsilon \qquad (6\text{-}10)$$

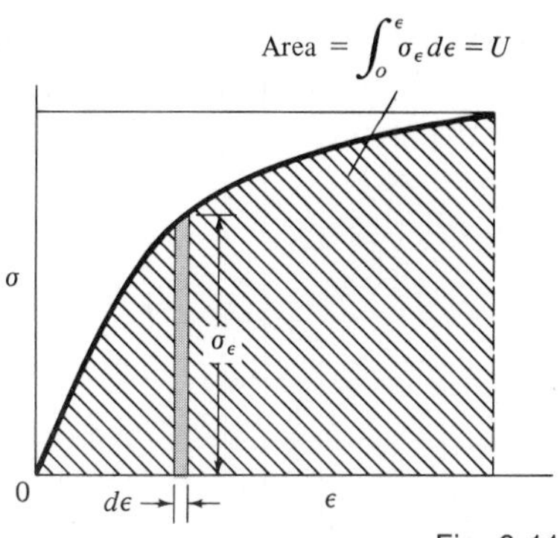

Fig. 6-11

where σ_ϵ denotes the fact that the axial stress σ varies with the axial strain ϵ. For a uniaxial stress-strain curve, such as the one depicted in Fig. 6-11, the integral in Eq. (6-10) corresponds to the *area* under the stress-strain curve up to whatever strain (or stress) level is reached. Hence we arrive at the following:

Theorem 6-1 The strain-energy density or stress work per unit volume for a uniaxial test is measured by the *area* under the stress-strain curve. Its units are commonly given as inch-pounds of energy per cubic inch of material.

As a mechanical parameter, the strain energy is a measure of the ability of a material to absorb energy. This property is of particular importance in situations involving dynamic forces resulting from an impact between two deformable bodies.

Modulus of Resilience

Given a uniaxial stress-strain curve for a material, the area under the *linear* portion of the curve is called the *modulus of resilience R*. It is considered to be a measure of the *elastic* strain energy of the material.

Modulus of Toughness

Given a complete uniaxial curve all the way to rupture, the area under the *entire* curve is called the *modulus of toughness T*. It is considered to be a measure of the total strain energy the material can withstand before rupture. Its value is often obtained within sufficient accuracy by multiplying some approximate average stress by the strain at rupture. A comparison of T to R is often an indication of the ductile or brittle character of the material, with a large T to small R indicating relatively ductile behavior (see Fig. 6-12).

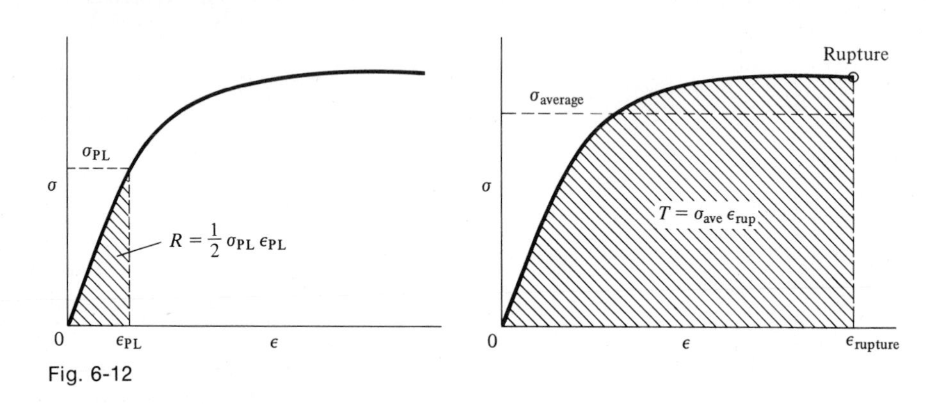

Fig. 6-12

The unique feature of strain energy as a material parameter is that it takes into account both stress and strain simultaneously rather than individually. The stress-strain curves in Fig. 6-13 illustrate that a material with a high ultimate stress might not have as much energy-absorption capacity (toughness) as another material with a substantially lower ultimate strength.

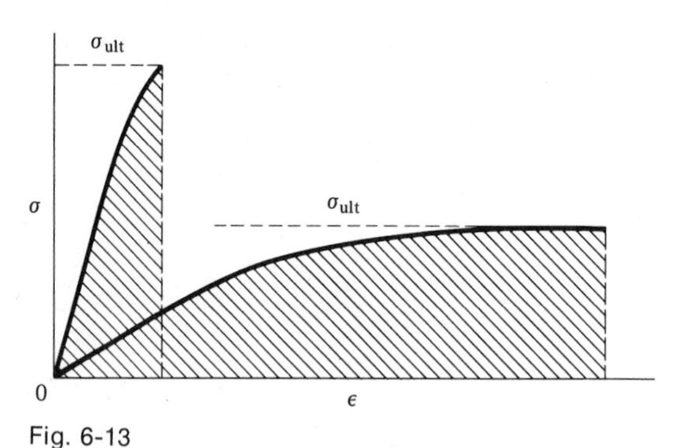

Fig. 6-13

EXERCISE PROBLEMS

6-11 At some instant during a tensile test on the specimen shown, the lengths A and B were measured to be 0.4994 and 0.5025 in, respectively. Originally, these lengths were both 0.5000 in. What is the value of Poisson's ratio for this instance?

PROB. 6-11

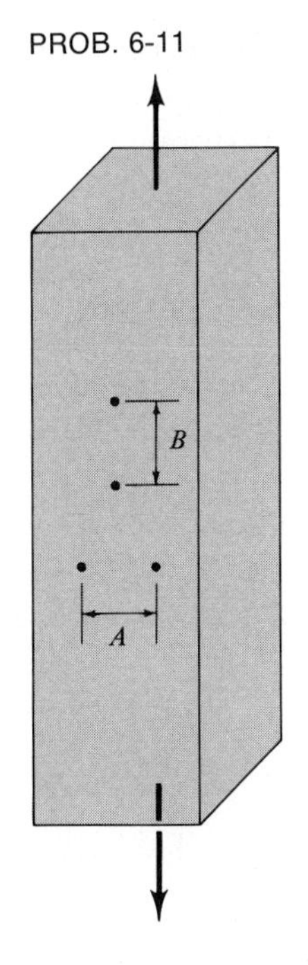

PROB. 6-12

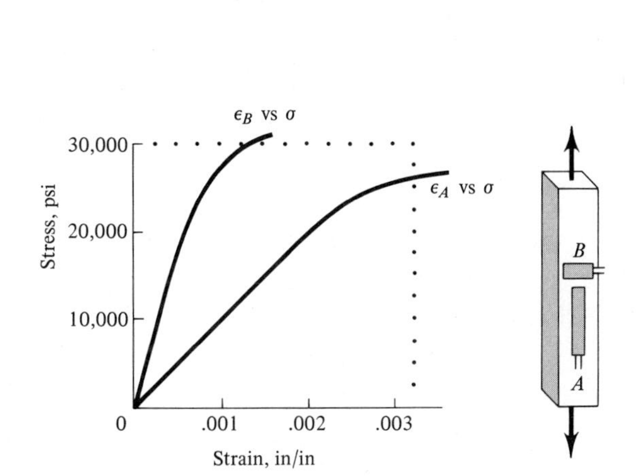

6-12 Two electrical strain gages are attached to a test specimen as shown and the corresponding stress-strain data recorded. From these data, determine (*a*) the modulus of elasticity and (*b*) the Poisson ratio.

6-13 Suppose we have a specimen of material with a Poisson ratio of $\frac{1}{3}$. In terms of the axial strain ϵ, what would be the percent change in cross-sectional area of such a specimen due to the Poisson effect during a uniaxial test?

6-14 Given a stainless-steel test specimen with original gage length of 8 in and rectangular area section $1 \times \frac{1}{4}$ in, what will be the deformed dimensions when the test load reaches 15,000 lb? (See Table 1 in Appendix B.)

6-15 For the material of Prob. 6-1, determine the modulus of resilience and the modulus of toughness.

6-16 For the material of Prob. 6-4, determine the modulus of resilience and the modulus of toughness.

6-17 For the material of Prob. 6-7, determine the modulus of resilience and the modulus of toughness.

6-18 For the material of Prob. 6-10, determine the modulus of resilience and the modulus of toughness.

6-19 Estimate the modulus of resilience and the modulus of toughness for the materials of Fig. 6-4(*a*). Comment on the relative ductility and/or brittleness of these materials.

6-20 Estimate the modulus of resilience and the modulus of toughness for the material of Fig. 6-4(*b*). Comment on the relative ductility and/or brittleness of these materials.

6-21 Estimate the modulus of resilience and the modulus of toughness for the materials of Fig. 6-4(*c*). Comment on the relative ductility and/or brittleness of these materials.

6-22 For the curve in Fig. 6-4(*d*), estimate the modulus of toughness. What about the modulus of resilience?

6-6 FAILURE AND DESIGN CRITERIA

At this point you are probably saying, "All right, now that we have all these assorted mechanical properties and parameters, what do we do with them and how do we use them in solving problems?" It is the purpose of this and the subsequent sections and chapters to put these ideas to use in solving some very fundamental problems in structural mechanics. Before we do this, however, let us briefly discuss the notion of failure.

We have said throughout this text that the primary purpose of most machines and structural elements is to transmit or support loads. When a member or device ceases to perform its function properly, we say the member or device fails. Such failure can manifest itself in several ways, which we call *modes of failure*, the most common of which for static loading are:

1. *Fracture or rupture due to an excessive static load.* We usually attribute this type of failure to a tensile, compressive, or shear stress which exceeds the corresponding ultimate stress of the material. From the discussion in Sec. 6-4, ductile materials will deform significantly before rupture, whereas brittle materials often rupture with little or no such indication of imminent failure. Accordingly, for members made of brittle materials, the ultimate strength is often used as the design criterion for determining their load-carrying capacity, whereas for ductile materials other factors might be more significant.
2. *General yielding or loss of structural stiffness.* This type of failure can occur only in a member or device made of a ductile material whose stiffness can decrease considerably before it ruptures. (Recall the definition of the tangent modu-

lus.) Such failure is usually manifested by some sort of collapse or excessively large overall deformation, with rupture being a possible secondary effect. An example of such failure would be the collapse of a steel automobile bumper during impact or the progressive collapse of an aluminum carport roof during an unexpectedly heavy snowfall. For structural members made of ductile materials, the yield strength (or yield-point stress) is usually used as the design criterion for failure due to general yielding, although, in some situations, particularly those involving dynamical loads, the modulus of resilience or modulus of toughness might be the significant failure criterion.

3 *Small but excessive deformation.* This type of failure can occur in situations in which the deformation (elastic or plastic) produced by the applied loads must be within some prescribed tolerances. For example, you would not want the crankshaft of your automobile engine to deform to any significant amount during operation, or the floors of an office building to sag significantly due to the weight of the furniture or people. All material bodies deform under applied loads, but structural elements can be designed to keep this deformation within prescribed limits. To do this, the engineer must determine the load-deformation relationship for the particular structural element involved.

4 *Buckling or sudden collapse due to structural instability.* A loaded structure is said to be *stable* if a small change in the loading results in a corresponding small change in the deformation. On the other hand, a structurally *unstable* situation is one in which a small change in load results in a sudden and dramatically significant large deformation which usually leads to a catastrophic failure of the structure. The sudden buckling of the hull of a submarine due to the ocean water pressure or the collapse of a column supporting the floor of a building are examples of this type of failure. Various types of unstable situations will be discussed in Chap. 10. Until then, we merely state that most buckling failures occur when a relatively slender structural element is required to support a relatively large compressive load. Such situations should be avoided as much as possible and otherwise carefully analyzed in the design of structures and machines.

While there are other modes of failure, these four are the most common for situations involving statical loading. Some additional factors which can be of significant importance in other situations will be considered in Sec. 6-8.

From the above discussion of modes of failure, it is apparent that the design criterion for a structural element is usually some *allowable* stress or deformation. Therefore, the engineer must be able to relate the applied loads to the resulting stresses and deformations. Then, by comparing these actual stresses and deformations to the allowable stresses and/or deformations, the engineer can make a judgment as to the probable performance of the structural element. Values for the allowable stresses are usually obtained from a stress-strain curve for the material, with the ultimate stress being the significant parameter insofar as rupture is concerned, the yield strength (or yield-point stress) the significant parameter for general yielding, and the elastic limit the significant parameter for the initiation of plastic (permanent) deformation.

We will now illustrate the application of these ideas to the design of centroidally loaded two-force members. These same ideas will be applied to other types of members, namely, shafts in Chap. 7 and beams in Chap. 9.

For a *centroidally loaded two-force member*, Theorem 5-1 gave the maximum normal stress as

$$\sigma_N = \frac{P}{A} \tag{5-13}$$

and the maximum shear stress as

$$\sigma_S = \frac{1}{2}\frac{P}{A} \tag{5-14}$$

Also, by Eq. (6-2), the axial strain is given by

$$\epsilon = \frac{e}{L} \tag{6-2}$$

Thus, Eqs. (5-13) and (5-14) relate the load to the maximum normal and shear stresses. To relate the load to the deformation, we need to relate the stress to the strain. For *elastic* (linear) behavior, Eq. (6-3) states that

$$E = \frac{\sigma}{\epsilon} \tag{6-3}$$

which, when combined with Eqs. (5-13) and (6-2), yields

$$E = \frac{P/A}{e/L} = \frac{PL}{Ae}$$

Rearranging, we have

$$e = \frac{PL}{AE} \tag{6-11}$$

which, for elastic behavior, relates the applied axial load P to the resulting axial deformation e. This is the basic load-deformation relationship for an elastic, homogeneous, axially loaded two-force member of constant cross section.

For inelastic (nonlinear) behavior, we must discard Eq. (6-3) and use the stress-strain curve itself to determine the strain corresponding to a given value of stress (or vice versa), and then use Eq. (6-2) to determine the deformation.

Equations (5-13), (6-2), and (6-11) (or its inelastic counterpart) enable us to relate the load to the resulting stresses and deformations for uniaxially loaded members. Thus, for a given two-force structural element, we can determine its mode of failure by comparing the allowable stresses and deformations with the actual stresses and deformations in the member as given by the above-noted equations.

Finally, before giving an illustrative example, we define a so-called "factor of safety."

Definition 6-3 The *factor of safety* is the ratio of the design load to the actual applied load.

A factor of safety of 2 implies that a member is designed to carry a load twice as large as the actual applied load. A factor of safety is often simply a "factor of ignorance" in that quite often the engineer must guess as to the actual loads which a device must support or how the device will be used. For example, how many ways might a screwdriver be used other than for driving screws? What loads will be exerted on the agitator of a household washing machine? The design engineer usually makes an intelligent estimate and then throws in a reasonable safety factor based on a mixture of experience, economics, and probability. We comment that overdesign can be as big a sin as underdesign.

EXAMPLE 6-1

The members of the truss shown in Fig. 6-14(*a*) are made of mild ASTM A-7 structural steel having an ultimate tensile stress of 65,000 psi and an ultimate compressive stress of 50,000 psi. Based on a factor of safety of 2.5, determine the minimum required cross-sectional area for member *eb*. How much will this member of minimum cross section deform under the actual loads?

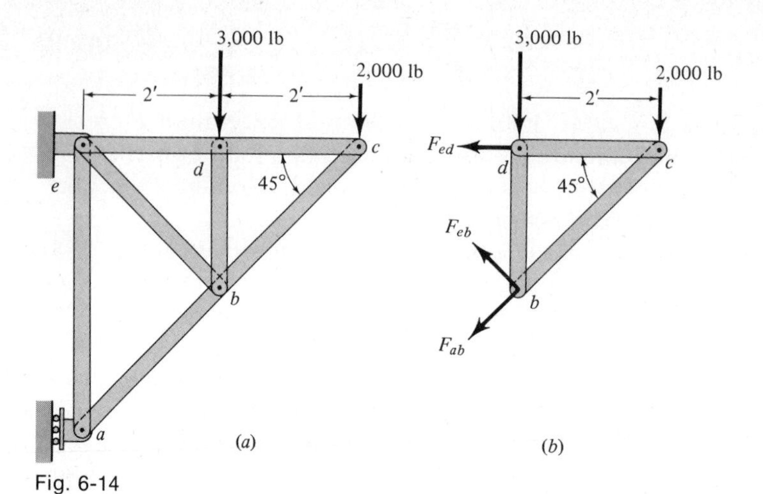

Fig. 6-14

Solution: Recall the method of sections for trusses in Sec. 4-7. Then, using the free-body diagram shown in Fig. 6-14(*b*) and taking moments with respect to point *c*, equilibrium requires

$$\Sigma \mathbf{M}_c = \mathbf{0}$$
$$= \mathbf{k}(3{,}000)(2) - \mathbf{k}(F_{eb})(2\sqrt{2})$$
$$\therefore F_{eb} = \frac{3{,}000}{\sqrt{2}} \text{ lb} \qquad \text{(as shown)}$$

Thus, the *actual* load in member *eb* is a tensile force of 2,120 lb. From the limited amount of information available, we will assume the mode of failure to be rupture, and, therefore, the failure criterion will be the *ultimate tensile stress* (since the member is in tension rather than compression). Applying the factor of safety, we will design for a load of

$$F_{\text{design}} = \text{F.S.} \times F_{\text{actual}}$$
$$= 2.5 \times \frac{3{,}000}{\sqrt{2}} = \frac{7{,}500}{\sqrt{2}} \text{ lb}$$

Then, using Eq. (5-13) or (6-1), we obtain

$$\sigma = \frac{P}{A}$$
$$\therefore \text{Area} = \frac{F_{\text{design}}}{\sigma_{\text{ult}}}$$
$$= \frac{7{,}500/\sqrt{2} \text{ lb}}{65{,}000 \text{ psi}} = 0.0815 \text{ in}^2$$

which is the minimum required area.

To find the deformation of this member, we first evaluate the *actual* stress in the member:

$$\sigma = \frac{P}{A} = \frac{3{,}000/\sqrt{2}}{0.0815} = 26{,}000 \text{ psi}$$

which is probably below the elastic-limit stress [see Fig. 6-4(*a*)]. Therefore, we can use Eq. (6-11) to determine the deformation of member *eb*. Hence,

$$\begin{aligned} e &= \frac{PL}{AE} \\ &= \frac{(3{,}000/\sqrt{2}\text{ lb})(2\sqrt{2}\text{ ft})(12\text{ in/ft})}{(0.0815\text{ in}^2)(30 \times 10^6\text{ psi})} \\ &= 0.029\text{ in} \end{aligned}$$

Remark: Although a cross-sectional area of 0.0815 in^2 is adequate for the tensile member *eb*, it is highly doubtful that such a thin member could be used in compression, such as member *bc*, because of the likelihood of buckling.

6-7 STATICALLY INDETERMINATE TWO-FORCE MEMBERS

Up to this point in the text, including the previous Example 6-1, the necessary requirements of equilibrium (sum of forces and sum of moments equal to zero) have been sufficient to determine the unknown forces exerted on the various structures and machines and the members thereof. In many real structures, however, these necessary requirements are often *not* sufficient to determine the unknown forces and the corresponding stresses and deformations (recall the discussion in Sec. 4-5). Fortunately, many of these statically indeterminate situations can be solved by adjoining to the equilibrium requirements the additional kinematical requirements of *structural compatibility*. In simple terms, structural compatibility means that the overall deformation of a structure must conform to whatever external constraints may exist and that the deformation of one part of a structure must be compatible (agreeable) with those deformations of the other adjoining parts of the structure so that the parts can fit together as a unit. Generally, the particular conditions of compatibility are peculiar to each individual problem, and consequently each statically indeterminate situation requires its own careful analysis. The following examples are intended to illustrate the basic ideas in handling some statically indeterminate situations. These examples deal with axially loaded members, while later chapters will deal with other types of statically indeterminate situations. However, the basic procedure is essentially the same, namely:

1. Draw a free-body diagram and write the corresponding equations of equilibrium. If they are not *sufficient* to solve for the unknown loads, then
2. Examine the kinematical requirements for the deformation(s) of the member(s). Then
3. Assuming elastic (or inelastic) behavior of the material, write the corresponding load-deformation relations. When enough of these are adjoined to the equilibrium requirements, the problem is determinate and can be solved.
4. Finally, any assumptions which were made in the above procedure should be verified to complete the solution.

This procedure will vary somewhat from problem to problem, but these ingredients form a good guideline.

EXAMPLE 6-2

Three pieces of $\frac{1}{2}$-in-diameter high-strength steel cable are attached to a small steel ring as shown in Fig. 6-15(a). What will be the stress in each of these cables if a vertical load of 8 tons is applied at the ring?

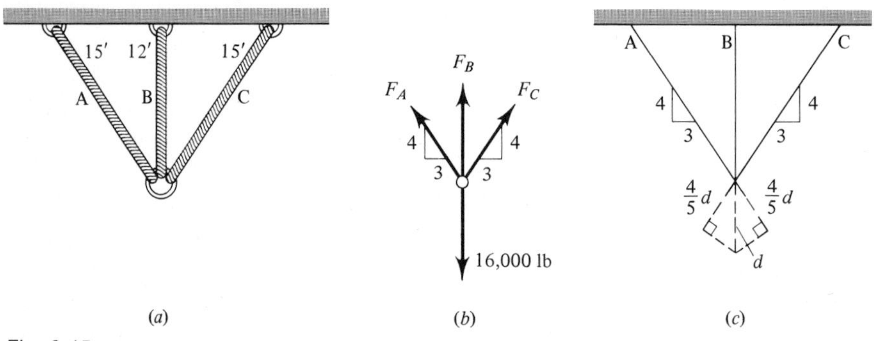

Fig. 6-15

Solution: A free-body diagram of the ring is shown in Fig. 6-15(b), from which we obtain the necessary requirements for equilibrium

$$\text{Sum of forces} = \mathbf{0}$$

$$\mathbf{i}(-\tfrac{3}{5}F_A + \tfrac{3}{5}F_C) + \mathbf{j}(\tfrac{4}{5}F_A + \tfrac{4}{5}F_C + F_B - 16{,}000) = \mathbf{0}$$

$$\therefore F_A = F_C \tag{a}$$

$$F_B + \tfrac{8}{5}F_A = 16{,}000 \tag{b}$$

These conditions are *not sufficient* to solve for the unknown forces F_A, F_B, and F_C. Accordingly, we must look for some additional requirements.

In this particular problem, *because of the symmetry of the cable arrangement,* we can expect that the applied vertical load will cause a *vertical* displacement d of the ring, as greatly exaggerated in Fig. 6-15(c). Hence, cable B will stretch by an amount d, while cables A and C will each stretch approximately by an amount $\frac{4}{5}d$. Assuming elastic behavior and using Eq. (6-11) we have, for cable B,

$$d = \frac{F_B L_B}{A_B E}$$

and for cable A

$$\tfrac{4}{5}d = \frac{F_A L_A}{A_A E}$$

Eliminating d, we obtain

$$F_B L_B = \tfrac{5}{4} F_A L_A$$

or

$$F_B = \frac{5}{4}\frac{L_A}{L_B} F_A = \frac{5}{4}\frac{15}{12} F_A \qquad (c)$$
$$= \tfrac{25}{16} F_A$$

Combining this result with Eqs. (b) and (a) gives

$$\tfrac{25}{16} F_A + \tfrac{8}{5} F_A = 16{,}000$$
$$F_A = F_C = 5{,}060 \text{ lb}$$
$$F_B = \tfrac{25}{16} F_A = 7{,}900 \text{ lb}$$

The stresses in the cables would then be

$$\sigma_A = \frac{F_A}{A} = \frac{5{,}060}{(\pi/4)(1/2)^2} = 25{,}800 \text{ psi}$$
$$\sigma_B = \frac{F_B}{A} = \frac{7{,}900}{(\pi/4)(1/2)^2} = 40{,}200 \text{ psi}$$
$$\sigma_C = \sigma_A = 25{,}800 \text{ psi}$$

These stresses appear to be low enough to justify our assumption of elastic behavior.

Remark: If the cable arrangement were *not* symmetric, a vertical load would *not* necessarily produce a purely vertical displacement, and the kinematical compatibility conditions would be much more complicated than in this example.

EXAMPLE 6-3

A mounting device shown in Fig. 6-16(a) consists of a solid stepped circular plug loosely inserted into a comparatively *rigid* socket. The bottom of the plug rests on the base of the socket and the clearance between the shoulder of the plug and the top of the socket is 0.10 in when the device is not loaded. If the plug is to carry an axial compressive load of 10 tons applied at the top, what will be the bearing stress between (*a*) the bottom of the plug and the bottom of the socket and (*b*) the shoulder of the plug and the top edge of the socket? A partial compressive stress-strain curve for the relatively soft plug material is shown in Fig. 6-16(*b*).

Solution: When the load of 10 tons is applied to the top of the plug, either the bottom of the plug will support all the load, as indicated in Fig. 6-16(*c*), or the bottom and the shoulder will each support part of the load, as indicated in Fig. 6-16(*d*). Before the shoulder can support any part of the load, the lower 1-in-diameter portion of the plug would have to be compressed 0.10 in in order to take up the clearance. The question is whether or not this occurs.

If the bottom of the plug supports all the load as in Fig. 6-16(*c*), the stress in the lower portion of the plug would be

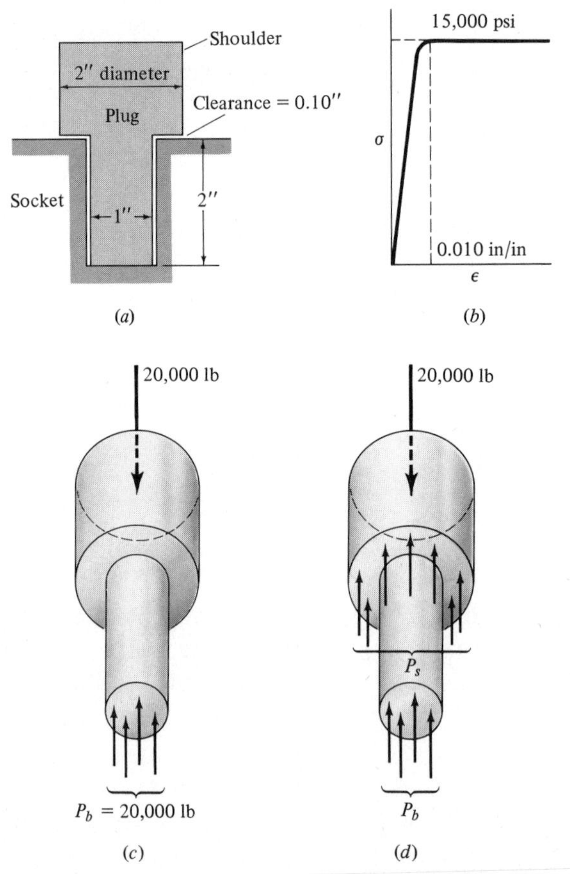

Fig. 6-16

$$\sigma = \frac{P}{A} = \frac{(10)(2{,}000)}{(\pi/4)(1)^2} = 25{,}500 \text{ psi}$$

which greatly exceeds the yield-point stress of 15,000 psi for the material. Hence, it is quite likely that the lower portion of the plug will yield enough for the shoulder to make contact with the top edge of the plug, and we will have the situation depicted in Fig. 6-16(*d*), with the equilibrium requirement being

$$P_b + P_s = 20{,}000 \text{ lb}$$

For this situation, the strain in the lower portion of the plug will be

$$\epsilon = \frac{e}{L} = \frac{0.10}{2} = 0.05 \text{ in/in}$$

For this amount of strain, by the stress-strain curve in Fig. 6-16(*b*), the corresponding compressive stress in the lower portion of the plug will be the yield stress of 15,000 psi. Thus, the force carried by the lower portion of the plug will be that necessary to produce the yield stress of 15,000 psi. This force P_b is given by

$$\sigma_{ys} = \frac{P_b}{A}$$

$$P_b = \sigma_{ys} A = (15{,}000)\left(\frac{\pi}{4}\right)(1)^2$$

$$= 11{,}800 \text{ lb}$$

Hence, if the total load on the top of the plug is 20,000 lb, the bottom will support 11,800, and the shoulder will support

$$P_s = 20{,}000 - 11{,}800 = 8{,}200 \text{ lb}$$

Finally, by the definition of bearing stress given in Sec. 5-6,

$$\sigma_B(\text{bottom}) = \frac{P_b}{A_b} = \frac{11{,}800}{(\pi/4)(1)^2} = 15{,}000 \text{ psi}$$

$$\sigma_B(\text{shoulder}) = \frac{P_s}{A_s} = \frac{8{,}200}{(\pi/4)(2^2 - 1^2)} = 3{,}480 \text{ psi}$$

EXAMPLE 6-4

A 6061-T6 aluminum rod 8 ft long and 2 in square is firmly anchored at each end while unstressed at a temperature of 90°F. The temperature of the rod is later decreased to −10°F. Assuming the anchors do not move, how much force must the anchors exert, and what will be the stress in the rod at this temperature?

Solution: From Table 1 of Appendix B, 6061-T6 aluminum has a coefficient of thermal expansion of 13.1×10^{-6} in/in of strain per degree Fahrenheit, and $E = 10 \times 10^6$ psi. When the aluminum rod cools, it would tend to contract by an amount

$$e = \alpha \times L \times T \qquad (a)$$

where α is the coefficient of thermal expansion, L is the length of the rod in inches, and T is the temperature *change* in degrees—in this case 100°F. Therefore, the anchors must exert enough force to prevent this contraction and effectively must *stretch* the rod by the same amount that it tends to contract. If we assume elastic behavior, the force-deformation relationship is as given by Eq. (6-11)

$$e = \frac{PL}{AE} \qquad (b)$$

Combining (a) and (b), we have

$$\alpha T = \frac{P}{AE}$$

or

$$\sigma = \frac{P}{A} = \alpha TE \qquad (c)$$

For our case the stress in the rod will be

$$\sigma = (13.1 \times 10^{-6}\ \text{in/in/°F})(100\text{°F})(10 \times 10^{6}\ \text{psi})$$
$$= 13{,}100\ \text{psi} \qquad \text{(tension)}$$

which is well below the yield strength of 35,000 listed in Table 1. Hence our assumption of elastic behavior appears to be reasonable. Finally, the force exerted by the anchors will be that necessary to produce the above tensile stress of 13,100 psi. Consequently, each anchor must exert a pull of

$$\sigma = \frac{P}{A}$$

$$\therefore P = \sigma A = (13{,}100)(2 \times 2)$$
$$= 52{,}400\ \text{lb}$$

EXERCISE PROBLEMS

6-23 Rework Example 6-1 for member dc.

6-24 The pin-connected truss shown is made of an aluminum alloy with $E = 10 \times 10^6$ psi. Its design is to be based on an allowable tensile stress of 10,000 psi and an allowable compressive stress of 8,000 psi. In addition, the overall axial deformation of member bc is not to exceed 0.1 in. Find the required minimum cross-sectional area of member bc.

6-25 Two pieces of high-strength $\frac{1}{4}$-in-diameter steel cable are fastened to a steel ring as shown. The ultimate tensile strength

PROB. 6-24

PROB. 6-25

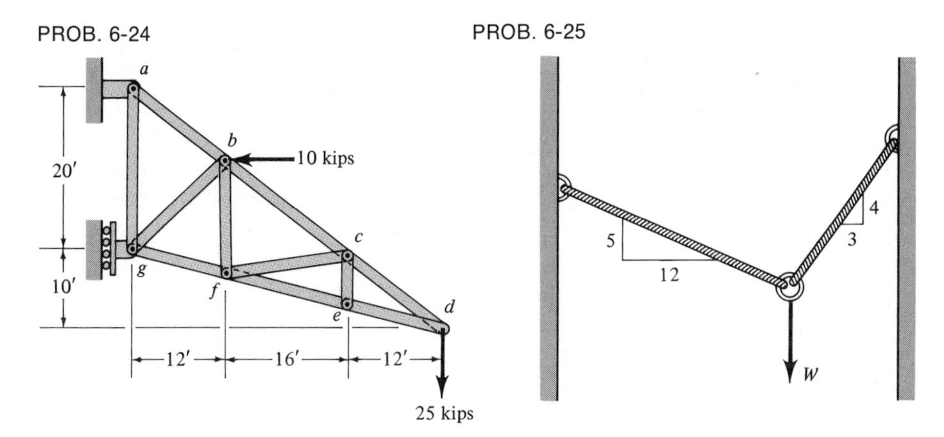

of the cable is 110,000 psi. Based on a factor of safety of 4, what would you say is the maximum weight which can safely be supported at the ring by this cable arrangement?

6-26 Consider a strip of 2024-T4 aluminum $\frac{1}{4} \times 2 \times 100$ in. How much tensile force would be required to stretch this strip 4 in? Refer to Fig. 6-4(*b*) for the stress-strain curve.

6-27 A small block of birchwood $2 \times 4 \times 10$ in standing on end is used to shim (support) one corner (out of four) of a 40-ton platform. In your judgment, neglecting the possibility of buckling, is this support safe? Refer to Fig. 6-4(*c*).

6-28 A homogeneous isotropic bar is made of a material whose axial stress-strain curve is approximated by the relation $\sigma = K\epsilon^{2/3}$, with $K = 10^5$. The original dimensions of the bar are $1 \times 2 \times 20$ in. If the bar is subjected to a longitudinal tensile load of 16 tons, what will be the corresponding change in length? If $\mu = \frac{1}{4}$, what will be the corresponding change in volume of the rod?

6-29 A thin circular aluminum ring with a room-temperature inside diameter of 2 in is to be heated and then slipped onto a solid steel cylinder with an outside diameter of 2.003 in. Treating the steel cylinder as rigid, what will be the tensile stress in the ring when it cools to room temperature? A partial σ-ϵ curve for the aluminum is shown.

PROB. 6-29

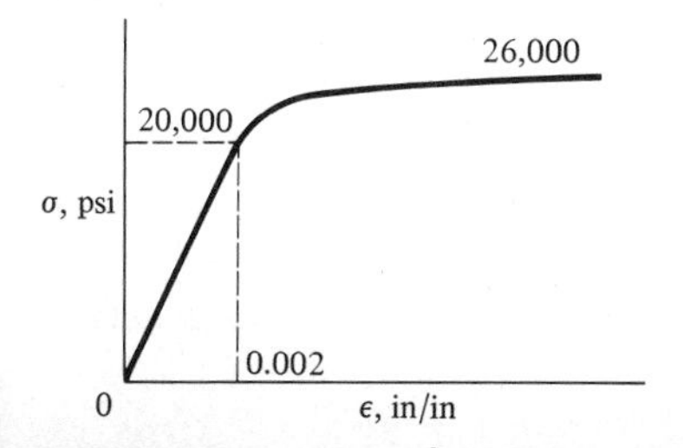

6-30 Same as Prob. 6-29 except the steel cylinder is of 2.008 in diameter.

6-31 A piece of steel pipe (OD = 2.375 in, ID = 2.067 in) is coupled to a piece of aluminum pipe (OD = 2.875 in, ID = 2.469 in) as shown. Find the stress in the steel and aluminum. Find the overall elongation of the coupled system due to the load.

PROB. 6-31

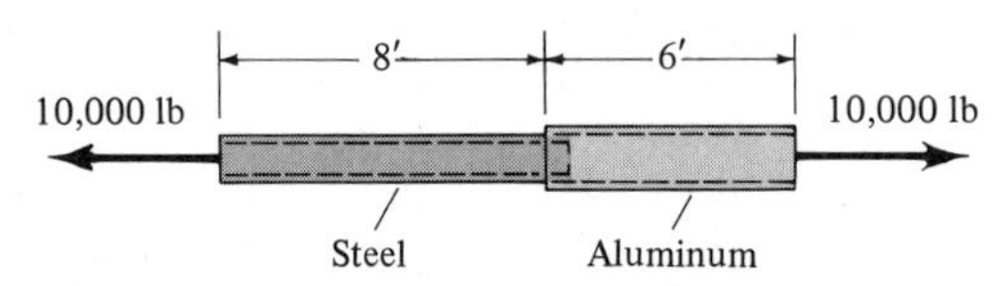

6-32 Rework Example 6-2 with cables A and C being 20 ft long rather than 15 ft.

6-33 Rework Example 6-3 for a clearance of 0.01 in instead of 0.10 in.

6-34 A uniform homogeneous steel rod 8 ft long and 1 in square is rigidly anchored at each end in a vertical position while unstressed. An axial downward load of 16,000 lb is then applied 3 ft from the top. Determine the forces exerted by each anchor.

6-35 The cylindrical steel bar shown is initially unstressed between its rigid supports, and then a load of $P = 12{,}000$ lb is applied. What are the reactions at each end, and how far will mn move downward due to P?

PROB. 6-35

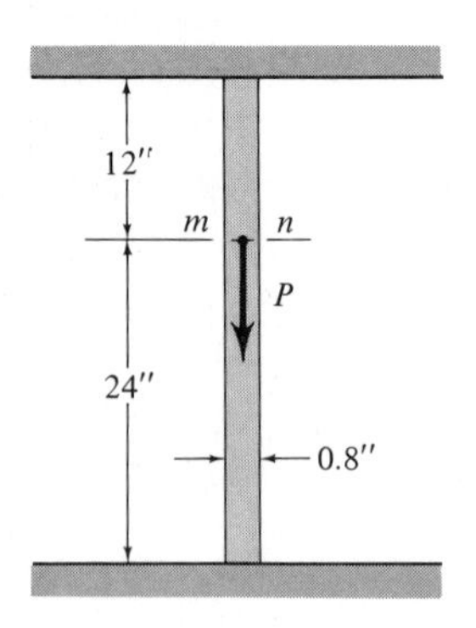

PROB. 6-36

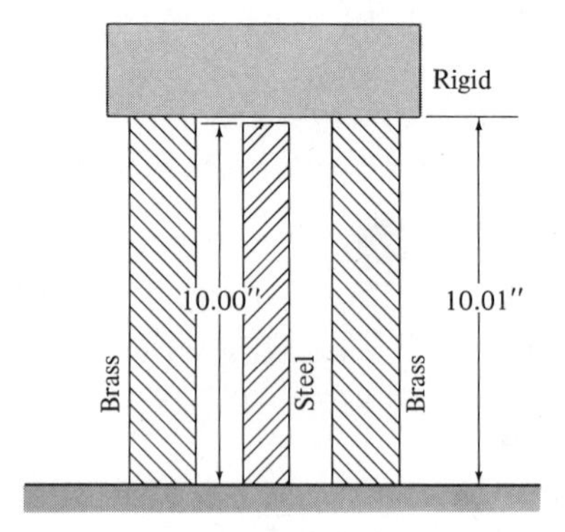

6-36 A mounting device consists of two brass bars, a steel bar, and a thick (rigid) flat plate. Each brass bar has a cross section of $1\frac{1}{2}$ in^2 and the steel bar, 1 in^2. If a compressive load of 20 tons is applied to the top of the plate, what will be the stress in the steel and brass bars?

6-37 Same as Prob. 6-34 except that the temperature of the unstressed rod is initially 40°F and is later raised to 80°F for the loaded rod. Use $\alpha = 5 \times 10^{-6}$ in/in/°F.

6-38 A mounting device consists of a solid circular brass plug ($E = 15 \times 10^6$) surrounded by a steel cylinder ($E = 30 \times 10^6$). Initially, the plug is 0.005 in longer than the cylinder. If a compressive load of 10 tons is applied to the rigid plate, what will be the stress in the plug? Assume elastic behavior. On the basis of your answer, is this assumption reasonable?

PROB. 6-38

Rigid plate
0.005″
1″ dia.
Steel cylinder
$1\frac{1}{4}$″ dia.
10.000″
Brass plug
2″ dia.

PROB. 6-40

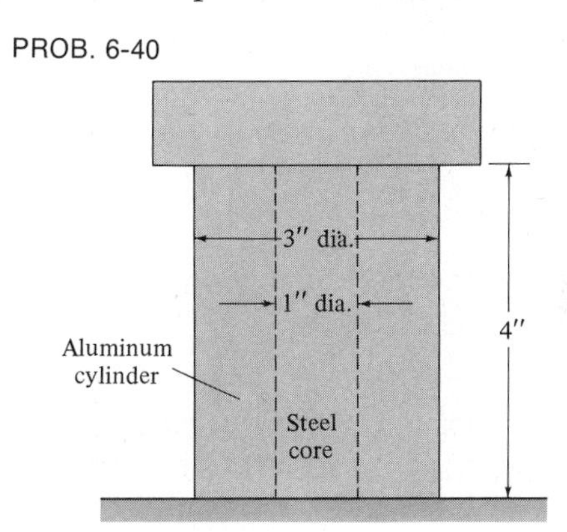

6-39 Same as Prob. 6-36 except the load is 30 tons.

6-40 A mounting device consists of an outer cylinder of aluminum ($E = 10 \times 10^6$) and an inner core of steel ($E = 30 \times 10^6$) topped by a thick (rigid) plate. The device is to carry a total compressive load of P lb distributed over the ends of the cylinder and core. If the allowable compressive stresses in the steel and aluminum are 30,000 psi and 15,000 psi, respectively, what is the allowable applied load P? Assume the cylinder and core to be intimately bonded and have the same Poisson ratio so as to act as a solid unit.

6-41 A very stiff bar is free to pivot at a and is supported horizontally by $\frac{1}{2}$-in-diameter steel cables at b and c. If a vertical load of 2 tons is applied at d, what will be the stress in each cable due to this load?

PROB. 6-41

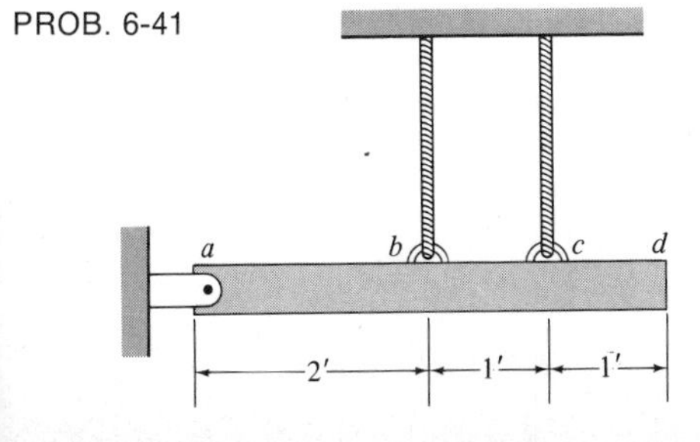

6-42 A $\frac{3}{4}$-in-diameter threaded stud is part of a heavy steel jig fixture as shown. A loose-fitting brass ($E = 12 \times 10^6$ psi) sleeve of 2 in OD and 8 in long is slipped onto the stud and a washer-and-nut combination is initially tightened until it just touches the sleeve. If the nut is then tightened a quarter turn, estimate the tensile stress in the shank of the stud and the compressive stress in the sleeve. The thread size is NF-16 (16 threads/in). Make whatever assumptions seem to be reasonable and justifiable.

PROB. 6-42

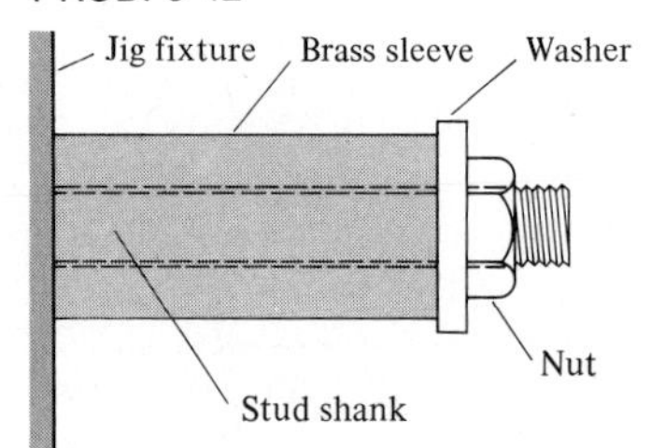

6-8 STRESS CONCENTRATIONS, COLD WORKING, AND OTHER CONSIDERATIONS

Thus far in our discussion of stress, strain, mechanical properties, etc., we have restricted our attention to what we have called "statical loading," which we vaguely distinguish from dynamical loading by saying that statical loads refer to dead loads (weight), fixed loads, and "slowly" applied loads. On the other hand, low-velocity and high-velocity impact loads, cyclic (repeated) loading, and inertia loading due to high acceleration are loosely referred to as dynamical loading. It is not our intention in this text to discuss in any depth the various dynamical material parameters and design considerations. However, no treatment of the statical theory of materials would be satisfactory without the student engineer being made aware of some of the nonstatical design considerations.

Probably, the most fundamental consideration is to recognize the dependence of the mechanical properties of materials on such factors as the rate of loading, the temperature of the material, and the previous history of the loading of the material. For example, Fig. 6-17(*a*) shows stress-strain curves for bone, one for quasi-static (slow loading with $\dot{e} = 0.001$/sec) and others for more rapid rates of loading. The difference in behavior of the material is self-evident and fairly typical.

Similarly, Fig. 6-17(*b*) shows stress-strain curves for a stainless steel alloy from room temperature to 800°F. Once again, the difference in response is rather typical for most structural materials.

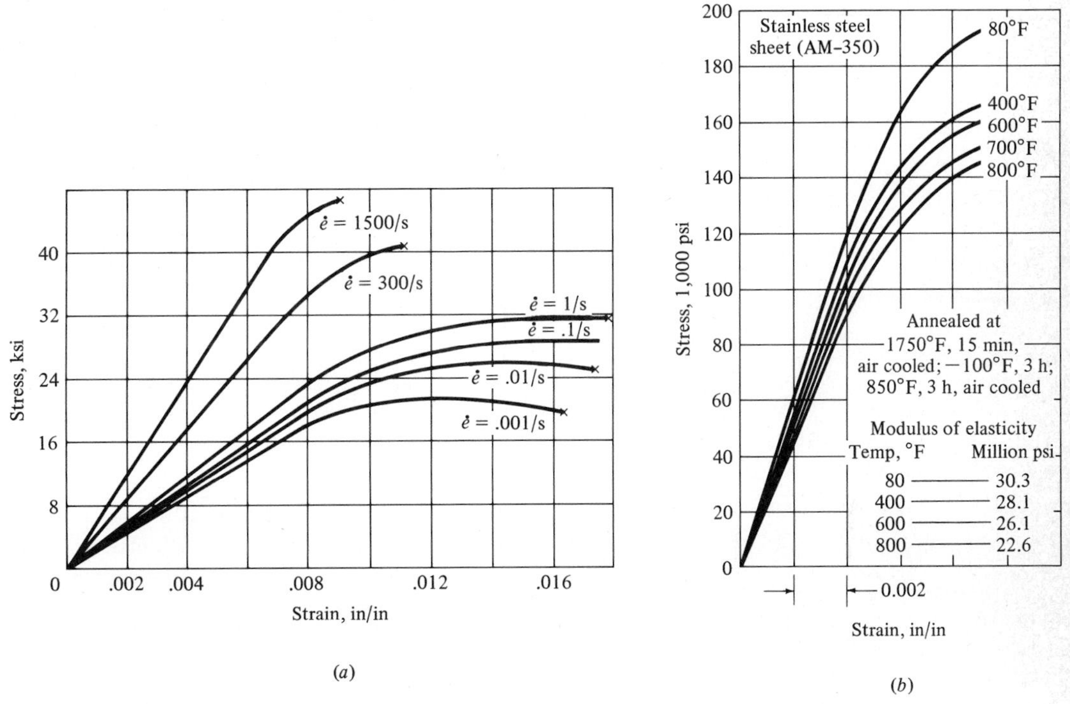

Fig. 6-17 (*a*) Stress-strain curves for human femur bone specimens; (*b*) stress-strain curves for a stainless-steel alloy.

As another consideration, the present mechanical response of a material might very well depend upon how it had been loaded or deformed previously. Figure 6-18 gives a good example of this phenomenon, usually referred to as *cold working.* Cold working is the process by which the mechanical properties of a material are significantly altered by working (rolling, drawing, pressing, etc.) the material at normal temperature. For some materials, this alteration can be quite pronounced, as indicated by the curves in Fig. 6-18 for low-carbon steel. Thus, cold-rolled steel and hot-rolled steel are considered to be two distinctly different materials although their chemical makeup is quite similar.

While most cold working is usually done to the material during fabrication, cold working can also be done to the material during its normal use with or without the engineer's knowledge. This is particularly true in dynamical machinery in which stresses and strains are applied to the structural parts many many times over the life span of the machine, causing the material to progressively crack due to *fatigue,* and eventually resulting in failure of the machine.

In Sec. 6-4, we classified materials as ductile and brittle ac-

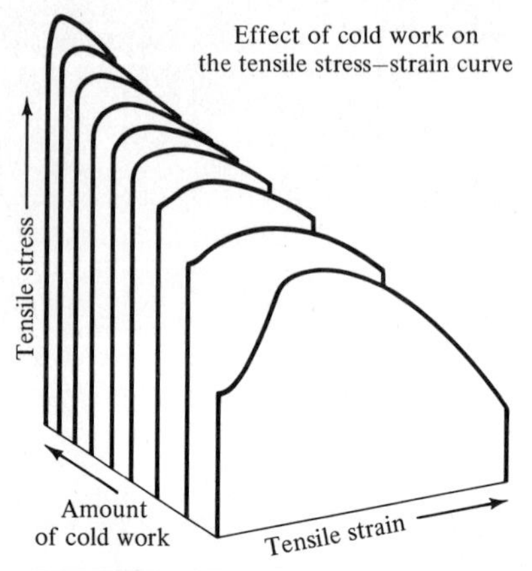

Fig. 6-18

cording to the amount of plastic deformation they could withstand before rupture. But, as you can see by the above discussion, a material which is brittle at room temperature might become quite ductile at elevated temperatures. A material that is ductile under static loads might behave as a brittle one under dynamical loads or after considerable cold working. The engineer must take these factors into consideration when selecting materials and design configurations for a particular application. Quite often this is done by incorporating into the design some experimentally determined empirical factors related to the material and/or the geometry involved.

The most common example of such an empirical design factor is the *stress-concentration factor*. In our study of stresses and deformations in two-force members in this and the previous chapter, we have assumed the members to be straight and to have a uniform cross section, and the material to be reasonably homogeneous and isotropic. However, in real structural or machine elements, it might be quite apparent that any one or several of these assumptions would be inapplicable. For example, many bars and rods are tapered rather than straight, or have holes, notches, or fillets. In such situations the use of Eq. (6-1) for determining the axial stress must be questioned. To see this, consider the bar of width D with a hole of radius r, as shown in Fig. 6-19(a). Suppose we were to use Eq. (6-1),

$$\sigma = \frac{P}{A} \tag{6-1}$$

to determine the tensile stress in the bar. In the solid part of the bar, the cross-sectional area is

$$A_{\text{total}} = t \times D$$

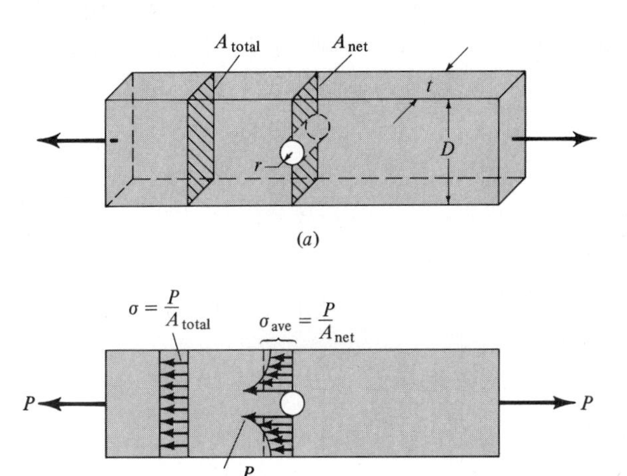

Fig. 6-19

while at the hole location the area of the bar is

$$\begin{aligned} A_{net} &= A_{total} - A_{hole} \\ &= t \times D - t \times 2r \\ &= t(D - 2r) \end{aligned}$$

Both theoretical and experimental analyses have shown that this abrupt reduction in effective cross section causes a concentration of stress and strain in the immediate vicinity of the hole. That is, while the stress in regions far removed from the hole is a uniform value of

$$\sigma = \frac{P}{A_{total}}$$

near the hole the stress is quite nonuniform, with a peak value σ_{max}, which can greatly exceed the average value, given by

$$\sigma_{ave} = \frac{P}{A_{net}}$$

as illustrated in Fig. 6-19(*b*). How much the peak value exceeds the average value depends upon the geometry of the concentration (in this case the relative values of D and r) and the material of the bar. This concentration of stress with its associated peak value can cause a local failure of the material in the form of a small crack which can eventually propagate across the cross section and subsequently cause complete failure of the member.

Definition 6-4 The ratio of the maximum (peak) stress to the average stress in the vicinity of an abrupt change in geometry is called the stress-concentration factor and is denoted by k:

$$k = \frac{\sigma_{\max}}{\sigma_{\text{ave}}} \tag{6-12}$$

This definition is applicable to various types of structural elements, such as shafts, beams, plates, etc., as well as the simpler two-force member being discussed here.

The crucial question, of course, is, "How do we determine the value of the stress-concentration factor for a particular situation under consideration?" For linear (elastic) behavior of the material, the stress-concentration factor is dependent primarily upon the geometry involved, and can be determined by various analytical and experimental techniques, notably the theory of elasticity and the photoelastic technique. Values obtained in this manner are called *theoretical* stress-concentration factors and are denoted by k_t. Figure 6-20 gives values of k_t for some axially loaded members with grooves, holes, and fillets. Generally, the stress-concentration factor becomes larger as the change in geometry becomes more severe, i.e., the sharper the radius, the higher the stress concentration. Similar types of stress-concentration factors have been determined for shafts, beams, and a great variety of other geometries, and an extensive tabulation of them can be found in R. E. Peterson, "Stress Concentration Design Factors," John Wiley & Sons, Inc., 1953. Note from Fig. 6-20 that the peak stress can be as much as three times the value of the average stress. Therefore, there should be no doubt that this can indeed be a significant design consideration.

The next question is, "How does inelastic behavior affect the value of the stress concentration?" First of all, a brittle material behaves elastically almost to rupture, and, therefore, the theoretical value of the stress-concentration factor is applicable for all levels of stress. On the other hand, ductile materials have the ability to deform and flow in the vicinity of the stress concentration to such an extent that they can adjust themselves so as to redistribute the stress in such a manner that the peak stress is only slightly higher than the average value. Therefore, the *effective* stress concentration factor k_e approaches the value of unity when significant plastic behavior occurs.

Summarizing, for brittle materials and elastic behavior of ductile materials, the peak stresses are given by the theoretical stress-concentration factors, while for plastic behavior of ductile materials, the peak stresses eventually will not greatly exceed the average values. However, in the light of our previous discussion of cold working, even a ductile material can become brittle under

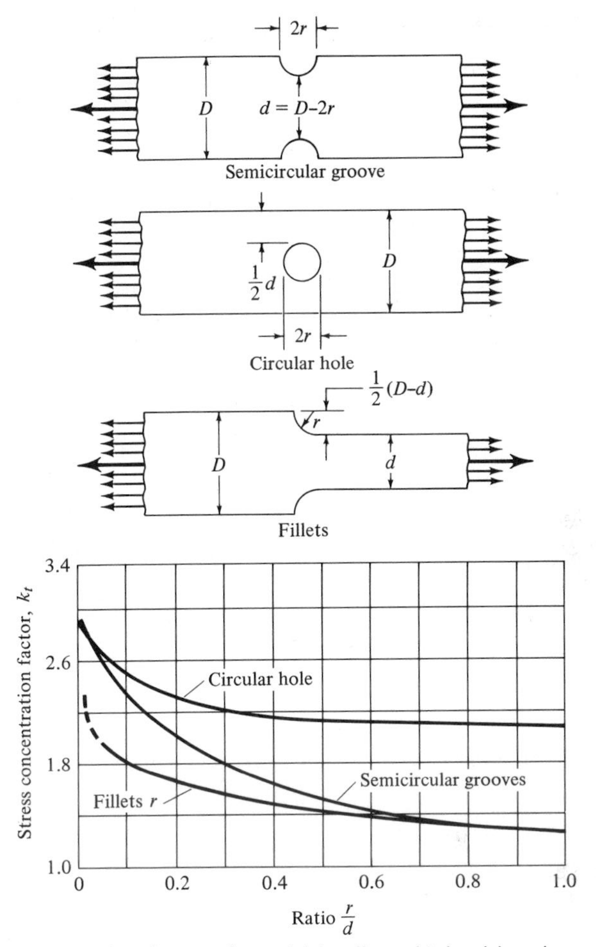

Fig. 6-20 Theoretical stress-concentration factors for axial loading obtained by photoelasticity. (*After Frocht.*)

repeated (fatigue) loading and unloading. Accordingly, for such situations the theoretical stress-concentration factors are used even though the original material might be a ductile one.

Before closing this chapter with an example problem, we remark that the concept of a stress-concentration factor is a typical scheme employed by engineers to get a simple answer to a complex problem. That is, although the engineer might be quite aware that a simple formula such as

$$\sigma = \frac{P}{A}$$

is not valid for this particular situation, by modifying the simple formula by the use of some "design factor" he hopes to obtain a reliable answer to his problem. Needless to say, experience and good judgment play an important role in such design problems.

EXAMPLE 6-5

The two-force member shaped as shown in Fig. 6-21(*a*) is subjected to a static tensile force of 30 tons. Find the maximum tensile stress in the member if it is made of (*a*) hot-rolled steel or (*b*) an alloy steel whose partial stress-strain curves are as shown in Fig. 6-21(*b*).

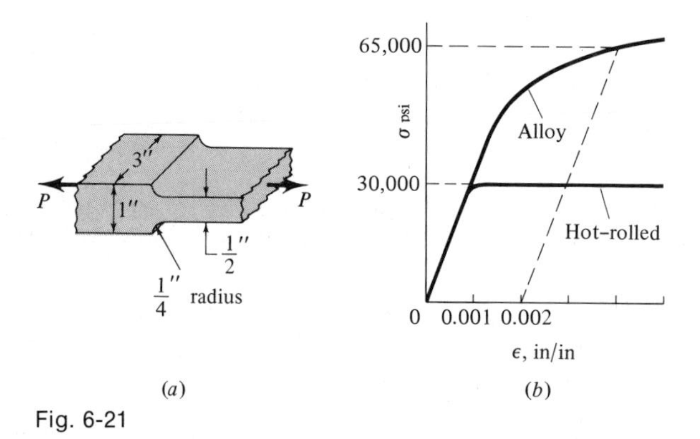

Fig. 6-21

Solutions: From the given geometry and Fig. 6-20,

$$\frac{r}{d} = \frac{1/4}{1/2} = \tfrac{1}{2}$$

so that the value of the theoretical stress-concentration factor is

$$k_t \approx 1.4$$

The average tensile stress in the smaller section is

$$\sigma_{\text{ave}} = \frac{P}{A} = \frac{(30)(2{,}000)}{(3)(1/2)}$$
$$= 40{,}000 \text{ psi}$$

(*a*) For the hot-rolled steel, σ_{ave} is well above the yield stress, so that undoubtedly the material will yield. Hence, the *effective* stress-concentration factor will be close to unity, so that

$$\sigma_{\max} \approx \sigma_{\text{ave}} = 40{,}000 \text{ psi}$$

(*b*) For the alloy steel, 40,000 psi is well below its 0.2 percent yield strength. Therefore, we will use the theoretical stress-concentration factor of 1.4. Hence

$$\sigma_{\max} = k_t \sigma_{\text{ave}}$$
$$= (1.4)(40{,}000) = 56{,}000 \text{ psi}$$

which is still below the yield strength of 65,000 and also below the ultimate strength of the material.

EXERCISE PROBLEMS

6-43 Find the maximum stress which would occur in an axially loaded plate whose cross section is $4\frac{1}{2} \times \frac{3}{8}$ in, with a $\frac{7}{8}$-in-diameter hole in the center, if the plate were made of a linearly elastic material and the axial load is 10,000 lb.

6-44 Find the axial load that can be applied to a long alloy-steel plate if the maximum stress may not exceed 72,000 psi. The cross section is $\frac{1}{4} \times 2\frac{1}{2}$ in and there is a $\frac{3}{8}$-in-diameter hole in the center.

6-45 A plate has a cross section of $1\frac{1}{2} \times \frac{1}{8}$ in and is loaded axially with a static load of 2,500 lb. If a $\frac{5}{16}$ in-diameter hole is drilled in the center of the plate, find (*a*) the maximum stress if the material is mild steel with a yield point of 30,000 psi and an ultimate stress of 72,000 psi, and (*b*) the maximum stress if the material is brittle with a linear stress-strain relationship to rupture at 48,000 psi.

6-46 The fixture in the figure is made of a brittle material whose stress-strain curve is linear to rupture at 12,000 psi. Find the allowable applied load P if a factor of safety of 2.2 is used.

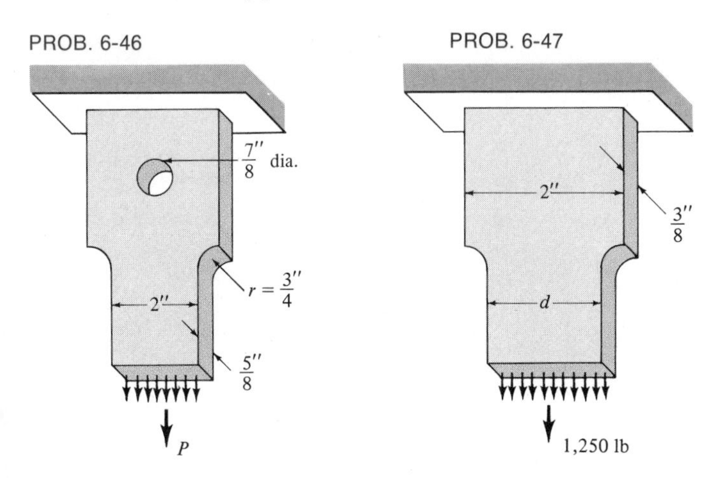

6-47 The material of the fixture in the figure is brittle with a linear stress-strain relationship to rupture at 14,000 psi. Approximately how narrow can the plate be cut down (find d) and still carry the load of 1,250 lb? Use a factor of safety of 3 based on a rupture mode of failure.

6-48 A brittle tensile plate $\frac{1}{4}$ in thick and 2 in wide must transmit 6,000 lb without exceeding a tensile stress of 34,000 psi in the material. Semicircular grooves must be machined across the $\frac{1}{4}$-in edges at the center of the plate for clearance with another member. Approximately how deep can these grooves be made?

6-49 Same as Prob. 6-48 except the plate is very ductile and the yield point is 24,000 psi. The plate is assumed to perform satisfactorily until it becomes fully yielded.

6-50 How much load can a 24-in-wide by $1\frac{1}{2}$-in-thick alloy-steel plate transmit if semicircular grooves 4 in deep are cut across the edges? The stress may not exceed 46,000 psi at any point.

chapter 7 TORSION

7-1 INTRODUCTION

In the previous chapter, the entire discussion of stress, strain, and mechanical properties was confined to centroidally loaded two-force members subjected to so-called "pure tension" or "pure compression." The axial (normal) stress and the accompanying axial strain and transverse strain implied by the dashed lines and Fig. 7-1(*a*) were defined and calculated according to Eqs. (6-1), (6-2), and (6-6), respectively. But from our introduction to the stress concept in Sec. 5-4, we know that other types of stresses, namely, shear stresses, can exist within a loaded body of material, and we expect that such a shear stress will be accompanied by some shear-type of strain, as illustrated in Fig. 7-1(*b*).

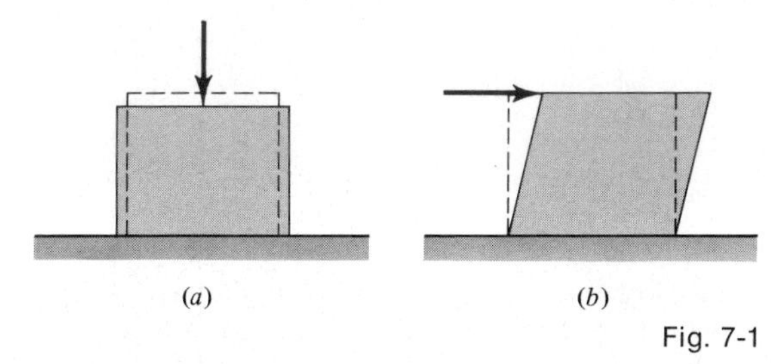

Fig. 7-1

This chapter will deal with one of the most fundamental applications of the concept of shear stress and shear strain insofar as structural materials are concerned, namely, the torsion or twisting of shafts. Our objective is to quantitatively relate the applied loading (twisting couple) to the resulting stresses and deformation in the shaft.

7-2 BASIC TORQUE-STRESS REQUIREMENTS

Let us consider a solid circular cylindrical shaft of some structural material which is being twisted by the couples as indicated by

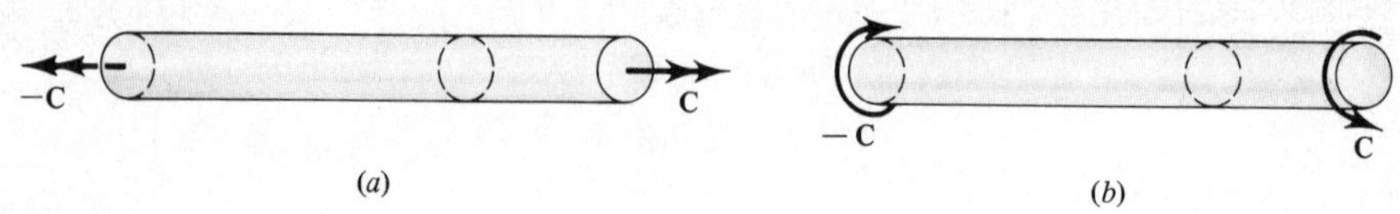

Fig. 7-2

either of the schemes in Fig. 7-2(*a*) or (*b*). A free-body diagram of a section of this shaft would show that the net internal reaction would be a twisting couple $\mathbf{C}_T$, as shown in Fig. 7-3, with no net *internal* normal force, no net internal shear force, and no net internal bending moment. Hence, by Eqs. (5-6) and (5-7), whatever stress vector distribution $\boldsymbol{\sigma}$ exists on the area A must satisfy

$$\iint_{\text{area}} \boldsymbol{\sigma}\, dA = \mathbf{R}_N + \mathbf{R}_S = \mathbf{0} \tag{7-1}$$

$$\iint_{\text{area}} \mathbf{p} \times \boldsymbol{\sigma}\, dA = \mathbf{C}_T + \mathbf{C}_B = \mathbf{C}_T \tag{7-2}$$

where $\mathbf{p}$ is the position vector relative to the center of the circular cross section.

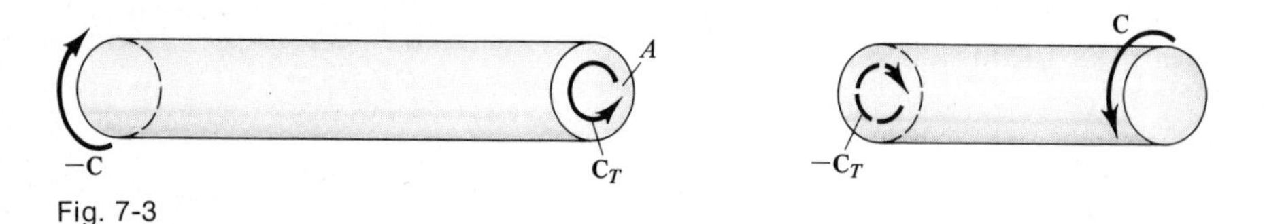

Fig. 7-3

We now face the problem of determining the stresses $\boldsymbol{\sigma}$ on the cross section A which will satisfy the load-stress requirements given by Eqs. (7-1) and (7-2). Let us first consider the possibility of normal stresses $\boldsymbol{\sigma}_N$ on the surface A, as shown in Fig. 7-4. Since their sum would have to be $\mathbf{0}$, some of these stresses would have to be tensile while others would have to be compressive. But such equal but opposite stress vectors might form $\mathbf{j}$ or $\mathbf{k}$ bending couples and thus violate Eq. (7-2). Hence, to simplify matters, let us speculate that there are *no significant normal stresses* on the area A.

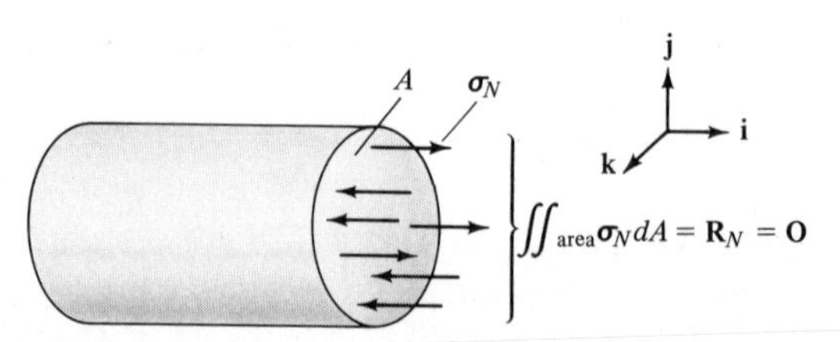

Fig. 7-4

What about shear stresses? From Eqs. (7-1) and (7-2), the shear stresses $\boldsymbol{\sigma}_S$ on area A would have to satisfy

$$\iint_{\text{area}} \boldsymbol{\sigma}_S \, dA = \mathbf{R}_S = \mathbf{0} \tag{7-3}$$

$$\iint_{\text{area}} \mathbf{p} \times \boldsymbol{\sigma}_S \, dA = \mathbf{C}_T \tag{7-4}$$

Since the cross section is circular, for convenience, let us use cylindrical polar coordinates x, r, θ and their associated longitudinal, radial, and tangential orthogonal unit vectors $\mathbf{i}$, $\boldsymbol{\mu}_r$, $\boldsymbol{\mu}_\theta$, respectively, as shown in Fig. 7-5(a).

The shear stress vector $\boldsymbol{\sigma}_S$ lies in the plane of the circular cross section and, therefore, can be expressed in terms of orthogonal components $\boldsymbol{\sigma}_{Sr}$ and $\boldsymbol{\sigma}_{S\theta}$ as

$$\begin{aligned} \boldsymbol{\sigma}_S &= \boldsymbol{\sigma}_{Sr} + \boldsymbol{\sigma}_{S\theta} \\ &= \sigma_{Sr}\boldsymbol{\mu}_r + \sigma_{S\theta}\boldsymbol{\mu}_\theta \end{aligned} \tag{7-5}$$

as shown in Fig. 7-5(b).

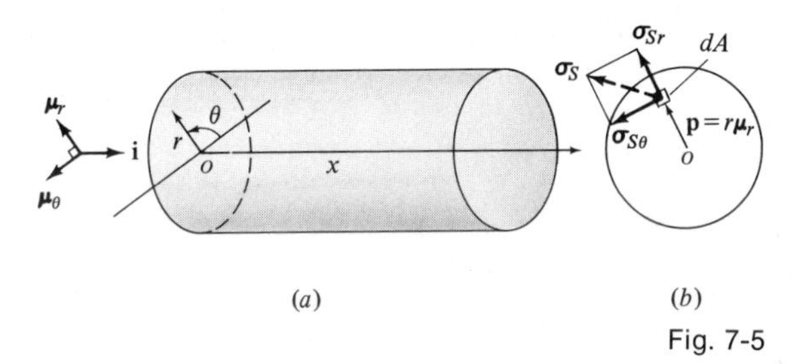

Fig. 7-5

Substituting Eq. (7-5) into the load-stress requirements (7-3) and (7-4) yields

$$\iint_A (\sigma_{Sr}\boldsymbol{\mu}_r + \sigma_{S\theta}\boldsymbol{\mu}_\theta) \, dA = \mathbf{0} \tag{7-6}$$

$$\iint_A (r\boldsymbol{\mu}_r) \times (\sigma_{Sr}\boldsymbol{\mu}_r + \sigma_{S\theta}\boldsymbol{\mu}_\theta) \, dA = \mathbf{C}_T \tag{7-7}$$

where we have used $\mathbf{p} = r\boldsymbol{\mu}_r$ as the position vector of dA relative to point o. By noting that $\boldsymbol{\mu}_r \times \boldsymbol{\mu}_\theta = \mathbf{i}$ and recalling that $\mathbf{C}_T$ is an $\mathbf{i}$ couple, Eq. (7-7) becomes

$$\mathbf{i} \iint_A r\sigma_{S\theta} \, dA = C_T \mathbf{i}$$

or

$$\iint_A r\sigma_{S\theta} \, dA = C_T \tag{7-8}$$

Thus, we observe that the component shear stress σ_{Sr} does not enter into the load-stress relationship of Eq. (7-8).

We now make the assumption that because of the circular symmetry in the problem, the magnitudes of the shear stresses $\sigma_{S\theta}$ and σ_{Sr} depend only upon r and not on θ. Consider, then, the stresses on two diametrically opposite elemental areas as shown in Fig. 7-6(a). The stress vectors $\boldsymbol{\sigma}_{Sr}$ (and $-\boldsymbol{\sigma}_{Sr}$) will "cancel out," while the stresses $\boldsymbol{\sigma}_{S\theta}$ (and $-\boldsymbol{\sigma}_{S\theta}$) will produce a net couple, as required by Eqs. (7-6) and (7-8). Therefore, in the light of these results and observations, we will *assume that the radial shear stress* $\boldsymbol{\sigma}_{Sr}$ *vanishes and that the tangential shear stress* $\boldsymbol{\sigma}_{S\theta}$ *is the primary stress produced by the twisting couple* $\mathbf{C}_T$. To simplify our notation, we will denote this shear stress by τ, so that Eq. (7-8) takes the form

$$C_T = \iint_A r\tau(r)\, dA = T \tag{7-9}$$

where $\tau(r)$ denotes the dependence of τ upon the variable r. This tangential shear stress is distributed over the entire cross section, as in Fig. 7-6(b). It is quite common to call the twisting couple the *torque* T. Equation (7-9) is then called the *fundamental torque-shear stress relation* for a circular shaft.

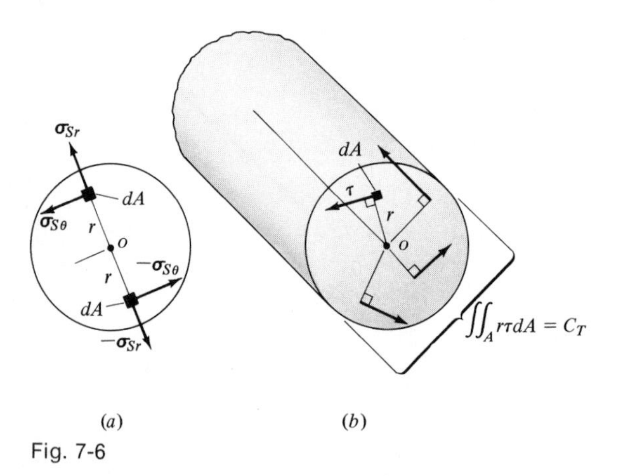

Fig. 7-6

Assuming our arguments and analysis thus far have been correct, the basic question now confronting us is: How does the shear stress τ vary with the radius r? If we can answer this question, we can then integrate Eq. (7-9) to get some kind of algebraic relationship between the torque load and the shear stress in the shaft. The next section will attempt to answer this question, but for now we will look at an example problem to illustrate what is involved in using Eq. (7-9).

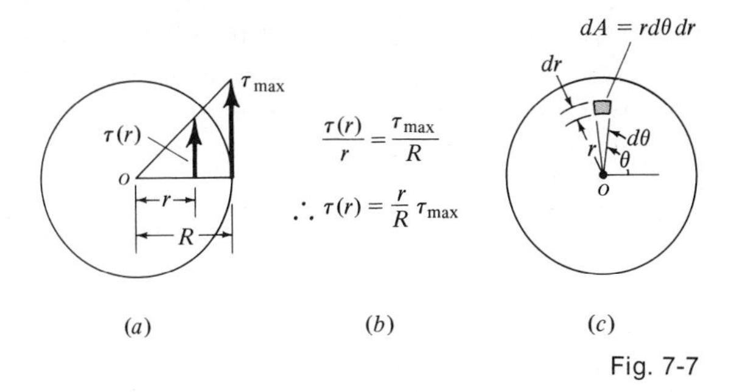

Fig. 7-7

EXAMPLE 7-1

A solid circular shaft of radius R is subjected to a pure torque T. Suppose that the shear stress τ varies linearly with the radius from zero at the center to τ_{max} at the outside radius R. Find the relationship between T and τ_{max}.

Solution: The linear stress variation is indicated in Fig. 7-7(a), and its algebraic representation by the equation in Fig. 7-7(b). By Fig. 7-7(c) the elemental area dA is given by

$$dA = r\, d\theta\, dr$$

Putting these results into Eq. (7-9), we obtain

$$\begin{aligned} T &= \iint_A r\tau(r)\, dA \\ &= \int_0^R \int_0^{2\pi} r\left(\frac{r}{R}\tau_{max}\right) r\, d\theta\, dr \\ &= \int_0^R 2\pi \frac{r^3}{R}\tau_{max}\, dr \\ &= 2\pi \frac{R^4}{4R}\tau_{max} = \frac{\pi R^3}{2}\tau_{max} \end{aligned} \tag{7-10}$$

A check on units shows

$$\frac{\pi R^3}{2}\tau_{max} = \text{in}^3 \frac{\text{lb}}{\text{in}^2} = \text{in-lb}$$

which appears to be okay. A more useful form of this result is given by the following:

Alternate Solution: Let us substitute the expression for $\tau(r)$ from Fig. 7-7(b) into Eq. (7-9),

$$\begin{aligned} T &= \iint_A r\left(\frac{r}{R}\tau_{max}\right) dA \\ &= \frac{\tau_{max}}{R}\iint_A r^2\, dA \end{aligned}$$

since τ_{max} and R are constants insofar as the integration is concerned. The integral represents a geometric property of the cross-sectional area A, called the *second polar moment of area* or *polar moment of inertia*, denoted by J:

$$J = \iint_{\text{area}} r^2 \, dA \tag{7-11}$$

(See Appendix A for properties of areas.) With this definition our result takes the form

$$T = \frac{\tau_{max}}{R} J \tag{7-12}$$

which is the basic torque-shear stress relation for a *linear variation* of the shear stress. As you will see in Sec. 7-4, this linear variation, and, consequently, Eq. (7-12), plays an important role in the stress analysis of circular cylindrical torsion members.

> ***Remark:*** You should realize that Eqs. (7-10) and (7-12) represent the same result, since, for a solid circular cross section,
>
> $$J = \frac{\pi R^4}{2}$$
>
> Substituting this into (7-12) yields (7-10), and vice versa.

EXERCISE PROBLEMS

7-1 A solid circular steel shaft of $1\frac{1}{2}$ in diameter and 16 in long is subjected to a pure torque of 500 ft-lb. What is J for this shaft? Assuming the shear stress varies linearly with the radius, what is the value of the maximum torsion shear stress? What happens to the values of J and τ_{max} if the length is 20 in rather than 16 in?

7-2 Same as Prob. 7-1 except the shaft is hollow with a 1-in diameter hole through it.

7-3 A solid circular shaft of diameter D is subjected to a pure torque T. Assume that the shear stress $\tau(r)$ varies linearly as in Example 7-1. How much does τ_{max} increase or decrease if (*a*) the diameter is doubled, (*b*) the torque is doubled, and (*c*) the diameter is halved?

7-4 Two shafts of equal length are made of the same material except one is solid and the other is hollow. If the two shafts weigh exactly the same (i.e., same amount of metal in each), which one has the larger J? Assuming a linear stress variation, which one will have the larger shear stress for a given applied torque T?

7-5 to 7-7 For a solid circular shaft subjected to a pure torque T, find the relationship between the torque T and the maximum shear stress τ_{max} for the stress variation shown in the figure.

PROB. 7-5

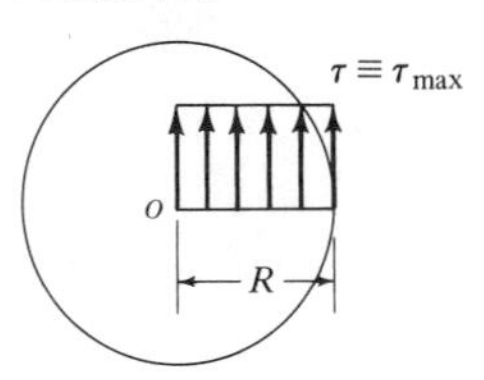

PROB. 7-6

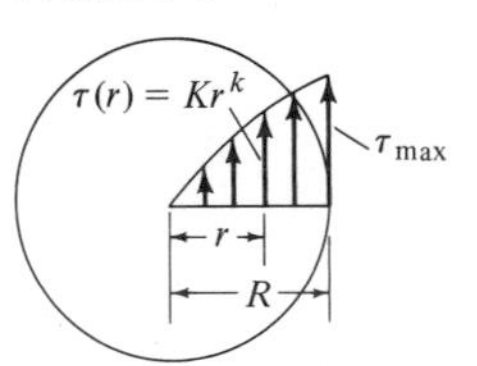

PROB. 7-7

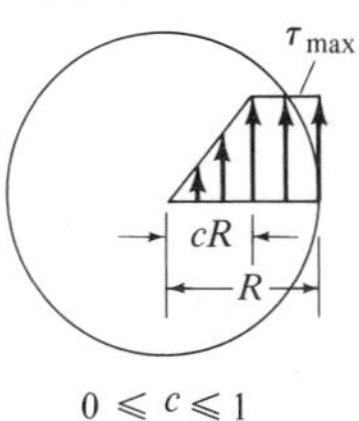

$0 \leqslant c \leqslant 1$

7-8 A solid 1-in-diameter steel shaft is twisted by a torque T until its shear stress is a uniform constant value of 20,000 psi, as depicted in the figure for Prob. 7-5. What is the corresponding value for T?

7-9 Rework Prob. 7-3 assuming the shear stress $\tau(r)$ is a constant throughout the shaft rather than a linear variation.

7-10 A hollow circular shaft of outside radius R and inside radius $cR(0 \leqslant c < 1)$ is subjected to a pure torque T. Find the relationship between the torque and the maximum shear stress if the stress variation is linear, as in Example 7-1.

7-11 Rework Prob. 7-10 for the case in which the shear stress is a constant throughout the shaft rather than a linear variation.

7-12 A solid circular shaft carries a torque T which, assuming a linear variation, produces a maximum shear stress τ_{max}. It is desired to achieve a 50 percent reduction in weight by making the shaft hollow while maintaining the same outside diameter D. What is the corresponding reduction in the allowable torque T if the value of τ_{max} is to be the same as for the solid shaft?

7-13 Same as Prob. 7-12 except the shear stress $\tau(r)$ is a constant, as in Prob. 7-5.

7-14 A solid steel shaft 16 in long and of 1 in diameter carries a torque of 200 ft-lb. Evaluate τ_{max} assuming (*a*) linear stress variation and (*b*) a constant stress throughout.

7-15 Two separate shafts are each carrying a torque of 100 ft-lb. One shaft is solid with a $\frac{3}{4}$ in diameter, and the other is hollow with an OD of $1\frac{1}{2}$ in and an ID of $1\frac{1}{4}$ in. Which shaft has the greater τ_{max} if (*a*) the stress variation is linear in both and (*b*) the stress variation is constant in both?

7-16 Which shaft can carry the larger torque – a solid 1-in-diameter shaft with a linear stress variation with $\tau_{max} = 18{,}000$ psi, or a hollow shaft with an OD of 1 in and an ID of $\frac{7}{8}$ in, having a constant stress throughout of 26,000 psi?

7-3 THE TORQUE-TWIST TEST

Now that we have seen how shear stresses might arise in a circular shaft subjected to twisting couples, let us look at this situation from the viewpoint of a laboratory experiment on materials.

Suppose we have a cylindrical specimen of a material with a gripping region and a test section, as shown in Fig. 7-8(*b*). This specimen is then placed in some type of torque machine, such as in Fig. 7-8(*a*), and the specimen is twisted by a reasonably pure torque with little or no bending or axial loading.

As with the uniaxial test, we will confine our observation to what occurs in the test region. In the previous section, we examined the loading situation and the accompanying load-stress requirements, with Eq. (7-9) being the fundamental torque-shear stress relation. We will now look at the kinematics of this test more closely.

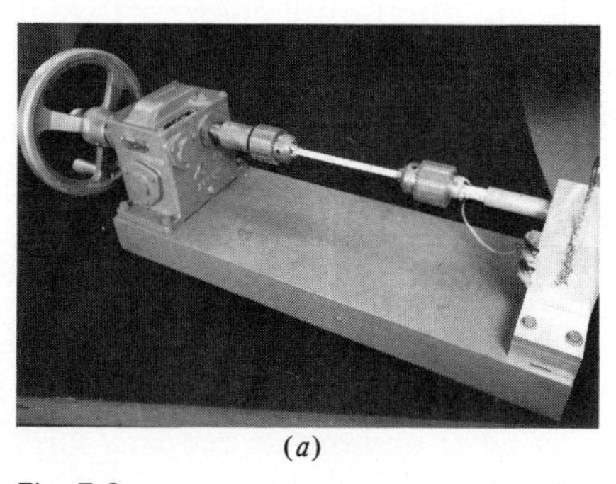

(*a*)

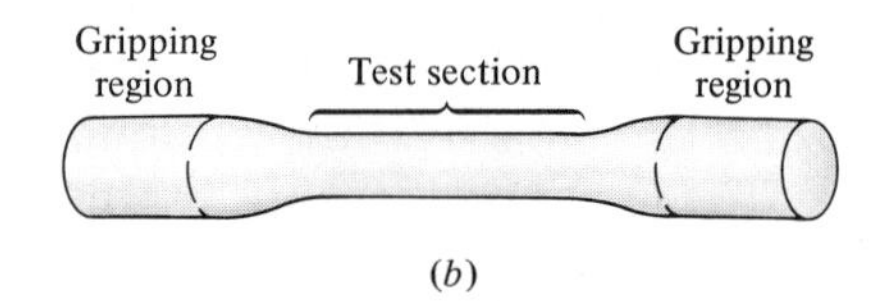

(*b*)

Fig. 7-8

For convenience, let us assume that the left end of the test section does not move. As the specimen is twisted, the most apparent data to record would be the increasing value of the applied torque T and the corresponding amount (angle) of relative twist ϕ between the two ends, as indicated in Fig. 7-9(*a*). These data are recorded as a torque-twist curve, such as the one in Fig. 7-9(*b*). It is quite common for the torque to be in in-lb and the angle in radians. Of course, just as with a uniaxial load-deformation curve, the torque-twist curve will be quite dependent upon the size of the test specimen as well as upon the material. For example, we would expect a

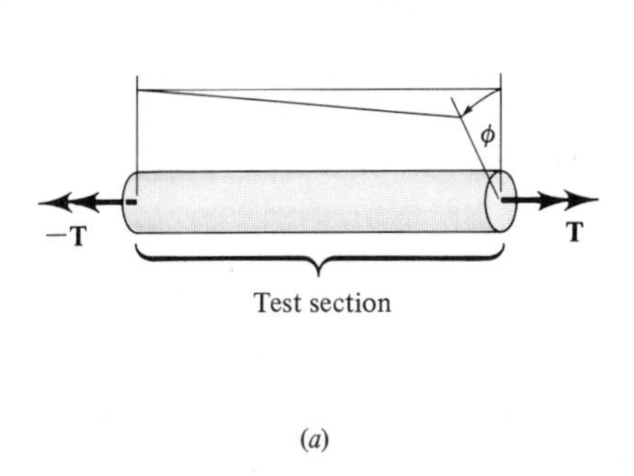

(a)

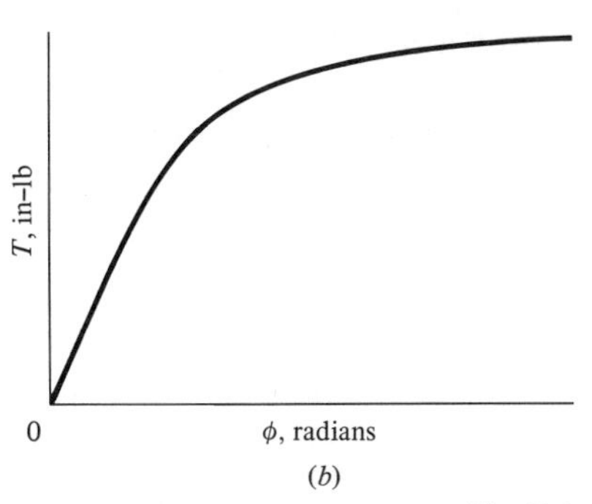

(b)

Fig. 7-9

long shaft to twist more than a shorter shaft and that it would take more torque to twist a 2-in-diameter shaft than a 1-in-diameter shaft. Also, from our study in Chap. 6, we might anticipate that an aluminum shaft would twist considerably more under a given torque than a corresponding steel shaft. Thus, we expect that the torque-twist curve is to some extent indicative of the mechanical response of a material to shear stresses and shear strains. Our problem is to determine how all these features of torque, angle of twist, shear stress, and shear strain are interrelated.

What do we know thus far? Equation (7-9) gives the integral relation between the shear stress τ and the applied torque T. But, while in the case of axial loading, for which the axial stress σ was assumed to be uniform over the entire cross-sectional area, we have no logical basis to expect that τ will be uniform over the circular cross section of the shaft.

The primary method of determining the stress variation in any solid body is by observing and analyzing the kinematics of the surface deformations. From such observations, one can often make rational conjectures as to how the stresses vary throughout the interior of the body. The torsion problem is an excellent example of this type of rational analysis.

We begin by assuming the material of the shaft to be *homogeneous* and *isotropic*, so that the material throughout the shaft is essentially the same regardless of location or orientation. Also, before we twist the shaft, let us scribe some longitudinal, radial, and circumferential lines on the surface, as shown in Fig. 7-10(*a*). Figure 7-10(*b*) shows the same shaft after the application of the torque T. This figure requires careful scrutiny as to the kinematics of the deformed lines.

1 The originally straight radial line will remain a straight radial line. To see why this is true, consider a portion of the shaft, as in Fig. 7-11(*a*), and assume that the radial lines on

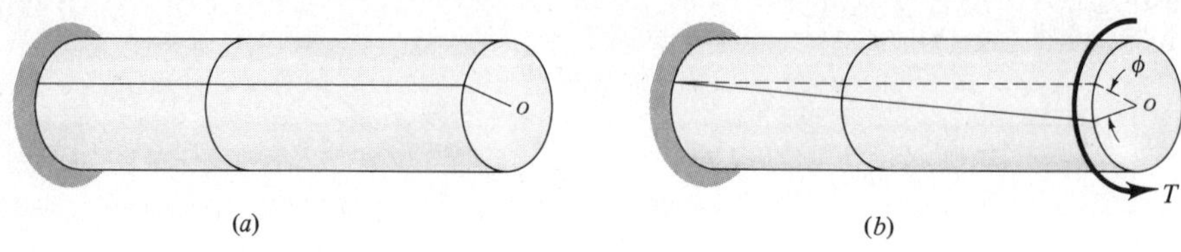

Fig. 7-10

the opposite ends curve in some manner. Symmetry would require that each of these opposing radial lines would have to curve into a shape similar to the other. This implies that the radial lines at the midsection would have to remain straight since it cannot curve both ways simultaneously. But this portion of the shaft under consideration is quite arbitrary, and, therefore, its middle could coincide with any location on the original shaft. Hence, all radial lines remain straight.

2 The originally straight longitudinal line becomes a helix (straight line advancing around a circular cylindrical surface). To see this, again consider an arbitrary portion of the shaft. Hold the left end stationary, twist the right end, and assume the longitudinal line curves in some manner as shown in Fig. 7-11(*b*). On the other hand, repeat the experiment by holding the right end and twisting the left end, as in Fig. 7-11(*c*). But Fig. 7-11(*b*) and (*c*) both represent the same portion of the shaft twisted by the same torque T, and, therefore, the curved lines in Fig. 7-11(*b*) and (*c*) should be the same shape. Carrying out this argument for any assumed shape of the curve, we eventually conclude that the only compatible shape is a straight line.

3 The circular circumferential line will remain circular and will not warp. The argument for this is essentially the same as those in (1) and (2) in that whatever warping occurs at the left end should also occur at the right end, as shown in Fig. 7-11(*d*). But then the middle could not warp both ways simultaneously, and so the conclusion is that neither end warps and all circumferential lines remain unwarped.

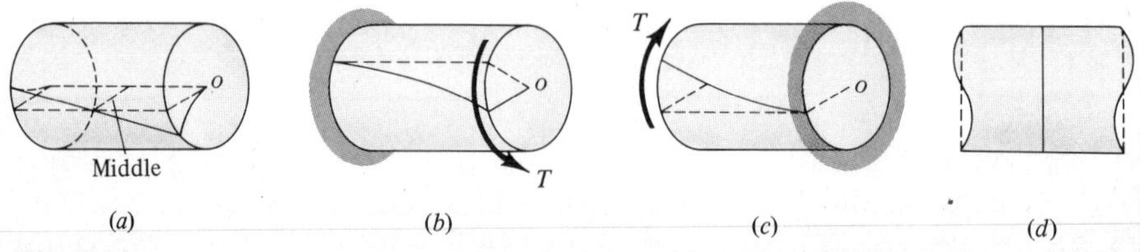

Fig. 7-11

Remark: All of the above kinematic arguments are called "compatibility conditions" in that they are based on the presumption that the deformation of one part of a body must be compatible (agreeable) with the deformations of neighboring parts of the body. Such *local* or *material* compatibility conditions are often an essential ingredient in the stress analysis of solids and structures.

An actual laboratory experiment would confirm each of these arguments and yield the deformation indicated in Fig. 7-10(*b*). Recalling the stress arguments of the previous section and the kinematical arguments of this section, the combination of the tangential shear stress τ and the resulting deformation is indicated in Fig. 7-12(*a*) for a section of the shaft of length L and radius R. Note that we are observing the deformation at the outer surface of the shaft. The corresponding value of the shear stress at the outer radius R is denoted by τ_R. Hence, at the *outer surface*, we have the situation depicted in Fig. 7-12(*b*), in which the longitudinal and circumferential lines on the cylindrical surface have been "flattened out" for convenience. Observe that what had been a right angle at o between the undeformed longitudinal and circumferential lines has been "distorted" by an amount γ_R. This angle of distortion γ_R is associated with the shear stress τ_R and is commonly called the *engineering shear strain* at the surface of the shaft. From Fig. 7-12(*a*) we see that *for small distortions* γ_R is related to the angle of twist ϕ by

$$L\gamma_R = R\phi$$

or

$$\gamma_R = R\,\frac{\phi}{L} \tag{7-13}$$

where ϕ and γ_R are measured in radians and R and L in some common unit of length. This relationship enables us to convert the angle of twist ϕ to the shear strain γ_R occurring at the *outer surface* of the shaft.

While γ_R is the shear strain at the outer surface, what is the shear strain in the interior of the shaft, say, at some radius r? From Fig. 7-12(*c*) we see that the angle of distortion γ_r at a radius r is given by

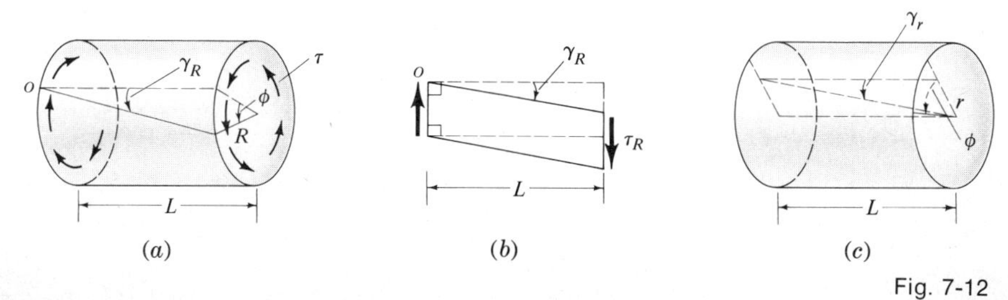

Fig. 7-12

$$L\gamma_r = r\phi$$

or

$$\gamma_r = r\frac{\phi}{L} \tag{7-14}$$

Hence, by Eqs. (7-13) and (7-14) we see that

$$\gamma_r = \frac{r}{R}\gamma_R \tag{7-15}$$

That is, the shear strain is directly proportional to the radial distance r, and the maximum strain occurs at the outer radius R. We formalize this observation with

Theorem 7-1 For a circular cylindrical shaft subjected to pure torsion, the *engineering shear strain* γ (angle of distortion) *will vary linearly* with the radius, with the maximum shear strain γ_{max} occurring at the outer surface of the shaft.

> ***Remark:*** Values of engineering shear strain are commonly of the order of 0.001 rad and accordingly are difficult to measure or observe without special instruments. However, laboratory experiments have verified Theorem 7-1 for shafts made of most structural materials subjected to reasonably large amounts of twist.

This theorem is the culmination of all the kinematical arguments and observations presented in this section and is the basis for analyzing the load-carrying behavior of circular (solid or hollow) shafts made of most engineering materials. Several of the arguments fail (for lack of symmetry) if the shaft is noncircular or nonconcentric, or if the material is very inhomogeneous or anisotropic; therefore, Theorem 7-1 does not hold for such cases. Although important, those nonsymmetric situations are beyond our purposes in this text.

7-4 TORQUE-STRESS AND TORQUE-TWIST RELATIONS

Recall the basic torque–tangential shear stress relationship from Sec. 7-2, namely, Eq. (7-9):

$$T = \iint_A r\tau(r)\, dA \tag{7-9}$$

and the basic shear strain-angle of twist relation (7-14)

$$\gamma_r = r\frac{\phi}{L} \tag{7-14}$$

How the torque T in Eq. (7-9) is related to the angle of twist ϕ in Eq. (7-14) depends upon how the shear stress τ is related to the shear strain γ. That is, the torque-twist relation for a shaft will depend upon the shear stress–shear strain properties for the material of the shaft.

In Chap. 6, we saw that the load-deformation relationship depended upon the stress-strain properties obtained from a uniaxial stress-strain curve. Unfortunately, shear stress-strain curves are not easily obtained for several reasons, not the least of which is the difficulty of measuring shear strains and calculating shear stresses. Torque-twist curves, like that in Fig. 7-9(*b*), are more common than shear stress–shear strain curves. However, on the basis of our knowledge of the mechanical response of structural materials gained in Chap. 6, let us conjecture that a typical shear stress–shear strain curve would take the form indicated in Fig. 7-13. The essential features of this curve are the initial linear portion up to the proportional limit and the rather flat top at the ultimate shear-stress level. This flattening effect would be most pronounced in a ductile material as it "flows" under the influence of large shear stresses, and would be far less pronounced or even nonexistent for a brittle material which undergoes little or no plastic deformation.

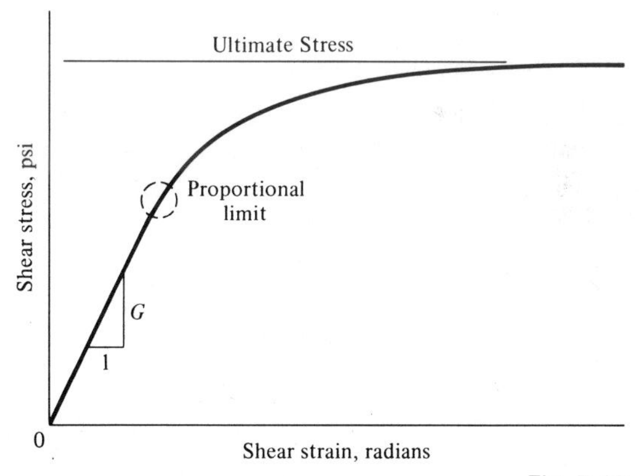

Fig. 7-13

The slope of the initial linear portion is called the *elastic shear modulus* or *modulus of rigidity* and is denoted by the letter G. Its units are usually given by

$$G = \frac{\text{shear stress, lb/in}^2}{\text{shear strain, rad}} = \text{psi} \tag{7-16}$$

since radian is a dimensionless unit. The shear modulus is a measure of a material's ability to resist distortion. Some typical values of G are

$$G = 12 \times 10^6 \text{ psi for steel}$$
$$= 3.8 \times 10^6 \text{ psi for aluminum}$$
$$= 10 \times 10^6 \text{ psi for cast iron}$$

Although the entire shear stress-strain curve might be of interest to an engineer, generally the elastic (linear) range and the flattened plastic range are the essential characteristics insofar as most structural applications are concerned. Accordingly, we will pursue further discussion of the torsion problem for these two situations, namely, linear elastic behavior and significantly ductile plastic behavior.

Elastic Behavior

For the engineer, elastic behavior of a material is generally characterized by a linear relationship between stress and strain. Therefore, if the material in a shaft is behaving elastically, the shear stress τ at any point in the shaft is related to the corresponding shear strain γ at that point by

$$\tau = G\gamma \tag{7-17}$$

Combining the stress-strain equation (7-17) with the strain variation given by Eq. (7-14), the stress variation $\tau(r)$ becomes

$$\tau(r) = G\gamma_r = Gr\frac{\phi}{L} \tag{7-18}$$

which shows that *the stress $\tau(r)$ varies linearly with the radius.* Recall Example 7-1, which dealt with the linear stress variation indicated in Fig. 7-7(*a*), and the corresponding relation of Fig. 7-7(*b*), namely

$$\tau(r) = \frac{r}{R}\tau_{\max} \tag{7-19}$$

where $\tau_{\max}$ is related to the torque T by the result in Eq. (7-12)

$$T = \frac{\tau_{\max}}{R}J \tag{7-12}$$

Hence, comparing Eqs. (7-18), and (7-19), we see that

$$\frac{\tau_{\max}}{R} = G\frac{\phi}{L} \tag{7-20}$$

or, by Eq. (7-12),

$$T = G\frac{\phi}{L}J \tag{7-21}$$

We summarize these results with

Theorem 7-2 For linear elastic behavior of a straight circular cylindrical shaft subjected to a pure torque T, the shear stress varies linearly with the radius, with the maximum stress τ_{max} occurring at the outer radius R and given by

$$\tau_{max} = \frac{TR}{J} \tag{7-22}$$

and the angle of twist given by

$$\phi = \frac{TL}{JG} \tag{7-23}$$

with J being the second polar moment of the cross section of the shaft.

Equations (7-22) and (7-23), which are rearrangements of Eqs. (7-12) and (7-21), are the *basic torque-stress* and *torque-twist relations* for a *linear elastic* shaft. We point out that, in using these relations, the torque T is usually given in inch-pounds, R and L in inches, J in inches^4, and G in pounds per square inch. Then the stress will be in psi and the angle ϕ in radians.

It is important to realize that these results are valid only so long as the entire shaft behaves elastically. This means that if the calculated shear stress τ_{max} given by Eq. (7-22) exceeds the elastic-limit (proportional-limit) shear stress for the material, or if the shear strain γ_R given by Eq. (7-13) exceeds the elastic-limit shear strain for the material, then the results are *not* valid. Hence there is a limit to the amount of torque and twist a shaft can undergo for elastic (linear) response, as is evident by the sample torque-twist curve of Fig. 7-9(*b*).

Fully Plastic Behavior

For a shaft made of a ductile material, it is sometimes possible to twist the shaft well beyond the proportional-limit strain without fear of immediate rupture of the shaft. For such a situation, the *stress* variation would no longer be linear as for elastic behavior but would become nonlinear in some fashion depending upon the shape of the shear stress–shear strain curve for the material. However, assuming this curve has a pronounced flattened portion as the shaft is twisted beyond its elastic limit, similar to the sample in Fig. (7-13), we can assume that the stress variation in the shaft will also progressively "flatten out," as indicated in Fig. 7-14, to the corresponding constant value which might be either the ultimate shear stress for most ductile materials or possibly the yield-point stress for mild steel (refer to Sec. 6-3). The idealizing assumption of a constant stress variation enables us to make a reasonable

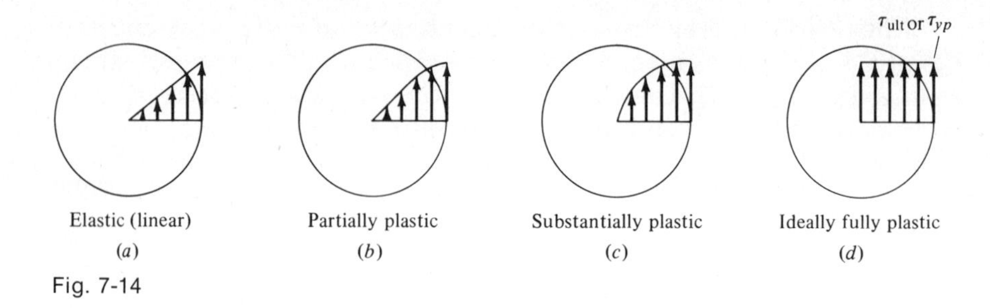

Fig. 7-14

guess as to the maximum torque a ductile shaft can carry before rupturing or failure due to excessive twisting. This limiting value for the torque is usually called the *fully plastic torque* T_{fp} and, by Eq. (7-9), is given by

$$T_{fp} = \iint_A r\tau_{fp}\, dA$$

$$= \tau_{fp} \iint_A r\, dA \tag{7-24}$$

Utilizing the polar coordinates of Fig. 7-7(*c*), we have, for a *solid* shaft,

$$T_{fp} = \tau_{fp} \int_0^R \int_0^{2\pi} r^2\, d\theta\, dr$$

$$= \tau_{fp} \frac{2\pi}{3} R^3 = \tfrac{4}{3}\tau_{fp} \frac{J}{R} \tag{7-25}$$

This result relates the limiting torque T_{fp} to the limiting shear stress τ_{fp} for a solid ductile shaft. If the shaft is hollow, the result will differ because the lower limit for the radius r will be some inside radius R_i rather than 0.

We remark that, for fully plastic behavior, the shear strain γ_r and the angle of twist will still be related by Eq. (7-14). However, when the stress-strain curve flattens, as in Fig. 7-13, the stress-strain relation is no longer single-valued in that the strain continues to increase while the stress value remains fixed. Therefore, the stress in Eq. (7-25), and thereby the fully plastic torque, cannot be uniquely related to the strain and angle of twist in Eq. (7-14). Accordingly, for the fully plastic condition, the amount of twist ϕ is not determined from the torque, except that both must be large enough to produce fully plastic behavior of the material. Generally, the fully plastic twist will be several times larger than the largest elastic twist, while the fully plastic torque will be roughly 30 percent larger than the limiting elastic torque.

We close this section with some illustrative examples of our theory and results.

EXAMPLE 7-2

A solid circular shaft 30 in long and of 1 in diameter is loaded in pure torsion. The applied torque causes an angle of twist of $(43.2/\pi)°$. What torque is necessary to do this if the shaft is made of the material whose idealized shear stress-strain curve is as shown in Fig. 7-15(*a*) and (*b*)?

Solutions: We will first determine the shear strain corresponding to the given angle of twist. Changing degrees to radians, by Eq. (7-13) the maximum shear strain is

$$\gamma_R = R\frac{\phi}{L} = \frac{1}{2}\frac{43.2}{\pi}\frac{\pi}{180}\frac{1}{30}$$

$$= 0.004 \text{ rad}$$

For the material in Fig. 7-15(*a*), this strain is below the proportional-limit strain of 0.005 rad, and, therefore, a shaft of this material will behave elastically for the given amount of twist and have the linear stress variation shown in Fig. 7-15(*c*). Consequently, by Eq. (7-23), the necessary torque is

$$T = \phi J\frac{G}{L}$$

$$= \frac{43.2}{\pi}\frac{\pi}{180}\left[\frac{\pi}{2}\left(\frac{1}{2}\right)^4\right][4 \times 10^6]/30$$

$$= 1{,}000\pi \text{ in-lb}$$

where $G = 4 \times 10^6$ psi is obtained from the stress-strain curve.

For the material in Fig. 7-15(*b*), the stress-strain curve is nonlinear, and our elastic formulas are not valid. Neither are our

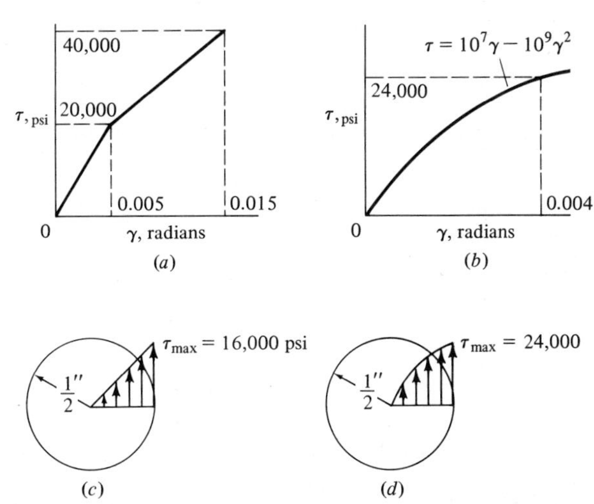

Fig. 7-15

fully plastic results, since the curve does not flatten off to any significant degree. Therefore, we return to our basic equations for all materials, namely, the torque-stress relation (7-9) and the strain-twist relation (7-14). Hence, for any radius r, by Eq. (7-14),

$$\gamma_r = r\frac{\phi}{L} = r\frac{43.2}{\pi}\frac{\pi}{180}\frac{1}{30}$$
$$= 0.008r \text{ rad}$$

Substituting this into the stress-strain relation for the material in Fig. 7-15(b) gives the stress variation

$$\tau(r) = 10^7\gamma_r - 10^9\gamma_r^2$$
$$= 80{,}000r - 64{,}000r^2$$

which is shown in Fig. 7-15(d). Thus, by Eq. (7-9), the torque is given by

$$T = \iint_A r\tau(r)\, dA$$
$$= \int_0^{1/2}\int_0^{2\pi} r\tau(r)r\, d\theta\, dr$$
$$= 2\pi\int_0^{1/2} \tau(r)r^2\, dr = 2\pi\int_0^{1/2} [80{,}000r - 64{,}000r^2]r^2\, dr$$
$$= 2\pi\left[80{,}000\frac{r^4}{4} - 64{,}000\frac{r^5}{5}\right]_0^{1/2}$$
$$= 2\pi[1{,}250 - 400] = 1{,}700\pi \text{ in-lb}$$

EXAMPLE 7-3

A hollow steel drive shaft with OD = 2 in and ID = $1\frac{1}{2}$ in is fastened to a coupling by means of three equally spaced $\frac{1}{2}$-in-diameter steel bolts around a 6-in-diameter circle, as shown in Fig. 7-16(a). The shear stress-strain curve for the steel is shown in Fig. 7-16(b). If the shaft rotates at 600 rpm, what is the maximum horsepower that it can transmit if failure occurs only if the system ruptures?

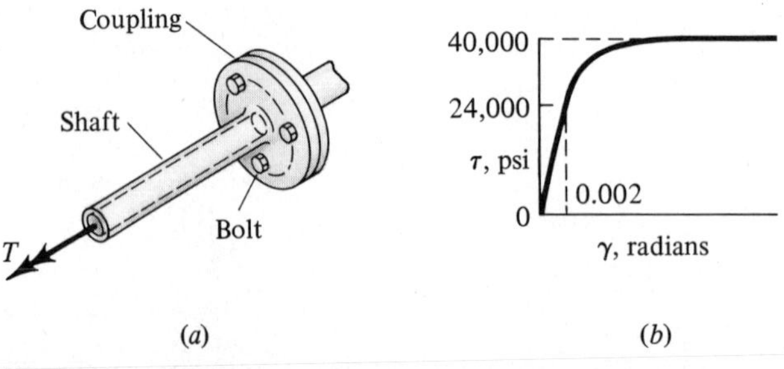

Fig. 7-16

Solution: Since either the shaft or bolts could fail, we will determine the failure torque for each. First, the shaft will become fully plastic before rupture, and by Eq. (7-24) with $\tau_{fp} = \tau_{ult}$,

$$\begin{aligned} T_{fp} &= \tau_{ult} \iint_A r\, dA \\ &= \tau_{ult} \int_{3/4}^{1} \int_0^{2\pi} rr\, d\theta\, dr \\ &= \tau_{ult}\, 2\pi \left[\frac{r^3}{3}\right]_{3/4}^{1} \\ &= 40{,}000\, \frac{2\pi}{3} \left[1 - \frac{27}{64}\right] = 15{,}400\pi \text{ in-lb} \end{aligned}$$

For the bolts, each can withstand a shear force of

$$\begin{aligned} F &= \tau_{ult} \times \text{area of one rivet} \\ &= 40{,}000 \left[\frac{\pi}{4}\left(\frac{1}{2}\right)^2\right] = 2{,}500\pi \text{ lb} \end{aligned}$$

Each bolt, therefore, can exert a torque of

$$\begin{aligned} T &= \text{Force} \times \text{lever arm} \\ &= 2{,}500\pi \times 3\text{-in-radius arm} \\ &= 7{,}500\pi \text{ in-lb} \end{aligned}$$

and so three bolts can exert a torque of

$$T_{total} = 3 \times 7{,}500\pi = 22{,}500\pi \text{ in-lb}$$

Hence, the shaft can withstand less torque than the coupling, and so the limiting failure torque is $15{,}400\pi$ in-lb.

To express our answer in terms of horsepower (rate of work), we have

$$33{,}000 \text{ ft-lb/min} = 1 \text{ hp}$$

Work is force times distance and power is force times velocity, or, in this case,

$$\begin{aligned} \text{Power} &= \text{torque} \times \text{angular velocity} \\ &= 15{,}400\pi \text{ in-lb} \times \frac{1}{12 \text{ in/ft}} \times 600\, \frac{\text{rev}}{\text{min}} \times 2\pi\, \frac{\text{rad}}{\text{rev}} \\ &= 15.4\pi^2 \times 10^5 \text{ ft-lb/min} \end{aligned}$$

Hence, the maximum transmitted horsepower is

$$\begin{aligned} \text{hp} &= \frac{15.4\pi^2 \times 10^5 \text{ ft-lb/min}}{33{,}000 \text{ ft-lb/min}} \\ &= 460 \end{aligned}$$

EXERCISE PROBLEMS

7-17 A solid steel shaft of 2 in diameter and 12 ft long is subjected to a pure torque which twists the shaft through an angle of 0.12 rad. Assuming elastic behavior, what is the maximum shear stress in the shaft? What is the applied torque?

7-18 Same as Prob. 7-17 except the shaft is hollow with an OD of 2 in and an ID of 1 in.

7-19 Rework Example 7-2 for a hollow shaft with an OD of $1\frac{1}{4}$ in and an ID of $\frac{1}{2}$ in.

7-20 Rework Example 7-3 if failure of the system is based on the *initiation* of inelastic behavior.

7-21 A laboratory torsion test on a solid 1-in-diameter shaft 18 in long produced the torque-twist curve indicated in the figure. From these data determine (*a*) the modulus of rigidity, (*b*) the elastic-limit shear stress, and (*c*) the ultimate shear stress.

7-22 Same as Prob. 7-21 except the shaft was hollow with an inside diameter of $\frac{7}{8}$ in.

PROBS. 7-21 and 7-22

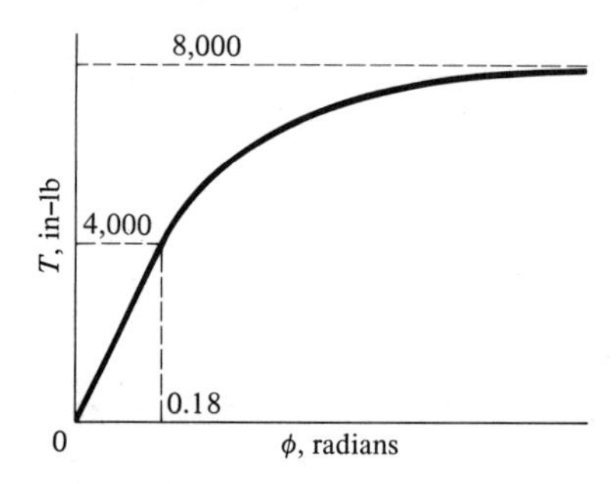

PROBS. 7-23 and 7-24

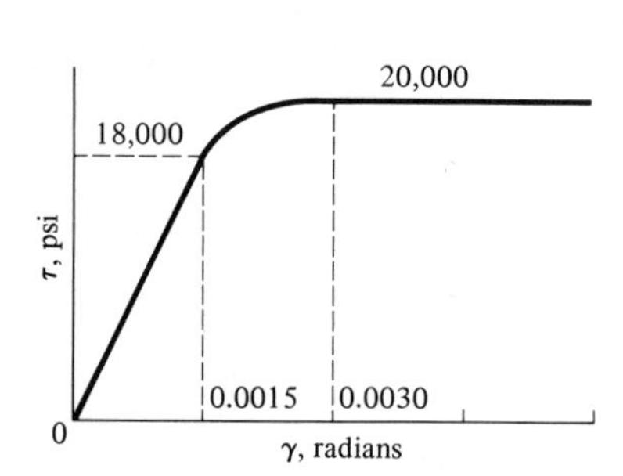

7-23 A solid 1-in-diameter shaft 15 in long is twisted through an angle of 2°. What is the applied torque? What is the maximum shear stress due to this torque? The partial shear stress-strain curve for the material is shown.

7-24 Same as Prob. 7-23 except the angle of twist is 10°.

7-25 A 20-in-long hollow circular shaft is rotating at 500 rpm. Its outside diameter is 4 in and its inside diameter is 2 in. Using a factor of safety of 2, what is the maximum horsepower the shaft may transmit if it is made of a brittle material with an ultimate shear stress of 60,000 psi and $G = 10 \times 10^6$ psi?

7-26 Same as Prob. 7-25 except the shaft is made of a very ductile material with an ultimate shear stress of 20,000 psi.

7-27 A 3-ft-long solid steel shaft is to transmit 5 hp at 315 rpm. The steel is fairly brittle with an ultimate shear stress of 60,000 psi. In addition, the angular twist of the shaft is not to exceed 0.036 rad. Using a factor of safety of 5, determine the minimum allowable diameter of the shaft.

7-28 A straight, hollow, concentric circular shaft is made of a material whose partial shear stress-strain curve is shown. The OD is 3 in and its length is 25 in. The shaft is twisted by the application of a pure torque. A very skilled experimental mechanician measured the shear strain at the outside and inside surfaces and found them to be 0.006 and 0.004 rad, respectively. Determine (*a*) the inside radius, (*b*) the angle of twist, and (*c*) the applied torque.

PROB. 7-28

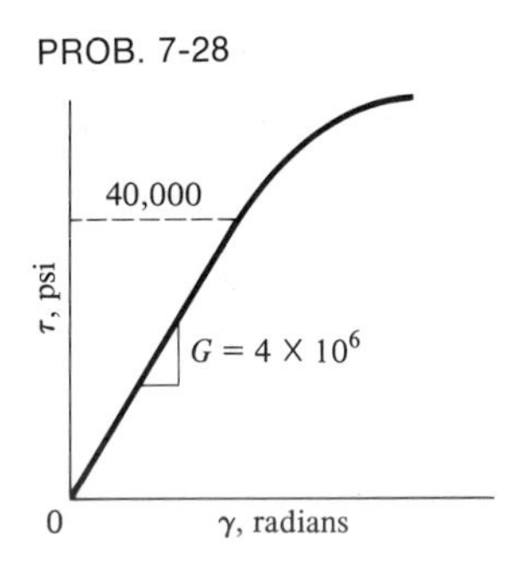

PROB. 7-29

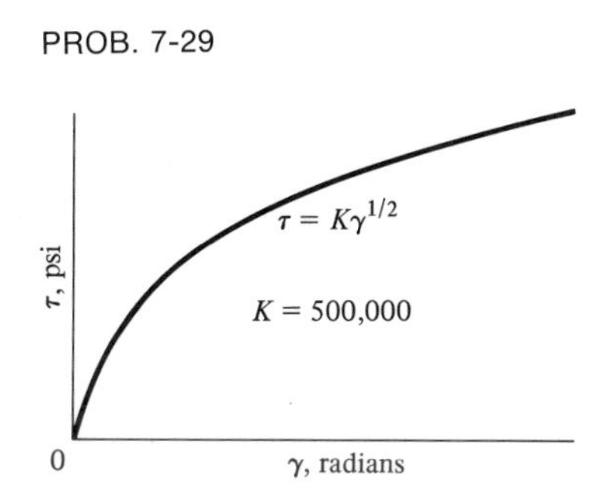

7-29 Same as Prob. 7-28 except the stress-strain curve is nonlinear and is approximated by the curve shown.

7-30 Consider a 1-in-diameter circular shaft made of the material whose partial shear stress-strain curve is shown in the figure for Prob. 7-23. Suppose the shaft is twisted by a torque through an angle of 5° and then released. What will be the residual amount of twist remaining after the torque has been removed if the shaft is (*a*) 30 in long and (*b*) 15 in long? Assume the material behaves linearly while unloading.

7-31 A serious skiing injury is the torsion fracture of the tibia (large bone in the lower leg). The tibia is of approximately 1 in outside diameter with a wall thickness of $\frac{1}{4}$ in. Assuming that the skier's leg is perpendicular to the ski and 4 ft from the ski tip, what lateral force on the ski tip would cause a shear stress of 5,000 psi in the tibia? Assume elastic behavior.

7-32 Rework Example 7-2 for a solid 2-in-diameter shaft.

7-5 CLOSURE

In this chapter we have considered a rather idealized type of structural member, namely, a straight shaft with a uniform circular cross section made of a homogeneous and isotropic material loaded with a pure twisting torque. Needless to say, there are many torque-carrying members which do not satisfy all of the above conditions, such as tapered shafts, shafts with grooves or fillets, members with square or some other noncircular cross section, composite shafts made of two or more materials, etc. Some of these situations can be handled with the theory developed in this chapter (for example, tables of stress-concentration factors for shafts with grooves and fillets are available), but the theory of noncircular shafts is beyond the scope of this text. Also, it is not uncommon to have members loaded with some combination of twisting, bending, and axial loads rather than just pure twisting. We will encounter some of these combined loading situations in subsequent chapters.

Just as for axially loaded two-force members, an idealized torsion problem can be statically indeterminate. Accordingly, it becomes necessary to invoke some kinematical compatibility condition to supplement the equilibrium requirements. Example 7-4 which follows shortly deals with a statically indeterminate torsion problem.

Finally, by this stage in your study of strength of materials, you should be acutely aware of the three primary factors which govern the load-carrying capacity of a structural element. First, the manner of loading (axial, twisting, etc.), second, the geometry of the member (straight, circular, etc.), and third, the mechanical properties of the material. All three must be taken into consideration before an engineer can make a rational analysis and judgment as to the performance of a load-carrying member.

EXAMPLE 7-4

A composite shaft shown in Fig. 7-17(*a*) is made up of a $1\frac{1}{2}$-in-diameter brass core ($G = 5.6 \times 10^6$ psi) intimately bonded to a 2-in-diameter steel sleeve ($G = 12 \times 10^6$ psi) so as to act as a solid shaft. If a torque of 10,000 in-lb is applied at the ends of this shaft, what will be (*a*) the largest shear stress in the steel, (*b*) the largest shear stress in the brass, and (*c*) the angle of twist?

Solution: The core and sleeve both transmit part of the total torque, and statical equilibrium requires that

$$T_s + T_c = T_{\text{total}} = 10{,}000 \text{ in-lb} \qquad (a)$$

Once we determine T_s and T_c, we can calculate the stresses in the sleeve and core by treating each one as an individual shaft. How-

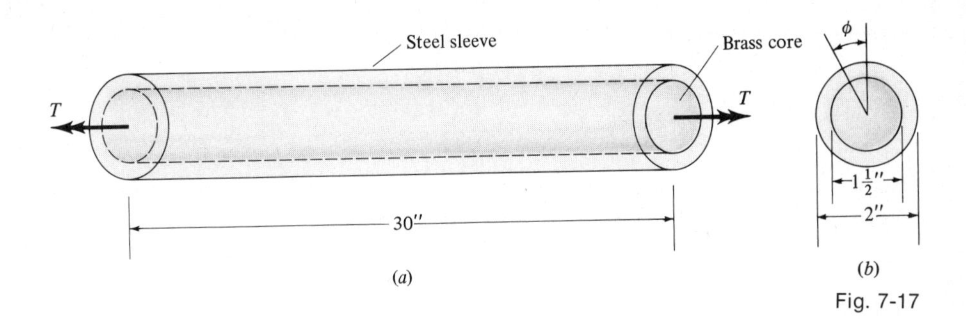

Fig. 7-17

ever, Eq. (*a*) is not sufficient to determine T_c or T_s, and the problem is statically indeterminate. Accordingly, we must appeal to some kinematical requirement to find another equation involving T_s and T_c.

Since the core and sleeve are bonded together as a solid unit, they must both twist through the same angle ϕ, as indicated in Fig. 7-17(*b*). Thus

$$\phi = \phi_s = \phi_c \tag{b}$$

Let us assume that both the sleeve and the core are behaving elastically, so that Eq. (7-23) is applicable to both. Then

$$\phi_s = \frac{T_s L_s}{J_s G_s} = \frac{T_c L_c}{J_c G_c} = \phi_c \tag{c}$$

or

$$T_s = \frac{J_s G_s}{J_c G_c} T_c \tag{d}$$

Now

$$J_s = \frac{\pi}{2}\left[1^4 - \left(\frac{3}{4}\right)^4\right] = \frac{175\pi}{512} \text{ in}^4$$

$$J_c = \frac{\pi}{2}\left(\frac{3}{4}\right)^4 = \frac{81\pi}{512} \text{ in}^4$$

Thus

$$T_s = \frac{175}{81}\,\frac{12 \times 10^6}{5.6 \times 10^6}\,T_c = 4.63T_c \tag{e}$$

Combining (*e*) with (*a*) yields

$$(4.63 + 1)T_c = 10{,}000$$

$$T_c = 1{,}775 \text{ in-lb}$$

$$T_s = 8{,}225 \text{ in-lb}$$

Still assuming elastic behavior, we use Eq. (7-22) to calculate the maximum shear stresses:

$$\tau_s = \frac{T_s R_s}{J_s} = \frac{(8{,}225)(1)}{175\pi/512} = 7{,}670 \text{ psi}$$

$$\tau_c = \frac{T_c R_c}{J_c} = \frac{(1{,}775)(3/4)}{81\pi/512} = 2{,}680 \text{ psi}$$

Both of these stresses appear to be low enough to justify our assumption of elastic behavior. Finally, by (*c*)

$$\phi = \frac{T_s L_s}{J_s G_s} = \frac{(8{,}225)(30)}{(175\pi/512)12 \times 10^6}$$

$$= 0.0191 \text{ rad}$$

$$= 1.1°$$

EXERCISE PROBLEMS

7-33 Rework Example 7-4 if the core is steel and the sleeve is brass.

7-34 The stepped shaft shown is made of steel and is to transmit a torque of 300 ft-lb. Neglecting the stress concentration at the fillet, what will be the maximum shear stress and the angle of twist?

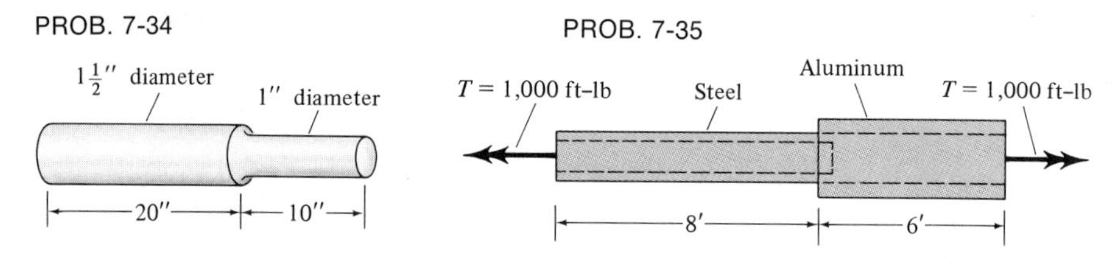

7-35 A piece of steel ($G = 12 \times 10^6$ psi) pipe (OD = 2.375 in, ID = 2.067 in) is coupled to a piece of aluminum ($G = 4 \times 10^6$ psi) pipe (OD = 2.875 in, ID = 2.469 in) as shown. Find the maximum shear stress in the steel and aluminum. Find the total angle of twist.

7-36 A shaft is made of a solid section of steel ($G = 12 \times 10^6$ psi) of 1 in diameter coupled to a solid section of brass ($G = 5.6 \times 10^6$ psi) of 2 in diameter. The combined length is 50 in. A torque of 5,000 in-lb applied to each end produces a total angle of twist of 5°. What is the length of the brass section?

7-37 The $1\frac{1}{4}$-in-diameter steel shaft carries an input torque of 500 ft-lb at pulley *B* and output torques of 300 and 200 ft-lb at *A* and *C*, respectively. Find the maximum shear stress in the shaft. What is the relative angle of twist between pulleys *A* and *C*?

PROB. 7-37

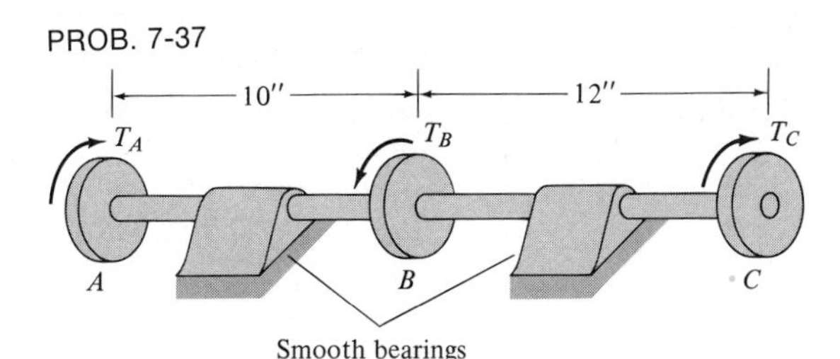

7-38 Same as Prob 7-37 except T_C is an input torque of 500 ft-lb and T_A (opposite to that shown) and T_B are output torques of 200 and 300 ft-lb, respectively.

7-39 The large stepped steel turbine shaft shown rotates at 5,500 rpm. Design specifications allow for a maximum shear stress of 30,000 psi and a total angle of twist of 3° and call for a factor of safety of 3. Assuming elastic behavior and neglecting the stress concentration at the fillet, determine the maximum horsepower the shaft may transmit.

PROB. 7-39

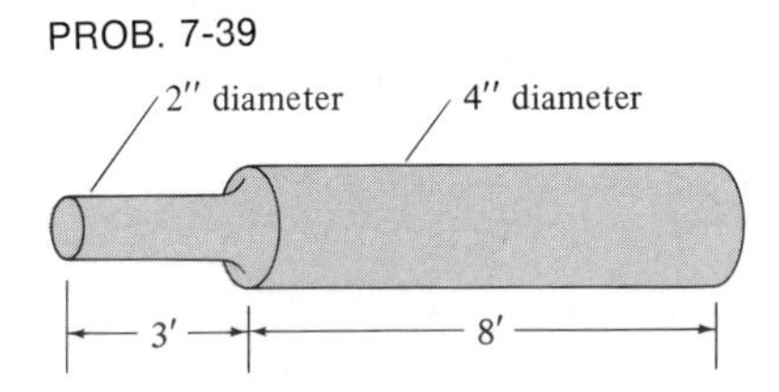

7-40 A hollow steel shaft with OD = 2 in and ID = 1 in is fastened between two rigid walls and is loaded as shown. A partial shear stress-strain curve is also shown. What is the torque reaction at each wall if the angular motion θ of the bar is 0.5°?

PROB. 7-40

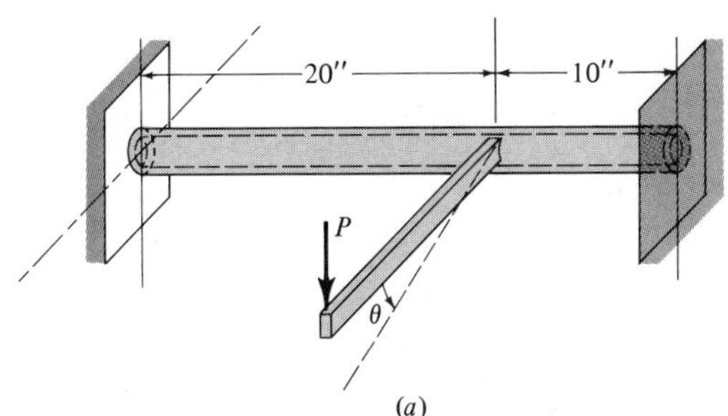

(a)

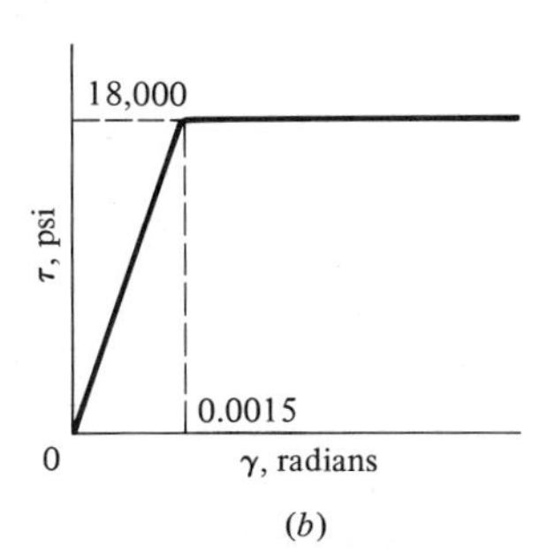

(b)

7-41 Same as Prob. 7-40 except $\theta = 5.0°$.

7-42 For the composite shaft shown, what is the ratio of the maximum shear stress in the aluminum to the maximum shear stress in the steel. Assume elastic behavior and that $G_s = 3G_a$.

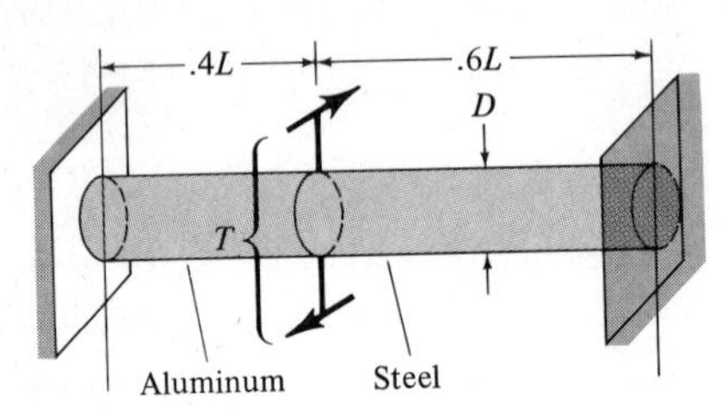

PROB. 7-42

chapter 8 GENERALIZED STRESS AND STRAIN

8-1 INTRODUCTION

In the preceding chapters we have considered the stresses, strains, and deformations in two types of relatively simple, but common, structural members, namely, axially loaded two-force members and circular cylindrical torsion members. However, in many, if not most, machines and structures, things are not quite so simple in that the members often have some complex geometry and/or are loaded with a combination of axial, twisting, bending, and shearing loads (recall the internal reactions of Chap. 5). Consequently, in such members it is highly probable that both normal and shear stresses will exist simultaneously and that the resulting deformations will involve both axial and shear strains.

It is our intention in this chapter to first look at situations involving combined normal and shear stresses. Then we will examine the kinematics of combined axial and shear strains. Finally, we will relate these combined stresses and strains in terms of a generalized stress-strain relation for solid structural materials. This study of generalized stress and strain will form the basis for the analysis and design of the more complex structural members, such as pressure vessels, which will be considered in this chapter, and beams, which will be considered in Chap. 9.

8-2 GENERAL STATE OF STRESS

Let us begin our discussion of combined loading by considering a solid circular cylinder subjected simultaneously to a centroidal axial load P and a twisting torque T, as shown in Fig. 8-1(a). We have already considered the stresses produced by each of these

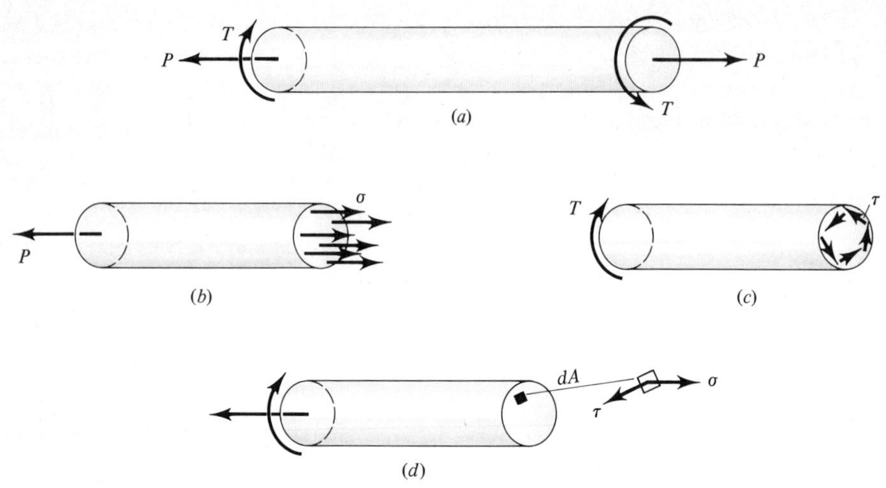

Fig. 8-1

loads individually: the axial tensile stresses indicated in Fig. 8-1(*b*) and the torsion shear stresses in Fig. 8-1(*c*). Recall that the normal stress σ in Fig. 8-1(*b*) will likely be uniform throughout the entire member, whereas the torsion shear stress τ will usually vary with the radial distance from the axis of the cylinder. Thus, combining these situations, we see in Fig. 8-1(*d*) that a normal stress σ and a shear stress τ will exist simultaneously on a typical elemental area dA of the cross section perpendicular to the longitudinal axis of the cylinder. This is a relatively simple example of *combined stresses.*

Remark: In this example it is quite likely that the value of σ could be calculated by using Eq. (6-1) [or Eq. (5-13)]

$$\sigma = \frac{P}{A}$$

and the values of τ by using Theorem 7-2 for elastic behavior

$$\tau = \frac{Tr}{J}$$

or some other similar relation for inelastic behavior. This procedure of calculating the individual stresses produced by individual loads and then combining the stresses, as we have done in Fig. 8-1(*d*), is called *superposition of stresses.* Superposition is usually valid so long as neither of the individual loads *nor their combined effect* severely deforms the member or causes a high degree of nonlinear inelastic behavior of the material. Such extreme situations are much more difficult to analyze and will not be treated in this text.

Now that we have seen how combined stresses can arise in a relatively simple situation, let us look at a much more general situation. Consider some type of embedded fixture of arbitrary size and shape loaded by some complex combination of loads, as illustrated in Fig. 8-2(*a*). Let us try to visualize the stresses which might exist in the material of this fixture at some arbitrary point such as point q. Recalling that the stress vector on any plane area can be resolved into a normal stress component and a shear stress component [Eq. (5-5)], let us agree to denote a normal stress component (tensile or compressive) by σ and a shear stress component (not necessarily due to torsion) by τ. Also, let us choose the x, y, z coordinates as shown.

First, let us visualize a cutting plane through point q parallel to the y–z plane (perpendicular to the x axis), as indicated in Fig. 8-2(*b*). The corresponding normal stress is denoted by σ_x and the shear stress by τ_x. Similarly, if we pass a cutting plane parallel to the x–z plane (perpendicular to the y axis), the corresponding

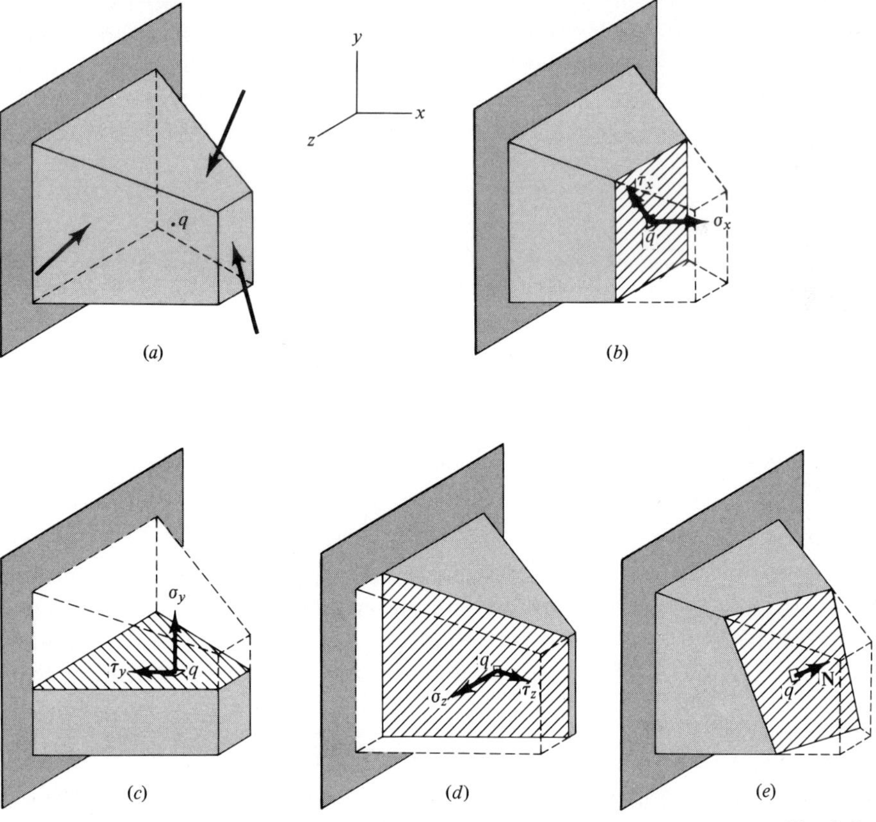

Fig. 8-2

stresses are denoted by σ_y and τ_y, as in Fig. 8-2(*c*). The same procedure is followed for the x–y plane and the stresses σ_z and τ_z indicated in Fig. 8-2(*d*). From this discussion it should be apparent that one cannot talk meaningfully about the stresses at a point without identifying the area on which those stresses act. The values and directions of both the normal stress and the shear stress depend upon the orientation of the cross-sectional area on which the stresses act (recall the discussion of Sec. 5-5); that is, at any arbitrary point such as point q,

$$\sigma = \sigma(\mathbf{N}) \qquad \tau = \tau(\mathbf{N}) \tag{8-1}$$

where **N** is the unit vector normal (perpendicular) to whatever cross-sectional area is considered at point q, as illustrated in Fig. 8-2(*e*). When we can determine σ and τ for *any* arbitrary **N**, we say that we know the *state of stress* or the *stress tensor* at point q.

Before continuing with this formal discussion of state of stress, you might perform a little experiment to demonstrate the importance of orientation when considering stresses and strength of materials. Take two pieces of notebook paper and mark a point q on each. Slit one with a razor blade or knife, as indicated in Fig. 8-3(*a*), and the other as indicated in Fig. 8-3(*b*). Pull on each of them as in Fig. 8-3(*c*) and (*d*) and observe the rather dramatic difference in the effect which the cuts have on the behavior of the paper.

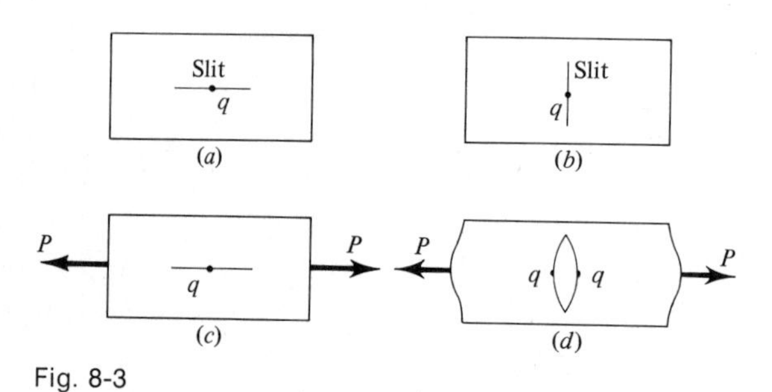

Fig. 8-3

Returning to Fig. 8-2, let **i**, **j**, and **k** be the usual unit vectors, so that the vector **i** is the unit vector normal to the plane area in Fig. 8-2(*b*), **j** is the unit normal for the plane area in Fig. 8-2(*c*), and **k** is the unit normal for that in Fig. 8-2(*d*). Then, in the notation of Eq. (8-1)

$$\begin{aligned} \sigma_x &= \sigma(\mathbf{i}) \qquad & \tau_x &= \tau(\mathbf{i}) \\ \sigma_y &= \sigma(\mathbf{j}) \qquad & \tau_y &= \tau(\mathbf{j}) \\ \sigma_z &= \sigma(\mathbf{k}) \qquad & \tau_z &= \tau(\mathbf{k}) \end{aligned} \tag{8-2}$$

Recall that any unit vector **N** can be written as

$$\mathbf{N} = N_1\mathbf{i} + N_2\mathbf{j} + N_3\mathbf{k}$$

Hence, for any **N**, Eq. (8-1) becomes

$$\begin{aligned}\sigma &= \sigma(\mathbf{N}) = \sigma(N_1\mathbf{i} + N_2\mathbf{j} + N_3\mathbf{k}) \\ \tau &= \tau(\mathbf{N}) = \tau(N_1\mathbf{i} + N_2\mathbf{j} + N_3\mathbf{k})\end{aligned} \tag{8-3}$$

In the next sections we shall see how Eqs. (8-2) and (8-3), when taken together with the requirements of equilibrium, imply that the stresses $\sigma(\mathbf{N})$ and $\tau(\mathbf{N})$ on any surface with orientation specified by the unit vector **N** are given by some linear combination of the stresses σ_x, τ_x; σ_y, τ_y; and σ_z, τ_z. We formalize this implication by

Theorem 8-1 The state of stress at any one point q in a body of solid material is completely specified by giving each of the normal and shear stresses on three mutually perpendicular plane sections passing through the point; i.e., the stress *tensor* is specified by giving the stress *vector* on each of three mutually orthogonal planes.

Quite often it is convenient to visualize a state of stress at some point q in a loaded body by using the scheme indicated in Fig. 8-4. In this scheme a small rectangular prism is used to visualize the planes at point q of the body, and the stresses on three mutually perpendicular faces of the prism are labeled in some manner.

The notation used in Fig. 8-4 is quite common (although not unique), whereby the normal stresses on the x, y, and z faces (or **i**, **j**, and **k** faces) are denoted by σ_x, σ_y, and σ_z, respectively, and the shear stress τ_x has been resolved into two components τ_{xy} and τ_{xz}; likewise, τ_y into τ_{yx} and τ_{yz}; and τ_z into τ_{zx} and τ_{zy}. In this way, the state of stress at point q is completely described by nine component

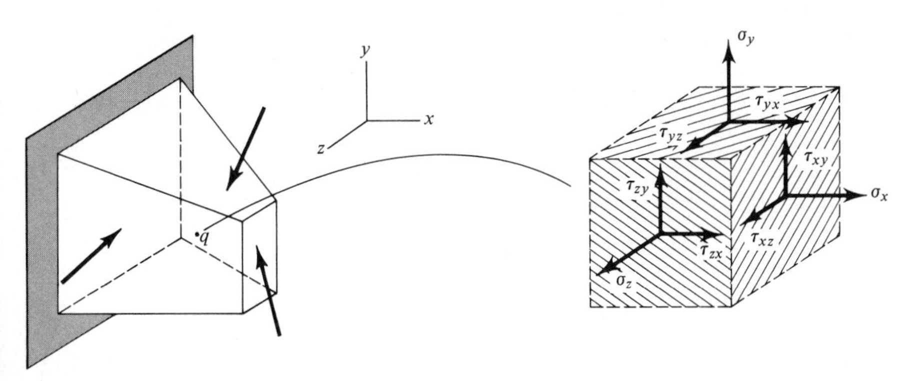

Fig. 8-4

stresses, three normal stresses, and six shear stresses. Figure 8-4(b) illustrates what is usually referred to as a *general state of stress.*

Remark: If the rectangular prism in Fig. 8-4 were actually a chunk of the material surrounding point q, stresses would likely exist on all *six* faces of the prism, the negative x, y, and z faces (with unit normals $-\mathbf{i}$, $-\mathbf{j}$, and $-\mathbf{k}$, respectively), as well as the positive x, y, and z faces. However, here we are using the prism merely as a convenient scheme for identifying three mutually perpendicular plane cross sections through point q, and it is on only those three planes that we have shown and labeled the stresses.

From this discussion you can see that the state of stress at any point in a loaded body can be a rather complex combination of normal and shear stresses. However, from an engineering viewpoint, the vital question is: What are the maximum normal and maximum shear stresses at this point? After all, the material of the body is quite unaware of our analysis and the manner in which we have labeled various planes, axes, and stresses. The material must be able to withstand the maximum stress (or stresses) *regardless* of the orientation of the plane on which it occurs.

For any given general state of stress as represented by Eq. (8-1), the problem of determining the maximum normal and shear stresses is a fundamental problem in linear algebra and tensor analysis. However, as a practical matter, this general three-dimensional problem often reduces to a much more tractable two-dimensional problem that will be considered in the next section; that is, in many situations the stresses on one or more of the faces of the prism in Fig. 8-4 will be zero or otherwise negligible. For example, in a uniaxially loaded two-force member, the state of stress

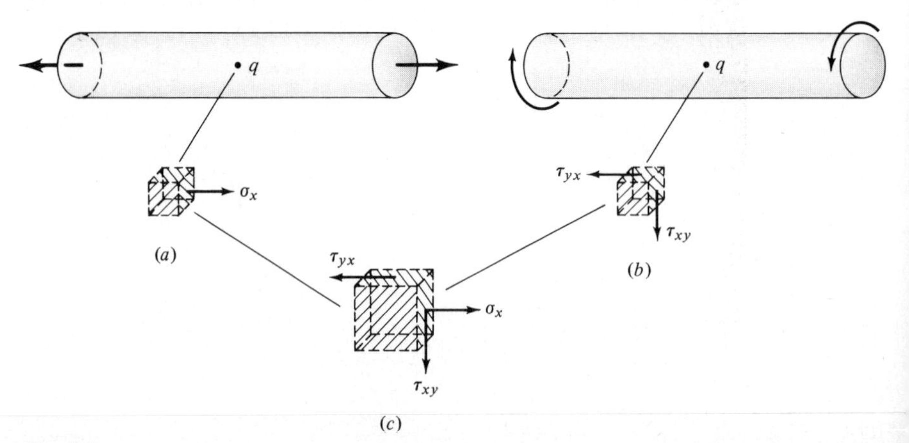

Fig. 8-5

is shown in Fig. 8-5(*a*), while that for a pure-torsion member is shown in Fig. 8-5(*b*), and their combination is shown in Fig. 8-5(*c*). Each of these is a much more simple state of stress than the general state of stress depicted in Fig. 8-4.

> ***Remark:*** For the pure-torsion member in Fig. 8-5(*b*), while the shear stress τ_{xy} can be anticipated from the work in Chap. 7, the simultaneous appearance of τ_{yx} will be explained in the next section.

EXERCISE PROBLEMS

8-1 A solid 2-in-diameter steel shaft carries an axial tensile load of 6 tons and a torque of 5,000 in-lb. Determine and sketch the state of stress existing (*a*) at a point on the outer surface of the shaft and (*b*) at a point on the longitudinal centroidal axis of the shaft.

8-2 A hollow steel shaft with an OD of 2 in and an ID of 1 in is loaded with an axial tensile force of 10,000 lb and a torque of 1,000 ft-lb. Determine and sketch the state of stress existing at a point (*a*) on the outside surface of the shaft and (*b*) on the inside surface of the shaft.

8-3 A scuba diver's air tank can be considered to be a hollow cylinder with spherical caps on the ends. The air pressure inside the tank produces stresses in the tank walls. Sketch what you believe to be the state of stress at a point (*a*) on the outer cylindrical surface, (*b*) on the outer spherical surface, and (*c*) on the inner cylindrical surface.

8-4 For the cantilever beam shown, sketch what you believe to be the state of stress at point *a*, point *b*, and point *c*.

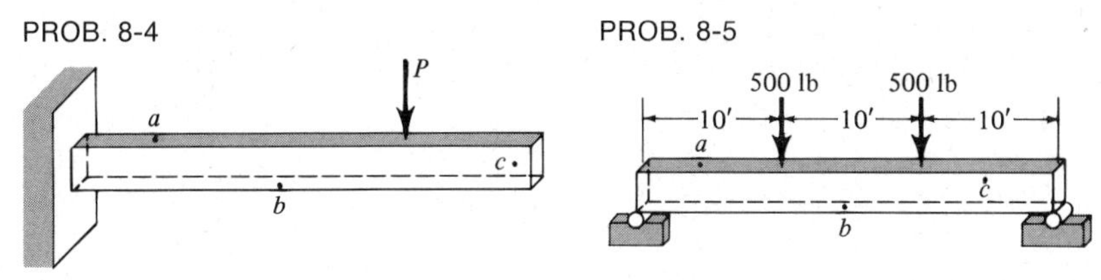

8-5 A 30-ft-long beam is effectively loaded and supported in the manner shown. Sketch what you believe to be the state of stress at point *a*, point *b*, and point *c*.

8-6 A piece of ¾-in thin-wall copper tubing carries a fluid under a hydrostatic pressure of 100 psi. Due to faulty installation, the tubing is also subjected to a torque of 80 in-lb. Sketch what you believe to be the state of stress in the wall of the tubing.

8-3 ANALYSIS OF PLANE STRESS

As we have said earlier, it is possible for the state of stress in a structural member to be the general triaxial (three-dimensional) state depicted in Fig. 8-4. However, for many common structural members such as beams, columns, shafts, plates, and shells, quite often the primary stresses, insofar as design considerations are concerned, occur at a free surface and, therefore, constitute a biaxial or *plane* (two-dimensional) state of stress.

Essentially, a plane state of stress is one for which all the stresses on one of the three mutually orthogonal faces of our rectangular prism are zero. Carrying this simplification a step further, a *uniaxial* state of stress is one for which all the stresses on two of the faces are zero. Thus, the state of stress in Fig. 8-5(*a*) is uniaxial, while that in Fig. 8-5(*b*) is a plane state, and their combination in Fig. 8-5(*c*) is also a plane state of stress. Thus, a uniaxial state is a special case of the plane (biaxial) state, and the plane state is a special case of the general (triaxial) state. We will now concentrate our attention on the plane state of stress.

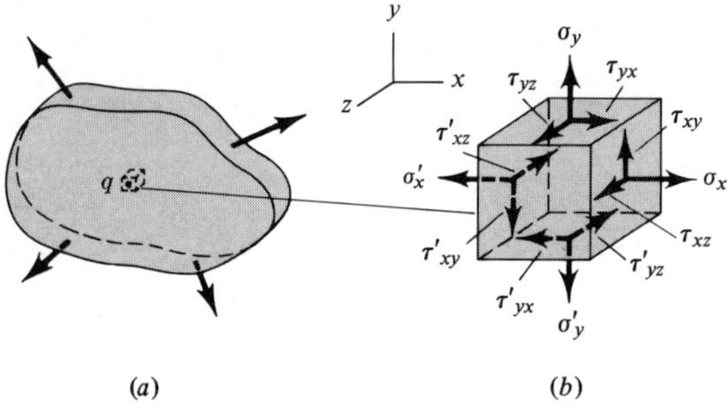

Fig. 8-6

Let us begin by considering in Fig. 8-6 a small rectangular *block of material* surrounding some point q in a loaded body, and let us suppose that the stresses σ_z, τ_{zx}, and τ_{zy} on the z faces are all zero. Notice that, since this block actually represents a piece of material, we have shown stresses on both the positive and negative x and y faces. We have denoted the stresses on the negative faces with primes merely to indicate that they need not have the same values (or directions) as the corresponding stresses on the positive faces.

If the original structural body in Fig. 8-6(*a*) is in structural equilibrium (recall Sec. 4-4), then the block in Fig. 8-6(*b*) must also be in equilibrium. We will now see what equilibrium requires in regard to the stresses on this block.

Remark: Strictly speaking, Fig. 8-6(*b*) is *not* a free-body diagram in that we have indicated stresses (force per unit area)

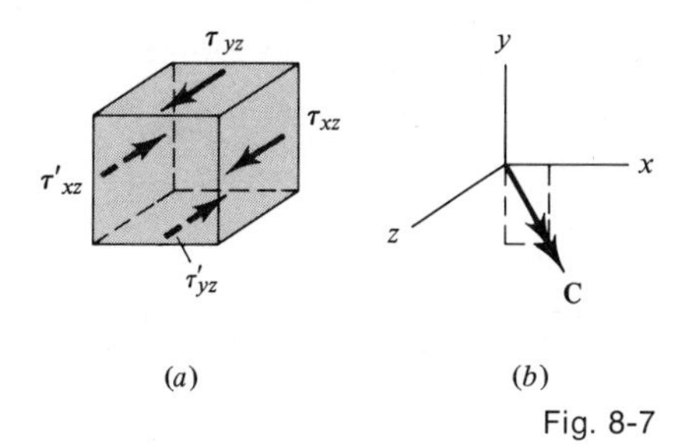

Fig. 8-7

rather than forces on the block of material. However, such *stress diagrams*, when properly interpreted, can be quite useful in discussing equilibrium requirements.

First of all, equilibrium requires that the sum of forces and the sum of moments must both be zero. Considering forces in the z direction, the four shear stresses τ_{xz}, τ_{yz}, τ'_{xz}, and τ'_{yz} shown in Fig. 8-7(a) must produce a zero net z force, and thus their only possible resultant effect can be a couple **C** whose vector lies in the x–y plane, as shown in Fig. 8-7(b). (This is plausible if you visualize the τ_{yz} and τ'_{yz} stresses as producing an x couple and the τ_{xz} and τ'_{xz} stresses as producing a y couple.) However, this couple must be zero since the block is in equilibrium, and none of the other remaining x–y stresses in Fig. 8-8(a) can produce an x–y couple. Hence, equilibrium requires that the z stresses τ_{xz}, τ'_{xz}, τ_{yz}, and τ'_{yz} must *all* vanish, and our state of stress reduces to that shown in Fig. 8-8(a). This is truly a *plane state* of stress in that all the stress vectors are parallel to the x–y plane, and this situation can be represented in a two-dimensional manner, as illustrated in Fig. 8-8(b). The half arrowheads on the shear stresses are commonly used in such representations.

Let us examine this plane state of stress more closely. As we have said, the stresses on the positive faces of this chunk of material need not be the same as those on the negative faces, since stresses can and usually do vary from point to point throughout a structural body. However, our intention here is to examine the

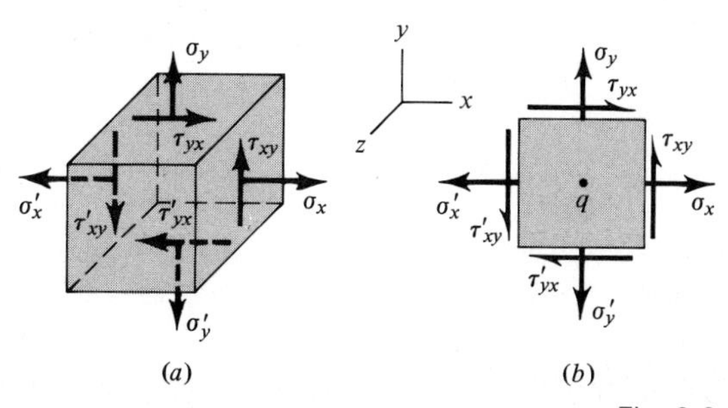

Fig. 8-8

state of stress at *one* particular point, in this case point q. Therefore, let us agree that this chunk of material surrounding point q can be made arbitrarily small. We symbolically indicate this assumption in Fig. 8-9(a) by labeling the dimensions of this block as dx, dy, and dz. Accordingly, if we denote the stresses on the positive faces as σ_x, τ_{xy}, σ_y, and τ_{yz}, then the stresses on the negative faces differ from these by some arbitrarily small amounts $d\sigma_x$, $d\tau_{xy}$, $d\sigma_y$, and $d\tau_{yx}$, as indicated in Fig. 8-9(b).

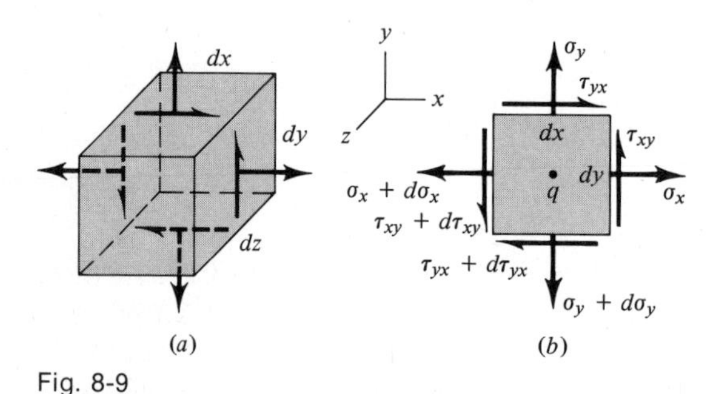

Fig. 8-9

Since this small block of material is assumed to be in structural equilibrium, let us consider moments with respect to point q. Realizing that stresses are forces per unit area and, therefore, must be multiplied by the area on which they act, we have

$$\mathbf{M}_q = \left[\tau_{xy}(dy)(dz)\,\frac{dx}{2} + (\tau_{xy} + d\tau_{xy})(dy)(dz)\,\frac{dx}{2}\right]\mathbf{k} - \left[\tau_{yx}(dx)(dz)\,\frac{dy}{2} + (\tau_{yx} + d\tau_{yx})(dx)(dz)\,\frac{dy}{2}\right]\mathbf{k} = \mathbf{0}$$

where $\mathbf{k}$ is the usual unit vector parallel to the z axis. Hence, rotational equilibrium requires

$$\tau_{xy} - \tau_{yx} + \frac{d\tau_{xy} - d\tau_{yx}}{2} = 0 \tag{8-4}$$

which, since $d\tau_{xy}$ and $d\tau_{yx}$ are arbitrarily small, implies

$$\tau_{xy} - \tau_{yx} = 0$$

or

$$\tau_{xy} = \tau_{yx} \tag{8-5}$$

This result is very important in that it says that if the shear stress τ_{xy} is nonzero, the shear stress τ_{yx} is also nonzero, and thus *these two shear stresses must always exist in simultaneous and equal pairs.* [This is the reason why the two shear stresses τ_{xy} and τ_{yx} were indicated in the state of stress for the torsion member in Fig. 8-5(b) and (c).] This result is generalized by the following theorem.

Theorem 8-2 The general state of stress at any point q in a body of solid material is *symmetric* in that the shear stresses on mutually perpendicular planes must exist in simultaneous and equal pairs; that is,

$$\begin{aligned}\tau_{xy} &= \tau_{yx} \\ \tau_{xz} &= \tau_{zx} \\ \tau_{yz} &= \tau_{zy}\end{aligned} \tag{8-5a}$$

Consequently, a state of stress is specified by giving values for six (rather than nine) stresses—three normal and three shear stresses—and a *plane* state of stress is specified by giving values for three stresses—two normal and one shear stress.

We close this article with some observations regarding the directions of shear stresses. Suppose the shear stress τ_{xy} on the positive x face of our small chunk of material is upward, as in Fig. 8-10(a). Then, vertical equilibrium would require the shear stress on the negative x face to be downward, as shown. (Note that *both* these arrows could be reversed and still satisfy vertical equilibrium.) Suppose also that the shear stress τ_{yx} on the positive y face is to the right. Then the shear stress on the negative y face would be to the left, as also shown in Fig. 8-10(a). (Again, note that both arrows could be reversed.) Then rotational equilibrium can be satisfied since the vertical shear stresses tend to produce a counterclockwise couple while the horizontal shear stresses tend to produce a clockwise couple. Thus, Fig. 8-10(a) and (b) shows permissible combinations of shear stresses, while Fig. 8-10(c) and (d) shows nonpermissible combinations, since both situations would violate rotational equilibrium. Study these figures carefully until you thoroughly understand them.

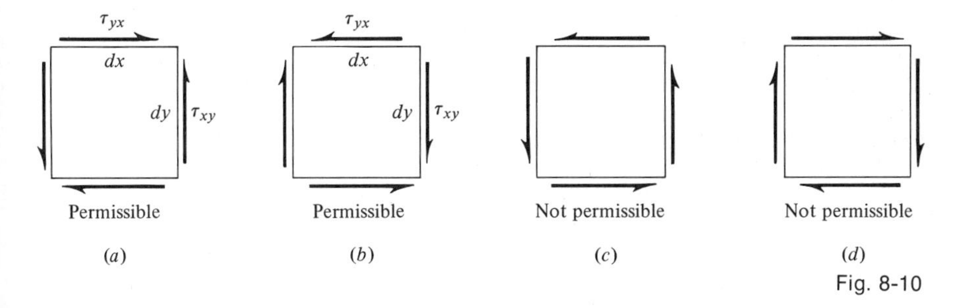

Fig. 8-10

8-4 MOHR's CIRCLE OF STRESS

We have just seen how a state of plane stress at a point can be specified by giving values of two normal stresses and one shear stress on mutually perpendicular planes, such as the vertical plane whose unit normal is the vector **i**, and the horizontal plane whose normal is **j**. We now return to the questions posed in Sec. 8-2, namely: If we know the stresses on the **i** and **j** planes through point q, what are the corresponding stresses on *any* other plane with normal **N**, as indicated in Fig. 8-11(a)? Furthermore, of all these possible planes, which is the one (or ones) with the largest normal and/or shear stresses?

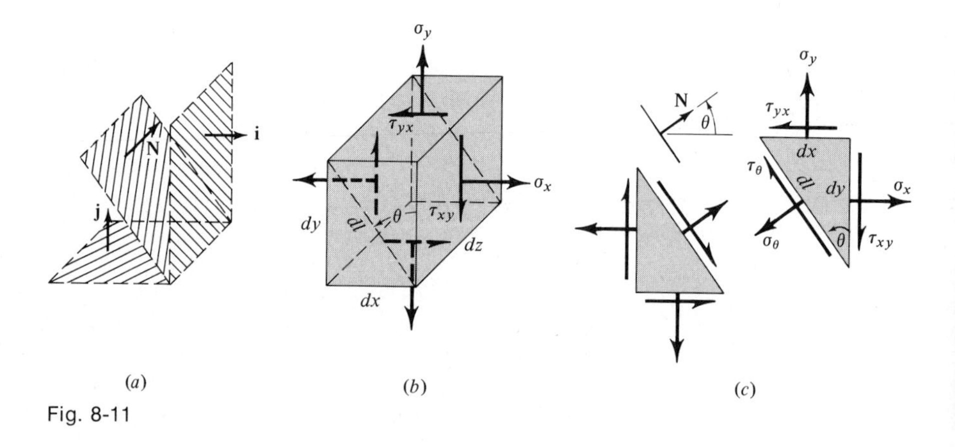

Fig. 8-11

Since our first objective is to determine the stresses on the **N** plane, let us proportion our small chunk of material in the manner shown in Fig. 8-11(b), whereby

$$dx = d\ell \sin\theta \qquad dy = d\ell \cos\theta \tag{8-6}$$

where $d\ell$ is the z-face diagonal of our block and θ is the angle of orientation of the inclined **N** plane relative to the **i** plane. Also, for reasons which will be apparent later, let us assume the normal and shear stresses on the vertical and horizontal faces (**i** plane and **j** plane, respectively) to be as shown. Now, let us cut this block into two wedges along the inclined plane and indicate the stresses acting on each of the two wedges, as shown in Fig. 8-11(c). The normal stress σ_θ and shear stress τ_θ represent the stresses acting on the inclined **N** plane. These wedges, like the original block, are chunks of material and are also in equilibrium. Realizing that the wedges are some dz units thick, if we sum forces (stress times area) in the **N** direction on the right wedge, equilibrium requires

$$\begin{aligned} -\mathbf{N}\sigma_\theta(d\ell)(dz) + \mathbf{N}\sigma_x(dy)(dz)\cos\theta - \mathbf{N}\tau_{xy}(dy)(dz)\sin\theta \\ + \mathbf{N}\sigma_y(dx)(dz)\sin\theta - \mathbf{N}\tau_{yx}(dx)(dz)\cos\theta = \mathbf{0} \end{aligned}$$

Rearranging, using Eqs. (8-6), then factoring out $(d\ell)(dz)$, we obtain

$$\sigma_\theta = \sigma_x \cos^2\theta + \sigma_y \sin^2\theta - 2\tau_{xy}\sin\theta\cos\theta \tag{8-7}$$

where we have used the result of Eq. (8-5), namely, $\tau_{xy} = \tau_{yx}$. Similarly, summing forces perpendicular to **N** yields

$$\tau_\theta = (\sigma_x - \sigma_y) \sin\theta \cos\theta + \tau_{xy}(\cos^2\theta - \sin^2\theta) \tag{8-8}$$

Observe that for $\theta = 0$

$$\sigma_0 = \sigma_x \qquad \tau_0 = \tau_{xy} \tag{8-9}$$

and for $\theta = 90°$

$$\sigma_{90} = \sigma_y \qquad \tau_{90} = -\tau_{xy} \tag{8-10}$$

Accordingly, it is convenient (and somewhat traditional) to adopt an algebraic sign convention for the normal stress σ_θ and shear stress τ_θ with the following physical interpretations:

1. Tensile stresses (pulling) will be assigned positive algebraic values, while compressive stresses (pushing) will be assigned negative values.
2. Shear stresses which tend to rotate our small block of material in a *clockwise* direction will be assigned *positive* algebraic values, while shear stresses which tend to rotate the block counterclockwise will be assigned negative values.

Thus, referring to Fig. 8-11(*b*), the indicated shear stress τ_{xy} would correspond to a positive value, while τ_{yx} would correspond to a negative value (of the same magnitude as τ_{xy}). Conversely, in Fig. 8-10(*a*) the indicated τ_{xy} would correspond to a negative value and τ_{yx} to a positive one. Generalizing these observations, the shear stresses on perpendicular planes will always have the same magnitude but opposite signs

$$\tau_{\theta+90°} = -\tau_\theta \tag{8-11}$$

This result [easily verified by Eq. (8-8)] is simply an algebraic restatement of the last paragraph of Sec. 8-3 in that if a shear stress tends to produce a clockwise rotation, then the simultaneous shear stress on the perpendicular plane must produce a counterclockwise rotation, and vice versa.

Equations (8-7) and (8-8) are called the *stress-transformation equations for plane stress*. If we know the values of σ_x, σ_y, and τ_{xy}, then these equations will enable us to determine the normal and shear stresses on any other plane oriented at angle θ relative to the original given planes. Thus we have answered the first question posed in the opening paragraph of this article.

Remark: The original given planes need not be vertical and horizontal, but can be any two mutually orthogonal planes.

The second question regarding the largest normal and shear stresses can now be answered by maximizing Eqs. (8-7) and (8-8). Before doing that, however, it is desirable to rewrite these equations as

$$\sigma_\theta = \frac{\sigma_x + \sigma_y}{2} + \frac{\sigma_x - \sigma_y}{2} \cos 2\theta - \tau_{xy} \sin 2\theta \tag{8-12}$$

$$\tau_\theta = \frac{\sigma_x - \sigma_y}{2} \sin 2\theta + \tau_{xy} \cos 2\theta \tag{8-13}$$

where we have utilized the double-angle identities

$$\cos 2\theta = \cos^2 \theta - \sin^2 \theta$$

$$\sin 2\theta = 2 \sin \theta \cos \theta$$

Now, to maximize, we differentiate Eqs. (8-12) and (8-13)

$$\frac{d\sigma_\theta}{d\theta} = -(\sigma_x - \sigma_y) \sin 2\theta - 2\tau_{xy} \cos 2\theta$$

$$\frac{d\tau_\theta}{d\theta} = (\sigma_x - \sigma_y) \cos 2\theta - 2\tau_{xy} \sin 2\theta$$

Thus, setting these equal to zero, the angle which maximizes (or minimizes) σ_θ is given by

$$\tan 2\theta = -\frac{\tau_{xy}}{(\sigma_x - \sigma_y)/2} \tag{8-14}$$

and that which maximizes (or minimizes) τ_θ is given by

$$\tan 2\theta = \frac{(\sigma_x - \sigma_y)/2}{\tau_{xy}} \tag{8-15}$$

Observe that Eqs. (8-14) and (8-15) are negative reciprocals, which physically means that the plane on which the maximum normal stress acts is not the same as the plane on which the maximum shear stress acts. In fact, these planes differ by 45° (recall Theorem 5-1).

All of these results given by Eqs. (8-7) to (8-15) can be summarized by a graphical representation called the Mohr[1] circle of stress. To develop this representation, let us transpose the first term on the right side of Eq. (8-12) to the left side and then square both this equation and Eq. (8-13), thus obtaining

$$\left[\sigma_\theta - \frac{\sigma_x + \sigma_y}{2}\right]^2 = \left[\frac{\sigma_x - \sigma_y}{2} \cos 2\theta - \tau_{xy} \sin 2\theta\right]^2$$

$$\tau_\theta{}^2 = \left[\frac{\sigma_x - \sigma_y}{2} \sin 2\theta + \tau_{xy} \cos 2\theta\right]^2$$

[1] Named after its developer, Otto Mohr, a professor of structural mechanics in Germany (1835–1918).

Now, adding these equations, expanding, and rearranging we obtain

$$\left[\sigma_\theta - \frac{\sigma_x + \sigma_y}{2}\right]^2 + [\tau_\theta - 0]^2 = \left(\frac{\sigma_x - \sigma_y}{2}\right)^2 + \tau_{xy}^2 \qquad (8\text{-}16)$$

which is the equation of a circle. To see this, recall that the equation of a circle in rectangular coordinates has the general form

$$(x - a)^2 + (y - b)^2 = r^2 \qquad (8\text{-}17)$$

where the center of the circle is located at coordinates (a,b) and the radius is r. Thus, if we let σ_θ and τ_θ correspond to the coordinate axes x and y, respectively, then the circle given by Eq. (8-16) will have its center at coordinates $[(\sigma_x + \sigma_y)/2,0]$, and its radius will be

$$r = \sqrt{\left(\frac{\sigma_x - \sigma_y}{2}\right)^2 + \tau_{xy}^2} \qquad (8\text{-}18)$$

as illustrated in Fig. 8-12(*a*). In this particular illustration, we have arbitrarily assumed that σ_x is a positive algebraic value, σ_y is negative, and τ_{xy} is a positive value. [Note that τ_{yx} corresponds to the negative value of τ_{xy}, as expected from Eq. (8-10) and our previously adopted sign convention.] This illustration embodies all the essential features of Mohr's circle of plane stress.

1 The coordinates of each point on the locus of the circle correspond to the stresses acting on some plane. For example, the point with coordinates (σ_x,τ_{xy}) in Fig. 8-12(*a*) corresponds to the stresses on the **i** plane in Fig. 8-12(*b*), the point

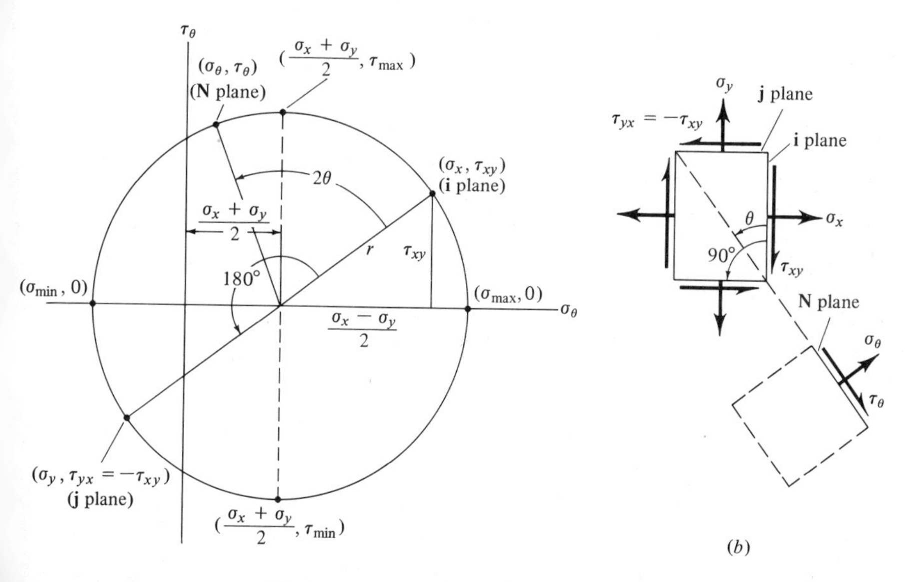

Fig. 8-12

with coordinates $(\sigma_y, -\tau_{xy})$ corresponds to the stresses on the **j** plane, and the coordinates of the arbitrary point $(\sigma_\theta, \tau_\theta)$ correspond to the stresses on the arbitrary **N** plane.

2 The angle θ between the **i** plane and the arbitrary **N** plane in Fig. 8-12(*b*) corresponds to the angle 2θ in Fig. 8-12(*a*) between the point representing the **i**-plane stresses (σ_x, τ_{xy}) and the point representing the **N**-plane stresses $(\sigma_\theta, \tau_\theta)$. In particular, since the **i** and **j** planes are 90° apart in Fig. 8-12(*b*), their corresponding points (σ_x, τ_{xy}) and $(\sigma_y, -\tau_{xy})$ are 180° apart on the circle in Fig. 8-12(*a*), i.e., diametrically opposite points.

3 The center of the circle is *always* on the horizontal σ_θ coordinate axis. Accordingly, the right extremity of the circle $(\sigma_{max}, 0)$ corresponds to the plane with the largest algebraic value σ_{max} of the normal stress σ_θ, and the left extremity $(\sigma_{min}, 0)$ corresponds to the plane with the minimum algebraic value σ_{min}. These maximum and minimum values of normal stress are called the *principal values of stress*, and the corresponding planes, the *principal planes*. Note that since the left and right extremities of the circle are diametrically opposite, the principal planes will always be perpendicular to each other. Also, there are *no* shear stresses on the principal planes.

4 Similarly, the upper extremity of the circle with coordinates $[(\sigma_x + \sigma_y)2, \tau_{max}]$ corresponds to the plane on which the largest shear stress acts, whose value is equal to the radius of the circle; and the lower extremity corresponds to the plane on which $\tau_{min} = -\tau_{max}$ acts. Observe that since these points on the circle are 90° away from the points $(\sigma_{max}, 0)$ and $(\sigma_{min}, 0)$ the planes of maximum shear stress will always be oriented at 45° from the principal planes, as previously observed in the paragraph following Eqs. (8-14) and (8-15).

In summary, then, the locus of the Mohr circle represents the complete *state of plane stress* at some particular point q in a loaded body. The coordinates of the points on the circle represent all possible combinations of normal and shear stresses which are acting on the corresponding planes through point q. Fortunately, the complete circle can be constructed by knowing the stresses on only two mutually orthogonal planes. Thus, for a plane state of stress, if the stresses are known on two mutually perpendicular planes, they are known on all planes.

We close this section with numerical illustrative examples. Before doing so, however, it would be worth your time and effort to review the previous discussion of the Mohr circle of stress.

EXAMPLE 8-1

A shaft is subjected to simultaneous torsion and compressive loads, which produce the state of stress shown in Fig. 8-13(*a*). Construct Mohr's circle for this state of stress and determine the principal stresses and principal planes. Also, find the value of the maximum shear stress.

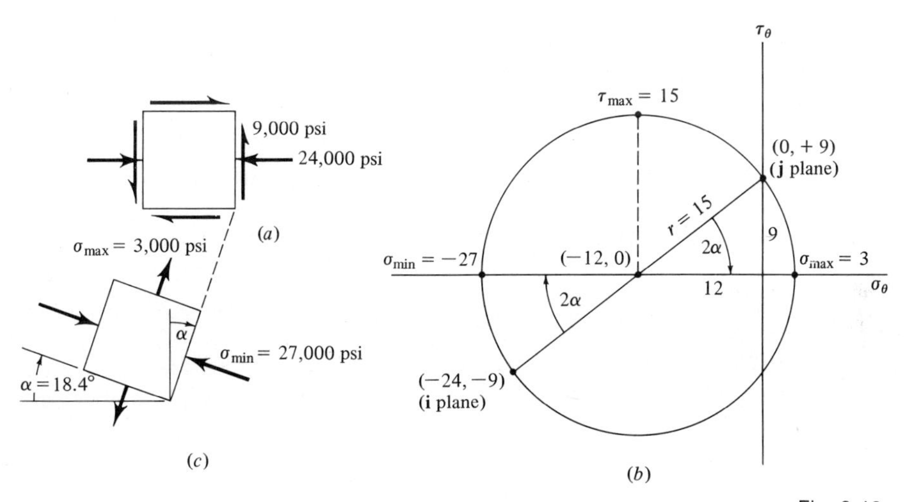

Fig. 8-13

Solution: Using our previously adopted notation and sign convention we have

$$\text{i-plane stresses:} \quad \sigma_x = -24{,}000 \qquad \tau_{xy} = -9{,}000$$
$$\text{j-plane stresses:} \quad \sigma_y = 0 \qquad \tau_{yx} = +9{,}000$$

which are the coordinates of two diametrically opposite points on the Mohr circle. Accordingly, after locating these two points and joining them with a diametral line, we can sketch the circle, as shown in Fig. 8-13(*b*). (Note that we have omitted the thousands from the numbers on the circle.) The center of the circle is located at (−12,0), and its radius is

$$r = \sqrt{(12)^2 + 9^2} = 15$$

Hence the value of the maximum shear stress is

$$\tau_{max} = 15{,}000 \text{ psi}$$

The values of the principal stresses are given by adding and subtracting the radius to the center location of the circle. Thus

$$\sigma_{max} = -12 + 15 = +3 \text{ or } 3{,}000 \text{ psi} \quad \text{(tension)}$$
$$\sigma_{min} = -12 - 15 = -27 \text{ or } 27{,}000 \quad \text{(compression)}$$

As for the orientation of the principal planes, let 2α be the angle on the circle measured *from* the point (0,+9) *to* the right extremity [also the angle from the point (−24,−9) to the left extremity].

$$2\alpha = \arctan \tfrac{9}{12} = 36.8^\circ \quad \text{(clockwise)}$$

Hence, the principal plane having the algebraic maximun stress of 3,000 psi is oriented at an angle of $\alpha = 18.4^\circ$ clockwise from the **j** plane, while the principal plane having the algebraic minimum stress of −27,000 psi is 18.4° clockwise from the **i** plane, as shown in Fig. 8-13(*c*).

Remark: In this example, the algebraic minimum of −27,000 psi (compression) is actually the physically largest stress for this state of stress and thereby would probably be the more significant design stress than the algebraic maximum of 3,000 psi (tension).

Remark: A carefully drawn and scaled sketch of the circle can be of great value as a reliable check on the analytical solution.

EXAMPLE 8-2

At some point in a timber beam, the state of stress is as shown in Fig. 8-14(*a*). Find the normal and shear stresses on a plane parallel to the indicated grain of the timber at this point.

Solution: Using our previously adopted sign convention

i-plane stresses:	$\sigma_x = 2{,}200$	$\tau_{xy} = 600$
j-plane stresses:	$\sigma_y = 0$	$\tau_{yx} = -600$

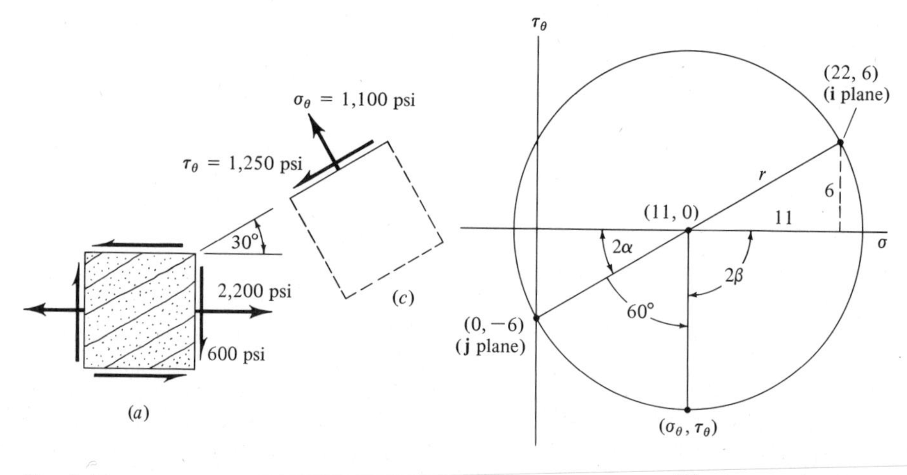

Fig. 8-14

which are the coordinates of two diametrically opposite points on the Mohr circle. The circle can then be drawn as in Fig. 8-14(*b*) with its center at (11,0) and a radius of

$$r = \sqrt{(11)^2 + (6)^2} = 12.5$$

Since we wish to determine the stresses on the plane oriented 30° counterclockwise from the **j** plane, we will locate the corresponding point on the circle at an angle of $2 \times 30° = 60°$ counterclockwise from the point (0,−6) on the circle. Although this can be done graphically by use of a protractor, we will do it analytically.

First, the angle 2α is given by

$$2\alpha = \arctan \tfrac{6}{11} = 28.6°$$

Then, from the circle we see that

$$2\alpha + 60° + 2\beta = 180°$$

from which

$$2\beta = 91.4°$$

Hence, for all practical purposes, the point with coordinates $(\sigma_\theta, \tau_\theta)$ coincides with the lower extremity of the circle whose coordinates are (11,−12.5). Thus, the stresses on the plane parallel to the grain of the timber are those indicated in Fig. 8-14(*c*). Note, in particular, the direction of the shear stress, which corresponds to the negative value obtained from the circle.

Remark: It is conceivable that this shear stress of 1,250 psi could cause failure of the timber along its grain, since wood is relatively weak in shear parallel to the grain.

EXERCISE PROBLEMS

8-7 Construct Mohr's circle for the uniaxial state of stress shown and, in effect, verify Theorem 5-1.

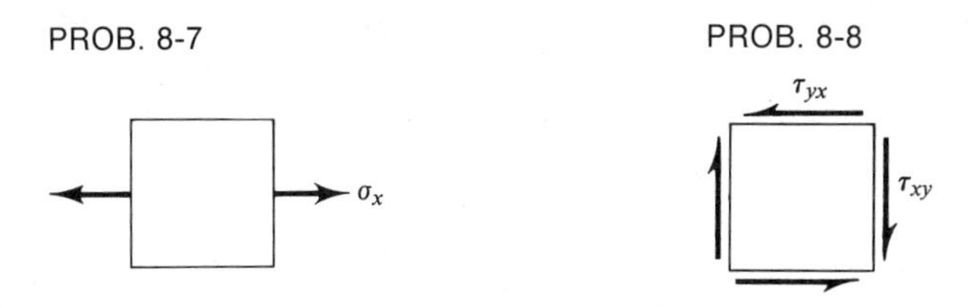

8-8 The state of stress in a circular cylindrical shaft subjected to a pure torque is shown in the figure and is called *pure shear*. Construct Mohr's circle for this state of stress and determine the principal values and principal planes of stress.

8-9 For the plane state of stress shown, construct Mohr's circle and determine the values and planes of maximum shear stress.

PROB. 8-9

$\sigma_y = -\sigma_x$

σ_x

PROB. 8-10

$\sigma_y = \sigma_x$

σ_x

8-10 Construct Mohr's circle for the state of plane stress shown. What are the principal values and principal planes?

8-11 Using Eq. (8-7) or (8-12), show that

$$\sigma_\theta + \sigma_{\theta + 90} = \sigma_x + \sigma_y$$

for any value of θ. Interpret this result geometrically in terms of Mohr's circle.

8-12 Using Eqs. (8-12) and (8-13), show that the angle which maximizes (or minimizes) σ_θ makes the shear stress τ_θ vanish. Interpret this result geometrically in terms of Mohr's circle.

8-13 to 8-18 Construct Mohr's circle for each of the states of plane stress shown. Determine the principal values and the corresponding principal planes. Finally, determine the values and planes of maximum shear stress. The magnitudes of the stresses are given in pounds per square inch.

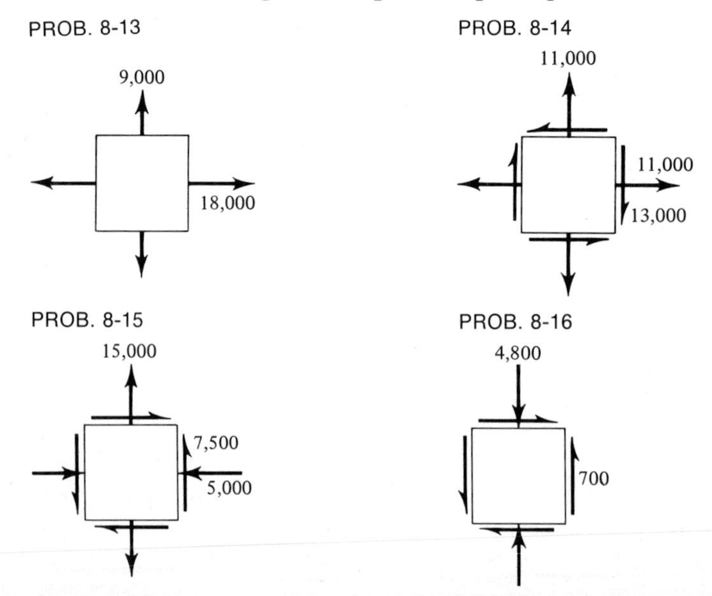

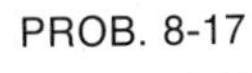
PROB. 8-17

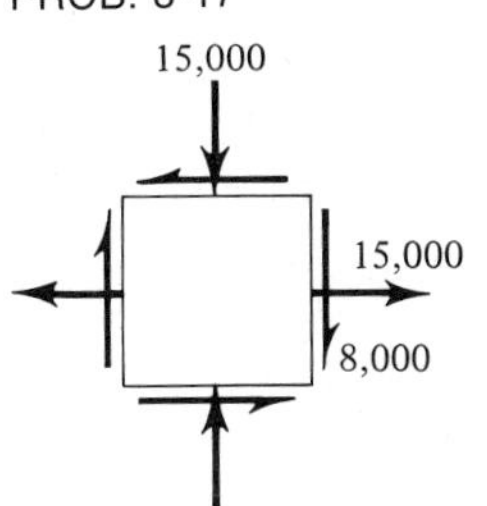

PROB. 8-18

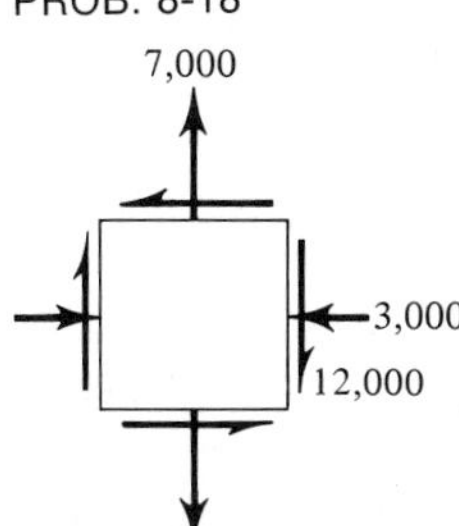

8-19 and 8-20 For the state of stress shown, find the normal and shear stresses on the planes indicated by dashed lines.

PROB. 8-19

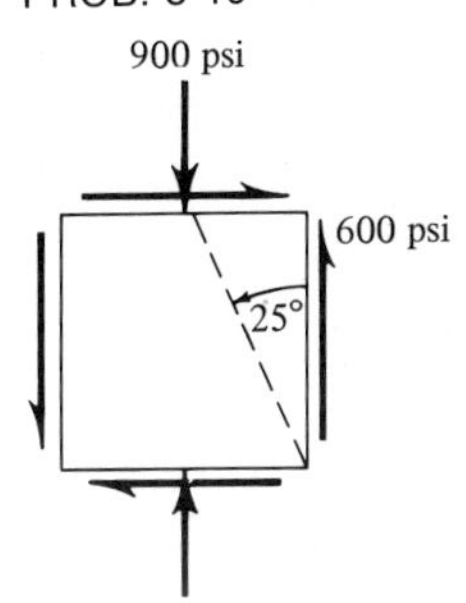

PROB. 8-20

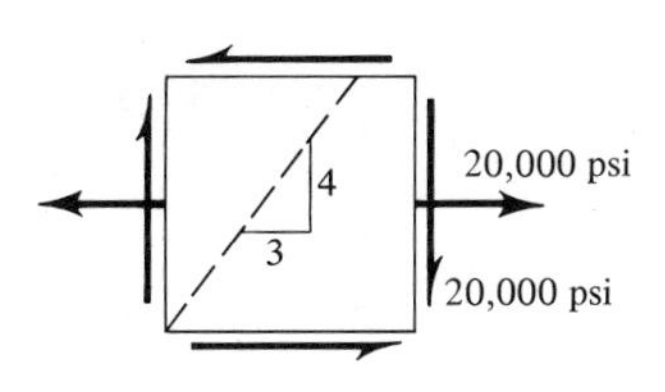

8-21 A circular shaft is subjected to a combined torsion and tension load producing the state of stress shown. Find the largest permissible value for σ_x if the maximum shear stress in the material may not exceed a design limit of 20,000 psi.

PROB. 8-21 and 8-22

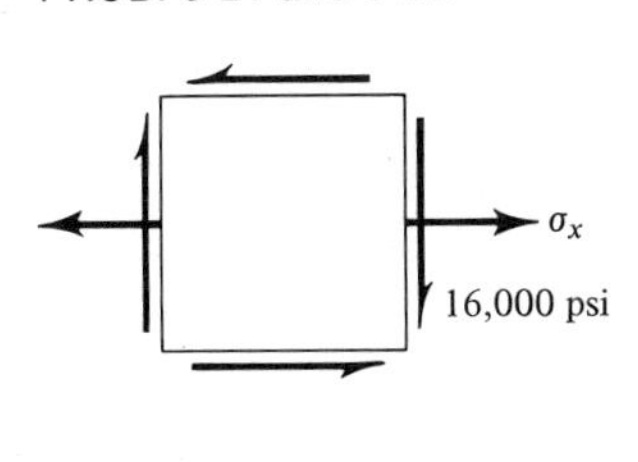

PROB. 8-23

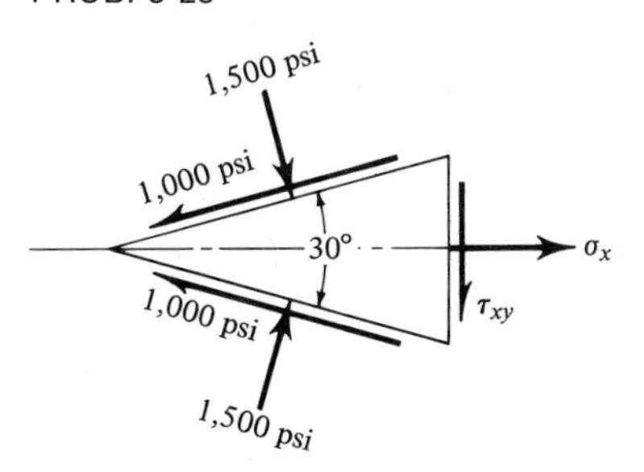

PROB. 8-24

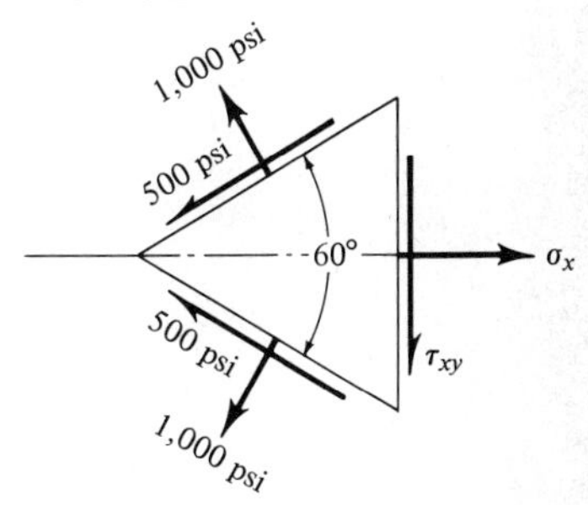

8-22 Same as Prob. 8-21 except the maximum tensile stress may not exceed a design limit of 30,000 psi.

8-23 and 8-24 By using Mohr's circle (or any other method you can devise) compute σ_x, τ_{xy}, $\sigma_{\max}$, $\sigma_{\min}$, and $\tau_{\max}$ for the state of stress shown.

8-5 SOME PROBLEMS IN COMBINED STRESSES

In the previous sections, we have seen how to analyze a plane state of stress at any point in a structural member to determine the values and directions of the maximum normal and shear stresses. The question that now arises is: What point or points in the structural member are the critical ones insofar as the possibility of failure is concerned? For example, for the simply supported beam shown in Fig. 8-15(*a*), at what point in the beam is failure likely to occur due to excessively large stresses? The same question is applicable to the circular shaft in Fig. 8-15(*b*) subjected to combined torsion and tensile loads.

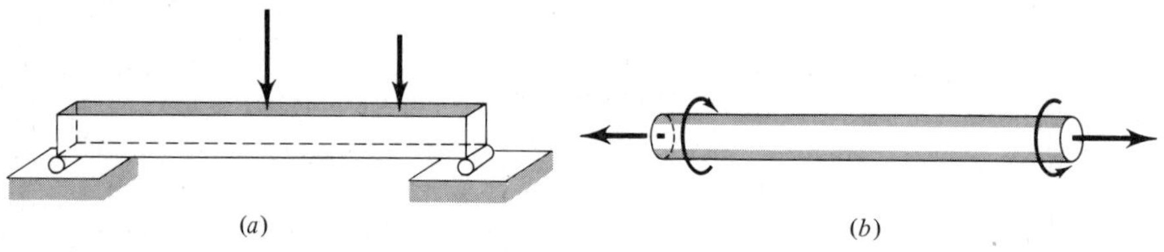

Fig. 8-15

In some situations, the answer to this question is intuitively obvious; in others, the answer can be obtained by comparatively simple analysis, while in more complex structures, the answer cannot be obtained except possibly by extensive analysis and/or experimentation. In this section, it is our intention to look at some relatively simple situations involving combined stresses to illustrate some of the questions and problems which arise in the design and stress analysis of structural members.

Axial and Torsion Loading

Consider a circular shaft subjected to simultaneous axial and torsion loading, as shown in Fig. 8-5 (repeated). The axial stress σ_x will have the same value at all points throughout the shaft, while the value of the shear stress τ_{xy} (and τ_{yx}) will vary with the distance from the center of the shaft (except for the idealized fully plastic situation). In any case, at any point in this member the state of stress is typified by the plane state depicted in Fig. 8-5(*c*). The Mohr circle for this state of stress is shown in Fig. 8-16(*a*), in which we have assumed σ_x to be a tensile (positive) stress and τ_{xy} to be a positive shear, although it is quite possible and permissible for either of these stresses to be negative. The analysis of this circle is given in Fig. 8-16(*b*), from which we can make the following observations.

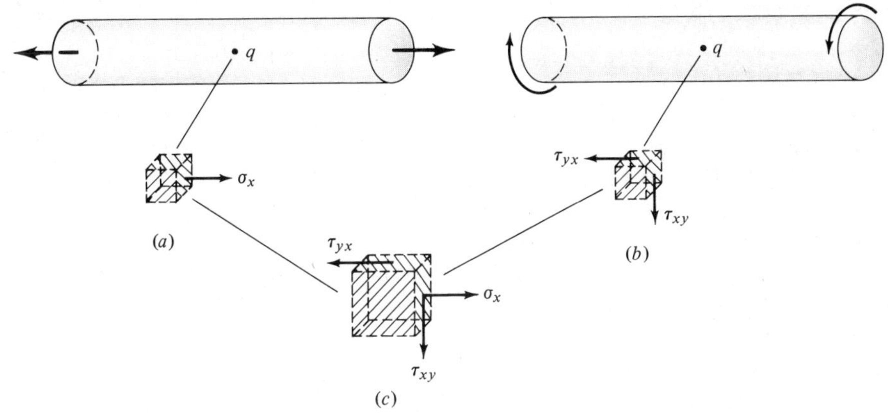

Fig. 8-5

1. Recalling that the normal stress σ_x is uniform throughout the shaft, the point(s) in the shaft which would have the largest Mohr's circle would be those points with the largest shear stress τ_{xy}, which are those points at the *outer surface* of the shaft. Therefore, the maximum shear stress τ_{max}, whose value is equal to the radius of the Mohr circle, occurs at the outer surface of the shaft.
2. Again, since σ_x is uniform throughout, the maximum normal stress σ_{max} will also occur at those points with the largest Mohr's circle. Accordingly, the maximum and minimum normal stresses occur at points on the outer surface of the shaft.
3. The observations in (1) and (2) remain valid even for the special cases in which either the torque is zero or the axial load is zero.

From these observations we conclude that for a circular shaft under combined axial and torsion loading, the maximum normal

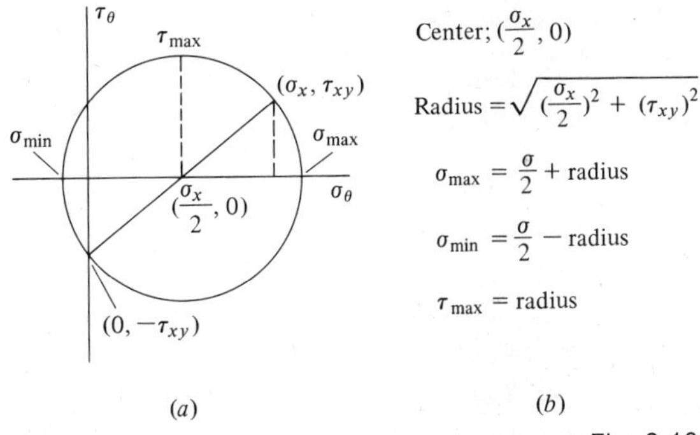

Fig. 8-16

and maximum shear stresses occur at the outer surface of the shaft. Their values and orientations depend upon the size of the axial load and the applied torque. Which of these—the maximum (or minimum) normal stress or the maximum shear stress—is more critical insofar as failure is concerned depends upon the mode of failure and the material of which the shaft is made. For example, for a brittle material the maximum tensile stress might well be the critical design factor based upon a rupture type of failure. Observe the 45° tensile-type fracture for the brittle torsion members in the photographs in Fig. 8-17(*a*). This can be anticipated by the fact that, for a state of pure shear ($\sigma_x = 0$), the maximum tensile stress will occur on a plane at 45° from the torsion-shear stresses.

On the other hand, if the material is rather ductile, then the maximum shear stress might well be the determining factor in predicting a shear type of failure of the shaft, as exemplified by the photographs in Fig. 8-17(*b*). Still another possible mode of failure is that of buckling due to excessively large compressive stresses. This is particularly possible in a thin-walled hollow shaft. The engineer must consider these and perhaps other possibilities in the design and stress analysis of such structural members.

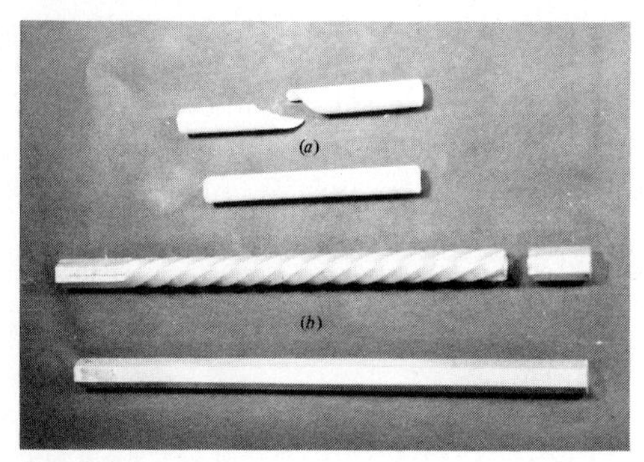

Fig. 8-17

Thin-Walled Pressure Vessels

We envision a pressure vessel as being some type of tube or closed container containing a fluid or gas under a hydrostatic gage pressure p (refer to Sec. 3-2). The essential feature that distinguishes a thin-walled pressure vessel from, say, a thick-walled pressure vessel is the assumption as to the stress variation through the wall of the vessel. To illustrate, consider in Fig. 8-18(*a*) a section of some tube subjected to an internal pressure p. A small piece of the wall is shown in Fig. 8-18(*b*), with σ_i denoting the tensile stress at the innermost point of the wall and σ_o the tensile stress at the outermost point of the wall. In a *thin-walled* tube or vessel the

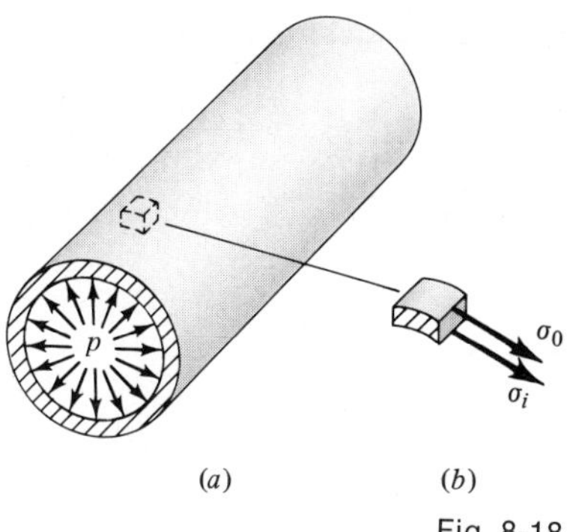

Fig. 8-18

difference between σ_i and σ_o is negligible compared to their magnitudes, so that we simply call this average wall stress *the circumferential* or hoop stress σ_c. On the other hand, in a thick-walled vessel, the variation between σ_i and σ_o can be of considerable importance in the design and analysis of the vessel. Accordingly, there is no clearly defined dividing line between thin-walled and thick-walled vessels, although a ratio of radius to wall thickness of about 10 to 1 serves as a reasonable rule of thumb, with larger ratios being thin walled and smaller ratios being thick walled.

With the assumption that the vessel is thin walled, the average stresses in the wall can usually be determined by a free-body-diagram analysis of carefully chosen sections of the vessel. To illustrate this procedure, consider a closed-ended cylindrical thin-walled pressure vessel as shown in Fig. 8-19(*a*) with an internal gage pressure p. Let us visually pass a transverse cut through this vessel and draw a partial free-body diagram of the left portion of this vessel with the pressure p pushing against the inside surface and the longitudinal tensile stress σ_L pulling on the cut wall of the vessel. Summing forces (stress times wall area and pressure times inside area) in the longitudinal direction, equilibrium requires

$$\sigma_L \pi D_i t - p\pi \frac{D_i^2}{4} = 0$$

where we have used $\pi D_i t$ as the approximate circumferential cross-sectional area of the wall. This is a good approximation if the thickness t is small compared with the inside diameter D_i, as is implicitly assumed by considering the vessel to be thin walled. Rearranging, we obtain

$$\sigma_L = \frac{pD_i}{4t} \tag{8-19}$$

as the relation between the pressure p and the longitudinal tensile stress σ_L.

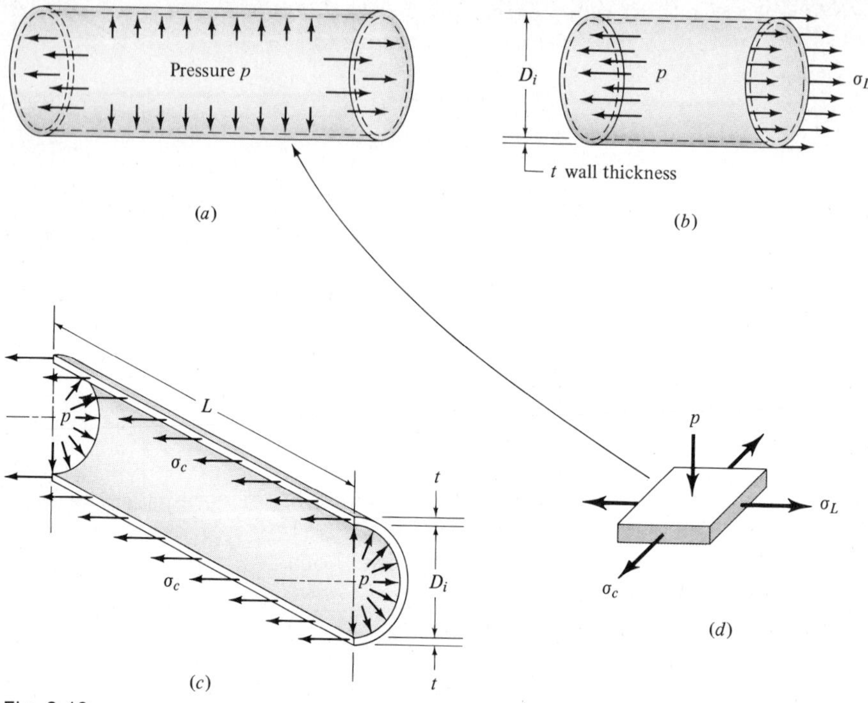

Fig. 8-19

In a similar manner, we can determine the circumferential stress previously referred to in Fig. 8-18. This time, let us pass a longitudinal cut through the vessel, as in Fig. 8-19(*c*), with the pressure p pushing against the inside curved surface and the circumferential stress σ_c pulling on the cut wall of the vessel. Summing forces [stress times wall area and pressure times *projected* inside area (recall Example 3-2)] parallel to σ_c, equilibrium requires

$$2\sigma_c Lt - pD_i L = 0$$

Rearranging, we obtain

$$\sigma_c = \frac{pD_i}{2t} \tag{8-20}$$

as the relation between the pressure p and the circumferential tensile stress σ_c. Equations (8-19) and (8-20) are typical of the pressure-stress relations obtained for many types of thin-walled pressure vessels. Note that for this cylindrical type of vessel, the circumferential stress σ_c is twice as large as the longitudinal stress σ_L.

Before leaving this example, let us examine a stress diagram of the wall. Figure 8-19(*d*) shows the stresses σ_c and σ_L and the pressure p acting on a typical small element or piece of the wall, this particular element being taken from the bottom side of the vessel.

At first glance, it appears that something is wrong or missing, since the pressure p would cause an unbalance of forces in the vertical direction. However, this dilemma is not disturbing for a number of reasons. For one, our stress diagram does not take into consideration the curvature of the wall. For another, quite often, in analyzing the state of stress in the wall of a thin-walled vessel, the compressive pressure stress on the inside surface is neglected. To see why this is reasonable, you only need to recall that for a thin-walled vessel

$$\frac{\text{Radius}}{\text{Wall thickness}} \geqslant 10$$

which when put into Eqs. (8-19) and (8-20) implies that

$$\sigma_L \geqslant \tfrac{20}{4}p$$
$$\sigma_c \geqslant \tfrac{20}{2}p$$

so that these tensile stresses are much more significant than the compressive stress due to the pressure.

Such dilemmas, paradoxes, and apparent inconsistencies like this one often arise when simplifying assumptions are made in analyzing stresses in structural members. However, there is usually a reasonable explanation to why the dilemma is acceptable within the limitations of the analysis.

Absolute Maximum Shear Stress

The previous discussion dealing with the state of stress in a thin-walled pressure vessel brings up an important situation in the use of Mohr's circle for plane stress. Consider a state of stress like that shown in Fig. 8-20(*a*), in which both σ_x and σ_y are considered to be the same sign, both tensile or both compressive. The corresponding Mohr circle has been drawn in Fig. 8-20(*b*) for the case $\sigma_x > \sigma_y \geqslant 0$. The radius of this circle is

$$\frac{\sigma_x - \sigma_y}{2} \tag{8-21}$$

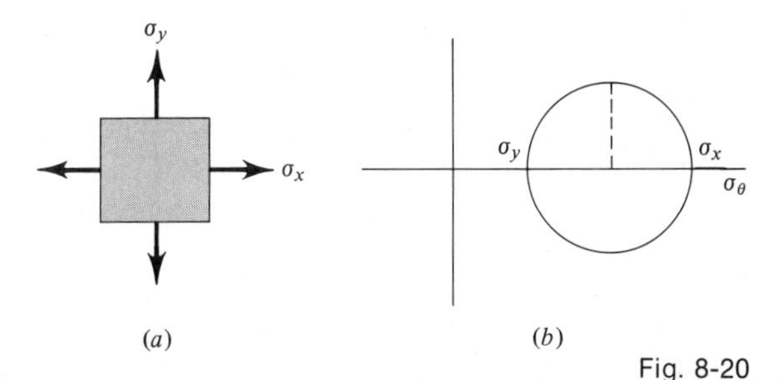

Fig. 8-20

The question that we raise is whether or not this is the value of the maximum shear stress. Before answering this question, let us first observe that for the state of stress in Fig. 8-20(*a*), as σ_y becomes larger, the value given by Eq. (8-21) becomes smaller, and in fact becomes zero when σ_y becomes equal to σ_x. This would imply that the maximum shear stress would be zero for the case in which $\sigma_x = \sigma_y$, which, as we shall see, is *not* correct.

Let us redraw the state of stress as a three-dimensional state in Fig. 8-21(*a*) with $\sigma_z = 0$ rather than the simplified two-dimensional representation of Fig. 8-20(*a*). Since there are no shear stresses on any of the indicated **i**, **j**, and **k** planes, these planes must be principal planes, and these stresses σ_x, σ_y, and $\sigma_z = 0$ must be principal stresses (extremities of Mohr's circle). With this in mind, we can think of this three-dimensional state of stress as being made up of three simultaneous two-dimensional states of stress—the state having σ_x and σ_y as its principal stresses, the state having σ_x and σ_z as its principal stresses, and the state having σ_y and σ_z as its principal stresses. Consequently, if we consider each of these two-dimensional states of stress separately, we can draw three separate Mohr circles, as shown in Fig. 8-21(*b*). In this case, with $\sigma_x > \sigma_y > \sigma_z = 0$, the largest Mohr circle is the x–z circle, and its radius is

$$\frac{\sigma_x - 0}{2} \tag{8-22}$$

which is larger than that for the x–y circle given by Eq. (8-21). Hence, Eq. (8-21) does not give the value of the maximum shear stress, but, for this case, Eq. (8-22) does. Furthermore, this maximum shear stress acts on a plane at 45° between the **i** plane and the **k** plane, as indicated in Fig. 8-21(*a*).

The situation which we have been considering in Fig. 8-21 is somewhat special in that σ_x, σ_y, and $\sigma_z = 0$ are all principal stresses, since no shear stresses are present on these planes. How-

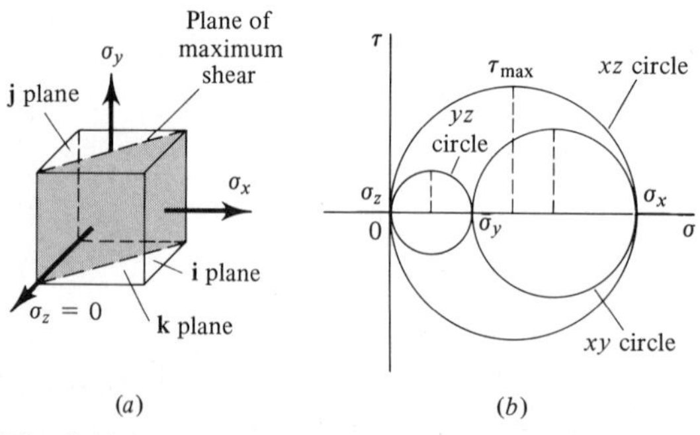

Fig. 8-21

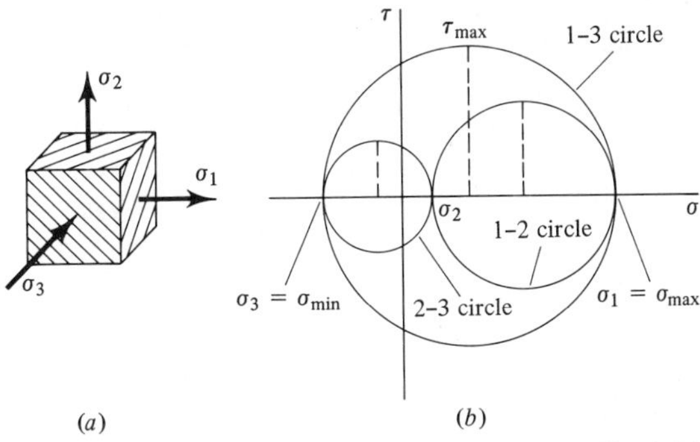

Fig. 8-22

ever, even the most general state of stress as depicted in Fig. 8-4 can, by suitable rotation of the planes, be reduced to a representation of the type in Fig. 8-22(a), in which σ_1, σ_2, and σ_3 are the three *principal* stresses acting on three mutually perpendicular principal planes. This general situation can be handled in the same manner as the previous one in that we can construct three Mohr circles with extremities σ_1, σ_2, and σ_3. Such construction is shown in Fig. 8-22(b) for the case where $\sigma_1 > \sigma_2 > 0$ and $\sigma_3 < 0$, although a similar construction can be made for any combination of values for the principal stress σ_1, σ_2, and σ_3.

From this construction, we see that the radius of the largest Mohr circle will always be

$$\frac{\sigma_{\max} - \sigma_{\min}}{2} = \tau_{\max} \tag{8-23}$$

where $\sigma_{\max}$ is the algebraic largest principal stress and $\sigma_{\min}$ is the algebraic minimum. Consequently, Eq. (8-23) also gives the value of the largest shear stress $\tau_{\max}$ for any general state of stress.

We close this article with example problems that illustrate some of the ideas discussed here. As a final remark, when analyzing stresses in real bodies, you should never lose sight of the three-dimensional aspects of the problem and the fact that you are dealing with real materials which can fail for a variety of reasons.

EXAMPLE 8-3

A thin-walled copper tube with a diameter of 1 in and a wall thickness of 0.020 in is required to transmit a torque T. The tube wall will buckle if the compressive stress in the wall reaches 12,000 psi. What is the maximum torque that can be applied?

Solution: For all practical purposes, the tube is a torsion member with all of its cross-sectional area concentrated at the outer diameter of 1 in. This area is approximately equal to the circumference

πD multiplied by the thickness t. Then, the polar second moment J is approximately (see Appendix A)

$$J = \left(\frac{D}{2}\right)^2 \times \text{area}$$

$$= \frac{D^2}{4}(\pi D t) = \frac{\pi D^3}{4} t$$

$$= \frac{(\pi)(1)^3}{4}(0.020) = 0.005\pi \text{ in}^4$$

The state of stress in the wall is pure shear, as indicated in Fig. 8-23(a), where τ is given by the torsion formula (7-22).

$$\tau = \frac{TR}{J}$$

$$= \frac{T(1/2)}{0.005\pi} = \frac{T}{0.010\pi}$$

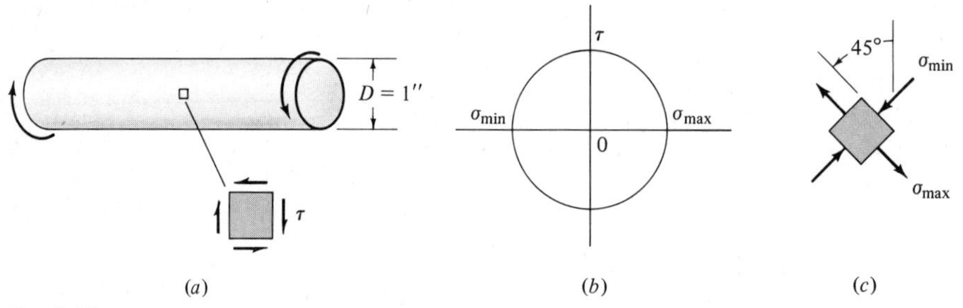

Fig. 8-23

Mohr's circle for pure shear is shown in Fig. 8-23(b), which is simply Fig. 8-16 with $\sigma_x = 0$. Hence, the maximum compressive stress corresponds to the value of σ_{min}, which in turn has the same value as the radius of the circle. Thus,

$$\sigma \text{ (compression)} = \tau = \frac{T}{0.010\pi}$$

or

$$T = (0.010\pi)\sigma$$

or

Since σ may not exceed 12,000 psi, the maximum allowable torque is

$$T = (0.010\pi)(12{,}000)$$

$$= 120\pi \text{ in-lb}$$

The principal planes of stress are indicated in Fig. 8-23(c).

EXAMPLE 8-4

A thin-walled tube with OD = 2 in and $t = \frac{1}{20}$ in carries a gas under a pressure p. Due to faulty installation, the pipe is subjected

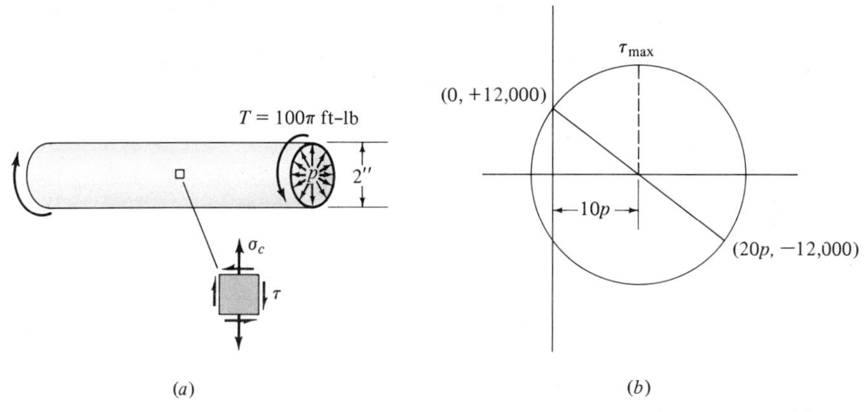

Fig. 8-24

to a constant torque of 100π ft-lb. If the maximum allowable safe shear stress is 20,000 psi, what is the largest pressure the pipe can carry?

Solution: We will treat the tube as an open-ended thin-walled pressure vessel subjected to a simultaneous internal pressure and external torque. Accordingly, the state of stress will be that shown in Fig. 8-24(a). The circumferential stress σ_c is given by Eq. (8-20):

$$\sigma_c = \frac{pD}{2t} = \frac{p(2)}{2(1/20)}$$
$$= 20p \text{ psi}$$

while the shear stress τ is given by Eq. (7-22) as

$$\tau = \frac{TR}{J} = \frac{TR}{2\pi R^3 t}$$
$$= \frac{(100\pi)(12)(1)}{2\pi(1)^3(1/20)} = 12{,}000 \text{ psi}$$

where we have used $2\pi R^3 t$ as an approximation for J (recall the previous example).

Mohr's circle for this state of stress is shown in Fig. 8-24(b). The radius of this circle is given by $\sqrt{(10p)^2 + (12{,}000)^2}$.

However, since the allowable maximum shear stress is 20,000, this value is also the maximum allowable radius. Hence,

$$(20{,}000)^2 = (10p)^2 + (12{,}000)^2$$
$$10p = 16{,}000$$
$$p = 1{,}600 \text{ psi}$$

which is the largest permissible gage pressure of the gas in the tube.

EXERCISE PROBLEMS

8-25 A solid $\frac{3}{4}$-in-diameter rod 12 in long is required to transmit a pure torque. The rod is made of a brittle material whose ultimate tensile stress is 16,000 psi and whose ultimate shear stress is 24,000 psi. Using a factor of safety of 2, determine the maximum permissible torque.

8-26 A thin-walled tube of 2 in OD and $\frac{1}{10}$ in wall thickness must transmit a pure torque T. The tube has a helical wall seam making an angle of 30° with the longitudinal axis of the tube. This seam will fail if the tensile stress on it exceeds 8,000 psi. What is the maximum allowable torque? Use a safety factor of 2.5.

8-27 A 10-in-diameter pipe with a wall thickness of 0.20 in carries simultaneous torsion and axial tensile loads of 2,000 ft-lb and 2,000 lb, respectively. Find the values and directions of the principal stresses. Also, find the value and direction of the maximum shear stress.

8-28 Under normal operating conditions, it was found that the state of stress at a typical point on the outer surface of a drive shaft was as indicated in the figure. If the allowable operating stresses are $\sigma_{max} \leqslant$ 24,000 psi and $\tau_{max} \leqslant$ 18,000 psi, was the shaft operating within these specifications?

PROB. 8-28

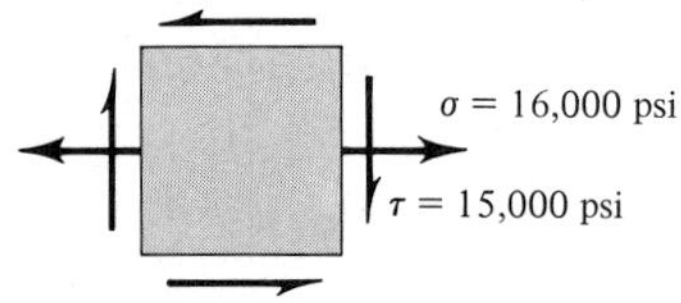

8-29 A hollow shaft is loaded simultaneously with a torque and an axial tensile load. The torque produces a torsion shear stress of $12Q$ psi, while the axial load produces a longitudinal tensile stress of $10Q$ psi. Using a factor of safety of 2, what is the maximum permissible value of Q if the allowable stresses are 25,000 psi (tensile) and 20,000 psi (shear)?

8-30 Same as Prob. 8-29 except the allowable stresses are 30,000 psi (tensile) and 15,000 psi (shear).

8-31 Rework Example 8-3 if an axial tensile load of 200 lb is also applied to the tube.

8-32 Same as Prob. 8-25 except a simultaneous axial tensile load of 2,000 lb is applied to the rod.

8-33 Derive a relationship between the gage pressure p and the tensile stress σ in the wall of a thin-walled spherical pressure vessel.

8-34 A thin-walled tube carries a refrigerant under a gage pressure of 80 psi. The tube has an elliptical-shaped cross section with a minor diameter of $\frac{1}{4}$ in and a major diameter of $\frac{1}{2}$ in and a uniform wall thickness of 0.010 in. Determine the maximum tensile stress in the tube wall.

8-35 A cylindrical water tank 30 ft high and 20 ft in diameter is made of $\frac{1}{8}$-in-thick steel plate. Estimate the maximum tensile stress in the tank wall when the tank is full.

8-36 A scuba diver's tank of diameter 6 in and length 18 in with a wall thickness of $\frac{1}{16}$ in can be considered to be a thin-walled cylindrical pressure vessel with spherical caps. Determine the maximum tensile and shear stresses in the tank wall for an internal gage pressure of 75 psi.

8-37 A cylindrical pressure vessel with spherical caps has a length of 30 in, a diameter of 8 in, and a wall thickness of 0.10 in. The tank is filled with a room-temperature (70°F) gas to a pressure of 100 psi. While in use, the gas becomes heated to a temperature of 400°F. What will be the maximum tensile and shear stresses in the tank wall? Use Gay-Lussac's law for a perfect gas.

8-38 A rectangular pressure chamber 8 × 8 × 8 in. has a flat lid held on by eight $\frac{1}{4}$-in-diameter bolts. Each bolt is tightened with a torque of 30 in-lb, and the chamber is then pressurized to 100 psig. What will be the shear stress in the bolts due to the torque? What will be the tensile stress in the bolts due to the pressure? What will be the maximum tensile stress in the bolts?

PROB. 8-38

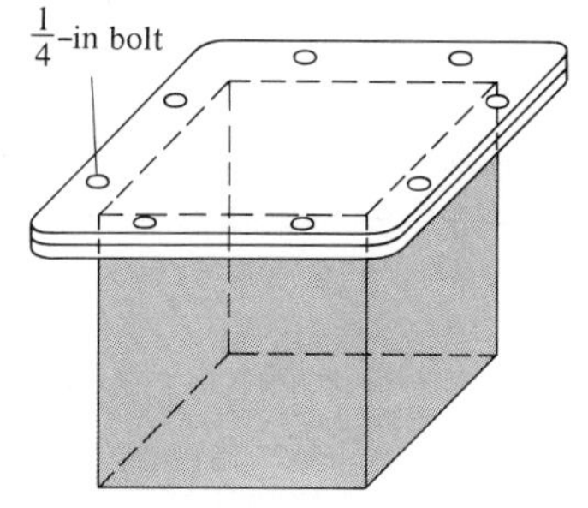

8-39 A closed-ended cylindrical pressure vessel has a radius of 3 in, a length of 25 in, and a wall thickness of $\frac{1}{8}$ in. The material of which it is fabricated has an allowable tensile stress of 10,000

psi and an allowable shear stress of 6,000 psi. What is the maximum allowable pressure the vessel may carry?

8-40 A thin-walled pipe with OD = 2 in and $t = \frac{1}{20}$ in carries a gas under a pressure p. Due to faulty installation, the pipe is under a constant torque of 800π in-lb. If the maximum allowable safe tensile stress is 16,000 psi, what is the largest pressure the pipe can carry?

8-6 GENERALIZED DEFORMATION AND STRAIN

Thus far in this chapter we have been dealing with some of the ideas and problems concerned with the *stress analysis* of load-carrying members. However, as has been said on a number of occasions, stress is actually a difficult, if not impossible, quantity to measure since it is, by definition, an internal force per unit area. For the most part, stresses are calculated by means of some mathematical relations which are derived on the basis of some rational assumption as to what is happening inside the body; the torsion formula is based on the assumption of linear stress variation in a circular torsion member, etc.

On the other hand, deformations and strains are to some extent physically measurable in that surface deformations are often observable and can often be quantified in some manner, e.g., the angle of twist in a torsion member or the overall elongation of a two-force tension member. Our immediate objective in this section is to develop a scheme to quantify and analyze an arbitrary surface deformation which might occur in some general load-carrying member.

In Chap. 6, we defined axial strain ϵ as an elongation or contraction e per unit original length L:

$$\epsilon = \frac{e}{L} \tag{6-2}$$

Also, in Chap. 7, we introduced shear strain γ as an angle of distortion resulting from some shear type of deformation. In general, an arbitrary deformation is made up of a combination of simultaneous elongations, contractions, and distortions resulting in a combination of axial and shear strains at each point in the deformed body. To see this, let us look more closely at the deformation which occurs at some point on the surface of some structural member.

Consider the unloaded cantilevered member shown in Fig. 8-25(a), and suppose we were to scribe two short perpendicular line segments on its surface at some arbitrary location q. Then, if this member were loaded as in Fig. 8-25(b), the member and the scribed

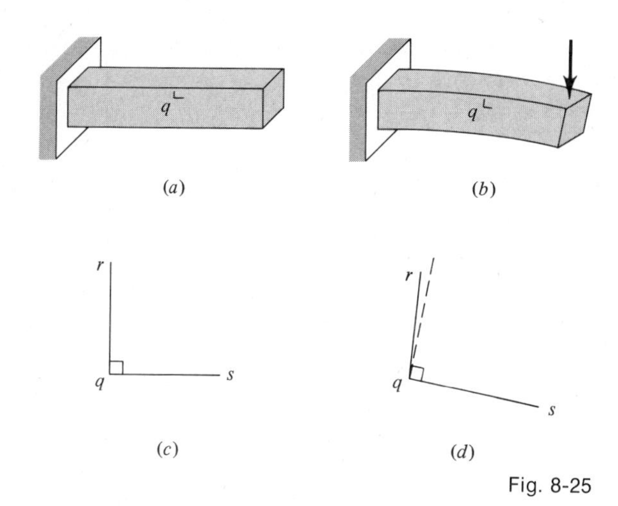

Fig. 8-25

lines would likely deform as shown (although perhaps exaggerated). As an overall description we can say that the member bent into some type of curve, but we desire to describe in greater detail the deformation that occurred locally at point q. The undeformed lines at point q are shown enlarged in Fig. 8-25(c) and are labeled qr and qs for identification purposes. The deformed lines are shown enlarged in Fig. 8-25(d).

First of all, since the whole member bent into some curve, it is quite likely that the original straight lines qr and qs would also be deformed into some curved shape. However, let us agree that if the original straight lines are "short enough," then the deformed lines will, for all practical purposes, also be straight lines, as shown in Fig. 8-25(d), and will be representative of the deformation which occurred "at point q." We observe that line qr was rotated and *contracted* during the deformation, while line qs was rotated and *elongated*. Furthermore, qs rotated slightly more than qr, so that they are no longer perpendicular, and, therefore, the right angle has become *distorted*. Hence, the deformation "at point q" is a combination of contraction, elongation, and distortion. We must now try to quantify this deformation in terms of axial and shear strains.

In view of what we have done thus far, let us redraw and relabel our undeformed and deformed line segments as shown in Fig. 8-26(a) and (b), respectively. The original horizontal line dx has been deformed into the line of length $dx + de_x$, and dy into $dy + de_y$, with the d's simply indicating that these lines conceptually can be made as short as we desire. Then, following the definition of axial strain as given by Eq. (6-2) and the concept of shear strain as a distortion, we give the following definition:

Definition 8-1 Axial strain is an elongation (considered positive) or contraction (considered negative) per unit original length. Accordingly

$$\epsilon_x = \frac{de_x}{dx} \qquad \text{and} \qquad \epsilon_y = \frac{de_y}{dy} \tag{8-24}$$

Also, shear strain is the angle of distortion (measured in radians) of two originally perpendicular line segments. Accordingly, the shear strain γ_{xy} is the angle by which the original right angle between the line segments dx and dy has been increased (considered positive) or decreased (considered negative).

The three strains ϵ_x, ϵ_y, and γ_{xy} give a quantitative description of the deformation occurring at some point q on the surface of a structural member. This is essentially a two-dimensional description. However, as has been said on numerous occasions, real bodies are three dimensional and deform in a three-dimensional manner. Consequently, the complete deformation at any point q in a body is described in terms of *three* axial strains ϵ_x, ϵ_y, and ϵ_z and *three* shear strains γ_{xy}, γ_{xz}, and γ_{yz} each of which has a definition and interpretation analogous to the two-dimensional considerations of Fig. 8-26. These six strains specify the *general state of strain* at a point in a body, just as the six stresses σ_x, σ_y, σ_z, τ_{xy}, τ_{xz}, and τ_{yz} specify the *general state of stress* (refer to Theorem 8-2).

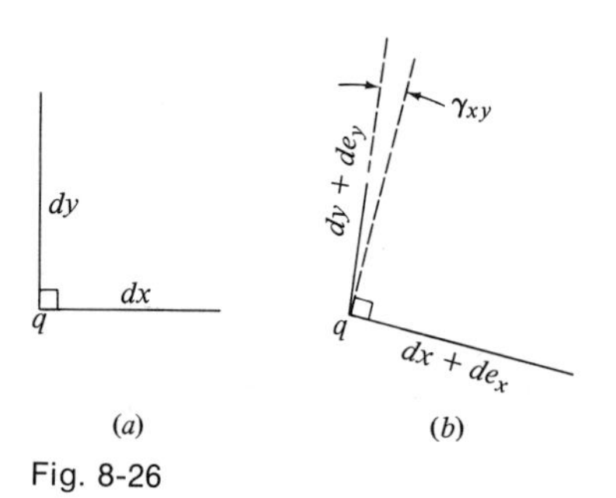

Fig. 8-26

As a practical matter, strains in structural members are usually of the order of 0.001 in/in and consequently are not easily discernible to the naked eye. However, by using special electrical and optical devices, deformations and average strains can be measured accurately and fairly easily at points on the surface of real bodies. Accordingly, in the next section we will pursue our analysis of two-dimensional surface strains and deformations such as those we have considered in Fig. 8-25.

At this point, it is important for you to have a clear understanding of axial and shear strains. Axial strain is an elongation or contraction of a line segment per unit original length. Shear strain is an angular distortion between two line segments. Accordingly, strains are dimensionless *measures* of deformation which are applicable to *any* real deformable body.

EXAMPLE 8-5

A 100-in-long steel wire is held taut in a vertical position as shown in Fig. 8-27. What strain will be induced in the wire if it is displaced horizontally 8 in at a point 60 in from the lower end?

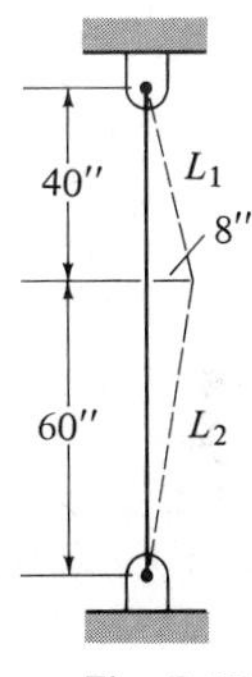

Fig. 8-27

Solution: The deformed length at the wire will be

$$L_d = L_1 + L_2$$

where

$$L_1 = \sqrt{40^2 + 8^2} = 40.79 \text{ in}$$

$$L_2 = \sqrt{60^2 + 8^2} = 60.53 \text{ in}$$

Therefore, the induced axial strain will be the overall elongation divided by the original overall length:

$$\epsilon = \frac{L_d - L}{L} = \frac{(40.79 + 60.53) - 100}{100}$$

$$= 0.0132 \text{ in/in} \qquad \text{(tensile)}$$

Remark: Quite often in problems of this type it is somewhat difficult (particularly with a slide rule) to accurately evaluate square roots, such as those above for L_1 and L_2, because the numbers inside the radical are of different orders of magnitude. Consequently, to evaluate a square root of the form

$$\sqrt{a^2 + b^2} \qquad \text{where} \qquad a \gg b$$

it is often desirable to use the following approximation. First observe that

$$\sqrt{a^2 + b^2} = a\left[1 + \left(\frac{b}{a}\right)^2\right]^{1/2}$$

Using the binomial expansion this becomes

$$\sqrt{a^2+b^2} = a\left[1+\frac{1}{2}\left(\frac{b}{a}\right)^2 - \frac{1}{8}\left(\frac{b}{a}\right)^4 + \cdots\right]$$

Now, since $b/a \ll 1$, this last expression is approximately

$$\sqrt{a^2+b^2} \approx a\left[1+\frac{1}{2}\left(\frac{b}{a}\right)^2\right]$$

In our case, $a = 40$, $b = 8$, and $a = 60$, $b = 8$. Hence

$$L_1 \approx 40[1+\tfrac{1}{2}(\tfrac{8}{40})^2] = 40 + \tfrac{64}{80} = 40.80 \text{ in}$$
$$L_2 \approx 60[1+\tfrac{1}{2}(\tfrac{8}{60})^2] = 60 + \tfrac{64}{120} = 60.53 \text{ in}$$

which agree quite closely with our previous values.

8-7 MOHR'S CIRCLE FOR TWO-DIMENSIONAL STRAIN

Let us suppose that at some point on the surface of a load-carrying member we were asked to determine the maximum axial strain occurring at that point, assuming, of course, that we had the necessary equipment to actually measure axial strain. Guided by our previous discussion, we conceivably might proceed by measuring the axial strains in some chosen **i** and **j** directions. But would either of these be the directions of the *maximum axial strain?* Probably not, and we would be faced with the problem of determining the direction as well as the magnitude of the maximum axial strain. This important engineering problem is very similar to the one we faced in the early part of this chapter when we tried to determine the values and directions of the maximum normal and shear stresses for a given state of stress. Analogously, we now seek to determine the values and directions of the maximum axial and shear strains for a given two-dimensional state of strain.

We begin by posing the following problem: Assuming we know or can measure the axial and shear strains for the **i** and **j** directions at some point q on the surface of a body, what are the corresponding strains for some other direction **N** as indicated in Fig. 8-28(a)? To answer this question, let us choose undeformed line segments dx, dy, and $d\ell$ in Fig. 8-28(b), proportioned such that

$$dx = d\ell \cos\theta \qquad \text{and} \qquad dy = d\ell \sin\theta \tag{8-25}$$

where θ is the initial angle between the **i** direction and the **N** direction. Assume that the line segments dx and dy undergo known positive (extensional) axial strains ϵ_x and ϵ_y and a positive shear strain γ_{xy} (increase in the right angle) as indicated (exaggerated) in Fig. 8-28(c). Note that we have utilized Eq. (8-24) in writing the defor-

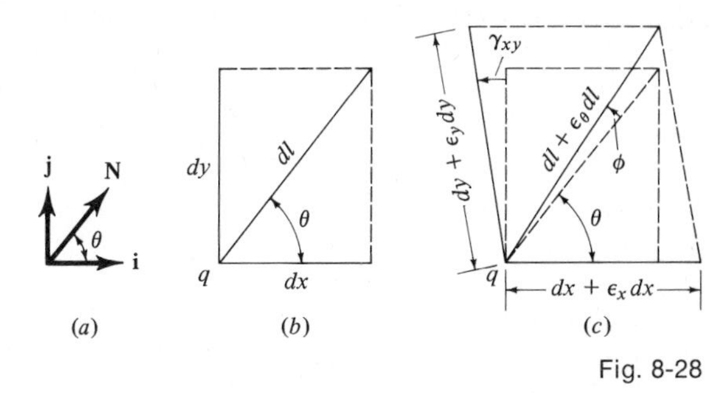

Fig. 8-28

mations de_x and de_y in terms of ϵ_x and ϵ_y. The line $d\ell$ has undergone an extensional deformation $de_\theta = \epsilon_\theta\, d\ell$ and a change of relative orientation specified by the angle ϕ.

It is apparent that the deformation of $d\ell$ is influenced simultaneously by all three known strains ϵ_x, ϵ_y, and γ_{xy}. However, for convenience and clarity we will consider the influence of each of the known strains ϵ_x, ϵ_y, and γ_{xy} separately and then add (superpose) these effects together to get the final result. This superposition procedure is valid so long as the strains are sufficiently small, of the order of 0.01 or less. Figure 8-29(a) to (c) shows the exaggerated effect of the strains ϵ_x, ϵ_y, and γ_{xy}, respectively. Each of these strains tends to change both the length and orientation of the line $d\ell$. Note that ϵ_x and ϵ_y produce an elongation of $d\ell$, while γ_{xy} causes a contraction. Adding these extensions algebraically, we see that the net elongation of $d\ell$ would be

$$\epsilon_\theta\, d\ell = \epsilon_x\, dx \cos\theta + \epsilon_y\, dy \sin\theta - \gamma_{xy}\, dy \cos\theta$$

Now, replacing dx and dy by using Eq. (8-25) and then factoring out $d\ell$, we obtain

$$\epsilon_\theta = \epsilon_x \cos^2\theta + \epsilon_y \sin^2\theta - \gamma_{xy} \sin\theta\cos\theta \qquad \text{(8-26)}$$

Thus, if we know the strains ϵ_x, ϵ_y, and γ_{xy}, this equation enables us

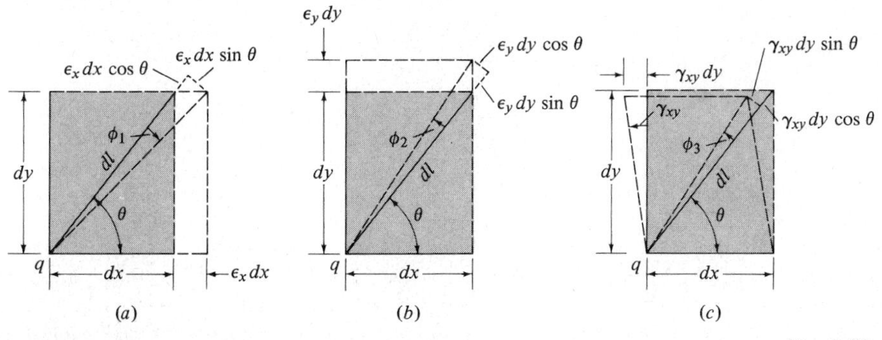

Fig. 8-29

to evaluate the axial strain ϵ_θ for any desired direction specified by the angle θ. Observe that for $\theta = 0$

$$\epsilon_0 = \epsilon_x$$

and for $\theta = 90°$

$$\epsilon_{90} = \epsilon_y$$

as anticipated.

Continuing with this analysis from Fig. 8-28(*c*), we see that the line $d\ell$ underwent a change in relative orientation ϕ as well as a change in length. The changes in relative orientation produced by ϵ_x, ϵ_y, and γ_{xy} are indicated by the angles ϕ_1, ϕ_2, and ϕ_3 in Fig. 8-29(*a*) to (*c*), respectively. Assuming these angles are *measured in radians*, we have

$$d\ell\ \phi_1 = \epsilon_x\, dx \sin\theta \qquad d\ell\ \phi_2 = \epsilon_y\, dy \cos\theta \qquad d\ell\ \phi_3 = \gamma_{xy}\, dy \sin\theta$$

Then, solving for the ϕ's and adding them algebraically, we obtain

$$\begin{aligned} \phi &= -\phi_1 + \phi_2 + \phi_3 \\ &= -\epsilon_x \sin\theta\cos\theta + \epsilon_y \sin\theta\cos\theta + \gamma_{xy}\sin^2\theta \end{aligned} \tag{8-27}$$

where we have again utilized Eq. (8-25). Equation (8-27) gives the angle of rotation which the line $d\ell$ undergoes *relative* to the line dx. However, ϕ is *not* the shear strain associated with the **N** direction since shear strain is defined as the distortion of a *right angle*. Consequently, we must consider the relative rotation of a second line segment dt originally perpendicular to $d\ell$, as shown in Fig. 8-30(*a*). But ϕ for dt would be given by Eq. (8-27) for θ replaced by $\theta + 90°$. Hence

$$\begin{aligned} \phi_{\theta+90°} &= (-\epsilon_x + \epsilon_y)\sin(\theta+90)\cos(\theta+90) + \gamma_{xy}\sin^2(\theta+90) \\ &= (\epsilon_x - \epsilon_y)\cos\theta\sin\theta + \gamma_{xy}\cos^2\theta \end{aligned} \tag{8-28}$$

where we have used the trigonometric identities

$$\sin(\theta+90) = \cos\theta \qquad \text{and} \qquad \cos(\theta+90) = -\sin\theta$$

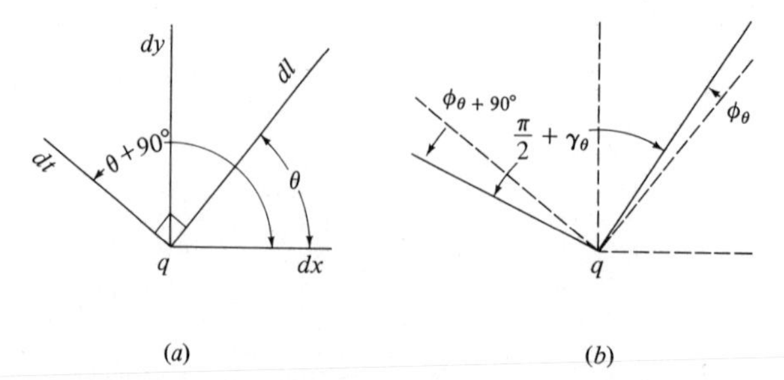

Fig. 8-30

Thus the distortion of the original right angle ($\pi/2$ rad) between $d\ell$ and dt would be

$$\phi_{\theta + 90°} - \phi_\theta = \text{Eq. (8-28)} - \text{Eq. (8-27)}$$

$$\gamma_\theta = 2(\epsilon_x - \epsilon_y) \sin \theta \cos \theta + \gamma_{xy}(\cos^2 \theta - \sin^2 \theta) \qquad (8\text{-}29)$$

as indicated in Fig. 8-30(*b*). This equation gives the shear strain γ_θ associated with the originally orthogonal line segments $d\ell$ and dt oriented at an angle of θ relative to the orthogonal line segments dx and dy, respectively.

Recapitulating, if we know the strains ϵ_x, ϵ_y, and γ_{xy}, Eqs. (8-26) and (8-29) enable us to determine the axial strain ϵ_θ and shear strain γ_θ for any desired direction specified by θ. These equations are called the *strain-transformation equations* and are quite analogous to the stress-transformation equations (8-7) and (8-8), respectively. In fact, these equations are identical if we make the following correspondences:

$$\sigma_x \sim \epsilon_x \qquad \sigma_y \sim \epsilon_y \qquad \tau_{xy} \sim \frac{\gamma_{xy}}{2} \qquad \sigma_\theta \sim \epsilon_\theta \qquad \tau_\theta \sim \frac{\gamma_\theta}{2} \qquad (8\text{-}30)$$

Hence, with these correspondences we can immediately construct *Mohr's circle of strain* with coordinate axes of axial strain ϵ_θ and one-half shear strain $\gamma_\theta/2$. Such a construction is shown in Fig. 8-31(*a*), where we have arbitrarily assumed the known strains ϵ_x, ϵ_y, and γ_{xy} in such a manner that $\epsilon_x > \epsilon_y$ and $\gamma_{xy} > 0$. This illustration embodies all the essential features of Mohr's circle for two-dimensional strain.

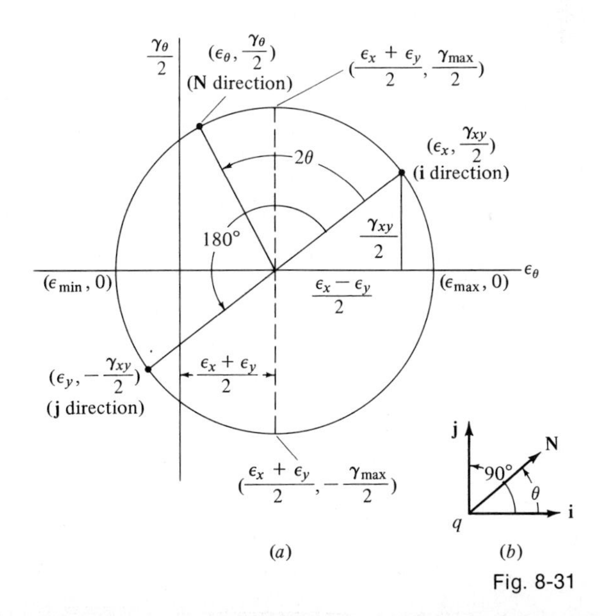

Fig. 8-31

1 The coordinates of each point on the locus of the circle correspond to the strains $(\epsilon,\gamma/2)$ associated with some direction at a point q on the surface of the body. For example, the point on the circle with coordinates $(\epsilon_x,\gamma_{xy}/2)$ in Fig. 8-31(*a*) corresponds to the strains associated with the **i** direction in Fig. 8-31(*b*); the point $(\epsilon_y,-\gamma_{xy}/2)$ in Fig. 8-31(*a*) corresponds to the strains associated with the **j** direction; and the arbitrary point $(\epsilon_\theta,\gamma_\theta/2)$ corresponds to the strains associated with the arbitrary **N** direction.

2 The angle θ between the **i** direction and the **N** direction in Fig. 8-31(*b*) corresponds to the angle 2θ on the circle in Fig. 8-31(*a*). Therefore, perpendicular directions such as **i** and **j** in Fig. 8-31(*b*) correspond to diametrically opposite points on the circle.

3 The center of the circle is *always* on the horizontal ϵ_θ axis, with coordinates $[(\epsilon_x+\epsilon_y)/2,0]$. The right extremity $(\epsilon_{\max},0)$ corresponds to the direction of the *largest* algebraic axial strain and the left extremity $(\epsilon_{\min},0)$, to the direction of the *smallest* algebraic axial strain. These maximum and minimum axial strains are called the *principal strains* and their corresponding directions, the *principal directions*. Note that there are *no shear strains* associated with the principal directions, so that line elements in the principal directions undergo *no distortion*—only elongations or contractions.

4 The upper and lower extremities of the circle with coordinates $[(\epsilon_x+\epsilon_y)/2,\gamma_{\max}/2]$ and $[(\epsilon_x+\epsilon_y)/2,-\gamma_{\max}/2]$, respectively, correspond to the directions associated with the *greatest* distortion i.e., shear strain. These directions are at 45° with the principal directions (90° in the circle).

In summary, the locus of Mohr's circle of strain represents the complete two-dimensional state of strain at some particular point q of a loaded body. The coordinates of the points on the circle represent all possible combinations of axial strain and one-half the shear strain associated with the corresponding line segments directed through point q. Thus, if we know the strain associated with two perpendicular directions, we can determine the strains for *any* direction.

EXAMPLE 8-6

A rectangular plate 20 by 20 in has two diagonal lines etched on its surface, as shown in Fig. 8-32(*a*). The plate is then stretched in such a way that it remains rectangular but with a vertical length of 20.2 in and a horizontal length of 20.4 in. How much did the angle

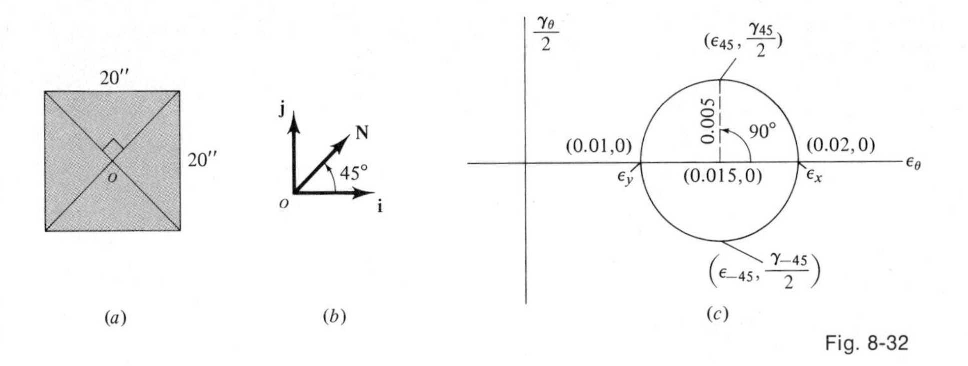

Fig. 8-32

at o change due to this stretching? How much did each diagonal stretch?

Solution: Since the deformed horizontal length is 20.4 in and the vertical, 20.2 in, then by definition of axial strain, with x and y being horizontal and vertical, respectively,

$$\epsilon_x = \frac{20.4 - 20}{20} = 0.02 \text{ in/in}$$

$$\epsilon_y = \frac{20.2 - 20}{20} = 0.01 \text{ in/in}$$

Also, since the plate remains rectangular,

$$\gamma_{xy} = 0$$

Mohr's circle for this two-dimensional state of strain is shown in Fig. 8-32(*c*). Its radius is 0.005, and its center is located at (0.015,0).

Initially, the diagonals intersect at a right angle. Therefore, the change in this right angle will be the shear strain γ_θ associated with a direction at 45° counterclockwise from the horizontal direction. On the circle, this corresponds to the point at 90° from the point $(\epsilon_x,0)$. Hence

$$\frac{\gamma_{45}}{2} = \text{radius} = 0.005 \text{ rad}$$

$$\gamma_{45} = 0.010 \text{ rad}$$

We interpret this result to mean that the indicated right angle in Fig. 8-32(*a*) will *increase* by (0.010)(180°/π) or 0.574°. Conversely, the other right angle at o will decrease by the same amount.

As for the stretching of the diagonals, the axial strain for $\theta = 45°$ and $\theta = -45°$ corresponds to the value given by the center of the circle. Thus

$$\epsilon_{45} = 0.015 \text{ in/in}$$

Recalling the definition of axial strain, the elongation of each diagonal is

$$\epsilon_{45} = \frac{e}{L} = \frac{e}{20\sqrt{20}}$$

$$e = \epsilon_{45} L$$
$$= (0.015)(20\sqrt{2}) = 0.42 \text{ in} \qquad \text{(elongation)}$$

8-8 STRAIN ROSETTES

Up to now we have specified a two-dimensional state of strain by giving two axial strains and the corresponding shear strain, such as ϵ_x, ϵ_y, and γ_{xy}. We have implied that it is possible to physically measure both axial and shear strains; but, as a practical matter, it is much more feasible to measure axial strains than it is to measure shear strains simply because small elongations and contractions are more easily and accurately measured than small angular distortions. Therefore, in experimental work, it is quite common to determine the state of strain at some point by measuring *three* axial strains at that point rather than two axial strains and one shear strain. To see how this can be done, suppose we have measured three axial strains ϵ_a, ϵ_b, and ϵ_c in three different directions specified by the angles θ_a, θ_b, and θ_c measured relative to some reference. Then, by Eq. (8-26),

$$\begin{aligned} \epsilon_a &= \epsilon_x \cos^2 \theta_a + \epsilon_y \sin^2 \theta_a - \gamma_{xy} \sin \theta_a \cos \theta_a \\ \epsilon_b &= \epsilon_x \cos^2 \theta_b + \epsilon_y \sin^2 \theta_b - \gamma_{xy} \sin \theta_b \cos \theta_b \\ \epsilon_c &= \epsilon_x \cos^2 \theta_c + \epsilon_y \sin^2 \theta_c - \gamma_{xy} \sin \theta_c \cos \theta_c \end{aligned} \qquad (8\text{-}31)$$

which represents three simultaneous equations with *known* measured values ϵ_a, ϵ_b, ϵ_c and θ_c, θ_b, θ_c and three *unknown* values ϵ_x, ϵ_y, and γ_{xy}. Thus, these three equations can be solved formally for the values of ϵ_x, ϵ_y, and γ_{xy} from which we can proceed to construct the Mohr circle of strain.

Because of the practical usefulness of this scheme for determining the state of strain, special devices called strain rosettes are available for measuring three simultaneous axial strains. Essentially these devices consist of three axial strain gages arranged in some configuration such as the 45° rectangular rosette in Fig. 8-33(*a*) or the 60° delta rosette in Fig. 8-33(*b*). For these particular configurations, the set of equations (8-31) can easily be solved and the Mohr circle constructed. The following example will illustrate the use of rosette data in determining principal strains.

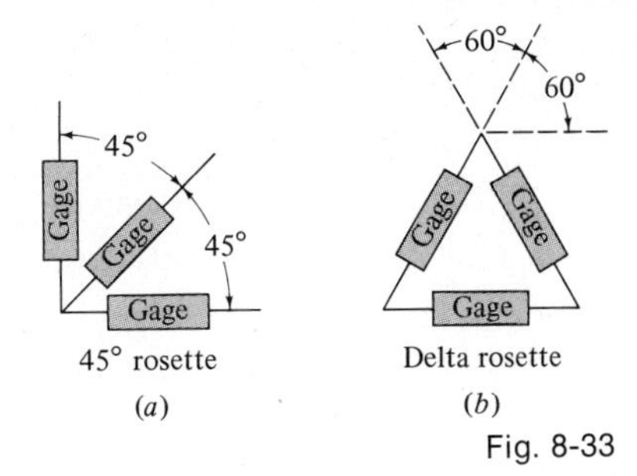

(a) (b)

Fig. 8-33

EXAMPLE 8-7

A delta rosette strain gage is attached at some point q on the surface of the fixture as shown in Fig. 8-34(a), and the gages indicate the following strains:

$$\epsilon_a = 1{,}600\ \mu\text{in/in} \qquad \epsilon_b = -700\ \mu\text{in/in} \qquad \epsilon_c = 1{,}200\ \mu\text{in/in}$$

where 1 μin/in = 0.000001 in/in. Determine the values and directions of the principal strains at point q.

Solution: The direction of gage a corresponds to the **i** direction. Then $\theta_a = 0°$, $\theta_b = 60°$, and $\theta_c = 120°$, as indicated in Fig. 8-34(b). Then Eqs. (8-31) become

$$1{,}600 = \epsilon_x \cos^2 0° + \epsilon_y \sin^2 0° - \gamma_{xy} \sin 0 \cos 0°$$
$$= \epsilon_x$$
$$-700 = \epsilon_x \cos^2 60° + \epsilon_y \sin^2 60° - \gamma_{xy} \sin 60° \cos 60°$$
$$1{,}200 = \epsilon_x \cos^2 120° + \epsilon_y \sin^2 120° - \gamma_{xy} \sin 120° \cos 120°$$

These last two become

$$-700 = 1{,}600\tfrac{1}{4} + \epsilon_y\tfrac{3}{4} - \gamma_{xy}\frac{\sqrt{3}}{4}$$

$$1{,}200 = 1{,}600\tfrac{1}{4} + \epsilon_y\tfrac{3}{4} + \gamma_{xy}\frac{\sqrt{3}}{4}$$

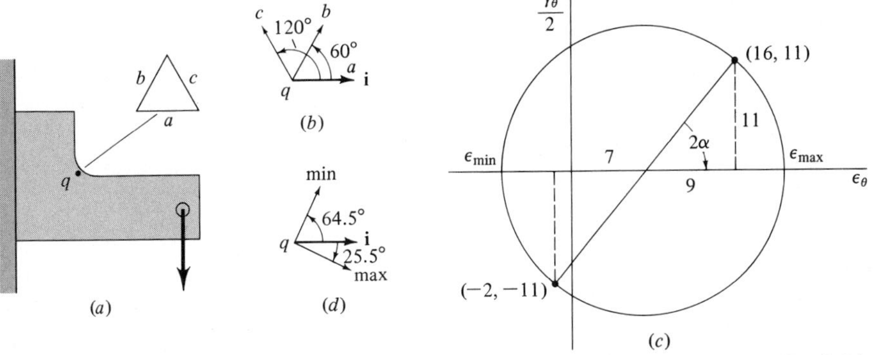

(a) (b) (c) (d)

Fig. 8-34

Solving these for ϵ_y and γ_{xy} we obtain

$$\epsilon_x = 1{,}600\ \mu\text{in/in} \qquad \epsilon_y = -200\ \mu\text{in/in} \qquad \gamma_{xy} = 2{,}200\ \mu\text{in/in}$$

We can now construct Mohr's circle as shown in Fig. 8-34(*c*). Its center is located at (7,0), and its radius is

$$r = \sqrt{9^2 + 11^2} = 14.2$$

Hence,

$$\epsilon_{\max} = 7 + 14.2 \quad \text{or} \quad 2{,}120\ \mu\text{in/in} \quad \text{(elongation)}$$
$$\epsilon_{\min} = 7 - 14.2 \quad \text{or} \quad -720\ \mu\text{in/in} \quad \text{(contraction)}$$

As for their directions, from the circle,

$$2\alpha = \arctan \tfrac{11}{9} = 51°$$
$$\alpha = 25.5°$$

Therefore, the direction of $\epsilon_{\max}$ is 25.5° clockwise from the **i** direction. Consequently, the direction of $\epsilon_{\min}$ is 64.5° counterclockwise from the **i** direction. These principal directions are indicated in Fig. 8-34(*d*). Note that $\epsilon_{\max}$ is roughly tangent to the fillet curve.

EXERCISE PROBLEMS

8-41 A rubber band 2 in long (4 in in perimeter) is slipped over a solid roll of paper 2 in in diameter. What is the axial strain in the rubber band?

8-42 A 20-ft-long guy wire is used to support a mast. The guy wire is made taut by using a turnbuckle. What strain would be induced in the taut guy wire if the turnbuckle were tightened two additional full turns? The turnbuckle screws have a thread pitch of 14 threads per inch.

8-43 The taut wire 80 in long is attached to a fixed anchor at o and a movable anchor at q. Determine the strain induced in the wire if q is moved X inches to the right and Y inches downward, where (*a*) $X = 4$ in, $Y = 0$, (*b*) $X = 0$, $Y = 4$ in, and (*c*) $X = 4$ in, $Y = 4$ in.

PROB. 8-43

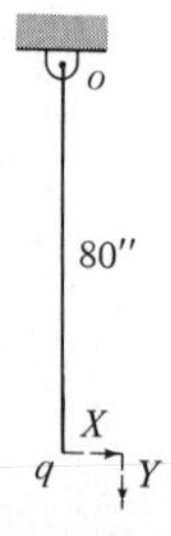

8-44 The rod ab is relatively rigid and is held horizontal by the vertical wire cd. Assuming the wire remains vertical as it slides on the rod, what strain will be induced in the wire if the rod is rotated about its pivot at a so that the end b undergoes a vertical movement of e inches downward? If e is small compared with L_R, give an approximation for the strain in the wire.

PROB. 8-44

PROB. 8-45

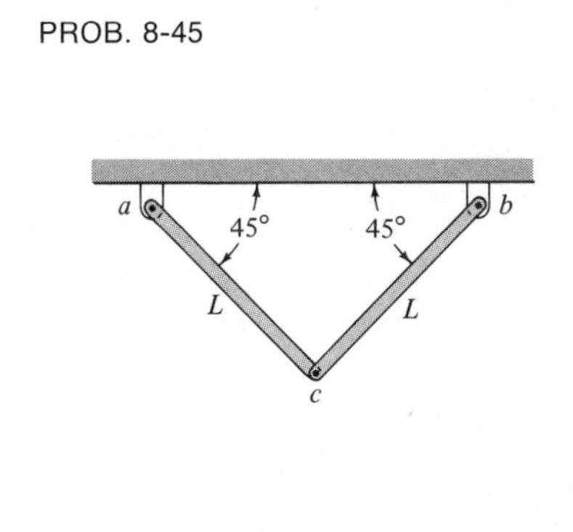

8-45 Two thin rods ac and bc each of length L units are pinned together as shown. The joint c is then displaced X units to the right and Y units downward. Express the strains in the rods in terms of X, Y, and L. If X and Y are small compared with L, give an approximation for the strains in the rods.

8-46 A rectangular rubber plate 20 by 20 in has two diagonal lines scribed on its surface. The top edge of the plate is translated 0.20 in to the right as indicated by the dashed lines, with the plate becoming a parallelogram. What is the change in the original indicated right angle at o? What is the change in length of each of these diagonals?

PROB. 8-46

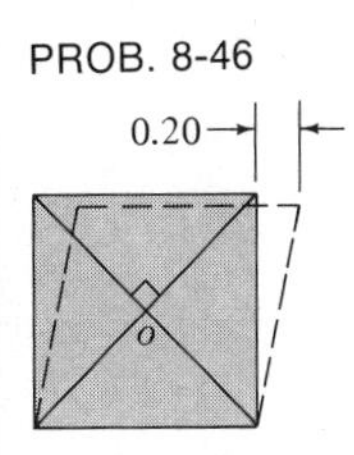

8-47 A rectangular block initially has dimensions of 4 by 4 by 8 in. The block undergoes a deformation such that the axial strains along its sides are 0.001 in/in/, 0.002 in/in, and 0.0015 in/in, respectively. What is the change in volume of the block due to the deformation?

8-48 A spherical weather balloon is usually filled with 20,000 ft^3 of helium. Inadvertently, the balloon was filled with 26,000 ft^3 of helium. What additional axial strain is induced in the skin of the balloon due to the additional 6,000 ft^3?

8-49 to 8-54 For each of the following two-dimensional states of strain, determine the values and directions of the principal strains. The strains are given in microinches per inch:

8-49 $\epsilon_x = -1{,}200$, $\epsilon_y = 200$, $\gamma_{xy} = -1{,}600$
8-50 $\epsilon_x = 300$, $\epsilon_y = 1{,}200$, $\gamma_{xy} = 800$
8-51 $\epsilon_x = 1{,}200$, $\epsilon_y = -300$, $\gamma_{xy} = 2{,}000$
8-52 $\epsilon_x = -120$, $\epsilon_y = -1{,}120$, $\gamma_{xy} = 1{,}000$
8-53 $\epsilon_x = 800$, $\epsilon_y = -200$, $\gamma_{xy} = -800$
8-54 $\epsilon_x = 750$, $\epsilon_y = 750$, $\gamma_{xy} = 1{,}500$

8-55 If $\epsilon_x = 0.00080$ in/in (compression), $\epsilon_y = 0.00020$ in/in (compression), and $\gamma_{xy} = 0.00080$ in/in, what value of strain would an axial strain gage indicate that was oriented 45° counterclockwise from the x direction?

8-56 If $\epsilon_x = 0.00240$ in/in, $\epsilon_y = -0.00060$ in/in, and $\gamma_{xy} = 0.00160$ in/in, what value would an axial strain gage indicate that was oriented 30° clockwise from the x axis?

8-57 to 8-60 The following are data from rosette strain gages. Determine the values and directions of the principal strains. The strains are in microinches per inch.

8-57 $\epsilon_o = 220$, $\epsilon_{45} = 120$, $\epsilon_{90} = 220$
8-58 $\epsilon_o = 400$, $\epsilon_{60} = 400$, $\epsilon_{120} = -600$
8-59 $\epsilon_o = -720$, $\epsilon_{60} = 300$, $\epsilon_{120} = -600$
8-60 $\epsilon_o = -120$, $\epsilon_{45} = -400$, $\epsilon_{90} = 1{,}200$

8-61 The cubical dilatation CD is defined as the change in volume per unit original volume. Using a rectangular prism with sides of lengths Δx, Δy, and Δz, show that for small strains the cubical dilatation is given by

$$CD = \epsilon_x + \epsilon_y + \epsilon_z$$

8-9 GENERALIZED HOOKE'S LAW

We have seen how the general state of stress at any point in a load-carrying body is described by six stresses—three normal stresses and three shear stresses. Similarly, the general state of strain is described by six strains—three axial strains and three shear strains. We now ask the question: For a given material, how are the six stresses at a point related to the six strains at the point? In more formal terminology, what is the mechanical constitutive equation for the material relating the state of stress to the state of strain?

In Chap. 6, we saw for an axially loaded member how the axial stress was related to the corresponding axial strain by means of a uniaxial stress-strain curve. Accordingly, a stress-strain curve can be interpreted as a one-dimensional constitutive equation relating one normal (axial) stress to the corresponding axial strain. From the shape of the curve, it was apparent that such one-dimensional stress-strain relationships can be linear and/or nonlinear, with the linear part quantitatively characterized by the modulus of elasticity E and the nonlinear portion being much more difficult to characterize in a quantitative manner. On the basis of such one-dimensional results, we can anticipate that the six stresses describing the general state of stress can be related to the six strains describing the general state of strain by means of some relationship which is initially linear and subsequently becomes nonlinear at higher levels of stress and strain. The problem of quantitatively describing such a linear-nonlinear relation between the general state of stress and the general state of strain for real materials has baffled and continues to baffle both theoreticians and experimental mechanicians. While a linear relation can be handled quite satisfactorily with some theories of linear algebra, efforts to describe nonlinear response have been far less successful, although some notable results have been achieved. In what follows, we will restrict our attention to the linear constitutive theory for solid materials, which is commonly referred to as the *generalized Hooke's law.*

Essentially, our problem is to relate the six variable stresses to the six variable strains in a linear manner. Although either set of variables could be considered the dependent and the other the independent variables, for convenience we will take the stresses as our independent variables and the strains as the dependent variables. Accordingly, the most general *linear* relation would take the form

$$
\begin{aligned}
\epsilon_x &= c_{11}\sigma_x + c_{12}\sigma_y + c_{13}\sigma_z + c_{14}\tau_{xy} + c_{15}\tau_{xz} + c_{16}\tau_{yz} \\
&\cdot \\
&\cdot \\
&\cdot \\
\gamma_{yz} &= c_{61}\sigma_x + c_{62}\sigma_y + c_{63}\sigma_z + c_{64}\tau_{xy} + c_{65}\tau_{xz} + c_{66}\tau_{yz}
\end{aligned}
\qquad (8\text{–}32)
$$

Thus, in general, 36 constants (material parameters) would be needed to describe the complete linear response of an arbitrary material. Fortunately, most structural materials exhibit certain characteristics which eliminate many of these constants and make the relationship much more tractable. We will not present a formal algebraic derivation, but rather we will appeal to some intuitive arguments for obtaining a more simplified linear relation. Our illustration will be for homogeneous and isotropic materials—isotropic meaning that the material exhibits the same response regardless of orientation. Most metals are isotropic, whereas wood, because of its grain, is not isotropic.

Consider a unit cube of material shown in Fig. 8-35(*a*), each side being one unit in length. Figure 8-35(*b*) illustrates the exaggerated deformations which the cube undergoes due to the application of the normal stress σ_x. Since the loading is uniaxial, we have, by the definition of the elastic modulus E given in Sec. 6-3 and the definition of Poisson's ratio μ in Sec. 6-4,

$$\epsilon_x = \frac{\sigma_x}{E} \qquad \epsilon_y = -\mu\epsilon_x = -\mu\,\frac{\sigma_x}{E} \qquad \epsilon_z = -\mu\epsilon_x = -\mu\,\frac{\sigma_x}{E} \tag{8-33a}$$

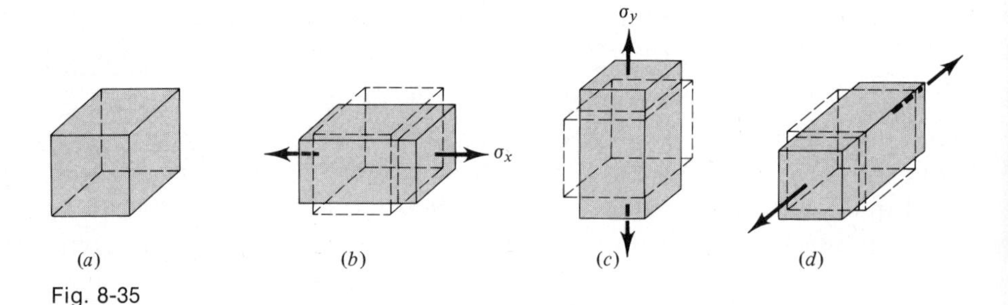

Fig. 8-35

Similarly, if the cube is loaded uniaxially in the y and z directions as illustrated in Fig. 8-35(*c*) and (*d*), respectively,

$$\epsilon_y = \frac{\sigma_y}{E} \qquad \epsilon_x = -\mu\,\frac{\sigma_y}{E} \qquad \epsilon_z = -\mu\,\frac{\sigma_y}{E} \tag{8-33b}$$

$$\epsilon_z = \frac{\sigma_z}{E} \qquad \epsilon_x = -\mu\,\frac{\sigma_z}{E} \qquad \epsilon_j = -\mu\,\frac{\sigma_z}{E} \tag{8-33c}$$

Note that the isotropy of the material has been employed by assuming that the E's and μ's in each of Eqs. (8-33*a*) to (8-33*c*) are the same E and μ, with no distinction being made because of the direction of the stress. Also, we have assumed that the axial stresses do not distort the cube.

Equations (8-33*a*) to (8-33*c*) give the axial strains caused by the individual stresses σ_x, σ_y, and σ_z, respectively. Since we are considering linear elastic behavior, we can anticipate that these strains will be relatively small. Consequently, we can superpose these individual strains to obtain the total axial strain due to the simultaneous application of all three stresses. Thus, adding the axial strains in the x direction caused by each stress, we obtain

$$\epsilon_{x(\text{total})} = \frac{\sigma_x}{E} - \mu \frac{\sigma_y}{E} - \mu \frac{\sigma_z}{E} \tag{8-34a}$$

Similarly, for the other directions

$$\epsilon_{y(\text{total})} = \frac{\sigma_y}{E} - \mu \frac{\sigma_x}{E} - \mu \frac{\sigma_z}{E} \tag{8-34b}$$

$$\epsilon_{z(\text{total})} = \frac{\sigma_z}{E} - \mu \frac{\sigma_x}{E} - \mu \frac{\sigma_y}{E} \tag{8-34c}$$

Consider now the same unit cube subjected to the shear stress τ_{xy} (and τ_{yx}), as in Fig. 8-36(*b*). Again, because of the isotropy of the material, we anticipate that this shear stress distorts the cube in such a manner as to produce a shear strain γ_{xy}, but no other strain; similarly for the stresses τ_{xz} (and τ_{zx}) and τ_{yz} (and τ_{zy}) in Fig. 8-36(*c*) and (*d*), respectively. Then, by the definition of the shear modulus G given in Sec. 7-4, we have

$$\gamma_{xy} = \frac{\tau_{xy}}{G} \qquad \gamma_{xz} = \frac{\tau_{xz}}{G} \qquad \gamma_{yz} = \frac{\tau_{yz}}{G} \tag{8-35}$$

where we have once again utilized the isotropy of the material by assuming the same G in each case.

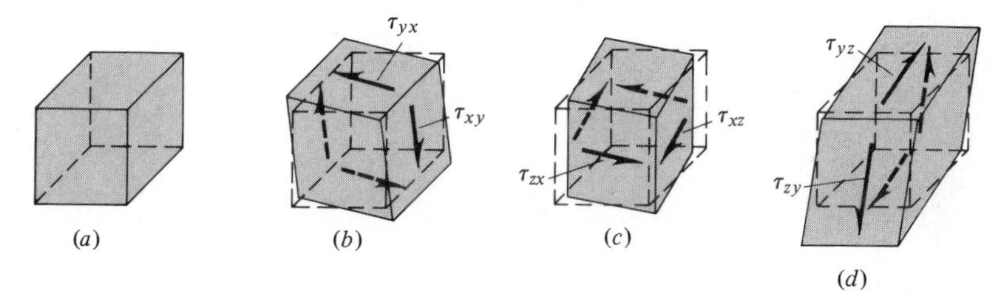

Fig. 8-36

The three equations (8-34*a*) to (8-34*c*) relating the axial stresses and strains and the three equation (8-35) relating the shear stresses and strains comprise the *generalized Hooke's law for linear elastic isotropic materials*. Note that the general linear relations for any material given by Eq. (8-32) involving 36 material constants have been reduced to relations involving only three material parameters, namely, E, G, and μ. We will see shortly how these three can further be reduced to two so that only two independent material parameters are needed to completely describe the linear elastic response of an isotropic solid material.

Before doing that, however, let us look at Hooke's law for the special but important case of *plane stress*. Recall that, for plane stress, the state of stress is essentially two dimensional, for example,

$$\sigma_x \neq 0 \qquad \sigma_y \neq 0 \qquad \tau_{xy} \neq 0$$
$$\sigma_z = 0 \qquad \tau_{xz} = 0 \qquad \tau_{yz} = 0$$

For this case, Eqs. (8-34) and (8-35) reduce to

$$\begin{aligned} &\epsilon_x = \frac{\sigma_x}{E} - \mu\frac{\sigma_y}{E} \qquad \text{and} \qquad \epsilon_y = \frac{\sigma_y}{E} - \mu\frac{\sigma_x}{E} \\ &\epsilon_z = -\frac{\mu}{E}(\sigma_x + \sigma_y) \qquad \gamma_{xy} = \frac{\tau_{xy}}{G} \qquad \gamma_{xz} = \gamma_{yz} = 0 \end{aligned} \tag{8-36}$$

These equations along with their inverse relations

$$\begin{aligned} \sigma_x &= \frac{E}{1-\mu^2}(\epsilon_x + \mu\epsilon_y) \\ \sigma_y &= \frac{E}{1-\mu^2}(\epsilon_y + \mu\epsilon_x) \\ \tau_{xy} &= G\gamma_{xy} \end{aligned} \tag{8-37}$$

play an important role in the stress analysis of many real load-carrying members.

Returning now to the question of reducing the three material parameters E, G, and μ to two independent parameters, consider the plane state of stress shown in Fig. 8-37(*a*), where σ is some given value. The corresponding Mohr stress circle is shown in Fig. 8-37(*b*), with $\tau_{\max} = \sigma$ occurring on a plane at 45° from the plane with σ_x. Now, by Hooke's law for plane stress [Eq. (8-36)], we see from the calculations in Fig. 8-37(*d*) that $\epsilon_x = \epsilon$ and $\epsilon_y = -\epsilon$, where $\epsilon = \sigma/E(1+\mu)$. Since $\gamma_{xy} = 0$, the Mohr strain circle is shown in Fig. 8-37(*e*), with $\gamma_{\max} = 2\epsilon$ occurring in a direction at 45° with the direc-

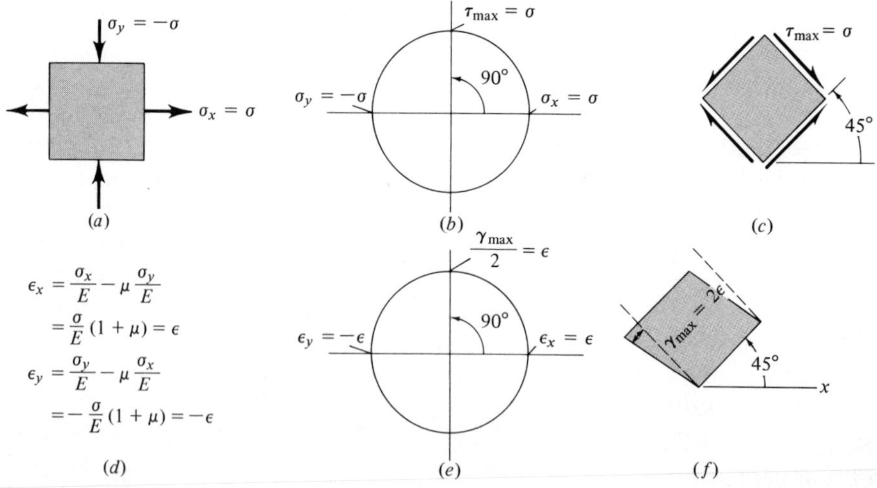

Fig. 8-37

tion of ϵ_x. Therefore, according to Hooke's law for shear stresses and strains [Eq. (8-35)], the shear stress τ_{max} and shear strain γ_{max} should be related by

$$\gamma_{max} = \frac{\tau_{max}}{G}$$

or

$$2\epsilon = \frac{\sigma}{G} = 2\,\frac{\sigma}{E}\,(1 + \mu)$$

Thus

$$G = \frac{E}{2(1 + \mu)} \tag{8-38}$$

which relates the three elastic constants E, G, and μ. Therefore, any one of these three parameters can be expressed in terms of the other two, and consequently only two are needed to characterize a linear elastic isotropic material. We remark that this relationship has been verified experimentally for many real materials.

We close this section and this chapter with illustrations of the application of Hooke's law and Mohr's circles to problems in stress analysis.

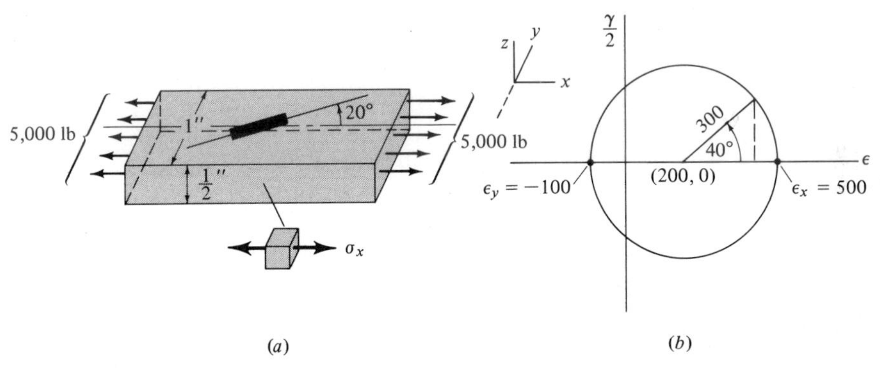

Fig. 8-38

EXAMPLE 8-8

A bar with the dimensions shown in Fig. 8-38(a) is stretched by a centrally applied load of 5,000 lb. An electrical axial strain gage is attached to the flat surface at an angle of 20° with the longitudinal axis of the bar. If the bar is linear elastic with $E =$ 20,000,000 psi and $\mu = 0.2$, what value of strain will the strain gage indicate?

Solution: Let the x axis correspond to the longitudinal dimension of the surface and the y axis to the transverse dimension. Since the bar is axially loaded, the longitudinal axial stress will be

$$\sigma_x = \frac{P}{A} = \frac{5{,}000}{(1/2)(1)} = 10{,}000 \text{ psi}$$

The state of stress is uniaxial, with all other normal and shear stresses associated with the x, y, and z axes being zero. Therefore, by Hooke's law [Eqs. (8-34) and (8-35)], we have

$$\epsilon_x = \frac{\sigma_x}{E} = \frac{10{,}000}{20 \times 10^6} = 500 \times 10^{-6} \text{ in/in}$$

$$\epsilon_y = -\mu \frac{\sigma_x}{E} = -(0.2)\frac{10{,}000}{20 \times 10^6} = -100 \times 10^{-6} \text{ in/in}$$

$$\gamma_{xy} = 0$$

It is not necessary to calculate the other three strains ϵ_z, γ_{xz}, and γ_{yz} since the x–y strains completely specify the two-dimensional surface strains. We can now plot the Mohr strain circle for these surface strains, which is shown in Fig. 8-38(*b*). The center of this circle is (200, 0) and its radius is 300.

The strain gage is oriented at an angle of 20° counterclockwise from the logitudinal axis. Hence, the corresponding strains are given by the coordinates of the point on the circle located at 40° counterclockwise from the point corresponding to $\epsilon_x = 500$. Thus

$$\begin{aligned}\epsilon_{20°} &= 200 + 300 \cos 40° \\ &= 200 + 230 = 430\end{aligned}$$

Therefore, the strain gage will indicate an axial strain of

$$\epsilon_{20°} = 430 \times 10^{-6} \text{ in/in}$$

EXAMPLE 8-9

At some point on the surface of a steel structural member, it was found that $\epsilon_x = 0.001150$ in/in, $\epsilon_y = -0.000250$ in/in, and $\gamma_{xy} = -0.000900$ in/in. What were the values of the principal stresses and the maximum shear stress?

Solution: Essentially, our problem is to convert the given strains to stresses by the use of Hooke's law and then find the principal stresses by the use of Mohr's stress circle. Alternatively, we could first determine the principal strains by the use of Mohr's *strain* circle and then convert these principal strains to principal stresses by the use of Hooke's law. Either procedure should result in the correct answer.

For the given strains, Mohr's strain circle is shown in Fig. 8-39(*a*) with its center at (450, 0) and a radius of

$$r = \sqrt{(450)^2 + (700)^2} = 832$$

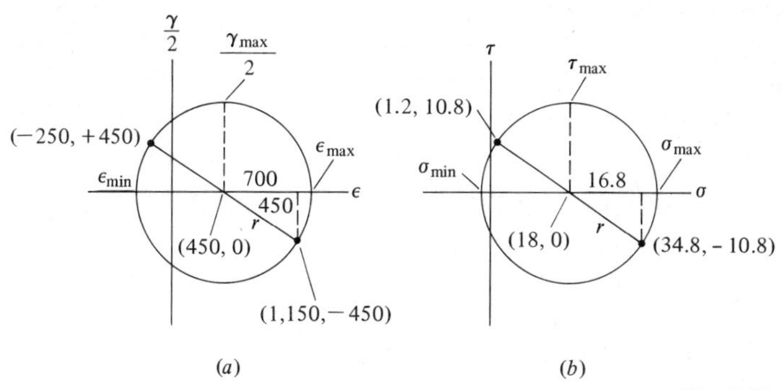

Fig. 8-39

Hence,

$$\epsilon_{max} = 450 + 832 = 1{,}282$$

or

$$\epsilon_{max} = 0.001282 \text{ in/in}$$

$$\epsilon_{min} = 450 - 832 = -382$$

or

$$\epsilon_{min} = -0.000382 \text{ in/in}$$

$$\gamma_{max} = 2r = 1{,}664$$

or

$$\gamma_{mas} = 0.001664 \text{ in/in}$$

At a free surface, the state of stress is plane, so that we can use Eqs. (8-37) to evaluate the principal stresses from the principal strains. Thus, using $E = 30 \times 10^6$ psi and $\mu = \frac{1}{4}$,

$$\sigma_{max} = \frac{E}{1-\mu^2}(\epsilon_{max} + \mu\epsilon_{min})$$

$$= \frac{30 \times 10^6}{1-(1/4)^2}(1{,}282 - \tfrac{1}{4}382) \times 10^{-6}$$

$$= 38{,}000 \text{ psi}$$

$$\sigma_{min} = \frac{30 \times 10^6}{1-(1/4)^2}(-382 + \tfrac{1}{4}1282) \times 10^{-6}$$

$$= -2{,}000 \text{ psi} \qquad \text{(compressive)}$$

$$\tau_{max} = G\gamma_{max}$$

$$= (12 \times 10^6)(1{,}664 \times 10^{-6})$$

$$= 20{,}000 \text{ psi}$$

On the other hand, by *first* using Eqs. (8-37), we have

$$\sigma_x = \frac{E}{1-\mu}(\epsilon_x + \mu\epsilon_y)$$

$$= \frac{30 \times 10^6}{1-(1/4)^2}(1{,}150 - \tfrac{1}{4}250) \times 10^{-6}$$

$$= 34{,}800 \text{ psi}$$

$$\sigma_y = \frac{30 \times 10^6}{1-(1/4)^2}(-250 + \tfrac{1}{4}1{,}150) \times 10^{-6}$$
$$= 1{,}200 \text{ psi}$$
$$\tau_{xy} = G\gamma_{xy} = (12 \times 10^6)(-900 \times 10^{-6})$$
$$= -10{,}800 \text{ psi}$$

For this state of stress, Mohr's stress circle is shown in Fig. 8-39(*b*) with its center at (18, 0) and a radius of

$$r = \sqrt{(16.8)^2 + (10.8)^2} = 20$$

Hence

$$\sigma_{\text{max}} = 18 + 20 = 38$$

or

$$\sigma_{\text{max}} = 38{,}000 \text{ psi}$$
$$\sigma_{\text{min}} = 18 - 20 = -2$$

or

$$\sigma_{\text{min}} = -2{,}000 \text{ psi} \qquad \text{(compressive)}$$
$$\tau_{\text{max}} = r = 20{,}000 \text{ psi}$$

which agree with our previous results.

> ***Remark:*** With a maximum stress of 38,000 psi, it is somewhat questionable as to whether or not the steel is behaving elastically. Therefore, our calculated maximum stresses may be somewhat higher than those actually existing in the member.

EXERCISE PROBLEMS

8-62 A bar of the dimensions shown is stretched by a centrally applied load of 10,000 lb. An axial strain gage is attached to the bar as shown. If the bar is elastic with $E = 10 \times 10^6$ psi and $\mu = \frac{1}{4}$, what value of strain will the gage indicate?

PROB. 8-62

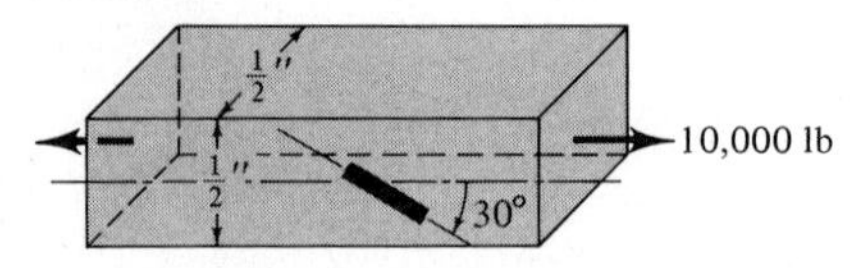

8-63 In a thin-walled closed-ended cylindrical pressure vessel the circumferential stress σ_c is twice the value of the longitudinal stress σ_L. If the vessel is made of steel with $E = 30 \times 10^6$ psi and $\mu = \frac{1}{4}$, what is the maximum axial strain in the wall of the vessel if $\sigma_c = 6{,}000$ psi? What is the maximum shear stress and corresponding shear strain?

8-64 At some point on the surface of a structural member, strain gages were used to measure the surface strains. The data indicated that $\epsilon_x = 100$, $\epsilon_y = 400$, and $\gamma_{xy} = 400$, all in microinches per inch. The structure is made of steel with $E = 30 \times 10^6$ and $\mu = \frac{1}{4}$. What were the principal stresses and the maximum shear stress?

8-65 Recall the fixture in Example 8-7 and Fig. 8-34. Assuming the fixture is made of steel with $E = 30 \times 10^6$ and $\mu = \frac{1}{4}$, what are the principal stresses at point q due to the applied load? What is the maximum shear stress at point q? Assume elastic behavior.

8-66 The bulk modulus K is defined as the ratio of the average normal stress divided by the sum of the axial strains (cubical dilatation) for any elastic state of stress. Show that

$$K = \frac{E}{3(1 - 2\mu)}$$

8-67 to 8-70 For the rosette strain gage data in Probs. 8-57 to 8-60, find the principal stresses and the maximum shear stress. Assume the material is steel with $E = 30 \times 10^6$ psi and $\mu = \frac{1}{4}$.

8-71 Consider a rectangular aluminum block $2 \times 4 \times 6$ in. Assume the block is elastic with $E = 10 \times 10^6$ psi and $\mu = \frac{1}{3}$. If each face of the block is subjected simultaneously to a compressive force of 20,000 lb, what will be the corresponding change in the volume of the block?

8-72 A solid circular steel ($E = 30 \times 10^6, \mu = \frac{1}{4}$) shaft of radius $1\frac{1}{2}$ in and length 16 in is subjected to a pure torque loading of 20,000 in-lb. An axial strain gage is bonded to the curved outer surface of the shaft. What strain will this gage indicate if its axis is (*a*) parallel to the axis of the shaft, (*b*) perpendicular to the axis of the shaft, and (*c*) at an angle of 30° to the axis of the shaft?

8-73 It is desired to "measure" the angle of twist and the torque being transmitted by a 6-ft-long hollow steel drive shaft with OD = 2 in and ID = $1\frac{1}{2}$ in. To accomplish this measurement, an axial strain gage is to be bonded to the outer surface of the drive shaft at an angle of 45° to the axis of the shaft. What will be the relationship between the strain ϵ indicated by this gage and the angle of twist? What will be the relation between ϵ and the torque? $E = 30 \times 10^6$ and $\mu = \frac{1}{4}$.

8-74 A thin-walled copper tube with OD = 2 in and $t = 0.10$ in carries a gas under a pressure 600 psi. An axial strain gage is attached to the surface of the tube. What strain will this gage indicate if its axis is (*a*) parallel to the axis of the tube, (*b*) perpendicular to the axis of the tube, and (*c*) at an angle of 25° with the axis of the tube? Use $E = 12 \times 10^6$ psi and $\mu = \frac{1}{3}$. Treat the tube as an open-ended pressure vessel; i.e., $\sigma_L = 0$.

8-75 A pressure gage is to be constructed by attaching two axial strain gages to the outer wall of a small closed-ended thin-walled aluminum tube having a diameter of $\frac{1}{2}$ in, wall thickness of 0.020 in, and length of 2 in. If the gages are at right angles to each other, what will be the relationship between the internal pressure p and the sum of the strains in both gages? Use $E = 10 \times 10^6$ and $\mu = 0.3$.

8-76 There is some question about the safety of a bridge across the Monongahela River. Under maximum load, inspectors measured the following strains: $\epsilon_x = 1{,}400$ μin/in, $\epsilon_y = -200$, and $\gamma_{xy} = 1{,}200$. The safety code says that the stresses cannot exceed the following values: $\sigma = 20{,}000$ psi, $\tau = 15{,}000$ psi. Is the bridge safe? Prove your answer. Use $E = 30 \times 10^6$, $G = 12 \times 10^6$, and $\mu = \frac{1}{4}$.

chapter 9 BEAMS

9-1 INTRODUCTION

Without doubt, beams are the most common of all simple structural members. Essentially, beams are relatively long members used to support and transmit bending and transverse shear loads (as distinguished in Fig. 5-5 from axial and twisting loads) across a span between some type of supports. For example, floor joists of a building are supported at the walls and carry the loads applied to the floor. Similarly you can observe the spanning beams supporting highway bridges, the curved beams supporting domed roofs, and the diving board of your community swimming pool. Less obvious are the nonhorizontal beam-like members of various structural frames and machines, such as those which were studied in Chap. 4. Consequently there are great varieties of types and shapes of beams each having their own peculiar structural characteristics. Generally, beams are classified according to their geometry (straight, curved, tapered, etc.) and the manner in which they are loaded and/or supported. Real beams are loaded in an infinite variety of ways, but generally the loadings are idealized as either concentrated loads at discrete locations or as easily described distributed loads (see Fig. 9-1).

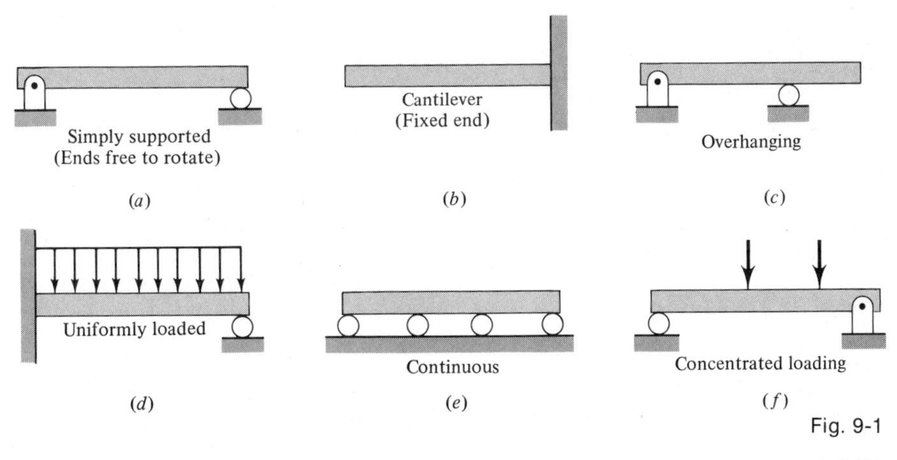

Fig. 9-1

In this chapter we will examine some of the basic considerations in the load analysis and design of beams. We will restrict our attention to straight beams of constant cross section and made of a homogeneous isotropic material. Our objective is to relate the applied loads to the resulting stresses and deformations produced by the loads.

9-2 SHEAR AND MOMENT DIAGRAMS

The primary structural function of a beam is to transmit or support bending and transverse shear loads. However, we can anticipate from the exaggerated illustrations in Fig. 9-2 that the amount and type of bending in a beam can vary significantly throughout the length of the beam. Therefore, one of the first things an engineer must be able to do is to analyze the bending and shear loads throughout the length of the beam so as to ascertain where the most severe bending and/or shear occurs in the beam; that is, an engineer must be able to determine where failure is most likely to occur. The primary tool for performing this type of load analysis of beams is the so-called *shear-and-moment-diagram* technique.

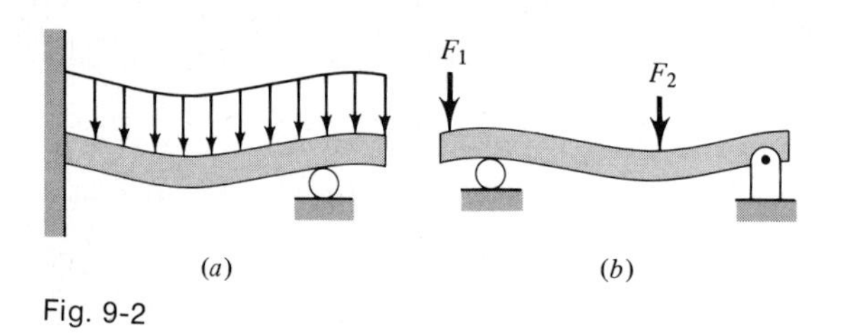

Fig. 9-2

Recall the method of sections from Chap. 5 in which a free-body diagram of a *section* of a structural member was drawn so as to determine the net *internal* reactions at some desired location in the structural member. Generally, the internal reactions might be a normal force, a shear force, a twisting couple, or a bending couple (recall Figs. 5-5 and 5-6). We will now apply this same technique for determining the shear force and bending moment (couple) at any and all locations throughout the length of the beam.

Consider a simply supported beam in Fig. 9-3(a) loaded with an arbitrarily distributed loading specified in terms of some loading function $w(\ell)$ per unit length of beam, for example, $w(\ell)$ pounds of load per foot of length of beam. Accordingly, the total load due to the loading $w(\ell)$ on the beam would be

$$F = \int_0^L w(\ell)\, d\ell \qquad (9\text{-}1)$$

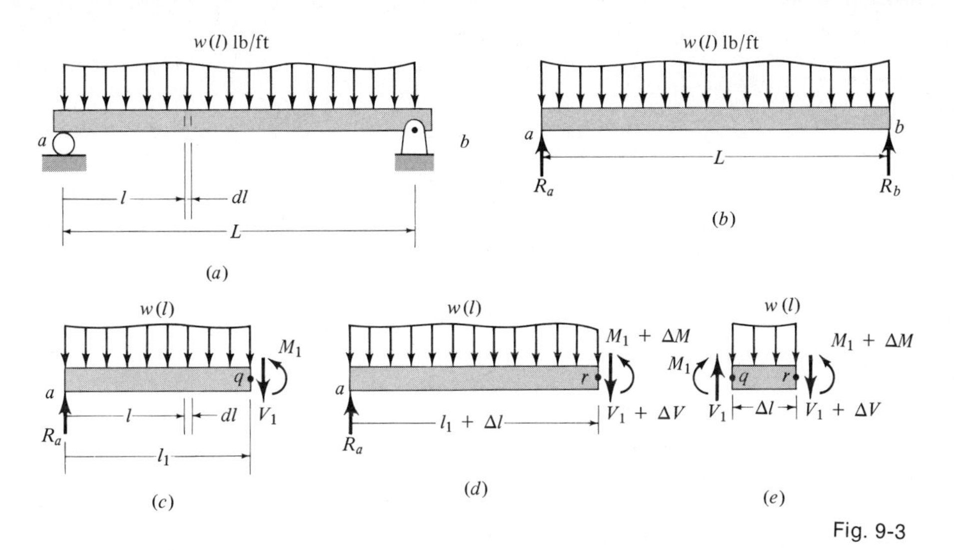

Fig. 9-3

We remark that in many situations the loading on a real beam might be reasonably described in terms of one or several discrete (concentrated) loads located at various points on the beam as, for example, in Fig. 9-1(*f*) or Fig. 9-2(*b*). In such situations the loading function $w(\ell)$ is zero between the concentrated loads and is undefined (mathematically discontinuous) at the loading points. Thus, with proper interpretation, what we do in the following general situation is applicable for the most part to all types of loading on a beam.

The first step in the load analysis of a beam is to determine the external reactions exerted by the supports on the beam. If the beam is statically determinate, these reactions can actually be evaluated from the requirements of statical equilibrium. Otherwise, if a beam is overrestrained and, therefore, statically indeterminate [as in Fig. 9-1(*d*) and (*e*)], the external reactions at the supports cannot be completely determined from equilibrium conditions alone. Our example beam in Fig. 9-3(*a*) is statically determinate, and we assume that the support reactions R_a and R_b have been determined as indicated in Fig. 9-3(*b*).

Recalling that our objective is to determine the *internal* shear force and bending moment at *any* location along the length of the beam, we now draw a free-body diagram of a *section* of the beam some arbitrary ℓ_1 units long as measured from the left end of the beam and indicated in Fig. 9-3(*c*). In this figure the internal shear force is labeled V_1, the subscript 1 indicating that this is the value of the internal shear force in the beam at location ℓ_1. Similarly, the bending moment (couple) is labeled M_1. By applying the requirements of statical equilibrium, V_1 and M_1 can be determined from the known loading function $w(\ell)$, the known reactions R_a and R_b, and the chosen length ℓ_1. Summing forces in the vertical direction yields

$$R_a - \int_0^{\ell_1} w(\ell)\, d\ell - V_1 = 0$$

$$V_1 = R_a - \int_0^{\ell_1} w(\ell)\, d\ell$$

Thus, in general, the internal shear force in the beam is a function of the location:

$$V_1 = V(\ell_1) \tag{9-2}$$

Note that in our present example, for $\ell_1 = 0$

$$V(0) = R_a$$

and for $\ell_1 = L$

$$V(L) = R_a - F = -R_b$$

where we have utilized Eq. (9-1). Thus the internal shear force as given by Eq. (9-2) can be algebraically positive or negative, and it is common practice in engineering to assign the positive value to a shear force, such as the one shown in Fig. 9-3(*c*); that is, a *positive* internal shear force is one that acts *downward* on the right-hand side of a section of a beam.

In a similar manner, summing moments in Fig. 9-3(*c*) with respect to point q yields

$$(R_a)(\ell_1) - \int_0^{\ell_1} w(\ell)(\ell_1 - \ell)\, d\ell - M_1 = 0$$

$$M_1 = R_a\ell_1 - \int_0^{\ell_1} w(\ell)(\ell_1 - \ell)\, d\ell$$

Thus, in general, the internal bending moment is also a function of the location:

$$M_1 = M(\ell_1) \tag{9-3}$$

In American engineering practice it is customary to assign a positive value to a bending moment, such as the one in Fig. 9-3(*c*). Physically, a positive bending moment is one that causes the beam to bend concave upward; that is, it compresses the upper side of the beam and stretches the lower side of the beam.

Recapitulating, both the internal shear force and the internal bending moment usually vary throughout the length of a beam, and these variations are formally given by Eqs. (9-2) and (9-3), which were obtained by applying the requirements of statical equilibrium to a free-body diagram of a typical section of the beam. A *shear diagram* is simply a graphical plot of how the internal shear force varies throughout the length of the beam. Similarly, a *moment diagram* is a graphical plot of the variation of the bending moment throughout the length of the beam.

Before giving some specific illustrations of such shear and moment diagrams, we will obtain some very useful relationships between the shear-force function $V(\ell)$, the bending-moment function $M(\ell)$, and the loading function $w(\ell)$. Figure 9-3(*d*) shows a free-body diagram of a section of our beam $\ell_1 + \Delta\ell$ units long, where $\Delta\ell$ is some relatively small length. The corresponding internal reactions on this section are labeled $V_1 + \Delta V$ and $M_1 + \Delta M$, the deltas indicating some relatively small changes in the values of the shear force and bending moment. Then, Fig. 9-3(*e*) is a free-body diagram of the $\Delta\ell$ portion of the beam. Note the application of Newton's third law for equal and opposite reactions between Fig. 9-3(*c*) and (*e*).

We now apply the requirements of equilibrium to this small portion of the beam. Summing vertical forces yields

$$V_1 - w(\ell')\,\Delta\ell - (V_1 + \Delta V) = 0$$

where $w(\ell')$ is the average value of the loading on this small portion of the beam. Rearranging and dividing by $\Delta\ell$ gives

$$\frac{\Delta V}{\Delta\ell} = -w(\ell')$$

If we let $\Delta\ell$ become arbitrarily small, this expression takes the form

$$\lim_{\Delta\ell \to 0} \frac{\Delta V}{\Delta\ell} = \frac{dV}{d\ell} = -w(\ell) \tag{9-4}$$

provided this limit exists, that is, provided the shear function $V(\ell)$ as given by Eq. (9-2) is differentiable.

Similarly, summing moments with respect to point r in Fig. 9-3(*e*) yields

$$M_1 + V_1\,\Delta\ell - w(\ell')\,\Delta\ell\,(\alpha\,\Delta\ell) - (M_1 + \Delta M) = 0$$

where α is some number between 0 and 1. Rearranging, dividing by $\Delta\ell$, omitting the subscripts, and letting $\Delta\ell$ become arbitrarily small, we have

$$\lim_{\Delta\ell \to 0} \frac{\Delta M}{\Delta\ell} = \frac{dM}{d\ell} = V(\ell) \tag{9-5}$$

provided this limit exists, that is, provided the moment function $M(\ell)$ as given by Eq. (9-3) is differentiable. Combining Eqs. (9-4) and (9-5), we see that the loading function $w(\ell)$, shear-force function $V(\ell)$, and bending-moment function $M(\ell)$ are related by

$$\frac{d^2M}{d\ell^2} = \frac{dV}{d\ell} = -w(\ell) \tag{9-6}$$

Finally, writing Eq. (9-5) in the form

$$dM(\ell) = V(\ell)\, d\ell$$

and integrating between two locations 1 and 2 in the beam specified by ℓ_1 and ℓ_2, we have

$$\int_1^2 dM(\ell) = \int_{\ell_1}^{\ell_2} V(\ell)\, d\ell$$

$$M(\ell_2) - M(\ell_1) = \int_{\ell_1}^{\ell_2} V(\ell)\, d\ell \tag{9-7}$$

so that the integral on the right gives the *difference* in the bending moment between two locations 1 and 2 in the beam.

The following examples will illustrate how Eqs. (9-4), (9-5), and (9-7) can be utilized in drawing the shear and moment diagrams for a beam.

EXAMPLE 9-1

Draw the shear and moment diagrams for the uniformly loaded cantilever beam shown in Fig. 9-4(*a*).

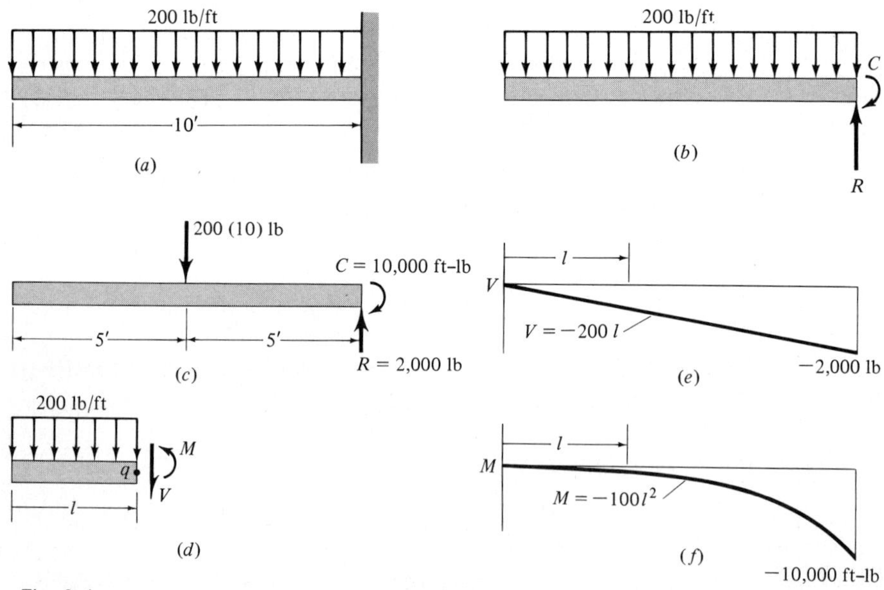

Fig. 9-4

Solution: The first step is to determine all the external forces and couples acting on the beam – in this case the reactions at the wall as shown in the free-body diagrams of Fig. 9-4(*b*). To find these *external* reactions, it is convenient to replace the uniformly distributed load of 200 lb/ft over the 10-ft length by a single statically equivalent concentrated load of (200)(10) = 2,000 lb acting at the centroid of the distribution – in this case at the middle of the beam as shown in Fig. 9-4(*c*). Then, applying the requirements of statical equilibrium to this free-body diagram yields

$$R = 2{,}000 \text{ lb} \qquad \text{and} \qquad C = (2{,}000)(5) = 10{,}000 \text{ ft-lb}$$

both as shown in Fig. 9-4(*b*) and (*c*).

Now that we have determined all the external reactions on the beam, we turn to the problem of finding the *internal* shear force and bending moment throughout the beam. While the 2,000-lb load in Fig. 9-4(*c*) is statically equivalent to the original distributed load, it does *not* produce the same *internal* effects. Therefore, we must draw a free-body diagram of a *typical* section of the original uniformly loaded beam, shown in Fig. 9-4(*d*), in which we have assumed both the shear force V and the bending moment M to be positive in accordance with the common engineering convention discussed earlier in the text. Summing forces yields

$$(200 \text{ lb/ft})(\ell \text{ ft}) + V = 0$$

Therefore, the shear function is

$$V(\ell) = -200\ell \text{ lb} \qquad (a)$$

Thus, the internal shear force varies *linearly* with the length ℓ measured from the left end of this cantilever beam, and its values are algebraically negative, indicating that its direction is actually opposite to that shown in Fig. 9-4(*d*). A graphical plot of equation (*a*) – the *shear diagram* – is shown in Fig. 9-4(*e*). Note that

$$V(0) = 0$$
$$V(10) = -200(10) = -2{,}000 \text{ lb}$$

which corresponds to the value of the vertical reaction at the wall, as expected.

Summing moments with respect to point q in Fig. 9-4(*d*) yields

$$(200 \text{ lb/ft})(\ell \text{ ft})(\ell/2 \text{ ft}) + M = 0$$

Therefore, the bending-moment function is

$$M(\ell) = -200\,\frac{\ell^2}{2} = -100\ell^2 \text{ ft-lb} \qquad (b)$$

Thus, the internal bending moment varies with the square of the length ℓ, and its values will be algebraically negative, indicating that its direction is opposite to that assumed in Fig. 9-4(d). A graphical plot of Eq. (b)—*the moment diagram*—is shown in Fig. 9-4(f). We note that

$$M(0) = 0$$

while

$$M(10) = -100(10)^2 = -10{,}000 \text{ ft-lb}$$

which corresponds to the external couple exerted by the wall. The shape of the moment curve is concave downward. To see why this is true, recall Eq. (9-5),

$$\frac{dM}{d\ell} = V(\ell)$$

which says that the *slope* of the moment curve is equal to the value of the shear force. Thus, since the shear force as given by Eq. (a) is negative and gets more negative as ℓ increases, the slope of the moment diagram should be negative and becomes more negative as ℓ increases. Thus the moment curve is concave downward rather than concave upward.

Finally, satisfy yourself that the shear-force function given by Eq. (a), the bending-moment function given by Eq. (b), and the uniform loading function of 200 lb/ft satisfy the relation given by Eq. (9-6).

EXAMPLE 9-2

A simply supported floor beam carries a distributed load which can reasonably be described by the loading shown in Fig. 9-5(a). Draw the shear and moment diagrams for this beam and determine the locations and values of the maximum shear force and bending moment.

Solution: A free-body diagram of the beam is shown in Fig. 9-5(b). In Fig. 9-5(c) the distributed load has been replaced by a statically equivalent concentrated load of (250 lb/ft)(12 ft) = 3,000 lb 6 ft from the left end, and the external reactions are then found to be

$$R_a = 1{,}875 \text{ lb} \qquad \text{and} \qquad R_b = 1{,}125 \text{ lb}$$

A free-body diagram of a typical section of the beam for $0 < \ell < 12$ ft is shown in Fig. 9-5(d), while a free-body diagram of a typical section of the beam for $12 < \ell < 16$ ft is shown in Fig. 9-5(e). It is important to note that neither of these diagrams is typical for the *entire* beam because of the discontinuity in loading which occurs at $\ell = 12$ ft.

Applying the requirements of statical equilibrium to the free-body diagram in Fig. 9-5(d) yields

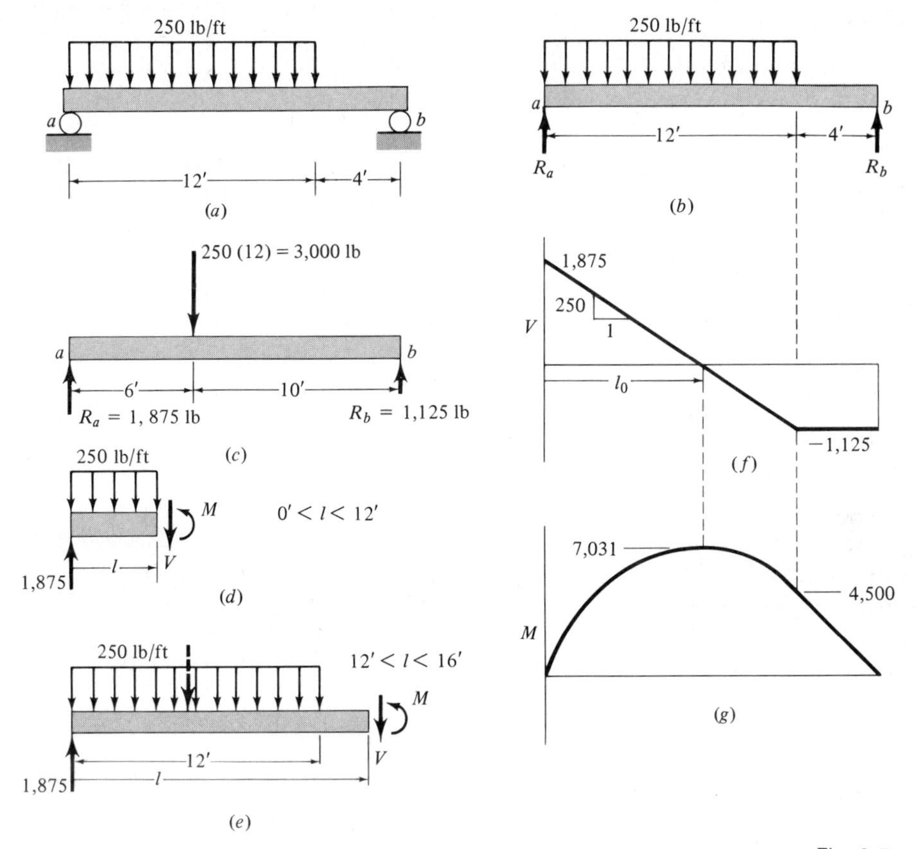

Fig. 9-5

$$\left.\begin{aligned} V &= 1{,}875 - 250\ell \text{ lb} \\ M &= 1{,}875\ell - 125\ell^2 \text{ ft-lb} \end{aligned}\right\} \quad \text{for } 0 < \ell < 12 \quad (a)$$

Similarly for the free-body diagram of Fig. 9-5(e), and recognizing that the distributed load in this case is statically equivalent to a single 3,000-lb load applied 6 ft from the left end of the beam, we obtain

$$\left.\begin{aligned} V &= 1{,}875 - 3{,}000 = -1{,}125 \text{ lb} \\ M &= 1{,}875\ell - 3{,}000(\ell - 6) \\ &= -1{,}125\ell + 18{,}000 \text{ ft-lb} \end{aligned}\right\} \quad \text{for } 12 < \ell < 16 \quad (b)$$

Hence, by Eqs. (a), the shear force has a value of 1,875 for $\ell = 0$ and *decreases* linearly with a slope of -250 lb/ft to a value of $-1{,}125$ at $\ell = 12$, as indicated in Fig. 9-5(f). The moment has a value of zero at $\ell = 0$ and varies parabolically with ℓ to a value of

$$M(12) = (1{,}875)(12) - (125)(12)^2 = 4{,}500 \text{ ft-lb}$$

at $\ell = 12$. Satisfy yourself as to the shape of this portion of the moment diagram in Fig. 9-5(g).

Similarly, by Eqs. (b), the shear force has a constant value of $-1{,}125$ lb for all ℓ between 12 and 16, while the moment decreases linearly from a value of 4,500 ft-lb at $\ell = 12$ to zero at $\ell = 16$. The complete shear and moment diagrams are shown in Fig. 9-5(f) and (g), respectively.

Recalling Eq. (9-5), $dM/d\ell = V$, the moment curve reaches its greatest value when its slope is horizontal; i.e., *the maximum moment occurs at the location in the beam where the shear is zero.* Accordingly, by ratio and proportion on the shear diagram of Fig. 9-5(f),

$$\frac{1{,}875}{\ell_o} = \frac{250}{1}$$

$$\therefore \ell_o = 7.5 \text{ ft}$$

Remark: This value could also have been obtained by setting the first of Eqs. (a), the shear equation, equal to zero.

Therefore, the maximum value of the bending moment occurs at $\ell = 7.5$ and is given by the second of Eqs. (a) as

$$M(7.5) = 1{,}875(7.5) - 125(7.5)^2$$

$$M_{\text{max}} = 7{,}031 \text{ ft-lb}$$

As a final observation we note that this value of the maximum moment is simply the increase in moment between $\ell = 0$ and $\ell = 7.5$. Accordingly, by Eq. (9-7), this *change* in moment is given by the triangular *area* of the shear diagram between $\ell = 0$ and $\ell_o = 7.5$. Thus

$$M(7.5) = \Delta M = \tfrac{1}{2}(1{,}875)(7.5) = 7{,}031 \text{ ft-lb}$$

which agrees with our previously obtained value.

EXERCISE PROBLEMS

9-1 to 9-10 For each of the beams shown, obtain the shear and moment equations and sketch the corresponding shear and moment diagrams, indicating the maximum values.

PROB. 9-1

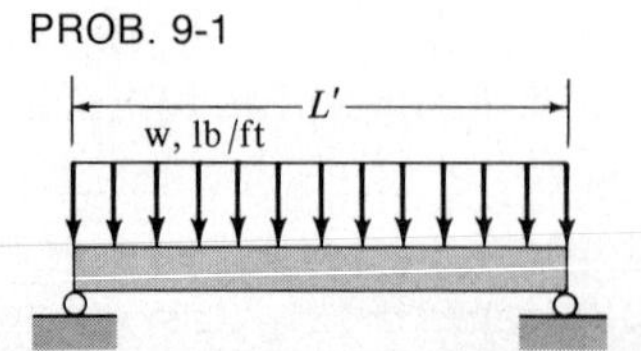

PROB. 9-2

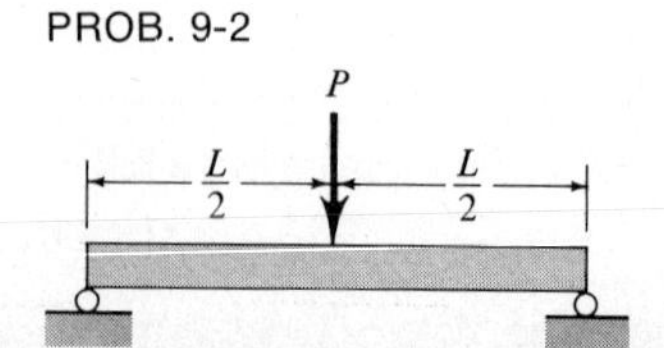

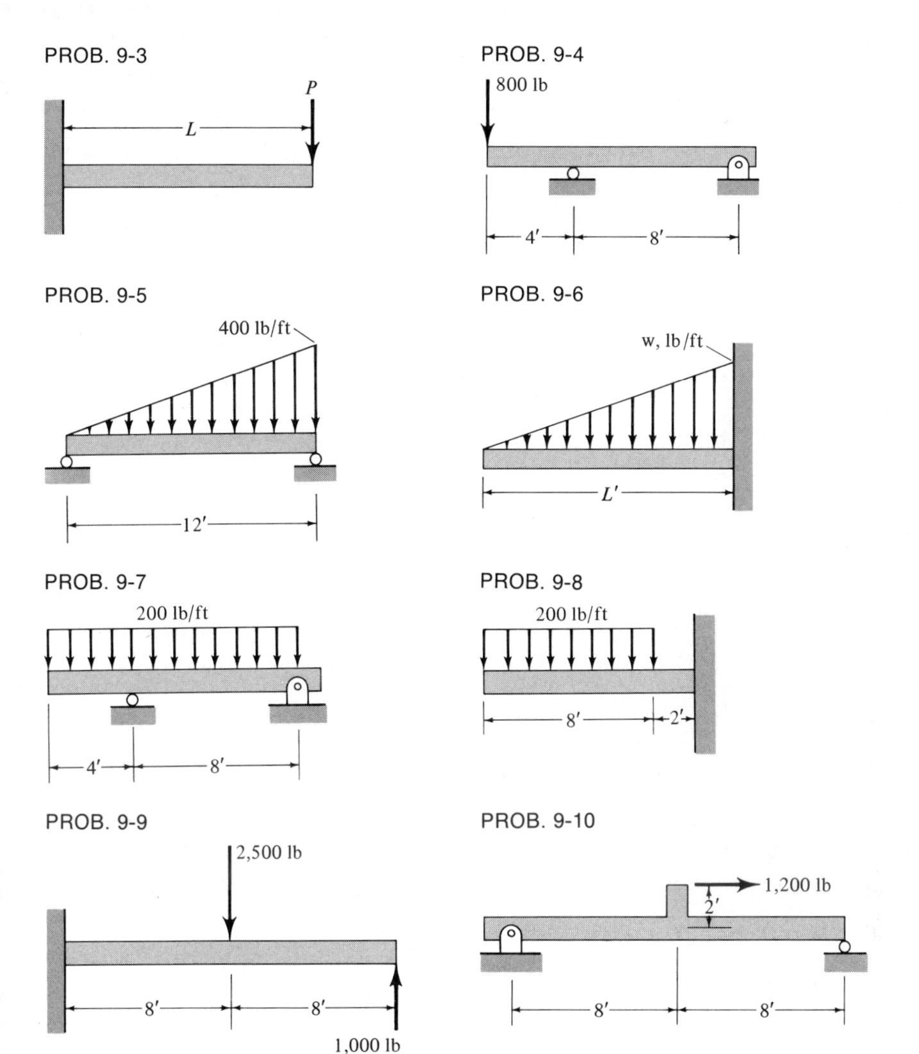

9-11 to 9-20 For the beams shown, sketch the shear and moment diagrams and determine the maximum values of the shear force and bending moment.

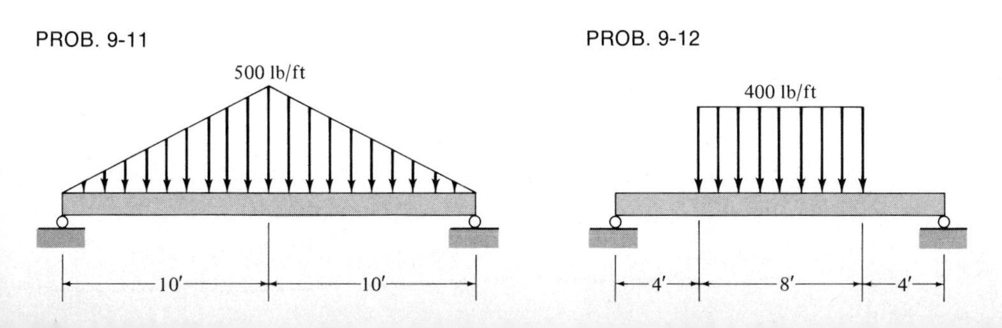

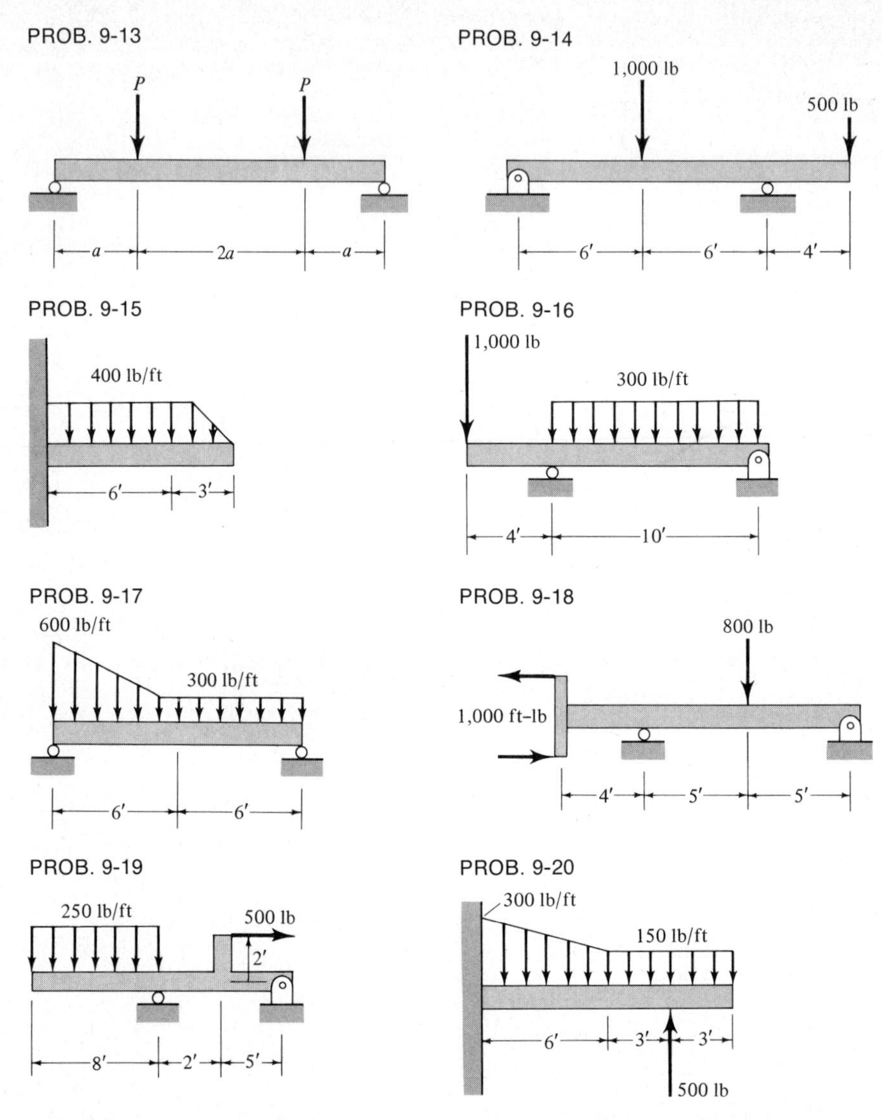

9-21 A simply supported beam has the shear diagram shown. Draw the moment diagram and indicate its maximum value. Sketch the beam and the manner in which it is loaded.

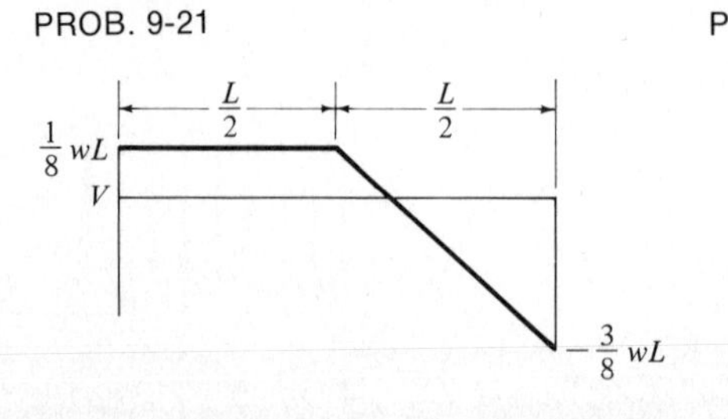

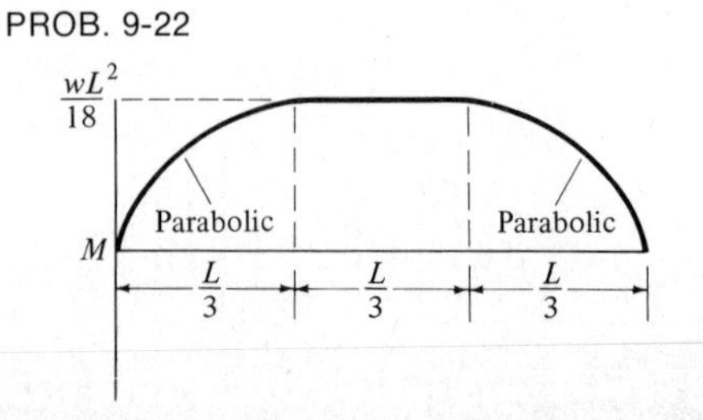

9-22 A simply supported beam has the moment diagram shown. Draw the shear diagram. Sketch the beam and the manner in which it is loaded.

9-23 A cantilevered beam has the shear diagram shown. Draw the moment diagram. Sketch the beam and the manner in which it is loaded.

PROB. 9-23

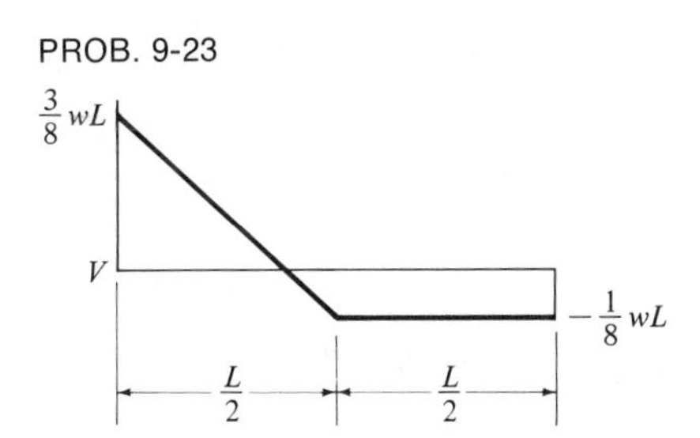

PROB. 9-24

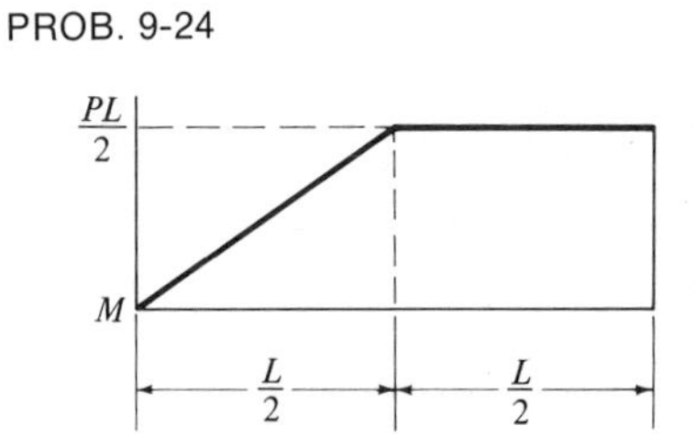

9-24 A cantilevered beam has the moment diagram shown. Draw the shear diagram. Sketch the beam and the manner in which it is loaded.

9-3 FLEXURE STRESSES

Now that we know how to determine the shear force and bending moment throughout a typical beam, we now turn to the question: How are the shear force and bending moment related to the stresses and deformation of the beam? Intuitively, we expect the bending moment to produce tensile and compressive stresses as it tends to bend the beam, whereas we expect the shear force to produce some type of shear stress. Thus we can anticipate that the state of stress at an arbitrary point in a beam would consist of some combination of normal and shear stresses. This is indeed the case, and, to simplify matters, we will first consider the bending (flexure) stresses due to the bending moment M and then consider the shear stresses due to the shear force V. Later on in the chapter we will consider the deformation (deflection) of the beam.

We begin our discussion of flexure stresses by making a few assumptions regarding the geometry of the beam and the manner of loading.

1 We assume the beam is straight and has a uniform cross section. Furthermore, we assume that the cross section of the beam has a longitudinal plane of symmetry, such as those indicated in Fig. 9-6.

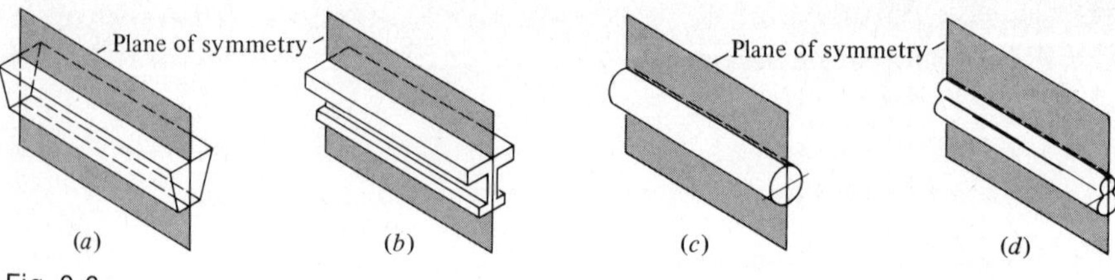

Fig. 9-6

2 We assume that the effective applied loads lie in the above-mentioned plane of symmetry; i.e., the beam is loaded symmetrically relative to its cross section.
3 We assume the material of the beam is homogeneous and isotropic but not necessarily linear or elastic.

Essentially these assumptions are made to minimize any tendency of the beam to *twist*. Figure 9-7 illustrates some situations which violate one or more of the above assumptions, resulting in possibly significant twisting effects.

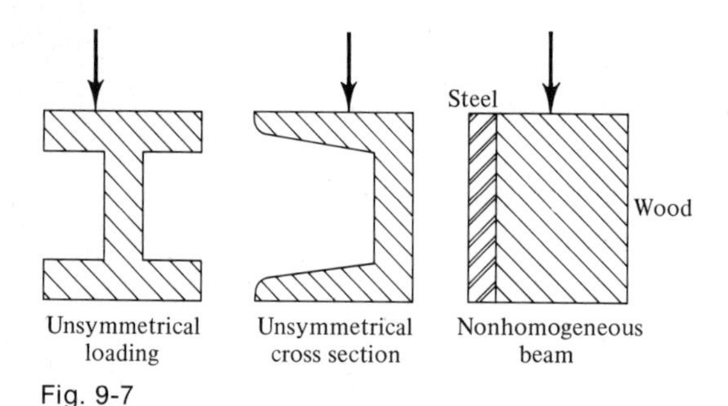

Fig. 9-7

Finally, since we are presently considering only the flexure (bending) stresses and not the shear stresses, we assume the beam is in *pure bending*, which is symbolized by the pure-couple loading depicted in Fig. 9-8(a). This idealized loading condition is merely for convenience in our analysis, although it can actually be achieved by the loading scheme in Fig. 9-8(b), for which the middle portion of the beam will have no internal shear force but will have a

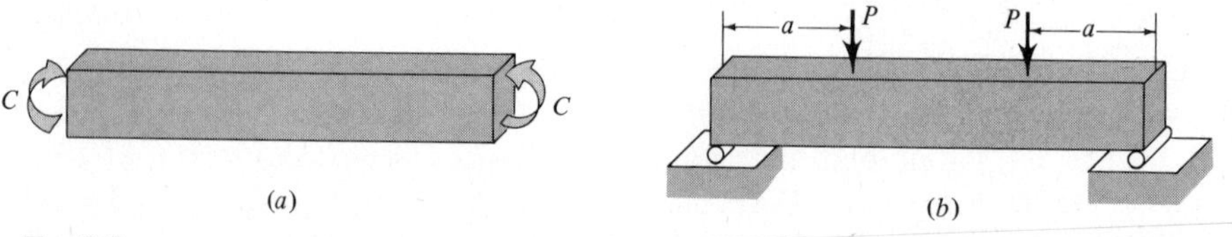

Fig. 9-8

constant internal bending moment of $P \times a$. (Satisfy yourself on this by sketching the shear and moment diagrams for this loading configuration.)

Kinematics of Pure Bending

To analyze the kinematics of pure bending, let us begin with an unloaded straight beam on whose surface we have scribed a pattern of longitudinal and vertical lines as shown in Fig. 9-9(*a*). Assume now that this beam is loaded in pure bending, so that a free-body diagram of a portion of the beam such as *a-b-d-c* would appear as in Fig. 9-9(*b*). In this figure we have "guessed" that the original straight line *a-c* deforms into the shape *a′-c′*. Consequently, since side *b-d* is loaded in the same manner as side *a-c*, the line *b-d* would deform into the similar shape *b′-d′*. But the middle line *e-f* cannot deform both ways at the same time, so that material compatibility would require that it remain straight as *e′-f′*. However, since the section *a-b-d-c* was quite arbitrary, line *e-f* can, in fact, be any one of the vertical lines on the beam, and thus we conclude that *all* the vertical lines *a-c*, *e-f*, *b-d*, etc., remain straight as indicated in Fig. 9-9(*c*). Also, while the vertical lines remain straight, the longitudinal lines *a-b*, *c-d*, etc., deform into circular arcs *a′-b′*, *c′-d′*, etc., with a center of curvature of, say, *o*.

Initially, in Fig. 9-9(*a*) the lines *a-b* and *c-d* were of the same length $\Delta \ell$, but we anticipate that the bending of the beam caused the line *a-b* to shorten, so that the arc length *a′-b′* is less than $\Delta \ell$, while the line *c-d* probably stretched, so that *c′-d′* is likely to be

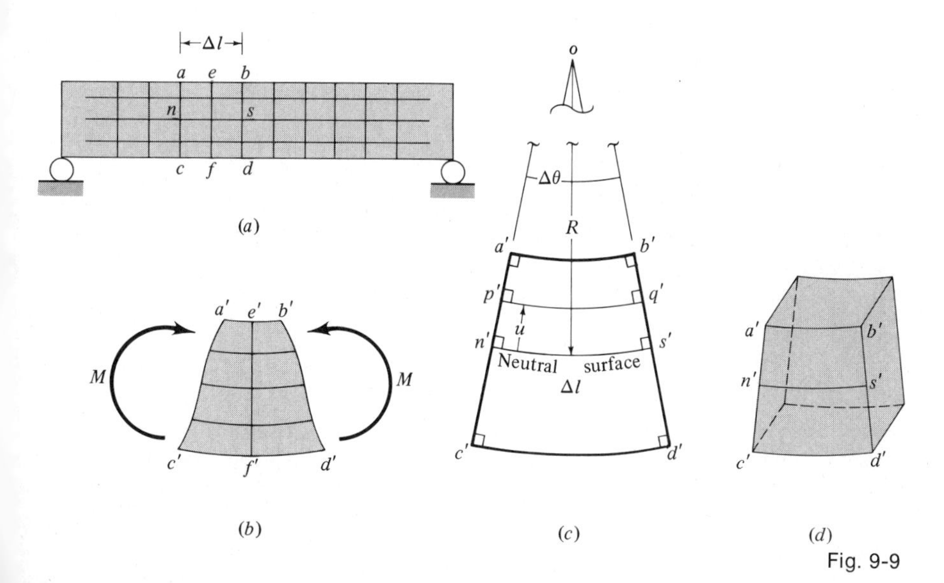

Fig. 9-9

longer than $\Delta\ell$. Consequently, there must be some horizontal line n-s between a-b and c-d that has deformed into an arc n'-s' which is still $\Delta\ell$ units long; that is, there is some line n-s which has not elongated or contracted. This line is called the *neutral line* or *neutral axis* of the beam, and its radius of curvature R is called the radius of curvature of the deformed beam.

> ***Remark:*** What we have called the neutral line is actually a *neutral surface* running through the thickness of the beam. The two-dimensional diagrams of Fig. 9-9(a) to (c) are representative of the deformation occurring throughout the thickness of the beam and should be interpreted as such. Accordingly, the above observation that vertical lines remain straight should be interpreted as meaning that *plane cross sections remain plane*, as illustrated in Fig. 9-9(d).

Returning to Fig. 9-9(c), let u be measured from the neutral surface. Then we see that

$$\Delta\ell = n's' = R\,\Delta\theta$$

$$p'q' = (R - u)\,\Delta\theta$$

Hence, the longitudinal axial strain $\epsilon(u)$ occurring at the distance u above the neutral surface is given by

$$\epsilon(u) = \frac{\text{final length} - \text{initial length}}{\text{initial length}}$$

$$= \frac{p'q' - \Delta\ell}{\Delta\ell} = \frac{(R-u)\,\Delta\theta - R\,\Delta\theta}{R\,\Delta\theta}$$

$$\epsilon(u) = -\frac{u}{R} \tag{9-8}$$

That is, the longitudinal axial strain varies *linearly* from the neutral surface of the beam. The minus sign in Eq. (9-8) indicates that a positive bending moment causes a compressive (negative) axial strain above (u positive) the neutral surface and a tensile (positive) axial strain below (u negative) the neutral surface.

We remark that up to this point we have argued the *existence* of a neutral surface and have not as yet determined its location. This will be done shortly. Also, the preceding kinematical arguments and the very important result of Eq. (9-8) all follow from the basic assumptions of geometry and loading made at the beginning of this section without regard to whether the beam material is linear, elastic, plastic, etc., so long as the deformation is not too severe.

Basic Moment-Stress Relationship

Consider a section of a beam with symmetrical cross section loaded in pure bending as in Fig. 9-10(*a*). Recalling the discussion of stresses in Sec. 5-4, we envision the net internal reaction M to be the accumulative effect of the stresses acting over the entire transverse cut cross section of the beam. In general, these stresses might be normal stresses (tensile or compressive) and/or shear stresses. However, the net shear force on the cross section is zero for pure bending. Also, from the kinematics of Fig. 9-9(*a*) and (*c*), we see that the right angles at a', b', c', d', etc., have all been preserved during the bending process, so that no distortion (shear) has occurred. Consequently, we are led to conjecture that the bending stresses on the transverse cut cross-sectional area are *normal* stresses rather than shear stresses, as indicated in Fig. 9-10(*b*) by the normal stress σ acting on the area dA.

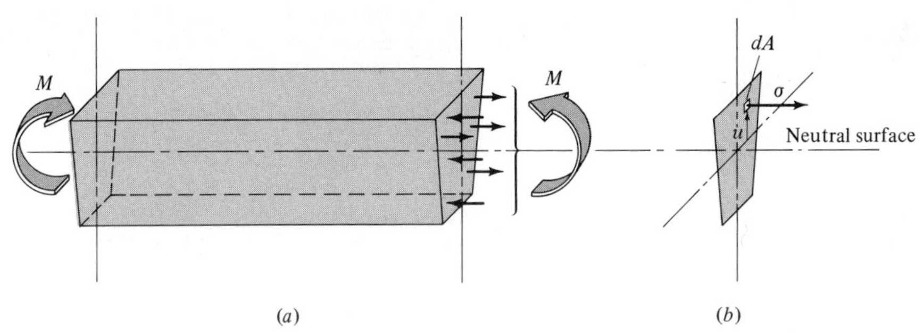

Fig. 9-10

In this figure, u is the vertical distance from the neutral surface of the beam. In general, the stress σ will vary over the cross-sectional area, but, because of the symmetry of the cross section, this stress variation can be expressed as a function of u, and we write

$$\sigma = \sigma(u) \tag{9-9}$$

Recall that the resultant effect of all the normal stresses over the cut cross section must be a pure couple equal to the bending moment M. Hence, the stress distribution $\sigma(u)$ must have a sum of zero

$$\iint_{\text{area}} \sigma(u)\, dA = 0 \tag{9-10}$$

while its moment effect must be

$$\iint_{\text{area}} \sigma(u) u\, dA = -M \tag{9-11}$$

where the minus sign arises because a positive (tensile) stress σ acting at a positive distance u above the neutral surface would produce a negative bending moment $-M$. Thus, whatever normal stresses are produced in the beam by the bending moment M, these stresses must satisfy both Eq. (9-10) and Eq. (9-11). Equation (9-11) is the *basic moment-stress relation* for pure bending of beams with symmetrical cross sections. Our problem now is to determine how the stress distribution $\sigma(u)$ varies with u so that Eq. (9-11) can be integrated to obtain an explicit relation between the normal stress σ and the bending moment M. Although many such relations can be found, depending upon the behavior of the material of the beam, we will examine two very important cases, namely, linear elastic behavior and ideally plastic behavior.

Linear (Elastic) Behavior

Let us briefly review what we know thus far. First, the axial *strain* due to bending varies linearly with the distance u from the neutral surface, as given by Eq. (9-8)

$$\epsilon(u) = -\frac{u}{R} \tag{9-8}$$

and indicated in Fig. 9-11(a). Second, from the preceding discussion, it appears that, for pure bending, the state of stress in the beam will be uniaxial, as indicated in Fig. 9-11(b). Now, if the material response is *linear*, the axial stress at any distance u from the neutral axis will be related to the axial strain at the same distance u according to Hooke's law for uniaxial stress, namely,

$$\sigma(u) = E\epsilon(u) \tag{9-12}$$

But, by Eq. (9-8), this takes the form

$$\sigma(u) = -E\,\frac{u}{R} \tag{9-13}$$

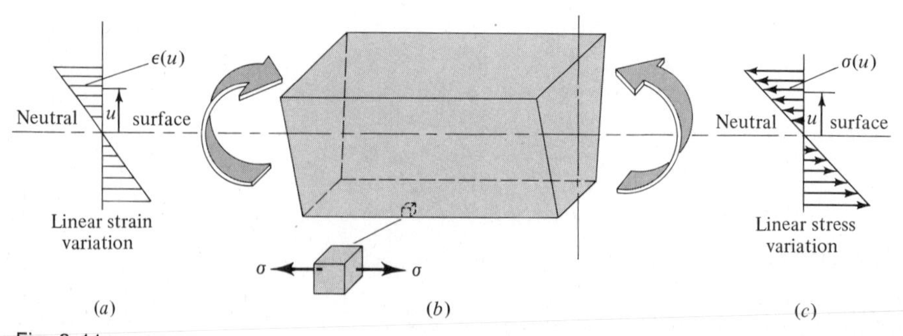

Fig. 9-11

so that for linear response the *stress varies linearly* from the neutral axis, as indicated in Fig. 9-11(*c*). Hence, with the linear stress variation given by Eq. (9-13), the basic moment-stress relation of Eq. (9-11) can now be integrated. Thus

$$-M = \iint_{\text{area}} \sigma(u)u \, dA$$

$$= \iint_{\text{area}} -E \frac{u}{R} u \, dA = -\frac{E}{R} \iint_{\text{area}} u^2 \, dA$$

or

$$M = \frac{E}{R} I \tag{9-14}$$

where

$$I = \iint_{\text{area}} u^2 \, dA$$

is the second moment of the cross-sectional area with respect to the neutral surface of the beam. (Refer to Appendix A for discussion of second moments of area.) Then, solving for E/R from Eq. (9-13) and substituting into Eq. (9-14) yields

$$M = -\frac{\sigma(u)}{u} I$$

or

$$\sigma(u) = -\frac{Mu}{I} \tag{9-15}$$

which is the *basic moment-stress relation for elastic (linear) behavior*. The minus sign indicates that a positive bending moment M produces negative (compressive) stresses in the upper fibers of the beam, whereas it produces positive (tensile) stresses in the lower fibers of the beam, which seems physically reasonable. Also, we see that the largest values of the stress occur at the outermost fibers of the beam; i.e., the maximum bending stresses occur at the top and bottom of the beam.

We return now to the question of the location of the neutral surface. To this end, recall that the stress variation $\sigma(u)$ must satisfy Eq. (9-10). Thus, substituting either Eq. (9-13) or (9-15) into Eq. (9-10) yields

$$\iint_{\text{area}} \sigma(u) \, dA = 0 \tag{9-10}$$

$$\iint_{\text{area}} -E \frac{u}{R} \, dA = -\frac{E}{R} \iint_{\text{area}} u \, dA = 0$$

which implies that

$$\iint_{\text{area}} u \, dA = 0 \tag{9-16}$$

But this integral is the *first moment* of the cross-sectional area with respect to the neutral surface of the beam. Its vanishing implies that the neutral surface of the beam must coincide with the *centroid* of the cross-sectional area.

Putting all these results together we have the following:

Theorem 9-1 For linear (elastic) behavior of straight symmetrical beams, the bending stress across the cross section varies linearly from the centroid of the cross section and is given by

$$\sigma(u) = -\frac{Mu}{I} \tag{9-15}$$

with the maximum (absolute) value on the cross section given by

$$|\sigma_{\max}| = \frac{|Mu_{\max}|}{I} \tag{9-17}$$

which occurs at the outermost fiber of the beam.

It is perhaps worthwhile to point out the similarity between the *elastic flexure formula* [Eq. (9-17)] and the *elastic torsion formula*

$$\tau_{\max} = \frac{TR}{J} \tag{7-22}$$

In using the flexure formula, the bending moment M is usually expressed in inch-pounds, the distance u in inches, and the second moment I in inch^4, so that the stress will be in pounds per square inch.

Fully Plastic Behavior

Suppose that the beam under consideration is made of a rather ductile material whose uniaxial stress-strain curve has one of the qualitative shapes depicted in Fig. 9-12. For example, the stress-strain curve for mild steel corresponds to that of Fig. 9-12(a), while a ductile aluminum would have a curve similar to that of Fig. 9-12(b). Both types of curves exhibit a pronounced flattening at

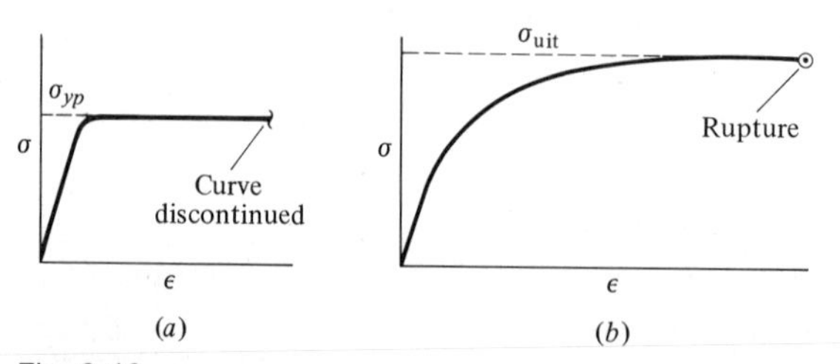

Fig. 9-12

some stress level such as the yield-point stress or ultimate stress, while the strain undergoes a comparatively large increase. Let us see what this type of behavior implies as to the moment-stress relation for a beam.

Recall that the arguments made in establishing the linear *strain* variation given by Eq. (9-8) were *not* restricted to linear elastic behavior of the material. Accordingly, we assume that the bending strain in the beam remains linear even though the material itself may begin to behave nonlinearly and undergo permanent plastic deformation. As the *strain* in the beam becomes larger, the corresponding *stress* distribution will progressively take the form indicated in Fig. 9-13. Initially, the variation is linear so long as

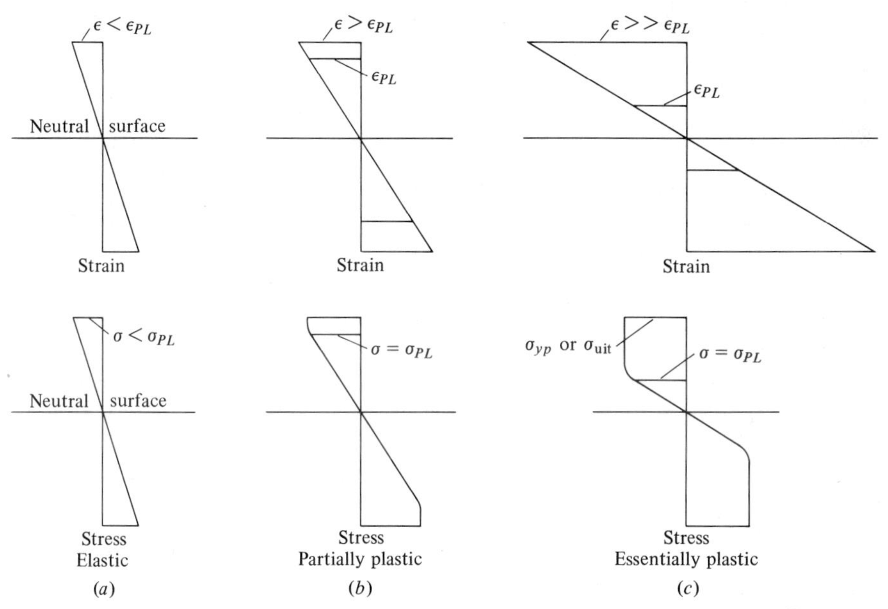

Fig. 9-13

the maximum stress and strain are below the proportional limit of the material; the stress variation then becomes partially flattened as the strain is increased, and the outer fibers of the beam begin to yield; and eventually the situation becomes essentially fully plastic as virtually all the material across the entire cross section reaches either the yield-point stress or the ultimate stress, depending upon the type of material involved. The situation is idealized by the illustration in Fig. 9-14 in which all the fibers above the neutral surface (which is not necessarily at the centroid of the cross section, as in the case for elastic behavior) are assumed to have a uniform compressive stress σ_c, while those fibers below the neutral surface have a uniform tensile stress σ_t.

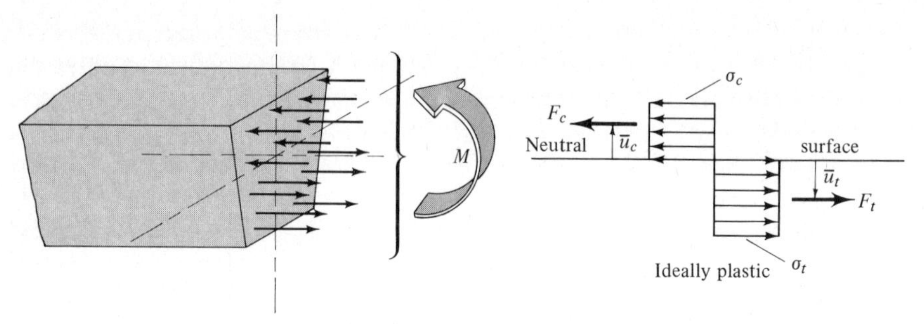

Fig. 9-14

With this idealized stress distribution, let us look at the basic bending stress relations of Eqs. (9-10) and (9-11). Essentially, these relations imply that the net effect of the stresses must be a pure couple M, so that the net compressive force F_c and the net tensile force F_t must satisfy

$$F_t = F_c$$
$$F_t\bar{u}_t + F_c\bar{u}_c = M \tag{9-18}$$

where

$$F_t = \sigma_t \times (\text{lower area})$$
$$F_c = \sigma_c \times (\text{upper area}) \tag{9-19}$$

and $\bar{u}_t$ and $\bar{u}_c$ are the distances from the neutral surface to the centroid of the lower and upper areas, respectively. Equations (9-18) and (9-19) are the basic requirements for the fully plastic condition in a beam. This condition is sometimes referred to as the formation of a *plastic hinge.*

The requirements of Eqs. (9-18) and (9-19) become somewhat more transparent for the case in which the tensile and compressive plastic stresses have the same value σ_p. Then, Eqs. (9-18) and (9-19) give

$$\sigma_p \times (\text{lower area}) = \sigma_p \times (\text{upper area})$$
$$\therefore \text{upper area} = \text{lower area} \tag{9-20}$$

and

$$\sigma_p \times (\text{lower area})\bar{u}_t + \sigma_p \times (\text{upper area})\bar{u}_c = M$$
$$\therefore M = \sigma_p[(\text{lower area}) \times \bar{u}_t + (\text{upper area}) \times \bar{u}_c] \tag{9-21}$$

That is, for this case, Eq. (9-20) implies that the neutral surface for the fully plastic condition divides the cross-sectional area in half, one half being in compression and the other half in tension. The relation between the bending moment M and the fully plastic stress σ_p is then given by Eq. (9-21).

Before giving some illustrative examples of elastic and plastic bending stresses in beams, some comments are appropriate at this time. First, in all of our discussions thus far, we have assumed the beam material to be homogeneous and isotropic. Thus, strictly speaking, our results would not be valid for, say, timber beams and concrete beams. However, in actual engineering practice, the elastic-flexure formula, Eq. (9-17), is widely used to estimate the bending stresses in most beams regardless of the homogeneity and isotropy of the material. Also, our discussion has assumed the loading on the beam to be pure bending, a situation which seldom occurs in real structures and machines. Again, engineers usually take the liberty of applying the pure-bending theory of this section to most loading situations. In the final analysis, the results obtained in this article can be utilized in a variety of situations so long as the engineer exercises a reasonable degree of good judgment and rational analysis.

EXAMPLE 9-3

A simply supported aluminum beam carries a load at its midpoint as shown in Fig. 9-15(a). Two electrical axial strain gages are attached 2 in from the top and bottom, respectively, directly under the loading point and indicate strains of $\epsilon_a = -0.0010$ in/in, $\epsilon_b = +0.0014$ in/in, respectively. The tensile and compression stress-strain curve for the aluminum is shown in Fig. 9-15(b). Determine the maximum tensile and compressive bending stresses in the beam.

Solution: From the discussions above we know that the strain always varies linearly from the neutral surface. Accordingly, we

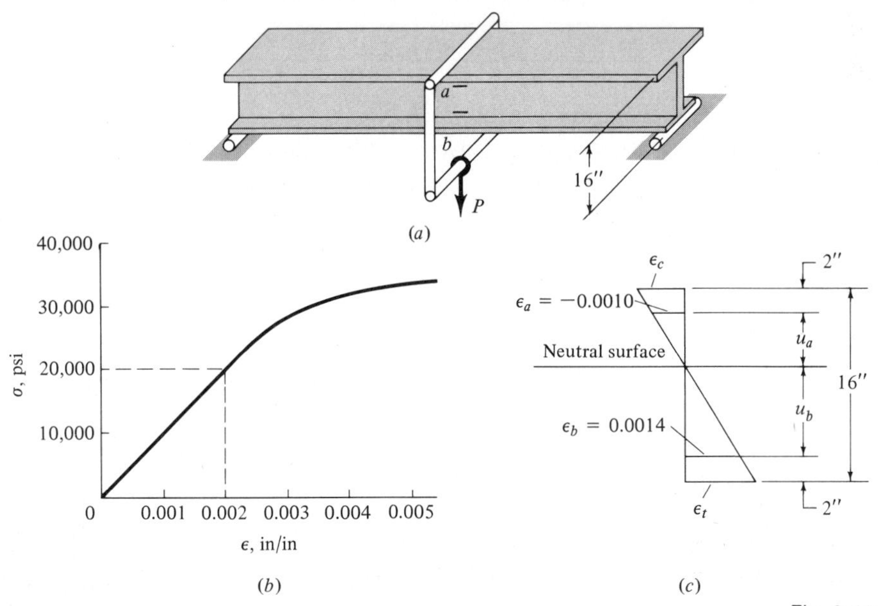

Fig. 9-15

can draw the *strain* distribution shown in Fig. 9-15(*c*). From this figure we see that

$$u_a + u_b = 12 \text{ in}$$

$$\frac{u_a}{0.0010} = \frac{u_b}{0.0014}$$

so that $$u_a = 5 \text{ in} \qquad \text{and} \qquad u_b = 7 \text{ in}$$

Consequently, the neutral surface is 7 in from the top edge of the beam or 9 in from the bottom. Also, by ratio and proportion the largest compressive strain at the top is

$$\frac{\epsilon_c}{7} = \frac{0.0010}{5}$$

$$\epsilon_c = 0.0014 \text{ in/in}$$

while the largest tensile strain at the bottom is

$$\frac{\epsilon_t}{9} = \frac{0.0014}{7}$$

$$\epsilon_t = 0.0018 \text{ in/in}$$

Both of these strains ϵ_c and ϵ_t are within the linear portion of the stress-strain curve. Hence, the corresponding compressive and tensile stresses are

$$\begin{aligned} \sigma_c &= E\epsilon_c \\ &= (10 \times 10^6)(0.0014) = 14{,}000 \text{ psi} \\ \sigma_t &= E\epsilon_t \\ &= (10 \times 10^6)(0.0018) = 18{,}000 \text{ psi} \end{aligned}$$

which represent the maximum compressive and tensile stresses at the midpoint of the beam. However, by sketching the moment diagram for a simply supported beam with a load at the midpoint, we know that the maximum bending moment also occurs at the midpoint. Hence, these stresses represent the maximum bending stresses throughout the entire beam.

EXAMPLE 9-4

The T beam shown in Fig. 9-16(*a*) is made of mild steel and is supposed to carry a uniformly distributed load of 1,000 lb/ft. Determine the maximum bending stress which this load would produce in the beam.

Solution: The first step is to determine the largest bending moment throughout the beam. A free-body diagram of a statically equivalent beam is shown in Fig. 9-16(*b*), from which the external fixed-end reactions are found to be

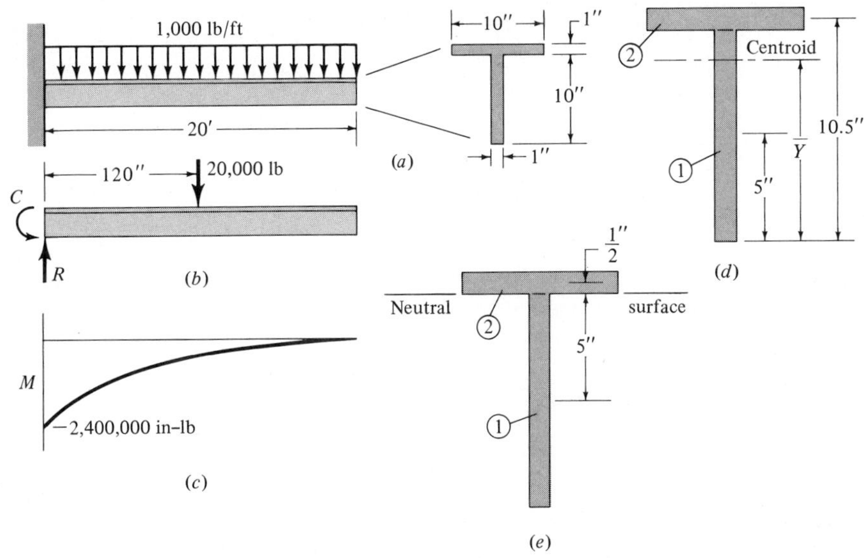

Fig. 9-16

$$R = 20{,}000 \text{ lb}$$
$$C = (20{,}000)(120) = 2{,}400{,}000 \text{ in-lb}$$

We can now sketch the moment diagram as in Fig. 9-16(*c*), from which we see that the largest bending moment occurs at the left end of the beam and has a value of

$$M = -2{,}400{,}000 \text{ in-lb}$$

At present we do not know whether the beam will behave elastically or plastically, but let us try the elastic case. Accordingly, to use the elastic-flexure formula, Eq. (9-17), we need to locate the centroid of the cross section and evaluate the second moment of the area. Referring to Fig. 9-16(*d*) and treating the cross section as a composite of two rectangles, the centroid will be located at

$$\bar{Y} = \frac{(10)(1)(5) + (10)(1)(10.5)}{(10)(1) + (10)(1)}$$
$$= 7.75 \text{ in from bottom of beam (or 3.25 in from top)}$$

Next, the second moment of the entire cross section will be the sum of the second moments of area 1 and area 2. Hence,

$$I_1 = \frac{bh^3}{12} + \text{area(distance)}^2$$
$$= \frac{(1)(10)^3}{12} + (10)(7.75 - 5)^2$$
$$= 83.3 + 75.6 = 158.9 \text{ in}^4$$

Similarly

$$I_2 = \frac{(10)(1)^3}{12} + (10)(10.5 - 7.75)^2$$
$$= 0.83 + 75.6 = 76.4 \text{ in}^4$$

Thus

$$I_{\text{total}} = 158.9 + 76.4 = 235 \text{ in}^4 \qquad \text{(rounded off)}$$

Now, the maximum tensile stress occurs in the top fibers for $u = 3.25$ in, while the largest compressive stress will occur in the bottom for $u = 7.75$ in. Thus

$$\sigma_{\text{tension}} = \frac{Mu}{I}$$
$$= \frac{(2{,}400{,}000)(3.25)}{235} = 33{,}200 \text{ psi}$$
$$\sigma_{\text{comp}} = \frac{(2{,}400{,}000)(7.75)}{235} = 79{,}200 \text{ psi}$$

both of which are above the yield point for mild steel, and our assumption of elastic behavior is incorrect. Since mild steel is quite ductile, we might anticipate that the beam will behave plastically and quite possibly may fail due to the yielding of the material across the entire cross section, i.e., formation of a plastic hinge.

To check this situation, we will determine the bending moment necessary to completely yield a cross section of the beam and compare this moment with the value of the maximum moment due to the loading which the beam is supposed to carry.

Recalling that for idealized fully plastic behavior the neutral surface will divide the cross-sectional area in half, in this case the neutral surface will be at the junction of the T section, as indicated in Fig. 9-16(*e*). Thus, using a yield-point stress of 30,000, Eq. (9-21) gives

$$M = \sigma_p[(\text{area 1})(5 \text{ in}) + (\text{area 2})(\tfrac{1}{2} \text{ in})]$$
$$= 30{,}000(50 + 5) = 1{,}650{,}000 \text{ in-lb}$$

as the bending moment which will cause the beam to yield completely across its cross section. Since this value is considerably less than the 2,400,000 in-lb which the loading will produce, we conclude that the beam cannot carry the indicated loading without failing due to yielding.

Remark: It is also possible that the large compressive stresses developed in the relatively thin web could cause this beam to buckle. Buckling-type failure will be discussed in Chap. 10.

EXERCISE PROBLEMS

9-25 Rework Example 9-3 for the case in which $\epsilon_a = -0.0018$ in/in and $\epsilon_b = 0.0030$ in/in.

9-26 to 9-29 A straight beam has the uniform cross section shown in the figure. Assuming the material behaves the same in tension and compression, determine the location of the neutral axis for (*a*) elastic behavior and (*b*) fully plastic behavior.

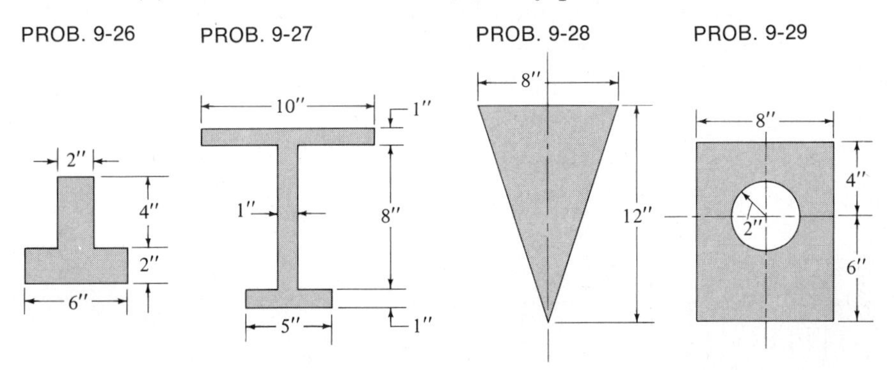

9-30 What is the longest span on which a simply supported 2 by 4 nominal-size timber can support a 200 lb load at its midspan if the allowable bending stress is 3,000 psi? Refer to the timber table in Appendix B.

9-31 Same as Prob. 9-30 except the timber is cantilevered. What is the maximum allowable distance between the load and the fixed end?

9-32 A simply supported W8 × 40 steel beam 16 ft long carries a uniformly distributed load of w lb/ft in addition to its own weight of 40 lb/ft. If the allowable bending stress is not to exceed 18,000 psi, what is the maximum permissible value of the additional loading w? Refer to Appendix B for properties of steel sections.

9-33 A cantilevered W10 × 100 steel beam 12 ft long carries a uniformly distributed load of w lb/ft in addition to its own weight of 100 lb/ft. If the allowable bending stress is not to exceed 24,000 psi, what is the maximum permissible value of the additional loading w? Refer to Appendix B for properties of steel sections.

9-34 Two American Standard steel channels are riveted together back to back to form an I-shaped beam. The beam is 16 ft long, simply supported, and carries a concentrated load of 8,000 lb at a point 10 ft from one end. Neglecting the weight of the beam itself, determine the minimum size channel required if the maximum bending stress is not to exceed 20,000 psi. Refer to the table in Appendix B and note that $S = I/(d/2)$ is the section modulus.

9-35 A standard steel S section 24 ft long is simply supported and carries a uniformly distributed load of 500 lb/ft over the left 12 ft and 1,000 lb/ft over the right 12 ft. If the maximum allowable bending stress is 20,000 psi, find the minimum size S shape required for a safety factor of 2.5. Refer to Appendix B and note that $S = I/(d/2)$ is the section modulus. Neglect the weight of the beam itself.

9-36 to 9-38 Determine the maximum tensile and compressive bending stresses in the beam shown. Assume elastic behavior.

PROB. 9-36

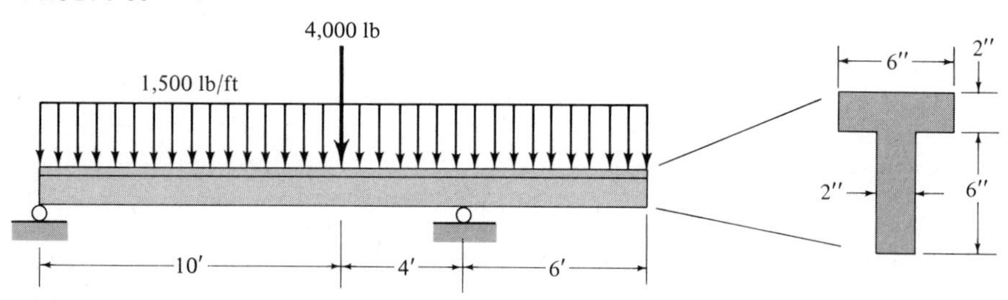

PROB. 9-37

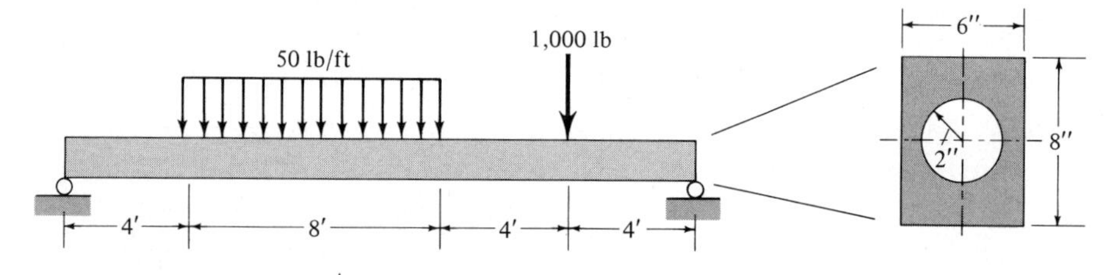

PROB. 9-38

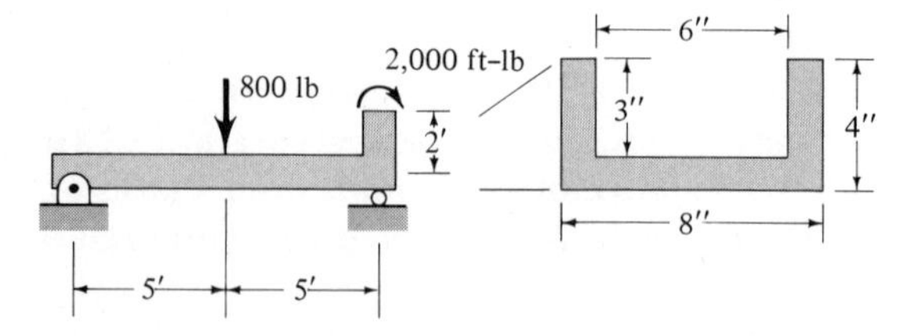

9-39 The beam-like weighing device shown is made of mild structural steel with a yield point of 30,000 psi. What would you specify as the weighing capacity of this device based on a factor of safety of 2? Use $L = 5$ ft.

PROB. 9-39

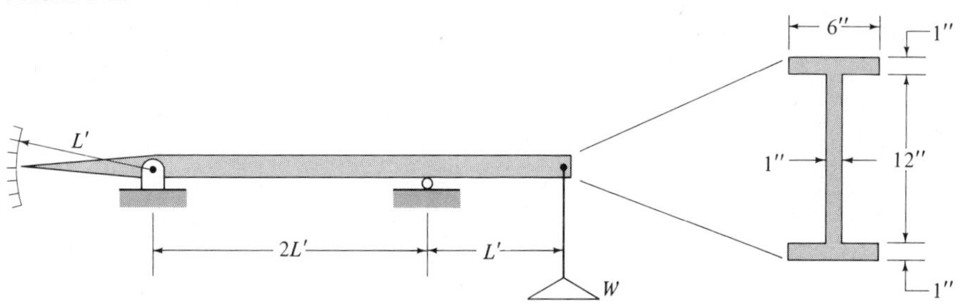

9-40 The beam shown is loaded in pure bending and is made of mild steel whose σ-ϵ curve is shown. Find the value of M if the axial strain at the bottom surface is (*a*) $\epsilon = 0.0002$ in/in, (*b*) $\epsilon = 0.002$ in/in, and (*c*) $\epsilon = 0.02$ in/in.

PROB. 9-40

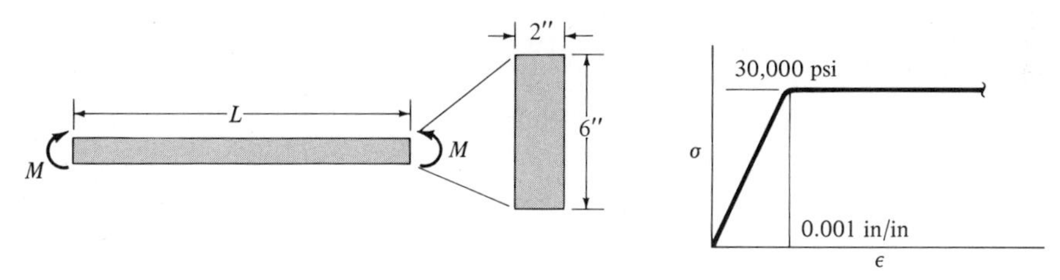

9-41 to 9-44 Consider a beam whose cross section is as shown. Find the ratio of the fully plastic bending moment to the maximum elastic bending moment if the beam is made of mild steel with a yield-point stress of 30,000 psi.

PROB. 9-41 PROB. 9-42 PROB. 9-43 PROB. 9-44

b″ h″ D″ 2″ 4″ 2″ 4″ 8″ 4″ 3″ 3″

9-45 to 9-48 Consider a beam whose cross section is as shown in Prob. 9-41 to 9-44, respectively. Find the ratio of the ultimate bending moment to the maximum elastic bending moment if the beam is made of a ductile aluminum with a proportional-limit stress of 24,000 psi and an ultimate strength of 40,000 psi.

9-49 A simply supported aluminum beam 8 ft long has the cross section shown in the figure for Prob. 9-26. The ductile aluminum has a proportional-limit stress of 18,000 psi and an ultimate stress of 34,000 psi. What is the largest concentrated load which may be applied at the midspan if (*a*) failure is due to initiation of inelastic behavior and (*b*) failure is due to total loss of structural integrity?

9-50 A cantilevered cast-iron beam 10 ft long has the cross section shown in the figure for Prob. 9-27. The cast iron is rather brittle with an ultimate tensile stress of 30,000 psi. Using a factor of safety of 3, what is the largest concentrated load which can safely be applied to the end of the beam?

9-4 TRANSVERSE SHEAR STRESSES

In the previous section we saw how the bending moment M at any location in a beam can be related to the corresponding tensile and compressive bending stresses across the cross section of the beam. We will now look at the shearing effects caused by the transverse shear force V which usually occurs in conjunction with the bending moment throughout the length of the beam. In the subsequent discussion we make the same assumption regarding the geometry and material of the beam as before; namely, the beam is straight, has a symmetrical cross section, is loaded in the plane of symmetry, and is made of a reasonably isotropic and homogeneous material. Although we no longer assume the loading to produce pure bending, we do assume the beam to be sufficiently long so that bending effects are still dominant; that is, the beam is a beam rather than a short shear block, as illustrated in Fig. 9-17. In a short shear block the shearing effects tend to dominate over the bending effects, and the applicability of the bending theory developed in the previous article is quite questionable.

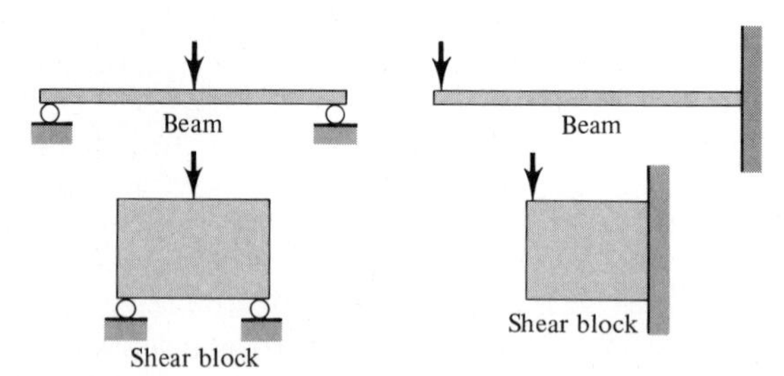

Fig. 9-17

Consider the simply supported beam in Fig. 9-18(*a*). At any location along the beam we envision the internal shear force V as being the cumulative effect of the shear stresses τ developed on the transverse cross section of the beam, as illustrated in Fig. 9-18(*b*). Our present purpose is to relate the shear force to these shear stresses.

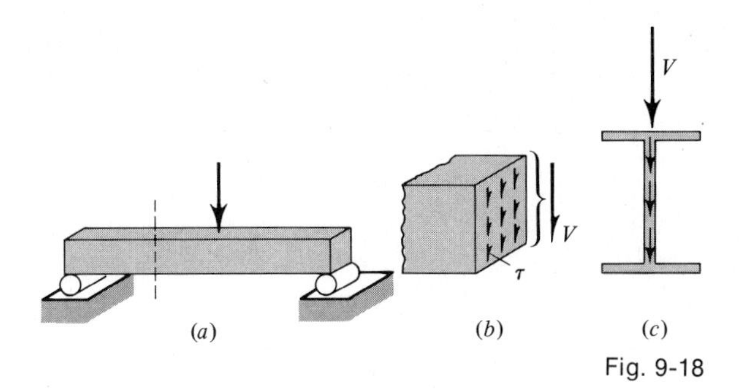

Fig. 9-18

As a crude approximation we could evaluate an average shear stress by dividing the shear force V by the cross-sectional area,

$$\tau_{\text{ave}} = \frac{V}{A} \tag{9-22}$$

and in some situations this value would give a rough indication of the relative size of the shear stresses as compared with the bending stresses produced by the bending moment. Quite often, the average shear stress will be very small (at least an order of magnitude) compared with the maximum bending stress, so that the shear effects may be considered to be relatively unimportant. However, in other situations the average value for the shear stress can be quite misleading. For example, for a thin cross section such as that depicted in Fig. 9-18(*c*), it is quite likely that most of the vertical shear force will be carried by the thin vertical web and very little carried by the thin horizontal flanges. Furthermore, the actual value of the largest shear stress is likely to be much larger than the average value given by Eq. (9-22). Consequently it is necessary that we develop a more refined relation between the shear force V and the corresponding shear stresses in a beam.

We begin this development by making an observation on a relatively simple experiment. Suppose several wooden planks are laid one on top of the other as shown in Fig. 9-19(*a*). If a bending force is applied as shown in Fig. 9-19(*b*), the stack will deform as indicated, with each plank behaving as a separate beam being compressed at the top side and stretched at the bottom side. Con-

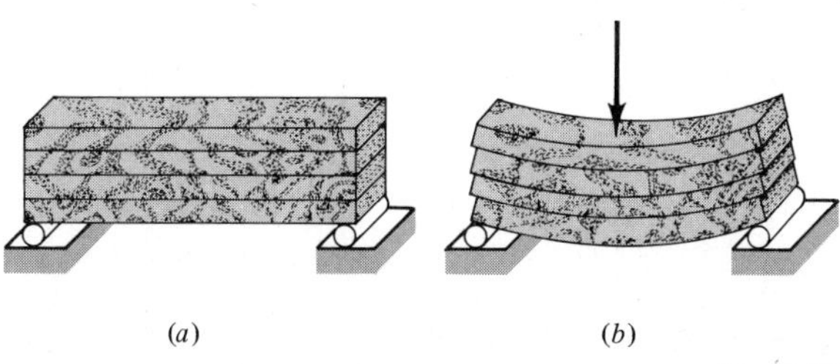

Fig. 9-19

sequently, the end of the stack becomes staggered, indicating that a relative sliding effect has occurred between the contact surfaces of the planks. This same tendency to slide or shear exists in any solid beam not loaded in pure bending and in effect causes the transverse cross section to warp rather than remain plane. This warping, exaggerated in Fig. 9-20, indicates the presence of shear strains and shear stresses on the transverse cross section, just as we have already anticipated.

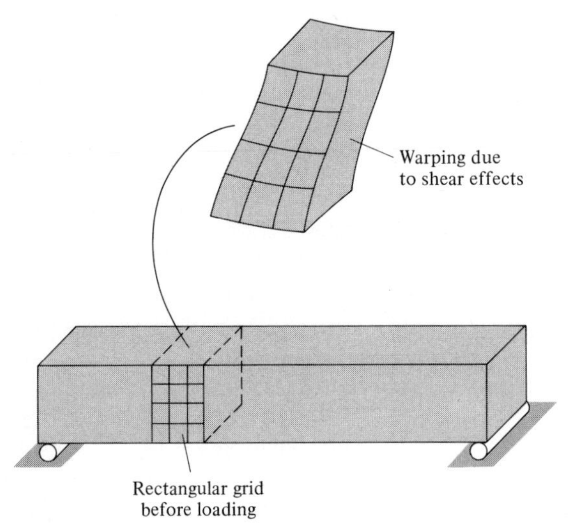

Fig. 9-20

Keeping in mind the experiment with the planks, consider a small longitudinal slice of the solid beam in Fig. 9-21. By considering the horizontal forces acting on this slice, the longitudinal force ΔF on the bottom surface of the slice arises because of the fact that the compressive bending stresses σ_1 on the left side of the slice will be smaller than the stresses σ_2 on the right side of the slice. This is true because the bending moment at ① is smaller than at ② due to the variation of the bending moment throughout the length of the beam (sketch the moment diagram for this beam). Hence, summing forces in the longitudinal direction yields

$$\iint_{\substack{\text{left}\\ \text{area}}} \sigma_1 \, dA + \Delta F - \iint_{\substack{\text{right}\\ \text{area}}} \sigma_2 \, dA = 0$$

$$\Delta F = \iint_{\substack{\text{cross section}\\ \text{of slice}}} (\sigma_2 - \sigma_1) \, dA \tag{9-23}$$

since the left area and right area are both the same size.

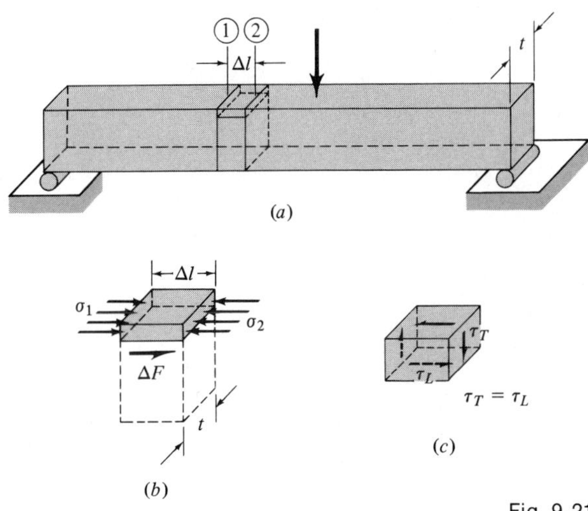

Fig. 9-21

Recall that ΔF represents the longitudinal shearing force on the bottom surface of the slice. We can formally convert this force to a *longitudinal shear stress* τ_L (force per unit area) by dividing this shear force ΔF by the bottom area of the slice. Thus

$$\tau_L = \frac{\Delta F}{\Delta \ell \, t} \tag{9-24}$$

Then, recalling from Sec. 8-3 that simultaneous and *equal* shear stresses always occur in mutually perpendicular surfaces, as illustrated in Fig. 9-21(*c*), we see that the transverse shear stress τ_T occurring on the vertical transverse cross section of the slice will be given by

$$\tau_T = \tau_L = \frac{\Delta F}{\Delta \ell \, t} \tag{9-25}$$

where ΔF is given by Eq. (9-23).

The discussion up to now has not considered whether the beam material is behaving elastically, plastically, or otherwise. Since ΔF by Eq. (9-23) depends upon the bending stresses in the beam, the behavior of the beam material will influence the shear stresses, as given by Eq. (9-25). Accordingly, let us *assume elastic behavior*, so that the compressive bending stresses σ_1 and σ_2 are given by the elastic-bending-stress relation (9-15).

$$\sigma_1 = \frac{M_1 u}{I} \qquad \text{and} \qquad \sigma_2 = \frac{M_2 u}{I} \tag{9-26}$$

where M_1 and M_2 are the bending moments at locations ① and ②, respectively. Substituting Eqs. (9-26) into Eq. (9-23) yields

$$\Delta F = \iint_{\substack{\text{cross section}\\ \text{of slice}}} \frac{(M_2 - M_1)}{I} u \, dA$$
$$= \frac{\Delta M}{I} \iint_{\substack{\text{cross section}\\ \text{of slice}}} u \, dA \tag{9-27}$$

where ΔM is the change in moment between locations ① and ②. Substituting this result in Eq. (9-25) yields

$$\tau_L = \tau_T = \frac{\Delta M}{I \, \Delta \ell \, t} \iint_{\substack{\text{cross section}\\ \text{of slice}}} u \, dA$$

which, as $\Delta \ell \to 0$, takes the form

$$\tau_L = \tau_T = \frac{dM}{d\ell} \frac{Q}{It} = \frac{VQ}{It} \tag{9-28}$$

where we have utilized Eq. (9-5),

$$V = \frac{dM}{d\ell} \tag{9-5}$$

and have defined

$$Q = \iint_{\substack{\text{cross section}\\ \text{of slice}}} u \, dA \tag{9-29}$$

Equation (9-28) gives the longitudinal and transverse shear stresses existing at any location throughout an *elastic beam.* The quantity V is the internal transverse shear force, I is the second moment of the *entire* cross-sectional area of the beam with respect to its centroid, and t is the thickness of the slice on which the stresses τ_T and τ_L are acting. The term Q as defined by Eq. (9-29) represents the *first moment of the cross-sectional area of the slice* with respect to the centroid of the entire cross-sectional area of the beam. Thus Q will vary with the cross section of the slice, the largest value of Q corresponding to a slice cut all the way down to the centroid of the beam. We note that Q will be zero at the very top and bottom of the beam. Consequently, the shear stress is zero at the top and bottom, but generally reaches a maximum value at the centroid of the beam. For a beam with a rectangular cross section, this variation of the shear stress will be parabolic throughout the depth of the beam (see Prob. 9-51).

The following example illustrates some of the ideas regarding the transverse shear stress and the application of Eq. (9-28).

EXAMPLE 9-5

A 24-ft-long simply supported steel I beam, S18 × 54.7, carries a uniformly distributed load of 1,000 lb/ft, which includes the weight of the beam. Determine (*a*) the maximum flexure stress in the beam, (*b*) the largest transverse shear stress developed at the junction of the flange and web, and (*c*) the largest transverse shear stress in the web.

Solutions: The beam and its corresponding shear and moment diagrams are shown in Fig. 9-22(*a*), the maximum shear force being

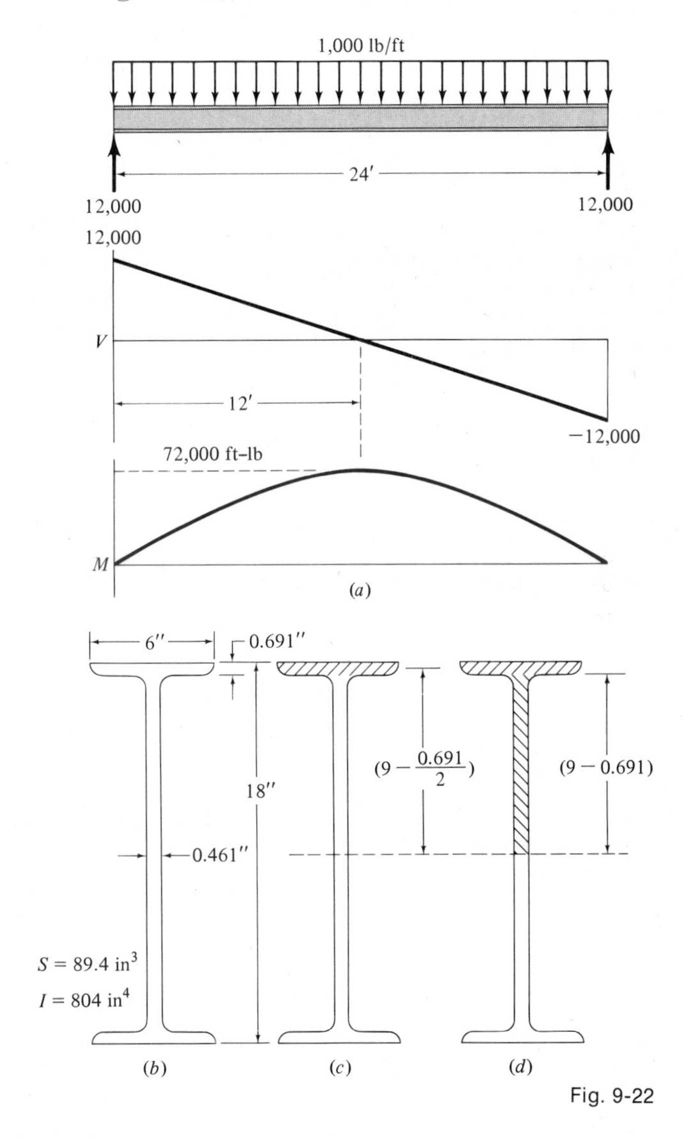

Fig. 9-22

12,000 lb and the maximum bending moment being 72,000 ft-lb.

(*a*) Referring to the table in Appendix B for standard I beams, the average cross-sectional dimensions for an S18 × 54.7 beam are shown in Fig. 9-22(*b*). Assuming elastic behavior, the maximum flexure stress would occur at the top and bottom surface ($u = 9$ in) at the midspan of the beam:

$$\sigma_{\max} = \frac{Mu}{I} = \frac{M}{S}$$
$$= \frac{(72{,}000)(12)}{89.4}$$
$$= 9{,}670 \text{ psi}$$

where $S = I/(d/2)$ is the section modulus. This value of stress is well below the proportional limit for mild steel, so that the beam is elastic.

(*b*) To determine the largest shear stress at the junction of the flange and web, we need to evaluate Q for the slice indicated by the cross-hatched area in Fig. 9-22(*c*). The quantity Q will be the first moment of this area with respect to the centroid of the entire cross section. Thus

$$Q = \text{area} \times \text{distance to centroid}$$
$$= (6)(0.691) \times \left(9 - \frac{0.691}{2}\right)$$
$$= 36 \text{ in}^3$$

With the maximum shear force of 12,000 lb, $I = 804$ in^4, and $t = 0.461$ in, the largest transverse stress shear at the junction will be, by Eq. (9-28),

$$\tau_T = \frac{VQ}{It} = \frac{(12{,}000)(36)}{(804)(0.461)}$$
$$= 1{,}170 \text{ psi}$$

(*c*) The largest transverse shear stress in the web will occur at the middle of the cross section. Consequently, Q must be evaluated for the shaded area indicated in Fig. 9-22(*d*); that is, for half the total cross section. For this composite area we have

$$Q = Q_{\text{flange}} + Q_{\text{half web}}$$
$$Q_{\text{half web}} = \text{area} \times \text{distance to centroid}$$
$$= (0.461)(9 - 0.691) \times \left(9 - \frac{0.691}{2}\right)$$
$$= 16 \qquad \text{(rounded off)}$$
$$Q = 36 + 16 = 52 \text{ in}^3$$

Thus, by Eq. (9-28) the largest transverse shear stress in the web will be

$$\tau_T = \frac{VQ}{It} = \frac{(12{,}000)(52)}{(804)(0.461)}$$
$$= 1{,}680 \text{ psi}$$

Remark: Notice that the value of the largest transverse shear stress is considerably smaller than the largest bending stress. This situation is usually true for most beams as opposed to short shear blocks. The longitudinal shear stress may, however, be significant in a relatively short-span timber beam, where the resistance to shear stress parallel to the wood grain may be low.

Remark: It was mentioned earlier in this section, for thin cross sections such as this I beam, that most of the vertical shear force is carried by the web. With this assumption, an *average* value of the shear stress in the web can be evaluated by dividing the shear force by the area of the web. Thus

$$\tau_{\text{ave}} = \frac{V}{A_{\text{web}}}$$
$$= \frac{12{,}000}{(0.461)[18 - 2(0.691)]}$$
$$= 1{,}570 \text{ psi}$$

which would be a reasonable approximation of the shear stresses in the web.

9-5 COMBINED STRESSES

We have seen how both normal stresses due to bending and shear stresses due to the shear force can, and usually do, exist simultaneously throughout a beam. Also, it is not uncommon for structural or machine members to be loaded by some combination of axial, torsion, bending, and shear loading, with a resulting combination of axial, torsion, bending, and shear stresses.

Generally, the stresses produced by each of the aforementioned types of loading can be calculated individually by using the appropriate theories and formulas developed in Chaps. 5, 6, 7, 8, and 9. Then these stresses can be *superposed* so as to obtain the combined state of stress existing at any particular point in a structural member. So long as all of the individual stresses are below the proportional limit of the material, the superposition principle for stresses is reasonably justifiable. However, when nonlinear inelastic behavior begins to occur, superposing the individual stresses becomes much more difficult to justify, primarily because of the larger deformations which usually accompany the inelastic behavior.

In summary, when considering the stresses in beams and other types of load-carrying members, the engineer must consider the effect of each type of load and their corresponding stresses and deformations. Furthermore, he must be fully aware of how the stresses vary throughout the member so that he can make a rational judgment as to when and where the critical stresses are likely to occur and initiate failure of the member.

EXAMPLE 9-6

The pulley-shaft-bearing assembly transmits the loads as shown in Fig. 9-23(a). The half-inch diameter shaft is made of steel

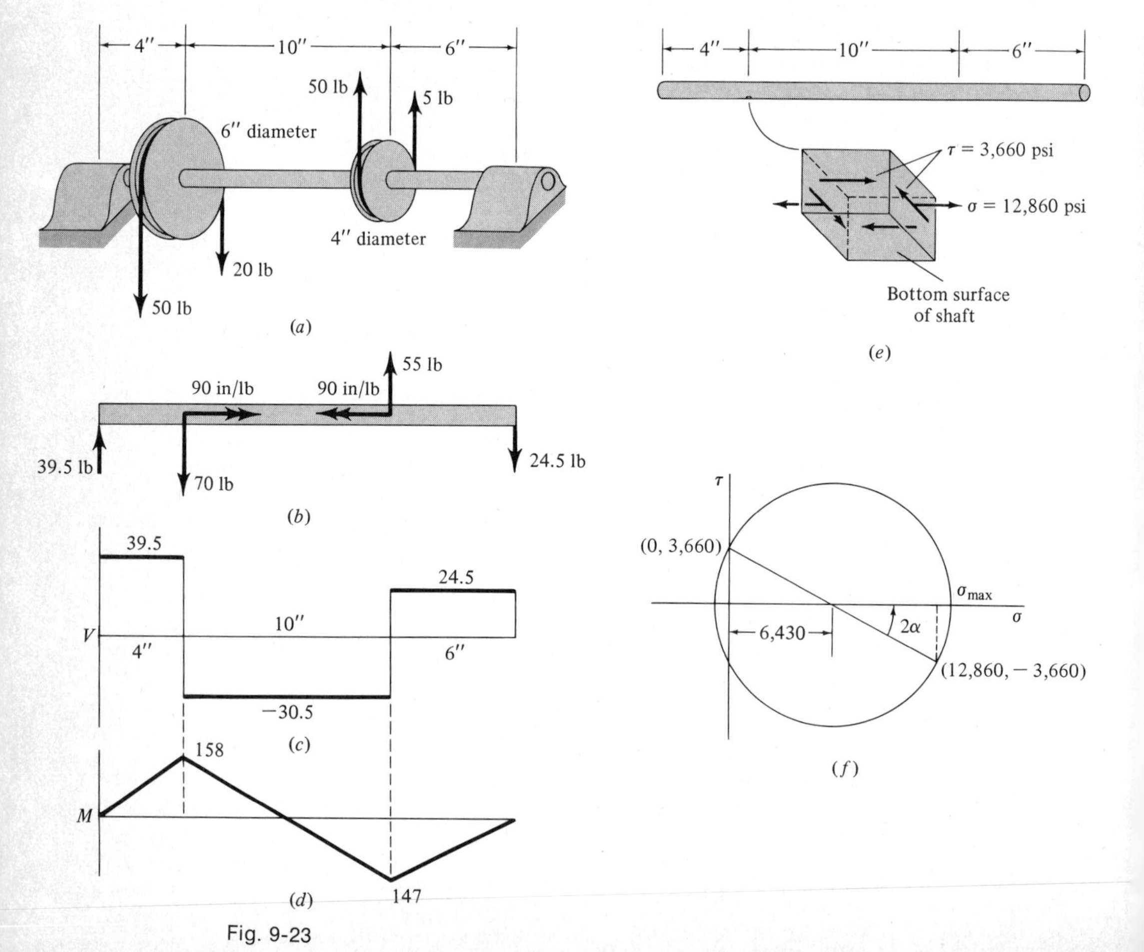

Fig. 9-23

and the bearings can be treated as simple supports. Determine (*a*) the location and value of the largest torsion stress in the shaft, (*b*) the location and value of the largest tensile bending stress in the shaft, and (*c*) the location, direction, and value of the largest normal stress in the shaft.

Solutions: (*a*) A free-body diagram of the shaft is shown in Fig. 9-23(*b*), from which it is apparent that the 10-in section between the pulleys carries a torque of 90 in-lb. Assuming elastic behavior and using the elastic-torsion formula (7-22) for evaluating the torsion shear stress, we obtain

$$\tau = \frac{TR}{J} = \frac{(90)(1/4)}{(\pi/2)(1/4)^4}$$
$$= 3{,}660 \text{ psi}$$

This stress occurs at the outside surface of the shaft anywhere in the 10-in section between the pulleys.

(*b*) The shear and moment diagrams for the shaft are shown in Fig. 9-23(*c*) and (*d*) respectively. The maximum value of the bending moment is 158 in-lb and occurs at the left pulley. Using the elastic-flexure formula (9-17) for evaluating the bending stress, we obtain

$$\sigma = \frac{Mu}{I} = \frac{(158)(1/4)}{(\pi/4)(1/4)^4}$$
$$= 12{,}860 \text{ psi} \qquad \text{(tensile)}$$

which occurs at the bottom outside surface of the shaft. We remark that this is the same value for the compressive bending stress at the top surface of the shaft.

(*c*) With the results of (*a*) and (*b*), the state of stress in the shaft between the left and right pulleys is a combination of a torsion shear stress and a tensile (or compressive) bending stress, with the largest values of both occurring at the outside surface and, in particular, at the left pulley. This combined state of stress is shown in Fig. 9-23(*e*), for which we can plot Mohr's stress circle to determine the maximum normal stress. From the stress circle in Fig. 9-23(*f*)

$$\text{Radius} = \sqrt{(3{,}660)^2 + (6{,}430)^2}$$
$$= 7{,}400$$

$$\sigma_{\max} = \text{center} + \text{radius}$$
$$= 6{,}430 + 7{,}400 = 13{,}800 \text{ psi}$$

$$2\alpha = \arctan \frac{3{,}660}{6{,}430}$$

$$\alpha = 14.8^\circ$$

Thus the maximum normal stress is 13,800 psi and acts in a direction of 14.8° to the axis of the shaft. This stress as well as all the other stresses in the shaft are small enough to justify the assumption of elastic behavior and the use of the superposition principle.

EXERCISE PROBLEMS

9-51 Consider an elastic beam with a rectangular cross section as shown. Show that the transverse shear stress varies parabolically with the distance u and that its maximum value is given by $3V/2A$.

PROB. 9-51

PROB. 9-53

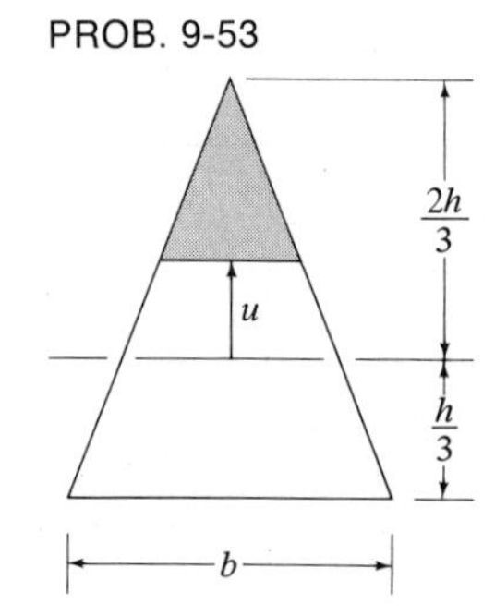

9-52 Consider an elastic beam with a solid circular cross section of radius R. Show that the transverse shear stress at the centroid is given by $4V/3A$.

9-53 Consider an elastic beam with a solid triangular cross section as shown. Show that the maximum transverse shear stress will occur at $u = h/6$ and is given by $3V/2A$.

9-54 and 9-55 For the cross section shown, find the value of Q/It at each of the levels indicated. Location ③ is the centroid.

PROB. 9-54

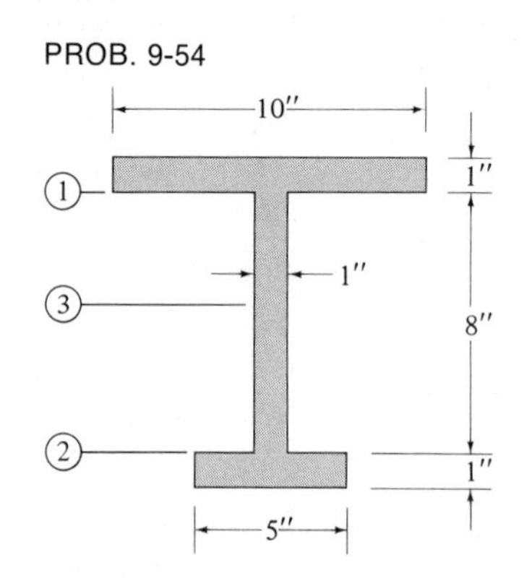

PROB. 9-55

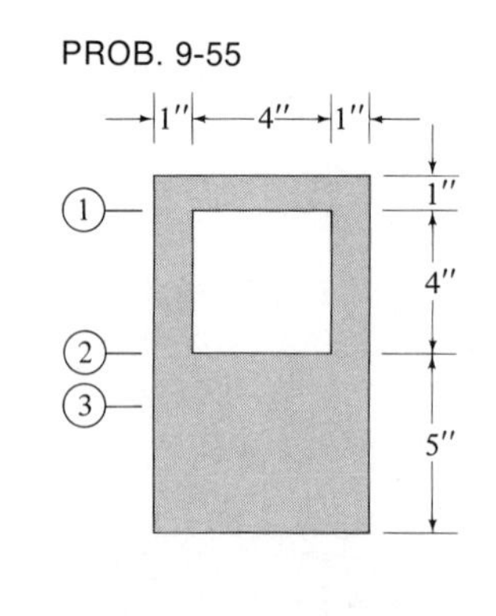

9-56 Two wooden planks 2 by 5 in are glued together to form a T beam. What is the shear stress on the longitudinal glued surface if the transverse shear force on the beam is 3,600 lb?

9-57 A simply supported, uniformly loaded timber beam 10 ft long has a full-size cross section 8 in wide and 12 in deep. If the maximum allowable bending stress is 1,200 psi and the maximum allowable longitudinal shear stress is 100 psi, what is the maximum allowable loading w, lb/ft?

9-58 A standard steel W10 × 72 beam is cantilevered and carries a concentrated load at its free end. How long may the beam be if the maximum bending stress is not to exceed five times the maximum transverse shear stress in the web? Neglect the weight of the beam.

9-59 A nominal 2 by 12-in spruce wood simply supported floor joist has an 18-ft span. The allowable tensile stress is 1,200 psi and the allowable longitudinal shear stress parallel to the grain is 200 psi. What is the largest effective concentrated load which may be applied at the midspan? Refer to Appendix B.

9-60 A timber beam 4 in wide (nominal size) is to be used as a floor joist on a span of 14 ft. The joist carries an effective uniformly distributed load of 250 lb/ft. What is the minimum nominal size timber required if the allowable tensile stress is 1,200 psi and the allowable longitudinal shear stress is 180 psi? Refer to Appendix B.

9-61 The beam shown in the figure has the cross section shown in the figure for Prob. 9-54. Find (*a*) the maximum tensile bending stress, (*b*) the maximum compressive bending stress, and (*c*) the maximum transverse shear stress. Assume elastic behavior.

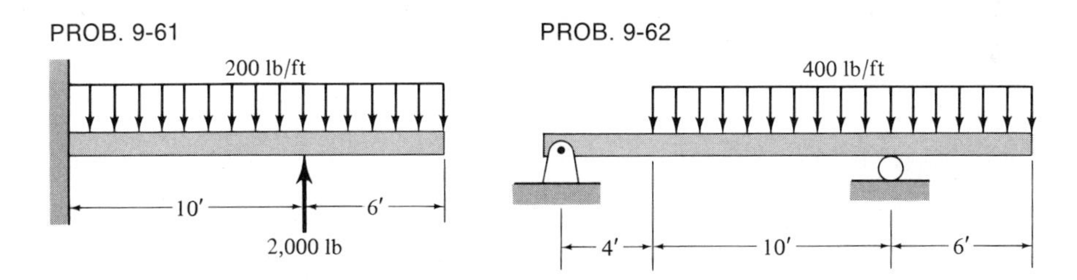

9-62 The beam shown in the figure has the cross section shown in the figure for Prob. 9-55. Find (*a*) the maximum tensile bending stress, (*b*) the maximum compressive bending stress, and (*c*) the maximum transverse shear stress. Assume elastic behavior.

9-63 The frame of a large C clamp has the dimensions and cross section shown in the figure. Assuming elastic behavior, determine the maximum tensile and compressive stresses developed across section *a-a* due to a load $P = 100$ lb.

PROB. 9-63

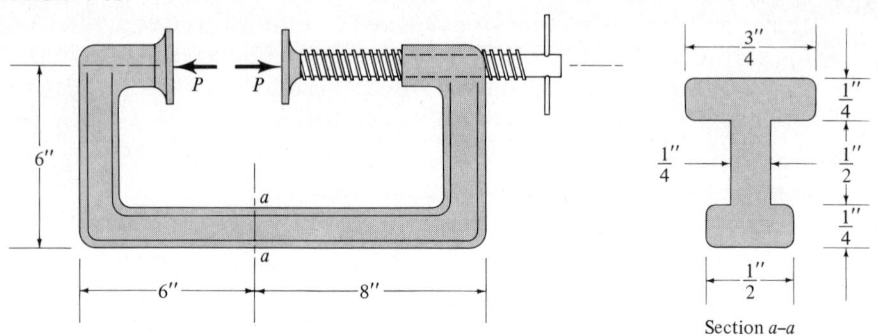

9-64 A large cylindrical steel tank has an inside diameter of 4 ft and a wall thickness of 1 in. The tank is effectively simply supported on its ends 30 ft apart and carries a liquefied gas under a pressure of 200 psi. The tank and its contents weigh 100,000 lb, which may be assumed to be uniformly distributed over its entire length. Find the maximum tensile stress in the tank wall. Use thin-walled approximations.

9-65 The fixture shown is to be made of a brittle material with an ultimate tensile stress of 40,000 psi and an ultimate compressive stress of 100,000 psi. The cross section is to be either a T or an inverted T as shown, whichever is better. Using a factor of safety of 2, what is the maximum allowable load P?

PROB. 9-65 and 9-66

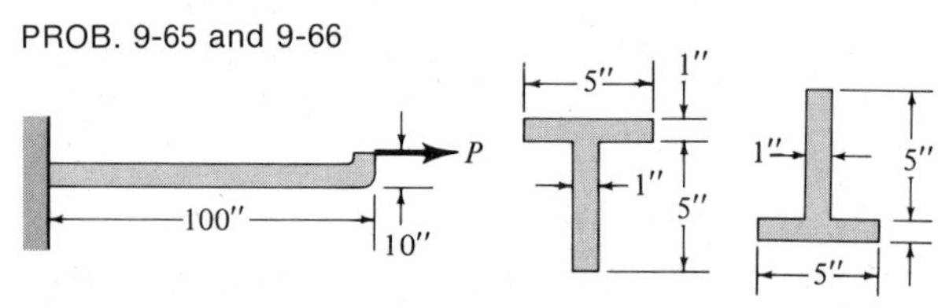

9-66 Same as Prob. 9-65 except the ultimate compressive stress is 50,000 psi rather than 100,000 psi.

9-67 A standard 4-in pipe (nominal size) is subjected to a torque of 2,000 ft-lb, an axial tensile thrust of 8,000 lb, and a bending moment of 6,000 ft-lb. If the proportional limit of the material is 32,000 psi, will the pipe remain elastic under these applied loads? Refer to Appendix B.

9-68 A standard 12-in pipe (nominal size) is installed in such a way that it is subjected to a torque of 22,000 ft-lb, a bending moment of 34,000 ft-lb, and an internal gas pressure of 600 psi. Determine the maximum principal stress and the maximum shear stress. Treat the pipe as an open-ended pressure vessel.

9-69 A relatively stiff plate is welded to the end of a cantilevered 1-in-diameter circular rod as shown. Determine the principal stresses produced at point a at the top surface of the rod. Also, find the maximum shear stress at point a.

PROB. 9-69 PROB. 9-70

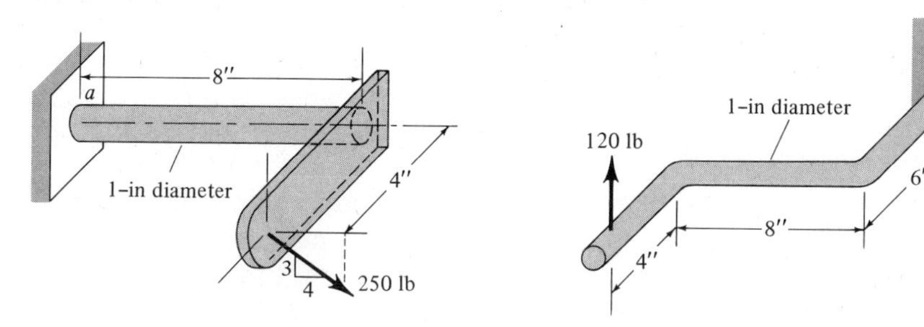

9-70 Find the principal stresses at section a-a of the crank and indicate their directions.

9-71 A simply supported 2 by 6-in (exact size) timber beam 8 ft long supports a uniformly distributed load of 200 lb/ft. The grain of the wood makes an angle of 15° with the longitudinal axis of the beam. What is the value of the largest shear stress parallel to the grain? Would an additional axial compressive load of 3,000 lb increase or decrease this shear stress, and by how much?

9-72 A 12-ft wooden ladder is leaning against a reasonably smooth wall. The bottom of the ladder is 3 ft from the wall. Neglecting the weight of the ladder, estimate the maximum tensile and compressive stresses developed in the side rails of the ladder due to a 180-lb man standing three-fourths of the way up the ladder. The side rails are 1 by $2\frac{1}{2}$ in.

9-73 A concrete pillow block has a square cross section as shown and is to support an effective concentrated compressive load. If the load is applied eccentrically rather than centroidally, it is possible to develop tensile stresses in the block due to bending effects. Prove that in order to prevent any net tensile stresses in the block, the load P must be applied in the middle third of the cross section.

PROB. 9-73

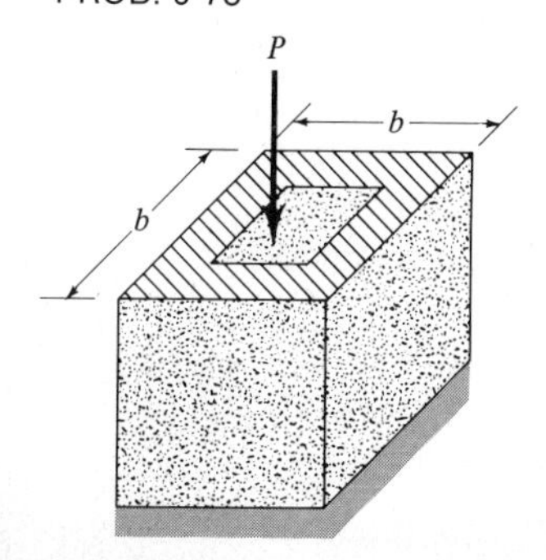

9-6 THE DEFLECTION CURVE

Now that we have seen how the loading on a beam can be related to the bending and shearing stresses throughout the beam, we turn our attention to the question of how the loading is related to the deformation of the beam. Generally, the loading will cause an originally straight beam to assume some deformed shape such as those exaggerated in Fig. 9-2. More specifically, the deformed shape of the *neutral axis* of the beam is called the *deflection curve*, and the deviation of this curve from its initial straight position at any location along the beam is called the *deflection* y at that location. Refer to Fig. 9-24. Our objective is to relate the applied loading to the deflection of the beam.

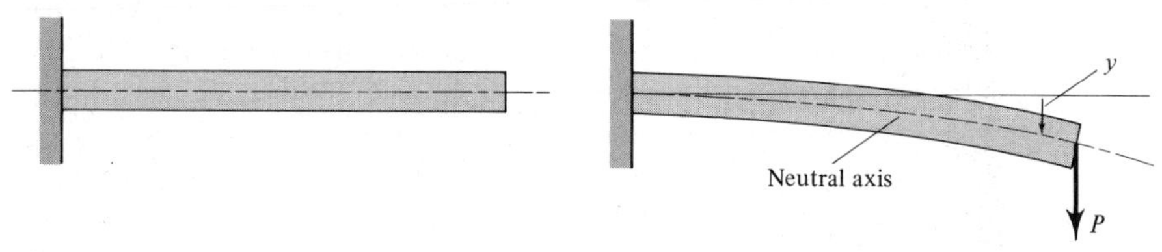

Fig. 9-24

We already know some things about the deflection of a beam; namely, for pure bending, Eq. (9-8) relates the bending *strain* ϵ to the radius of curvature R of the deflected neutral axis,

$$\epsilon(u) = -\frac{u}{R} \tag{9-8}$$

Consequently, if we can relate this strain to the bending stress, and then relate the stress to the applied loads, we will obtain a relationship between the radius of curvature of the deflected beam and the applied loads. Strictly speaking, such a relationship would be applicable only for pure bending, and the presence of additional transverse shearing effects would contribute to an additional deformation of the beam. However, as we have seen, in most beams the shearing effects will be considerably smaller than the bending effects, and the dominant deformation will be that due to the bending effects. Consequently, the following discussion will ignore the transverse shear effects on the deflection of the beam.

Before proceeding with the development of the deflection theory, a few comments are in order. In the design of beams and beam-like structures, the deflections sometimes play a significant role; but more often the stresses in the beam are the primary design criteria. In a well-designed structure the deflections will be unobservable even to the highly trained naked eye. Of course, there are situations in which even such small deflections can be objectionable; for example, if the floor beams in your house are too "springy" the floor (and everything on it) will bounce as you walk across it. Usually, relatively large deflections can occur only in beams made of ductile materials, and in such cases the fully plastic stress condition is likely to be taken as the failure criterion. Finally, one of the most important applications of deflection theory is in the solution of statically indeterminate beams, such as those of Fig. 9-1(*d*) and (*e*). Problems of this type will be considered in a later section.

In the light of the comments of the previous paragraph, we will develop the deflection theory assuming *elastic* (linear) behavior of the material of the beam. Accordingly, the bending stresses are given by Eq. (9-15) as

$$\sigma(u) = -\frac{Mu}{I} \tag{9-15}$$

and the bending stresses and bending strains are related by Hooke's law as

$$\epsilon = \frac{\sigma}{E} \tag{9-30}$$

Combining Eqs. (9-8), (9-15), and (9-30) yields

$$\frac{1}{R} = \frac{M}{EI} \tag{9-31}$$

This equation gives the curvature $1/R$ at any location in the beam in terms of the corresponding bending moment M at that location in the beam. Note that a positive bending moment M produces a positive (concave upward) curvature, and a negative moment produces a negative (concave downward) curvature.

Although the curvature as given by Eq. (9-31) is of some interest, we are much more concerned with determining the deflection y at any location along the beam. To this end we must relate the radius of curvature R to the deflection y. Consider an arbitrary plane curve $y = y(x)$ as is shown in Fig. 9-25(*a*), noting that positive y is measured upward. Then, from elementary calculus,

$$\frac{1}{R} = \frac{d^2y/dx^2}{[1 + (dy/dx)^2]^{3/2}} \tag{9-32}$$

which relates the curvature to the deflection y and the location x. Now, for elastic behavior, the deflection curve for a beam usually will be a very gentle curve whose slope θ at any location will be a very small angle. Consequently, for small slopes we have approximately

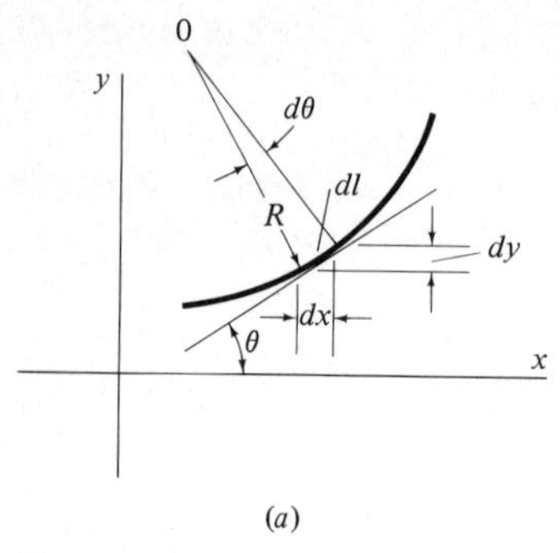

From elementary calculus

$$R d\theta = dl$$

$$dl = \sqrt{(dx)^2 + (dy)^2} = \sqrt{1 + (dy/dx)^2}\, dx$$

$$dy/dx = \tan\theta, \; d^2y/dx^2 = \sec^2\theta \; d\theta/dx$$

$$d\theta = \frac{d^2y/dx^2}{\sec^2\theta} dx = \frac{d^2y/dx^2}{1 + (dy/dx)^2} dx$$

$$\therefore \frac{1}{R} = \frac{d\theta}{dl} = \frac{d^2y/dx^2}{[1 + (dy/dx)^2]^{3/2}}$$

(b)

Fig. 9-25

$$\theta \approx \tan\theta = \frac{dy}{dx} \ll 1 \tag{9-33}$$

and Eq. (9-32) becomes

$$\frac{1}{R} \approx \frac{d^2y}{dx^2} \approx \frac{d\theta}{dx} \tag{9-34}$$

which is a linear approximation between the curvature R, the deflection y, and the slope θ in terms of the location x.

> ***Remark:*** The approximation (9-33) for small slopes implies that the deflection y will always be very small compared with the length of the beam. Also, it implies that there is no need to distinguish between the horizontal distance x and the length ℓ measured along the deformed beam.

Combining Eq. (9-34) with Eq. (9-31) yields

$$\frac{d\theta}{dx} = \frac{d^2y}{dx^2} = \frac{M}{EI} \tag{9-35}$$

which is the classical *Bernoulli–Euler Equation*[1,2] relating the slope θ and deflection y to the bending moment M. Recalling that M is usually a function of the location x, Eq. (9-35) is a linear second-order ordinary differential equation which usually can be successively integrated to give the slope function $\theta(x)$ and the deflection function $y(x)$.

The shape of the deflection curve depends upon the loading and the manner in which the beam is supported. For example, the deflection curve for the simply supported beam in Fig. 9-26(a) will have the qualitative shape shown, with the deflection being zero at

---

[1] James Bernoulli (1655–1705), a Swiss mathematician, developed the equation.
[2] Leonard Euler (1707–1783), a famous Swiss mathematician, extended its use to other problems.

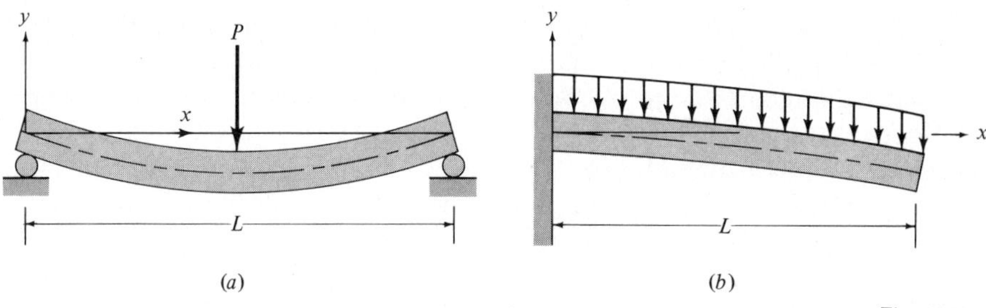

Fig. 9-26

both the left end ($x = 0$) and the right end ($x = L$). Similarly, for the cantilevered beam in Fig. 9-26(*b*), the deflection will be zero at $x = 0$, and the slope will also be zero at $x = 0$. These conditions are called *boundary conditions* and are necessary to uniquely specify the solution of the differential deflection equation (9-35).

We summarize these results with the following theorem:

Theorem 9-2 For a straight elastic beam, the deflection y, the slope θ, and the bending moment M are related by

$$y = y(x) \qquad \text{deflection curve}$$

$$\frac{dy}{dx} = \theta(x) \qquad \text{slope function}$$

$$EI\,\frac{d^2y}{dx^2} = EI\,\frac{d\theta}{dx} = M(x) \qquad \text{moment function}$$

Furthermore, the location of greatest deflection usually corresponds to the location at which the slope θ is zero. Otherwise, the greatest deflection occurs at one of the ends.

> ***Important Remark:*** From the discussion of shear and moment diagrams earlier in this chapter, we know that in many cases a single moment equation will not suffice for the entire beam, but, rather, a moment equation will be required for each portion of the beam having a different loading characteristic. Consequently, in such situations the deflection curve will not be described by a single algebraic function but by several functions each of which describes the deflection curve within some interval of the beam.

We close this section with an example which illustrates the so-called *direct method* of integrating the differential-deflection equation (9-35). There are numerous other analytical, semigraphical, and numerical integrating techniques, which are sometimes more convenient, and can be found in texts dealing primarily with strength of materials.

EXAMPLE 9-7

For the W8 × 10 steel beam shown in Fig. 9-27(a), derive the elastic-deflection curve and determine the maximum value of the deflections.

Solution: The appropriate free-body diagrams for the left and right portions of the beam are shown in Fig. 9-27(e) and (f), respectively, along with the corresponding shear and moment equations. The shear and moment diagrams are shown in Fig. 9-27(b) and (c), respectively, from which the maximum value of the moment is 9,000 ft-lb. Checking the maximum bending stress, from the W-shapes table in Appendix B, we have $I = 30.8\ \text{in}^4$ and $u = 3.95$, from which

$$\sigma_{\max} = \frac{Mu}{I} = \frac{(9{,}000)(12)(3.95)}{30.8}$$
$$= 13{,}850\ \text{psi}$$

which implies that the beam is elastic.

Recall Eq. (9-35),

$$EI\frac{d\theta}{dx} = EI\frac{d^2y}{dx^2} = M \qquad (9\text{-}35)$$

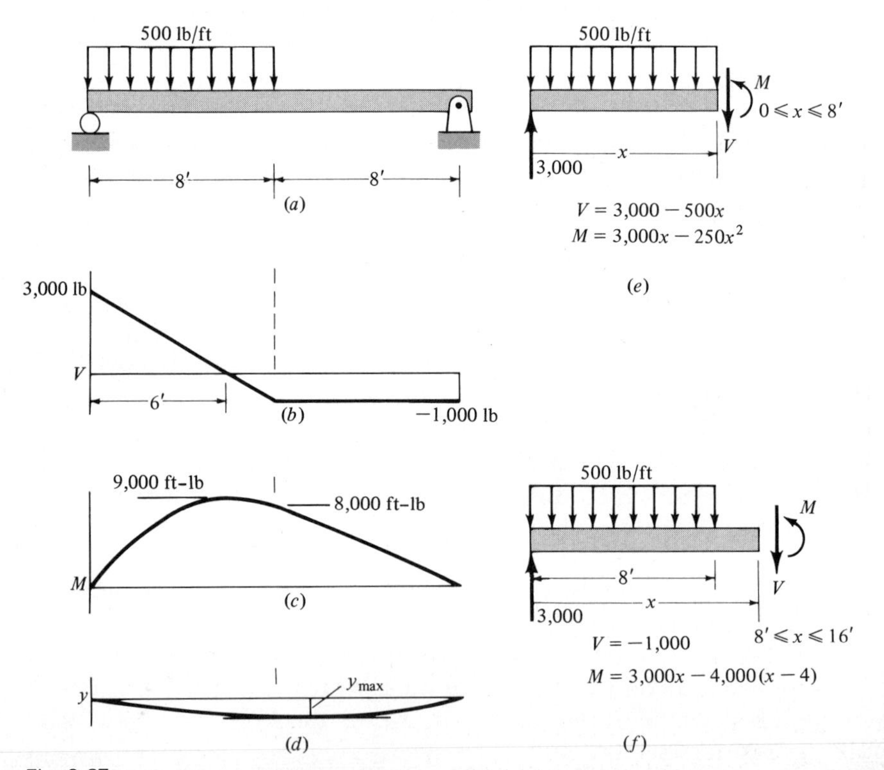

Fig. 9-27

The given E and I are in pounds per square inch and inch^4, respectively, while the moment equations of Fig. 9-27(*e*) and (*f*) are in foot-pounds. Accordingly, we must make the proper conversion of units, and, in this case, it is perhaps easier to convert the E and I to pounds per square foot and foot^4, respectively, rather than converting the moment equation. Then, integrating Eq. (9-35), for $0 \leq x \leq 8$ ft, we have

$$M(x) = 3{,}000x - 250x^2$$

$$EI\theta(x) = \int M(x)\,dx$$

$$= 1{,}500x^2 - \frac{250}{3}x^3 + C_1 \qquad (a)$$

$$EIy(x) = \int \theta(x)\,dx$$

$$= 500x^3 - \frac{250}{12}x^4 + C_1x + C_2 \qquad (b)$$

while for 8 ft $\leq x \leq$ 16 ft, we have

$$M(x) = 3{,}000x - 4{,}000(x - 4)$$

$$EI\theta(x) = \int M(x)\,dx$$

$$= 1{,}500x^2 - 2{,}000(x - 4)^2 + C_3 \qquad (c)$$

$$EIy(x) = \int \theta(x)\,dx$$

$$= 500x^3 - \frac{2{,}000}{3}(x - 4)^3 + C_3x + C_4 \qquad (d)$$

where C_1, C_2, C_3, and C_4 are constants of integration whose values must be determined from appropriate boundary conditions. Two of these boundary conditions are fairly obvious, namely,

$$y(0) = 0 \qquad (e)$$

$$y(16) = 0 \qquad (f)$$

The other two boundary conditions are less obvious, but nevertheless, quite necessary. The deflection given by Eq. (*b*) at $x = 8$ ft must have the same value as that given by Eq. (*d*) at $x = 8$ ft; that is, the deflection curve must be continuous at $x = 8$ ft. Similarly, the slope must be continuous at $x = 8$ ft. Thus

$$\text{Left portion } y(8) = y(8) \text{ right portion} \qquad (g)$$

$$\text{Left portion } \theta(8) = \theta(8) \text{ right portion} \qquad (h)$$

From Eqs. (*b*) and (*e*)

$$C_2 = 0 \qquad (j)$$

From Eqs. (*d*) and (*f*) we obtain

$$(2{,}000)(28)(16) + C_3 16 + C_4 = 0 \qquad (k)$$

From Eqs. (*a*), (*c*), and (*h*),

$$C_1 = C_3 + \frac{2{,}000}{3}(16) \tag{m}$$

From Eqs. (*b*), (*d*), (*g*), and (*m*)

$$C_4 = \frac{2{,}000}{3}(64) \tag{n}$$

Then, by Eqs. (*n*), (*k*), and (*m*)

$$C_1 = -\frac{2{,}000}{3}(72) \qquad \text{and} \qquad C_3 = -\frac{2{,}000}{3}(88) \tag{o}$$

Thus, Eqs. (*b*) and (*d*), with the values for the constants given by (*j*), (*n*), and (*o*), describe the elastic-deflection curve for this beam. A qualitative sketch of this curve is shown in Fig. 9-27(*d*).

The maximum deflection occurs where the *slope* of the deflection curve is zero, i.e., a horizontal tangent. Although we do not know where this occurs, let us guess that it occurs somewhere in the right half of the beam. Accordingly, setting Eq. (*c*) equal to zero yields

$$1{,}500x^2 - 2{,}000(x-4)^2 - \frac{2{,}000}{3}(88) = 0$$

$$x = 7.35 \text{ ft} \qquad \text{and} \qquad x = 24.60 \text{ ft}$$

both of which are outside the range $8 \leqslant x \leqslant 16$ for which Eq. (*c*) is valid. Thus our guess was incorrect and the maximum deflection occurs in the left half of the beam. Hence, setting Eq. (*a*) equal to zero yields

$$1{,}500x^2 - \frac{250}{3}x^3 - \frac{2{,}000}{3}(72) = 0$$

$$x = 7.3 \text{ ft} \qquad \text{(approximately)}$$

Finally, substituting this value into Eq. (*b*) yields

$$y_{\max} = \frac{[500(7.3)^3 - (250/12)(7.3)^4 - (2{,}000/3)(72)(7.3)]}{EI}$$

$$= -\frac{214{,}700}{(30 \times 10^6)(12)^2(30.8)(12)^{-4}}$$

$$= -0.0335 \text{ ft} \qquad \text{(downward)}$$

Remark: From this example it is apparent that it is quite easy to make a numerical error in solving problems of this type. Therefore, it is often desirable to formally work the entire problem in general terms using *w*, *L*, *P*, *E*, *I*, etc., rather than using the numerical values of the loading, length, force, etc. After the formal solution has been obtained, the numerical values can be substituted to obtain the final answer.

EXERCISE PROBLEMS

9-74 to 9-79 For the beam shown, obtain the solution for the elastic deflection curve. Also, determine the maximum value of the deflection.

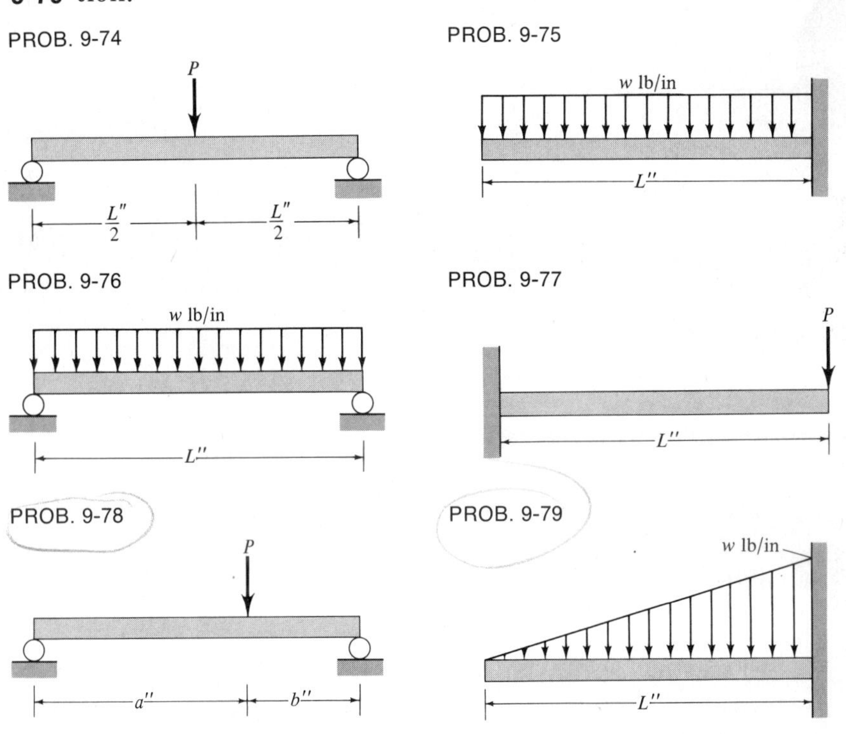

9-80 and 9-81 For the aluminum beam shown, find the deflection at the midspan. Use $E = 10.5 \times 10^6$ psi and $I = 24$ in^4.

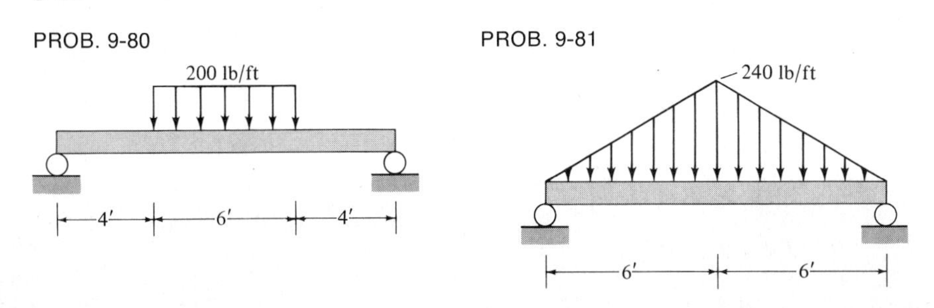

9-82 For the weighing device shown in the figure for Prob. 9-39, find the relation between the load W and the movement (in inches) of the pointer. Express your answer in terms of EI and L.

9-83 For the elastic beam shown, determine the maximum deflection in terms of P, a, E, and I. Also, find the value of the slope directly under the left load P.

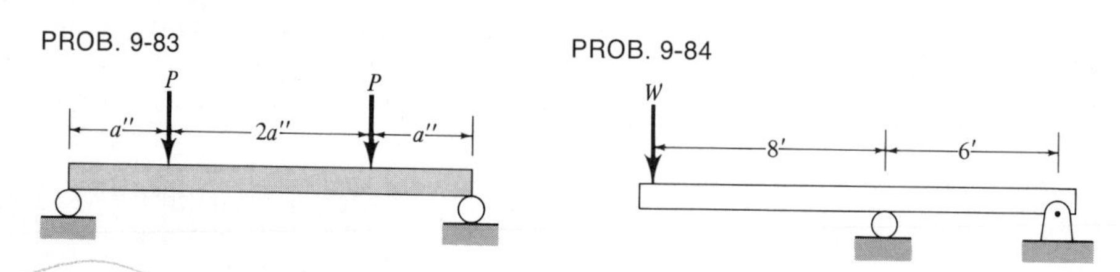

9-84 A diving board shown in the figure has a 2 by 18-in cross section. If the effective load W of an average man is 300 lb, what deflection will this produce at the left end of the board? Use $E = 1.8 \times 10^6$ psi.

9-85 A 2 by 10 (nominal size) timber beam 12 ft long and simply supported carries a uniformly distributed load. In an effort to brace the beam, the load is temporarily removed and a 2 by 4 is nailed lengthwise to the bottom surface of the beam to form an inverted T beam. How much (percentwise) will this bracing effort reduce (*a*) the largest bending stresses in the beam and (*b*) the maximum deflection of the beam?

9-86 The floor of a 16 by 24-ft room carries a total load of 9,000 lb which can be considered to be uniformly distributed over the floor. The floor is supported by eighteen 2 by 10-in (nominal size) floor joists 16 ft long and on 16-in centers. Neglecting any bridging effects of the floor, determine the maximum bending stress in the joists and the maximum deflection of the joists. Use $E = 1.4 \times 10^6$ psi.

9-87 Same as Prob. 9-86 except the floor is supported by sixteen 2 by 12-in (nominal size) joists on 18-in centers.

9-88 The second floor of a house is supported by 2 by 12-in (nominal size) floor joists across a 20-ft span. The owner is considering buying an 800-lb piano which would be placed in such a manner that its weight would be distributed across the middle of five joists. It is expected that the first-floor ceiling plaster will crack if the joists deflect more than 1/360 of their span. Will the piano cause the ceiling plaster to crack? Use $E = 1.4 \times 10^6$ psi.

9-89 For the steel beam shown, sketch the deflection curve. Determine the deflection of the free right end. Use $I = 108$ in^4.

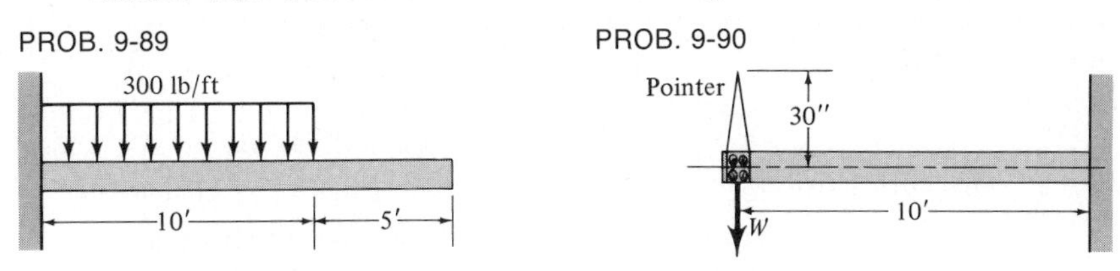

9-90 Find the horizontal and vertical displacement of the tip of the pointer for a weight W of 600 lb. Use $I = 88$ in^4, $E = 30 \times 10^6$ psi.

9-7 SUPERPOSITION OF DEFLECTIONS

The direct method of integration for obtaining the deflection curve of a loaded beam is usually a fairly straightforward process for a beam whose loading is not too complex. However, if a beam has numerous concentrated loads and/or distributed loads over its length, then the direct integration procedure can become quite tedious and error-prone. For such situations it is often possible and convenient to determine the deflection caused by the various individual loads and then add these deflections together to obtain the total deflection produced by the combined loads on the beam. For example, the loading on the beam in Fig. 9-28(a) can be thought of as the combination of the loadings depicted in Fig. 9-28(b) and (c).

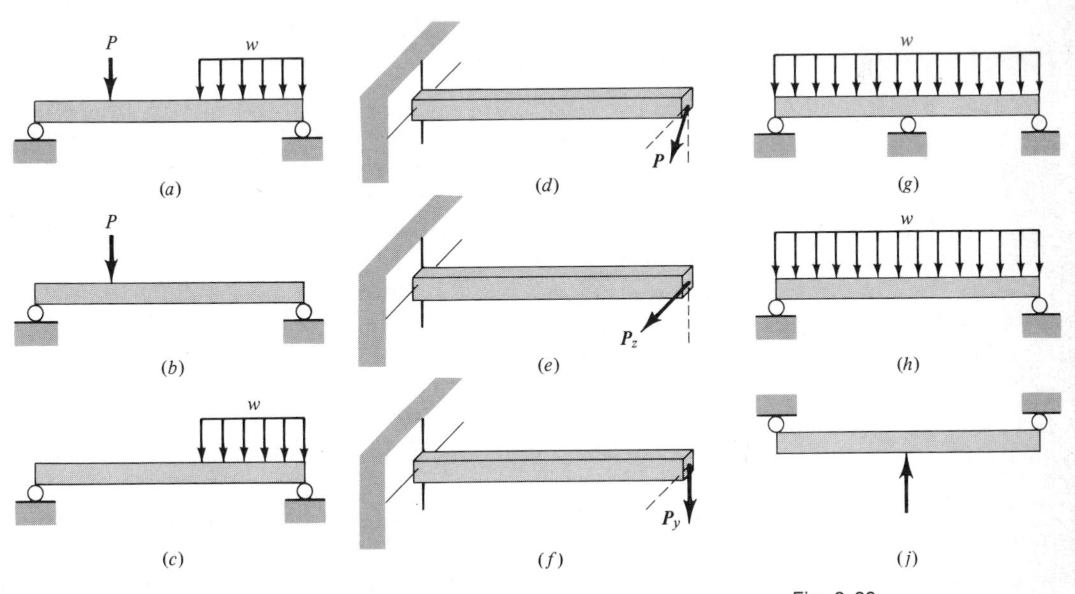

Fig. 9-28

Similarly, the loading **P** in Fig. 9-28(d) is the vector sum of the loadings $\mathbf{P}_z$ in Fig. 9-28(e) and $\mathbf{P}_y$ in Fig. 9-28(f). Consequently, the deflection of the beam in Fig. 9-28(d) would be the *vector* sum of the deflection produced by the loads in Fig. 9-28(e) and (f). A more subtle use of the superposition technique is indicated in Fig. 9-28(g) to (j), in which the loading and supports in Fig. 9-28(g) are visualized as the combination of the loadings in Fig. 9-28(h) and (j) in which the middle support is treated as a load rather than as a support. In all cases, the important thing to note is that the combination of loadings always produces the *same* loading at *each* point along the beam as does the original loading.

The elastic-deflection equation (9-35) is a *linear* differential equation. Therefore, if $y_1(x)$ is the elastic deflection due to the moment function $M_1(x)$ for some loading, and $y_2(x)$ is the elastic deflection due to the moment function $M_2(x)$ for some second loading, then for the individual loadings

$$\frac{d^2y_1}{dx^2} = \frac{M_1(x)}{EI}$$

$$\frac{d^2y_2}{dx^2} = \frac{M_2(x)}{EI}$$

Adding these equations yields

$$\frac{d^2(y_1 + y_2)}{dx^2} = \frac{M_1(x) + M_2(x)}{EI}$$

or

$$\frac{d^2y_{1+2}}{dx^2} = \frac{M_{1+2}(x)}{EI} \tag{9-36}$$

so that the deflection y_{1+2} due to the moment M_{1+2} for the combined loading is the sum of the individual deflections due to the individual loadings. Thus, in general, *elastic* deflections can be superposed to obtain the total deflection of a beam. However, a word of caution is necessary. It is possible that, while a beam would behave elastically due to individual loads, the total combined loading might result in inelastic behavior, thus invalidating Eq. (9-36) and the superposition principle. Consequently, before using the superposition principle, the engineer should analyze the stress to ensure the existence of elastic behavior.

9-8 STATICALLY INDETERMINATE BEAMS

Perhaps the most useful application of the superposition principle for deflections of beams is in the solution of statically indeterminate beams. A statically indeterminate beam is one for which the statical requirements of equilibrium are not sufficient to deter-

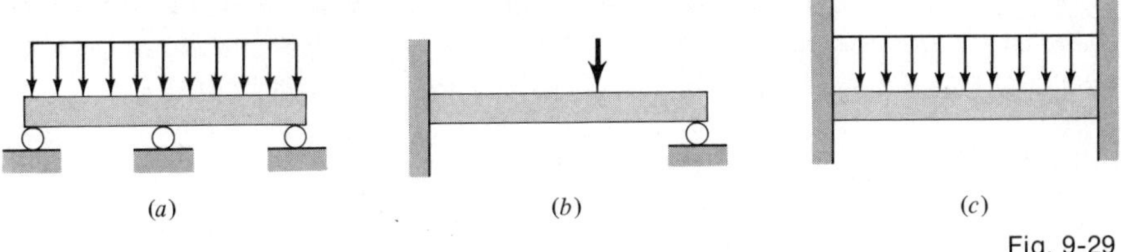

Fig. 9-29

mine the values of the external reactions at the various supports. Some examples were noted earlier in this chapter, and others are shown in Fig. 9-29. Just as in the statically indeterminate problems encountered in Chaps. 6 and 7 for axial and torsion members, respectively, the solution of such problems depends upon the kinematical restraints imposed on the deformation of the beam. In particular, the supporting reactions on a beam usually restrict the shape of the deflection curve in some manner. For example, in Fig. 9-29(*b*) the right support requires the deflection at that point to be zero, whereas the right support in Fig. 9-29(*c*) requires both the deflection and the slope to be zero at that point. When the appropriate kinematical requirements are adjoined to the equilibrium requirements, the problem becomes solvable.

The following example illustrates the solution of a statically indeterminate beam.

EXAMPLE 9-8

For the beam shown in Fig. 9-30(*a*), find all the support reactions and the maximum value of the deflection. Assume elastic behavior.

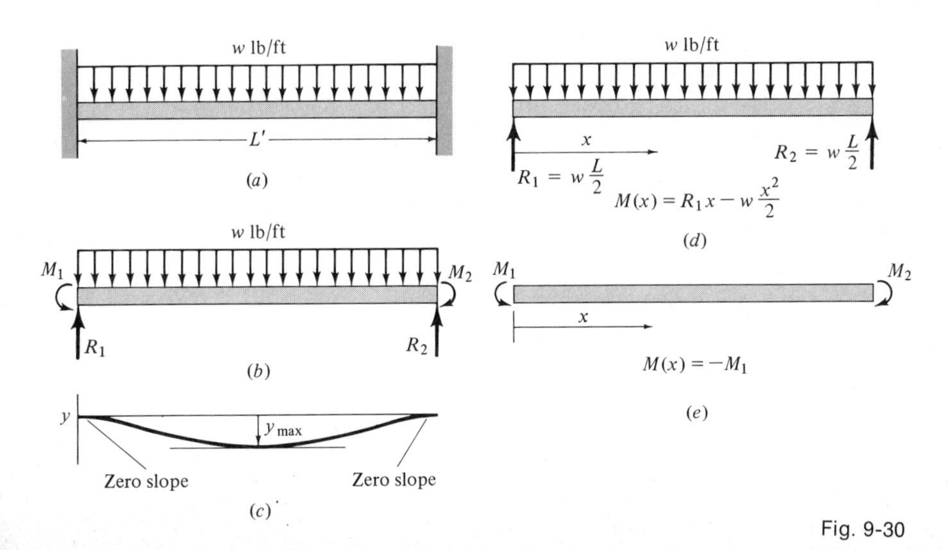

Fig. 9-30

Solution: Recognizing that the fixed supports are each capable of exerting a couple as well as a force, the free-body diagram of the beam is shown in Fig. 9-30(*b*). Equilibrium requires that

$$R_1 + R_2 - wL = 0 \tag{a}$$

$$R_1 L - M_1 + M_2 - w\frac{L^2}{2} = 0 \tag{b}$$

which are not sufficient to determine the external reactions R_1, R_2, M_1, and M_2. The beam is statically indeterminate.

A qualitative sketch of the deflection curve is shown in Fig. 9-30(*c*), from which we can recognize the kinematical restraints which the reactions impose on the deflection curve; namely, the slope and deflection are zero at the ends. Also, from symmetry, we anticipate that the slope is zero at the midpoint where the deflection will be greatest.

Using the superposition principle, we visualize the loading on the beam as being made up of two equilibrated loading situations depicted in Fig. 9-30(*d*) and (*e*), with the corresponding moment equation for each. Hence the moment equation for the original beam is

$$M(x) = \frac{wL}{2}x - \frac{wx^2}{2} - M_1 \tag{c}$$

Integrating, we obtain

$$EI\theta(x) = \frac{wLx^2}{4} - \frac{wx^3}{6} - M_1 x + C_1 \tag{d}$$

Now, $\theta = 0$ at $x = 0$ and $x = L$. Hence

$$C_1 = 0 \qquad \text{and} \qquad M_1 = \frac{wL^2}{4} - \frac{wL^2}{6} = \frac{wL^2}{12}$$

Thus, with the reactions

$$R_1 = R_2 = \frac{wL}{2} \qquad \text{and} \qquad M_1 = M_2 = \frac{wL^2}{12} \tag{e}$$

both the equilibrium requirements (*a*) and (*b*) and the kinematical requirements on the slope are satisfied. Finally, integrating Eq. (*d*), we obtain

$$EIy(x) = \frac{wL^3}{12} - \frac{wx^4}{24} - \frac{wL^2x^2}{24} + C_2$$

Since $y = 0$ at $x = 0$, $C_2 = 0$. The maximum deflection occurs at the midpoint $x = L/2$ and is given by

$$EIy_{\max} = \frac{wL^3}{96} - \frac{wL^4}{384} - \frac{wL^4}{96}$$

$$y_{\max} = -\frac{wL^4}{384EI}$$

EXERCISE PROBLEMS

9-91 to 9-96 For each of the beams shown, find all the support reactions. Assume elastic behavior.

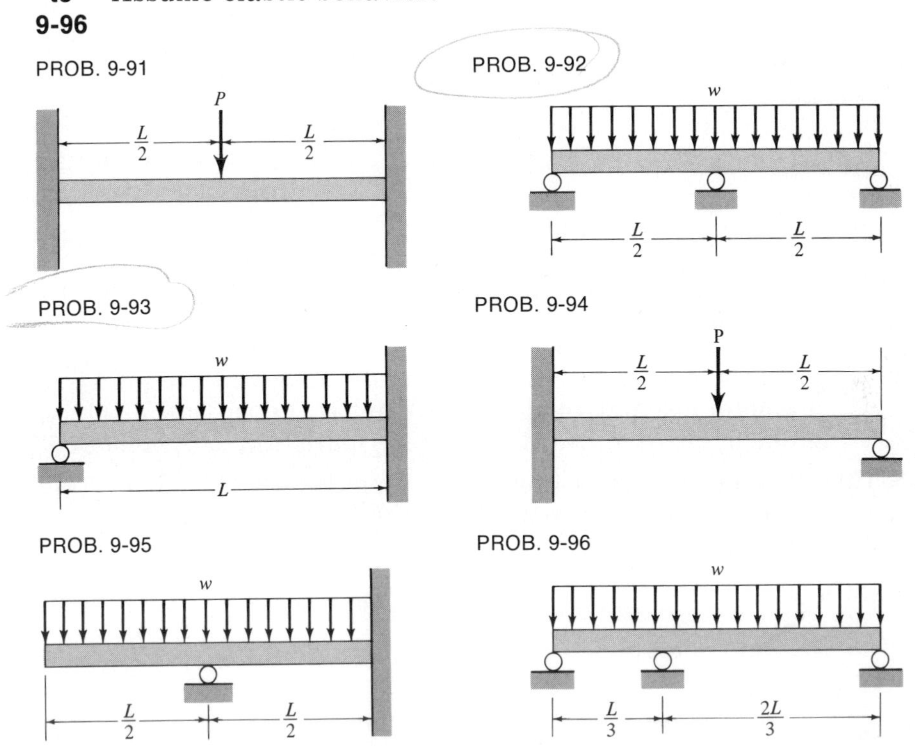

9-97 For the beam shown, find the support reactions on the beam and the deflection of the right end in terms of P, L, E, and I.

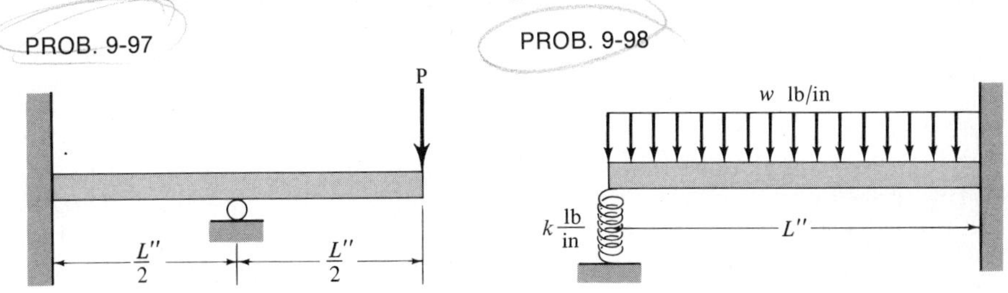

9-98 The left end of the cantilevered beam shown is supported by a linear elastic spring, with the spring constant k being the force required to compress the spring 1 in. Initially the unloaded beam is straight, and the spring is not compressed. How much will the spring be compressed when the uniformly distributed load of w lb/in is applied to the beam? Express your answer in terms of k, L, E, I, and w.

9-99 What ratio of couple M_o to the force P is necessary to keep the pointer horizontal? What will be the corresponding value of the horizontal displacement of the pointer in terms of E, L, and I?

PROB. 9-99

PROB. 9-100

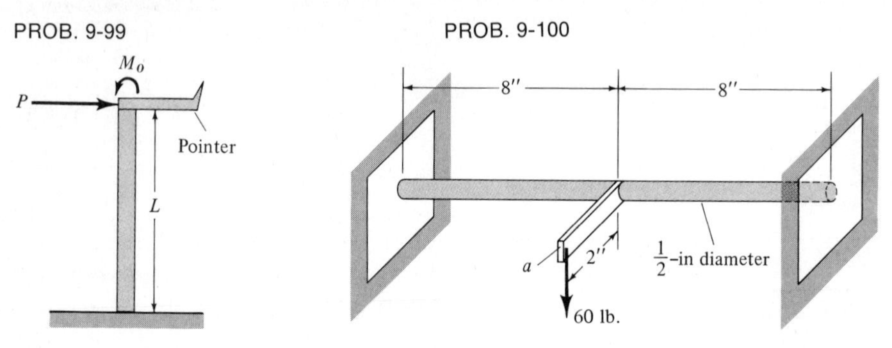

9-100 Find the maximum tensile stress developed in the solid $\frac{1}{2}$-in-diameter circular rod shown. What is the vertical displacement of point a due to the deformation of the steel rod?

9-101 The steel bar has a 1 by $\frac{1}{2}$-in rectangular cross section. Find the maximum tensile stress in the bar and the deflection of the right end.

PROB. 9-101

PROB. 9-102

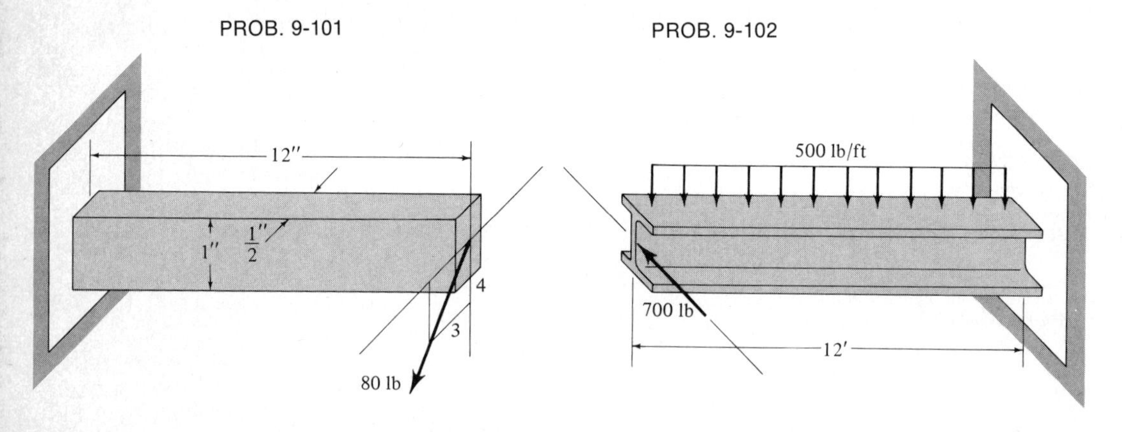

9-102 The beam shown is a W8 × 35 rolled steel section. Determine the maximum tensile stress in the beam. Also, find the deflection of the left end. The indicated loading includes the weight of the beam.

chapter 10 STABILITY

10-1 INTRODUCTION

Throughout this text we have dealt with bodies and structures acted upon by equilibrated force systems. With no information to the contrary, it was implied that the equilibrium configuration was a *stable* one. Essentially, a stable equilibrium situation is one for which a slight change in loading or geometry produces a correspondingly small change in the support reactions and/or configuration of the body or structure; i.e., a small disturbance results in a small response. This need not be the case, as Fig. 10-1 illustrates. The pendulum-like body in Fig. 10-1(a) is hanging in a stable configuration since a small disturbance, as indicated by the dashed lines, results in a small response, and the pendulum tends to return to its original vertical equilibrium configuration. However, the inverted pendulum configuration in Fig. 10-1(b) is *unstable* in that a small disturbance would result in a drastic change in the equilibrium configuration of the body.

While Fig. 10-1(a) and (b) are fairly obvious examples of a stable and unstable situation, respectively, the situation in Fig. 10-1(c) is not so apparent. Whether or not the vertical equilibrium configuration is stable or unstable depends upon the relative values of the weight W and the elastic spring constant k, the elastic spring constant being the force required to stretch or compress the spring one

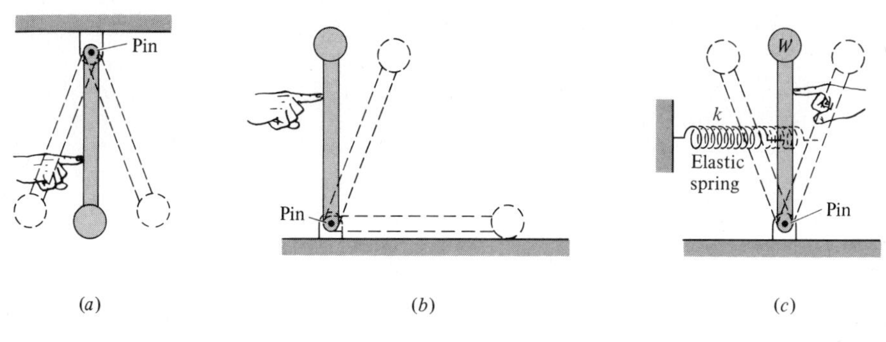

Fig. 10-1

unit of length: i.e., pounds per foot, pounds per inch, etc. If the spring is "strong enough" the configuration will be stable; otherwise, it will be unstable, with the question being how strong is "strong enough"? This is typical of the type of problem which we will consider later in this chapter.

The question of stability can arise in almost any type of structure or machine and quite often can be the criterion as to whether or not the device will or will not properly perform its function. Structural failures due to instability are usually catastrophic, such as the sudden collapse of a building or tower, the loss of control of an automobile on a slippery road, or the buckling of the hull of a submarine. The engineer must be aware of the possibility of this type of failure when he is designing a structure or machine element or when he is simply checking the performance of an existing structure or machine.

10-2 DRY (COULOMB)[1] FRICTION

Probably the most common everyday examples of a stable-unstable situation arise in connection with the frictional effect between two dry contact surfaces, e.g., the contact between the sole of your shoe and the floor, the contact between the automobile tires and the road, and the contact between a shaft and a sleeve-type bushing. In the first two cases the friction is highly desirable, while in the last case it can be considerably detrimental. These situations are variations of the basic problem depicted in Fig. 10-2(*a*). If P is small enough, the box will not slide; if P is big enough, the box will slide. What is the value of P which distinguishes these two situations, i.e., what value of P will cause *motion to impend?* This critical value of P may produce sudden initiation of motion of the box. The same question arises for the problem in Fig. 10-2(*b*), although in this case the box might tip before it slides. In Fig. 10-2(*c*), P might cause the upper box to slide on the lower one, or the stacked boxes might slide together as a unit across the floor.

Intuitively we expect that the limiting value of P is highly dependent upon the nature of the contact surface, e.g., wood on

[1] Charles Augustin Coulomb (1736–1800).

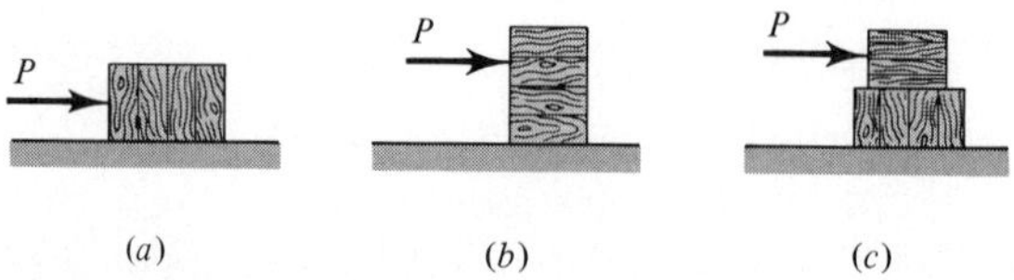

P P P

(*a*) (*b*) (*c*)

Fig. 10-2

wood, cardboard on tile, metal on concrete, etc. This qualitative idea can be made more precise by means of a simple experiment.

In Fig. 10-3(*a*), a block of weight W made of some material is resting upon an inclined plane made of another (or the same) material, where the angle of inclination θ can be increased or decreased as desired. A simplified free-body diagram of the block in Fig. 10-3(*b*) shows the weight vector **W**, the normal force **N**, and the friction force **F**.

> ***Remark:*** Each of the force vectors **W**, **N**, and **F** is the net resultant effect of a distributed force—the weight being distributed over the volume of the block and both the normal force and friction force being distributed over the entire contact surface between the block and the plane, as indicated in Fig. 10-3(*c*).

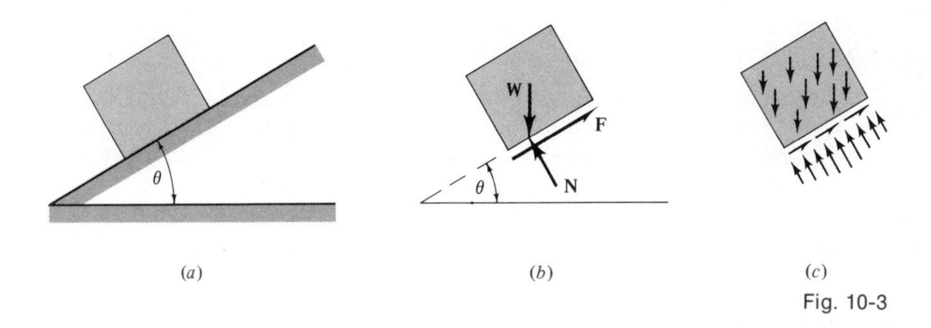

Fig. 10-3

Note that the friction force **F** is indicated to act up the plane in opposition to the *intended* downward motion of the block. Note, also, that if the block is in equilibrium, the three forces **W**, **N**, and **F** must form a concurrent system, so that the location of the net normal force **N** must change as the angle of inclination changes.

Summing forces parallel and perpendicular to the plane yields

$$F - W \sin \theta = 0$$
$$N - W \cos \theta = 0$$

which, when W is eliminated, yields

$$\frac{F}{N} = \tan \theta \tag{10-1}$$

As the angle of inclination is increased, there is a limiting critical value θ_c at which the block will *begin* to slide down the plane. Consequently, by Eq. (10-1), there is a limiting ratio of the value of the friction force F to the value of the normal force N. This limiting ratio

$$\frac{F_c}{N} = \tan \theta_c = f \tag{10-2}$$

is called the coefficient of static friction for the contact surface. Our experience leads us to anticipate that this value will depend upon the relative roughness of the contact surfaces of the block and plane. This phenomenological experiment is the basis for the following theorem:

Theorem 10-1 Theory of Dry Friction The *limiting* value F_c of the friction force between two dry contact surfaces depends directly upon the value of the normal force N between the contacting surfaces. The constant of proportionality f between F_c and N, called the *coefficient of static friction,* depends upon the nature of the contact surfaces but is otherwise independent of the size of the contact area or the manner in which the normal force is produced.

> ***Remark:*** It is important to note that $F_c = fN$ is generally *not* the friction force between the contacting surfaces but is the *upper bound* for the value of that friction force. This is true whether the bodies involved are sliding or at rest.

> ***Remark:*** There are many rational (and irrational) explanations for the validity of Theorem 10-1. We prefer to view this theorem as a useful engineering interpretation of a phenomenological observation.

Assuming that the contacting bodies involved are initially in stable equilibrium, this equilibrium configuration becomes precarious when the friction force (or forces) reaches the critical value F_c. This precarious condition, often referred to as *impending motion,* usually results in a dramatic transition from a static equilibrated situation to an unequilibrated dynamic situation. Once motion ensues, the theory of dynamic (kinetic) friction usually assumes that the sliding friction force is somewhat smaller (approximately 25 percent) than the limiting value of the static friction.

The table in Fig. 10-4 gives some representative values for the coefficient of static friction. Most of these are for dry contact surfaces. The presence of lubrication can significantly alter these values and render questionable the applicability of Theorem 10-1. The theory of lubricated surfaces can usually be found in texts dealing with machine design.

We close this section with a few example problems involving the application of the theorem of dry friction. The types of

VALUES OF THE COEFFICIENT OF LIMITING STATIC FRICTION

Materials	f
Wood on wood	0.25–0.5
Wood on metal	0.2–0.6
Leather on metal	0.6
Brake material on cast iron	0.4
Nylon on nylon	0.5
Nylon on steel	0.05–0.1
Brass or bronze on steel	0.3–0.4
Cast iron on steel	0.4
Rubber on concrete	0.6–0.9
Steel on steel	0.4–0.8

Fig. 10-4

problems which the engineer may encounter are quite diverse, but the guiding principles are basically the same. Above all, any assumptions that are made regarding impending motion or some other condition *must be verified* to ensure the validity of the solution.

EXAMPLE 10-1

A horizontal force P acts on the 200-lb wooden crate resting on the incline as shown in Fig. 10-5(a). Determine the range of values of P for which the crate will be in statical equilibrium. Use $f = 0.3$.

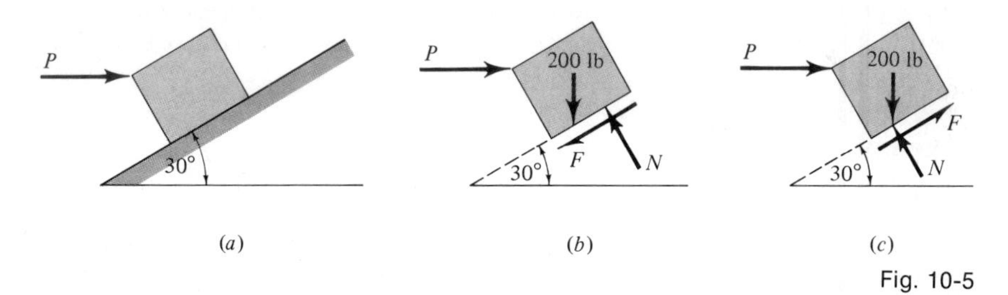

Fig. 10-5

Solution: If the crate tends to move up the plane, its free-body diagram will be as in Fig. 10-5(b), with the friction force F acting down the plane. For this case, summing forces parallel and perpendicular to the plane yields

$$F + 200 \sin 30° - P \cos 30° = 0$$
$$N - 200 \cos 30° - P \sin 30° = 0 \qquad (a)$$

The crate will be in stable equilibrium until F reaches its critical value $F_c = fN$. For this condition Eqs. (a) become

$$0.3N + 100 - 0.866P_c = 0$$
$$N - 173 - 0.5P_c = 0 \qquad (b)$$

where P_c is the critical value of P for motion *impending up the plane*. Hence, solving Eqs. (*b*) yields

$$N = 279 \text{ lb}$$
$$P_c = 211 \text{ lb} \qquad \text{(as shown)}$$

On the other hand, the crate might also tend to slide down the plane, for which case the free-body diagram will be as in Fig. 10-5(*c*). Summing forces as before yields

$$\begin{aligned} F - 200 \sin 30° + P \cos 30° &= 0 \\ N - 200 \cos 30° - P \sin 30° &= 0 \end{aligned} \qquad (c)$$

For *impending motion down* the plane, $F = F_c = fN$. Then, from Eqs. (*c*), the corresponding critical value for P is

$$N = 197 \text{ lb}$$
$$P_c = 47 \text{ lb}$$

Thus, the crate will be in stable statical equilibrium for values of P satisfying

$$47 \text{ lb} < P < 211 \text{ lb}$$

EXAMPLE 10-2

A 260-lb refrigerator is resting on wooden skids and has the dimensions shown in Fig. 10-6(*a*). Assuming the coefficient of friction between the skids and the floor is $f = 0.4$, can the refrigerator be in statical equilibrium if a man applies a horizontal force of 100 lb as indicated in the figure?

Solution: The 100-lb force could possibly cause the refrigerator to slide or tip, and we *must* check both possibilities. A free-body

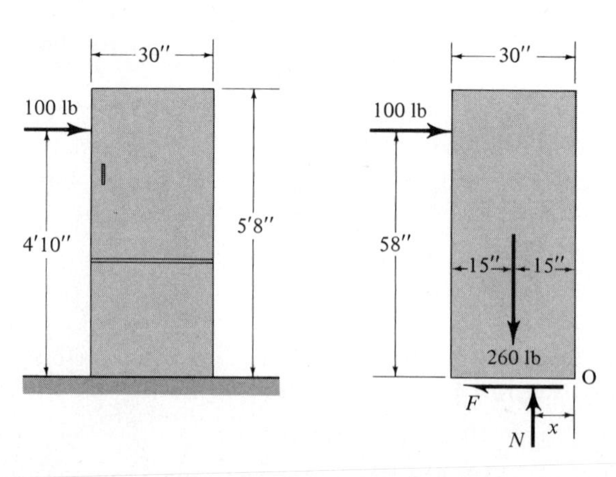

Fig. 10-6

diagram of the refrigerator is shown in Fig. 10-6(*b*). Note the indefinite location of the normal force N. Assuming the refrigerator to be in statical equilibrium, summing vertical and horizontal forces yields

$$260 - N = 0 \qquad N = 260 \text{ lb}$$
$$100 - F = 0 \qquad F = 100 \text{ lb}$$

The critical value for F is

$$F_c = fN = 0.4(260)$$
$$= 104 \text{ lb}$$

Hence $F = 100 < F_c = 104$, so that the refrigerator will *not* slide due to the 100-lb force.

Still assuming equilibrium and summing moments with respect to the lower right corner of the refrigerator yields

$$100(58) - 260(15) + N(x) = 0$$

Substituting the previously obtained value for N, we obtain

$$260x = -5{,}800 + 3{,}900$$
$$x = -7.3 \text{ in}$$

The minus sign means that the normal force N would have to act to the *right* of point o, which is impossible. Hence the refrigerator will tip.

EXAMPLE 10-3

A small steel wedge of neglible weight is used to hold an 800-lb machine up against a wall as shown in Fig. 10-7(*a*). Assuming the coefficient of friction is the same at all contact surfaces, what is its required minimum value for the wedge to stay in place without the aid of an additional force?

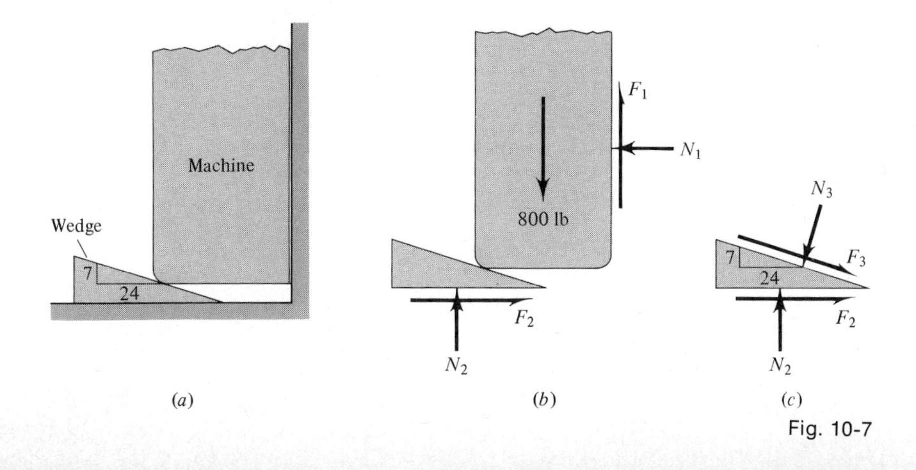

Fig. 10-7

Solution: A free-body diagram of the machine-wedge combination is shown in Fig. 10-7(*b*). Note carefully the directions of both friction forces F_1 and F_2, since the wedge would tend to move to the left as the machine tends to move downward. For equilibrium, summing vertical and horizontal forces yields

$$F_1 + N_2 - 800 = 0$$
$$F_2 - N_1 = 0 \qquad (a)$$

If motion impends, $F_1 = F_{1c} = fN_1$ and $F_2 = F_{2c} = fN_2$, which, when substituted into Eqs. (*a*), yields

$$f^2N_2 + N_2 = 800 \qquad (b)$$

This equation still contains two unknowns, and we must utilize another free-body diagram, such as the one for the wedge shown in Fig. 10-7(*c*). Note the direction of F_3 so as to be consistent with the intended leftward motion of the wedge. Summing forces yields

$$F_2 + \tfrac{24}{25}F_3 - \tfrac{7}{25}N_3 = 0$$
$$N_2 - \tfrac{7}{25}F_3 - \tfrac{24}{25}N_3 = 0 \qquad (c)$$

which, when solved simultaneously for the critical conditions $F_2 = F_{2c} = fN_2$ and $F_3 = F_{3c} = fN_3$, yield

$$f = \tfrac{1}{7}, -7$$

The value of $f = -7$ is discarded as an unacceptable value (recall the experiment used to define f). Hence, the value of $f = \frac{1}{7}$ will correspond to impending motion of the wedge, so that if $f > \frac{1}{7}$, the wedge-machine combination will be in stable equilibrium.

Finally, the complete solution of the problem yields

$$N_1 = 112 \text{ lb} \qquad N_2 = 784 \text{ lb} \qquad N_3 = 784 \text{ lb}$$

all of which are acceptable values. Thus, the solution appears to be correct.

EXERCISE PROBLEMS

10-1 Determine the range of values of the horizontal force P for which the 100-lb box will be in statical equilibrium. The coefficient of friction between the surfaces is 0.6.

10-2 Determine the range of values of the force P for which the 200-lb crate will be in statical equilibrium. The coefficient of static friction is 0.3.

10-3 Same as Prob. 10-2 except $f = 0.6$.

PROB. 10-1

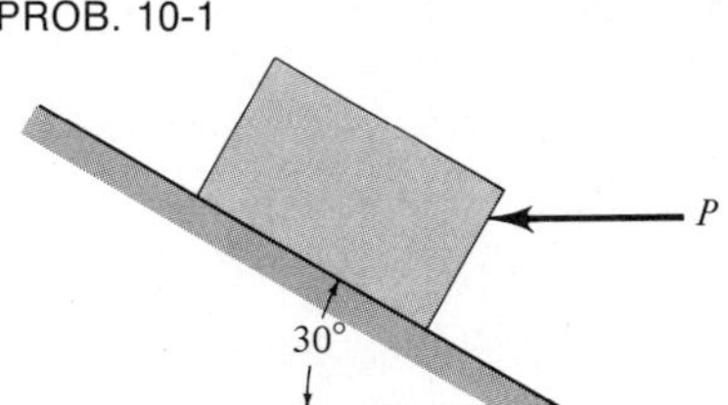

PROBS. 10-2 and 10-3

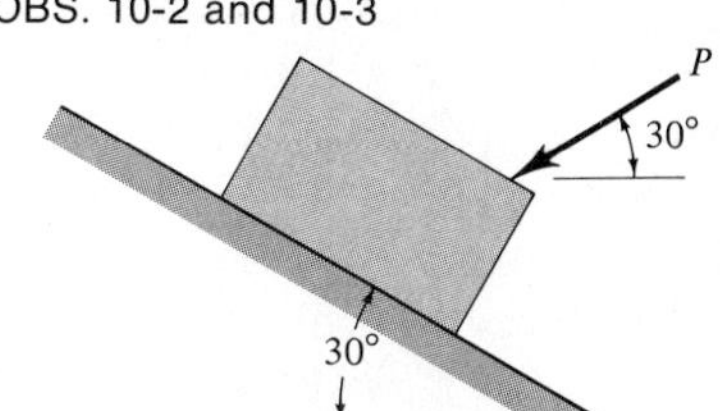

10-4 Determine the range of values of the force P for which the 250-lb crate can be held in statical equilibrium. The coefficient of friction between the surfaces is 0.3.

10-5 Same as Prob. 10-4 except $f = 0.6$.

PROBS. 10-4 and 10-5

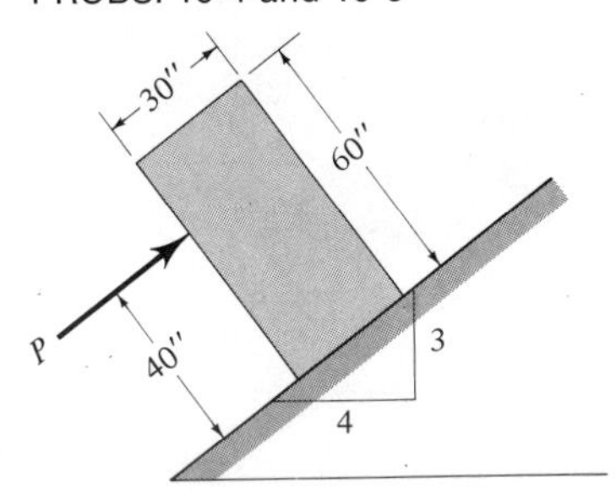

PROBS. 10-6 and 10-7

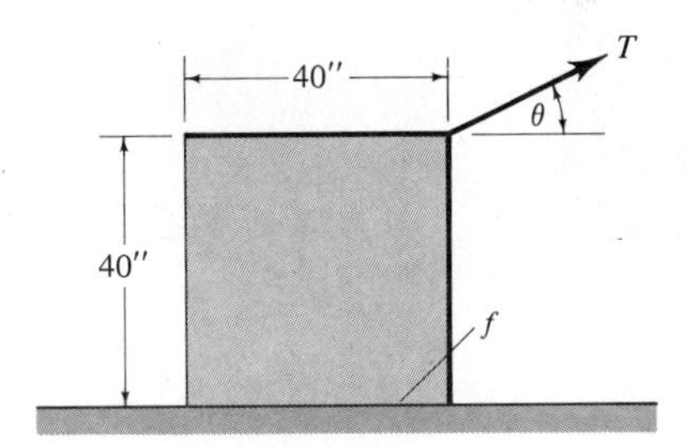

10-6 What minimum value of the tension T in the cable will cause the 150-lb box to move if (*a*) $\theta = 15°$, $f = 0.4$; and (*b*) $\theta = 15°$, $f = 0.6$?

10-7 Same as Prob. 10-6 except (*a*) $\theta = 75°$, $f = 0.4$; and (*b*) $\theta = 75°$, $f = 0.7$.

10-8 The large cabinet weighs 300 lb and rests on four casters which offer a rolling resistance equal to an effective coefficient of friction of $f = 0.25$. Find the value of P which will cause motion of the cabinet to impend. What is the load on each caster?

10-9 Same as Prob. 10-8 except the effective coefficient of friction is $f = 0.4$.

10-10 The coefficient of static friction between the 160-lb crate and the floor is 0.35. Determine the value of the force P and the distance h which will cause the crate to tip and slip simultaneously.

10-11 Same as Prob. 10-10 except $f = 0.3$.

PROBS. 10-8 and 10-9

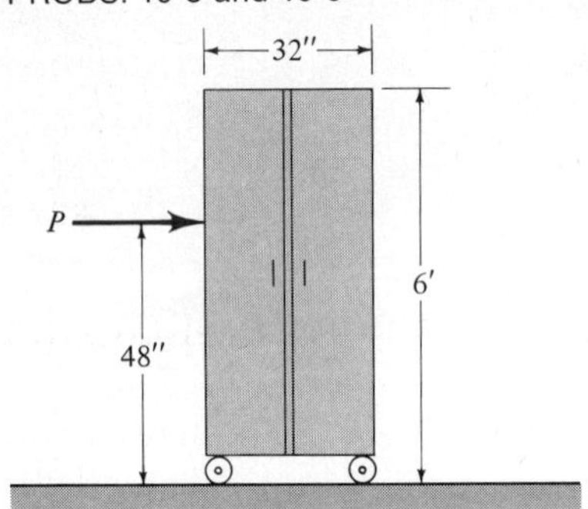

PROBS. 10-10 and 10-11

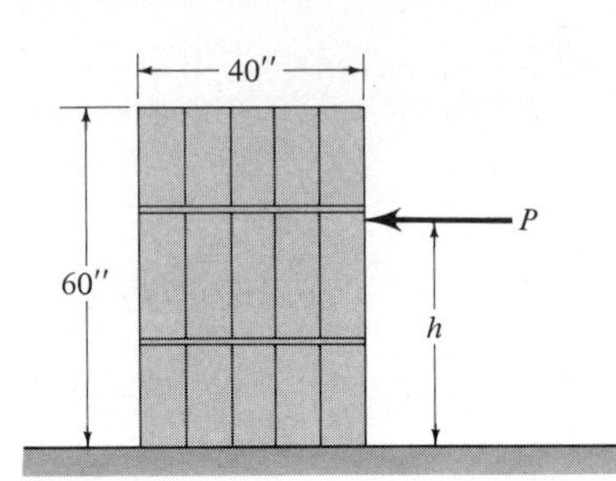

10-12 Can the 100-lb box shown be in statical equilibrium if P has a value of 150 lb? Use $f = 0.5$.

PROB. 10-12

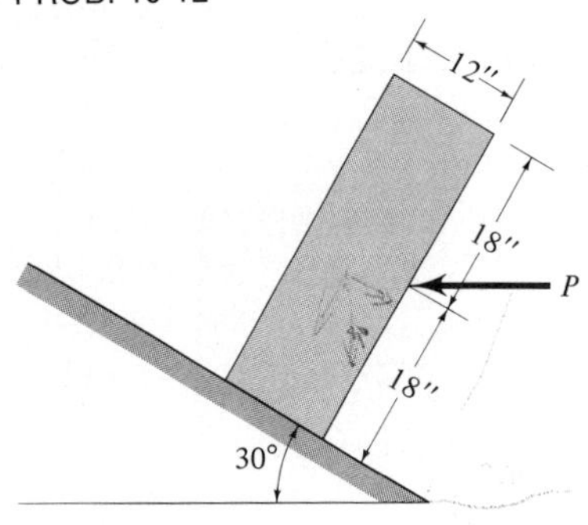

PROB. 10-13

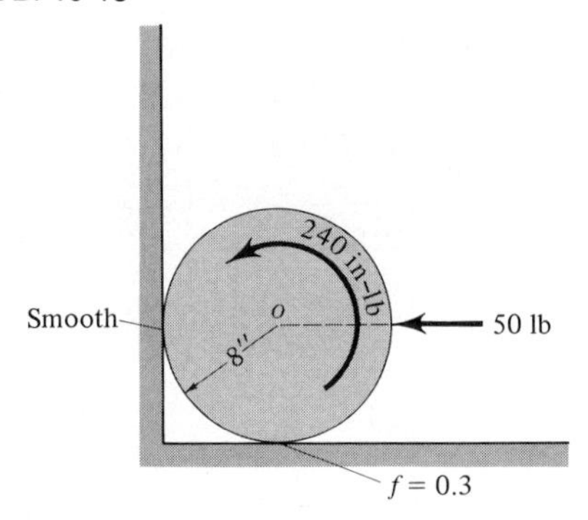

10-13 The 200-lb cylindrical disk is subjected to a 240-in-lb couple and a 50-lb horizontal force as shown. Is the disk in statical equilibrium? Find all unknown forces.

10-14 The automobile shown weighs 3,500 lb and the coefficient of friction between the tires and road is $f = 0.75$. What is the maximum angle θ of the incline before the driving wheels start to slip if (a) the car has rear-wheel drive, (b) the car has front-wheel drive, and (c) the car has four-wheel drive?

PROB. 10-14

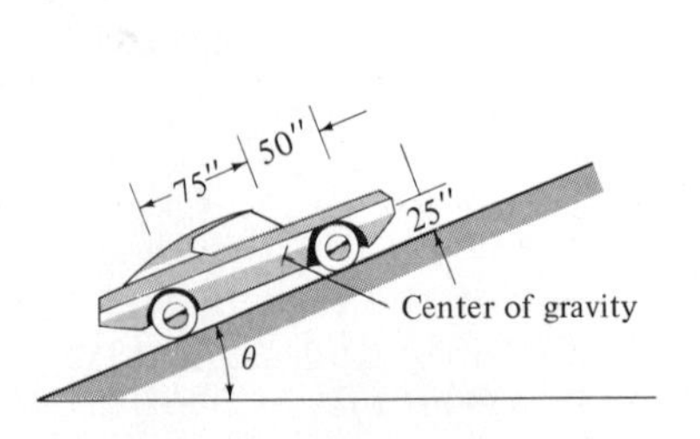

PROB. 10-15

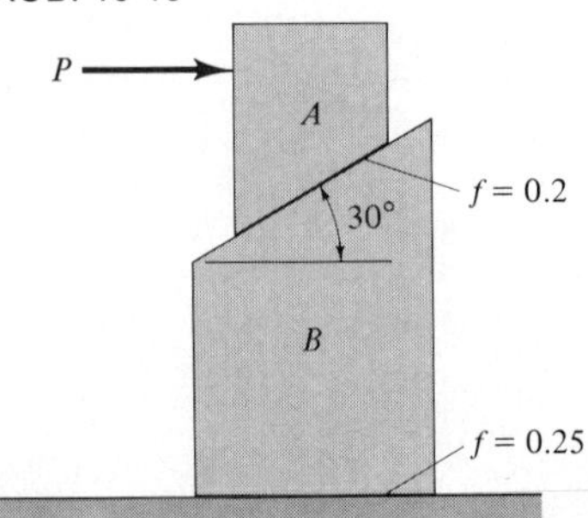

10-15 Block A weighs 50 lb and B weighs 100 lb. Determine the minimum value of P necessary to move A to the right.

10-16 The homogeneous cylinder weighs 100 lb and the weight B is 50 lb. If the system is in statical equilibrium, determine the angle θ and the necessary minimum value for the coefficient of friction between the cylinder and plane.

PROB. 10-16

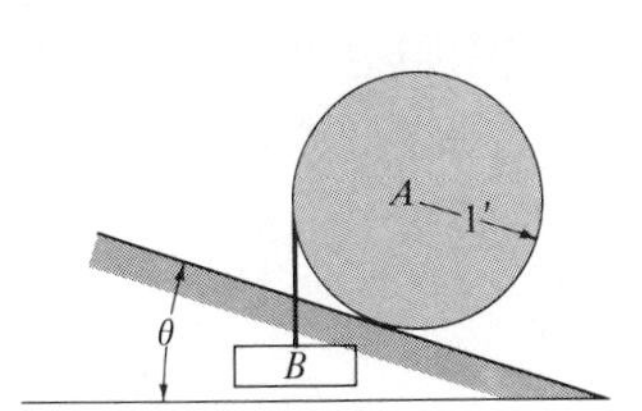

PROB. 10-17

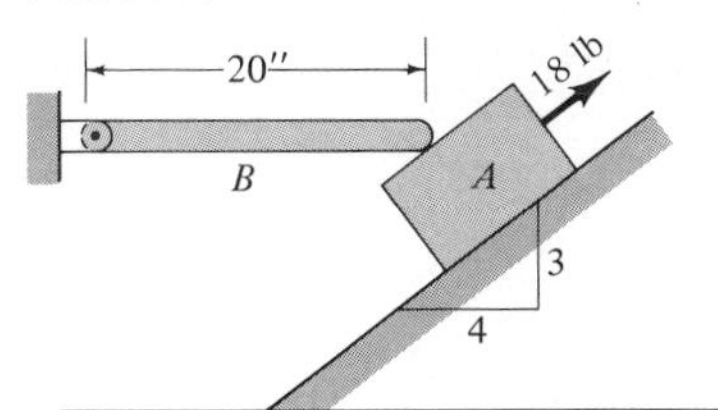

10-17 Block A weighs 20 lb and the homogeneous bar B weighs 25 lb. The friction between the block and plane can be considered negligible, but that between the bar and block has an effective coefficient of $f = 0.6$. Is the system in statical equilibrium?

10-18 The bar ab weighs 500 lb, bc weighs 200 lb, and block D weighs 50 lb. Determine the largest force P which can be applied without moving bar bc to the right.

PROB. 10-18

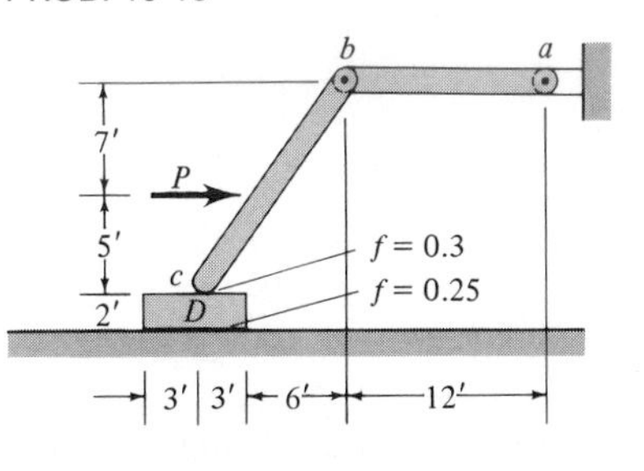

PROB. 10-19

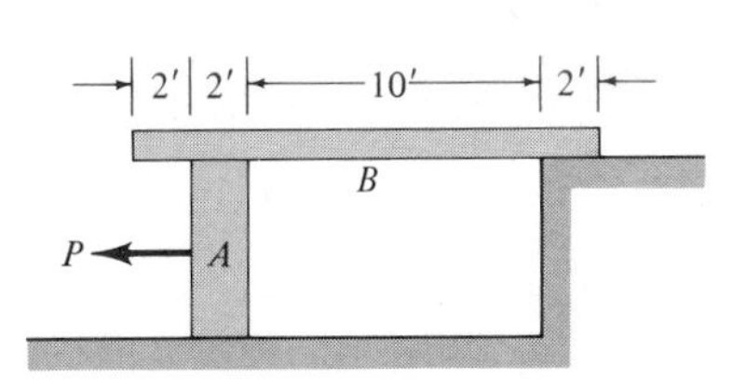

10-19 The 1,600-lb steel beam B is supported by a movable 100-lb wooden block A. What minimum force P is necessary to move the block? Assume $f = 0.3$ at all contact surfaces.

10-20 For the brake-drum mechanism shown, the coefficient of friction between the drum and brake shoe is 0.5. Would a force P of 125 lb be sufficient to hold the mechanism in equilibrium?

10-21 The ladder and man weigh 40 lb and 160 lb, respectively. The coefficient of friction of the wall is 0.2 and at the ground is 0.6. What is the smallest allowable value for the angle θ for statical equilibrium?

PROB. 10-20

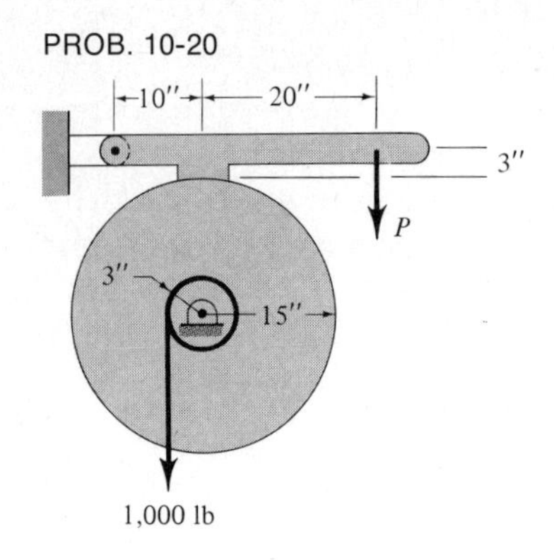

PROB. 10-21

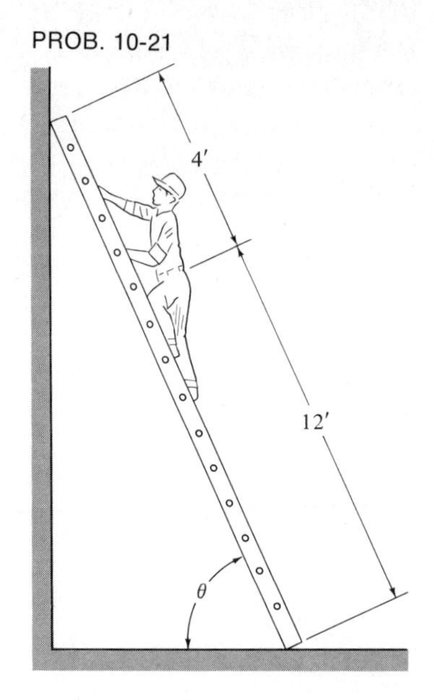

10-22 What value of the force P parallel to the edge of the roof will cause impending motion of the 80-lb block resting on the roof? The coefficient of friction is 0.6.

PROB. 10-22

PROB. 10-23

10-23 What minimum coefficient of friction between the shaft and sleeve is needed to keep the device from sliding downward? Assume contact is made at points a and b.

10-24 A small wedge is being used to raise a 1,000-lb machine. If the coefficient of friction at all contact surfaces is 0.3, what minimum force P is needed to move the wedge?

10-25 Can the 300-lb cylindrical disk shown be held in statical equilibrium by cable tension T if (a) the cable is wrapped as indicated by the solid line and (b) the cable is wrapped as indicated by the dashed line? The coefficient of friction between the cylinder and plane is 0.3.

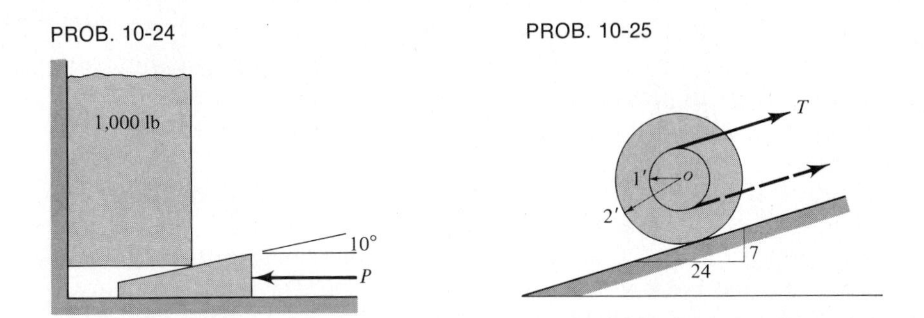

10-26 The roller-type belt buckle is supposed to be self-locking regardless of the tension. What is the minimum required coefficient of friction at each of the contact surfaces?

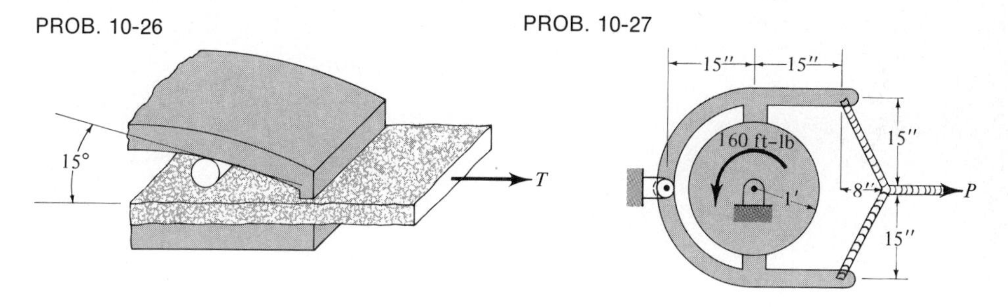

10-27 What minimum force P is required to keep the drum from turning? The coefficient of friction between the drum and the brake shoe is 0.4.

10-3 BELT AND DISK FRICTION

A useful application of the theory of dry friction deals with the frictional effect on ropes, belts, cables, etc., wrapped around pulleys, drums, etc. Experience tells us that if the box in Fig. 10-8 weighs, say, 100 lb, it will require a force F larger than 100 lb to raise the

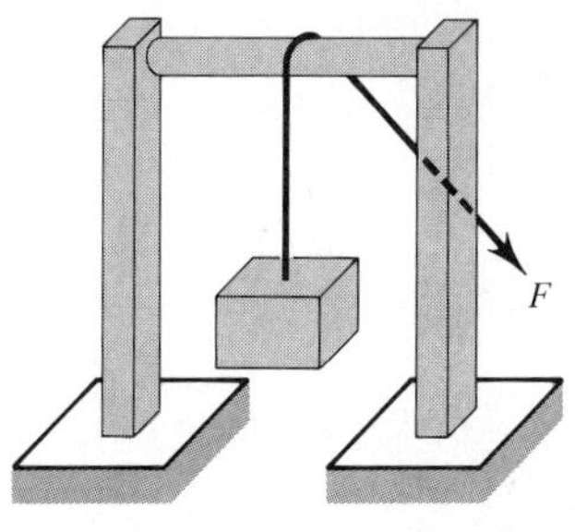

Fig. 10-8

weight due to the resistance of friction as the rope passes over the crossbeam. Conversely, it will require less than 100 lb to merely hold the weight in place due to the aid of the friction as the weight tends to drop. We will now make this qualitative observation more precise.

Figure 10-9(*a*) shows a free-body diagram of that portion of a rope, belt, or cable passing over a, say, fixed drum. The figure shows the normal and friction force on the belt distributed over the contact surface between the belt and drum. For convenience we assume that the tensions T_1 and T_2 satisfy $T_1 > T_2$, so that the belt is tending to move counterclockwise, with the *distributed* friction force tending to resist this intended motion. In this figure the total angle of contact between the belt and the drum is denoted by β. If T_1 becomes sufficiently larger than T_2, the belt will eventually slip on the drum. Our objective is to determine the limiting ratio between T_1 and T_2 for which motion will impend.

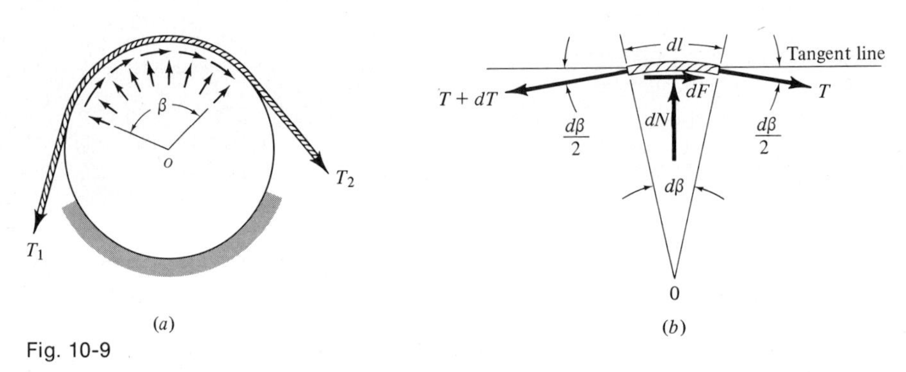

Fig. 10-9

Figure 10-9(*b*) shows a free-body diagram of an arbitrarily small portion $d\ell$ of the belt. Assuming equilibrium and summing forces parallel and perpendicular to the tangent line yields

$$T \cos \frac{d\beta}{2} + dF - (T + dT) \cos \frac{d\beta}{2} = 0 \qquad (10\text{-}3)$$

$$dN - T \sin \frac{d\beta}{2} - (T + dT) \sin \frac{d\beta}{2} = 0 \qquad (10\text{-}4)$$

Also, for *impending motion*,

$$dF = dF_c = f\, dN \qquad (10\text{-}5)$$

where f is the coefficient of static friction between the belt and drum. Then, as $d\beta \to 0$, Eqs. (10-3) and (10-4) become

$$\frac{dF}{d\beta} = \frac{dT}{d\beta}$$

$$\frac{dN}{d\beta} = T$$

which, when combined with Eq. (10-5) for impending motion, yields

$$\frac{dT}{d\beta} = fT$$

or

$$\frac{dT}{T} = f\,d\beta \qquad (10\text{-}6)$$

This expression gives the differential variation of the tension T with the angle β for impending motion. Integrating this relation yields

$$\int_{T_2}^{T_1} \frac{dT}{T} = \int_0^{\beta} f\,d\beta$$

$$\ln \frac{T_1}{T_2} = f\beta \qquad (10\text{-}7)$$

or

$$\frac{T_1}{T_2} = e^{f\beta} \qquad (10\text{-}8)$$

This is the *limiting* ratio of the tensions for impending motion. If T_1/T_2 is less than $e^{f\beta}$, the belt will *not* slip. For this expression, β is the total angle of contact measured in radians, and its value can be larger than 2π if the belt is wrapped completely around the drum or pulley. Note that if either f or β is zero, $T_1 = T_2$, as would be anticipated.

Remark: Although the drum in Fig. 10-9(*a*) was circular, the ensuing derivation of Eq. (10-8) did not utilize this fact. Hence, the final result, Eq. (10-8), is valid for any reasonably shaped drum or surface over which a belt or rope might slide so long as the angle of contact β can be well defined.

Remark: The above derivation neglected the weight of the belt or cable itself. In most statical problems this is quite reasonable. However, in a dynamical problem in which the belt might be moving at a relatively high speed, the centrifugal effects on the belt due to its mass might be rather significant and should be included in the analysis.

Disk Friction

Another interesting application of the theory of dry friction involves the transmission of torque between two flat disks as indicated in Fig. 10-10(*a*). Essentially the question is: What is the relationship between the axial pushing force P and the maximum *torque* T before slipping occurs between the disks? Figure 10-10(*b*) illustrates the basic mechanism by which the torque is transmitted across the disks. The bearing force (normal force) dN on the area

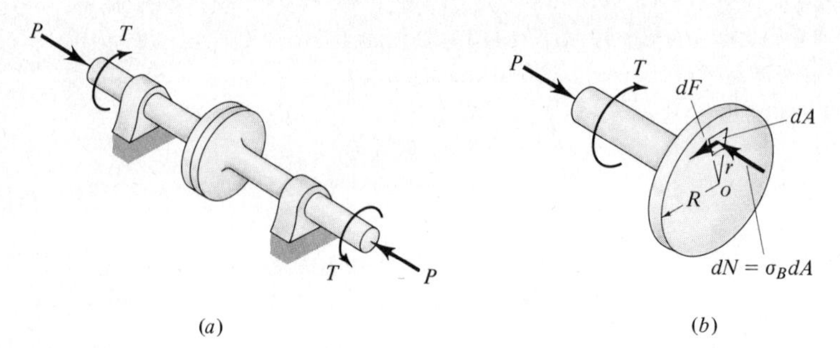

Fig. 10-10

dA produces a friction force dF, which in turn produces the torque on the shaft. *For impending motion,*

$$dF = dF_c = f\,dN \tag{10-9}$$

and

$$dN = \sigma_B\,dA \tag{10-10}$$

where σ_B is the bearing pressure (per unit area) between the disks caused by the axial compression force P. Thus, combining Eqs. (10-9) and (10-10), the maximum transmitted torque is given by

$$\begin{aligned} T &= \int r\,dF \\ &= \iint_{\text{area}} rf\sigma_B\,dA \end{aligned} \tag{10-11}$$

It is usually assumed that the coefficient of friction f and the bearing pressure σ_B are uniform over the contact surface. Then

$$T = f\sigma_B \iint_{\text{area}} r\,dA$$

which, when integrated over a *solid* disk of radius R, becomes

$$T = f\sigma_B \frac{2\pi R^3}{3} \tag{10-12}$$

Finally, with the *average* bearing pressure given by

$$\sigma_B = \frac{P}{A} = \frac{P}{\pi R^2}$$

Eq. (10-12) takes the form

$$T = \tfrac{2}{3} f\,PR \tag{10-13}$$

This expression gives the maximum torque T which can be transmitted across the solid disks *before slipping*.

Before looking at an illustrative example, a word of caution is in order. Both Eq. (10-8) for belt friction and Eq. (10-13) for disk friction give *limiting values* for the quantities involved. In general, the frictional effects on the tensions and torques are merely adequate to maintain equilibrium. However, if the limits given by Eqs. (10-8) and (10-13) are reached, then the equilibrium becomes unstable, and a dynamical situation may suddenly develop.

EXAMPLE 10-4

Find the minimum force P in Fig. 10-11(a) which will hold the 400-lb weight in statical equilibrium. The coefficient of friction between the brake drum and band is 0.4.

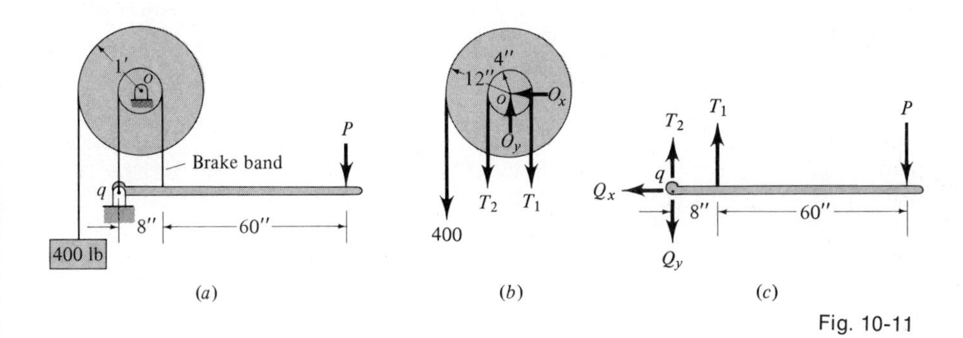

Fig. 10-11

Solution: A free-body diagram of the drum is shown in Fig. 10-11(b) with the weight, the band tensions T_1 and T_2, and the assumed smooth pin reactions at o. Assuming equilibrium and summing moments with respect to point o yields

$$400(12) + T_2(4) - T_1(4) = 0$$

$$T_1 - T_2 = 1{,}200 \qquad (a)$$

For impending counterclockwise motion of the drum, by Eq. (10-8),

$$\frac{T_1}{T_2} = e^{f\beta} = e^{0.4\pi} = 3.5 \qquad (b)$$

since the angle of contact is 180°. Substituting Eq. (b) into (a) gives

$$T_1 = 1{,}680 \text{ lb} \qquad \text{and} \qquad T_2 = 480 \text{ lb}$$

Summing moments on the free-body diagram of the lever shown in Fig. 10-11(c), we have

$$T_1(8) - P(68) = 0$$

$$P = 198 \text{ lb}$$

as the minimum force required to hold the weight in statical equilibrium.

EXERCISE PROBLEMS

10-28 and 10-29 A rope is wrapped around a stake as shown. If the coefficient of friction is 0.2, what is the ratio of the tensions T_1 and T_2 for impending motion?

PROB. 10-28 PROB. 10-29

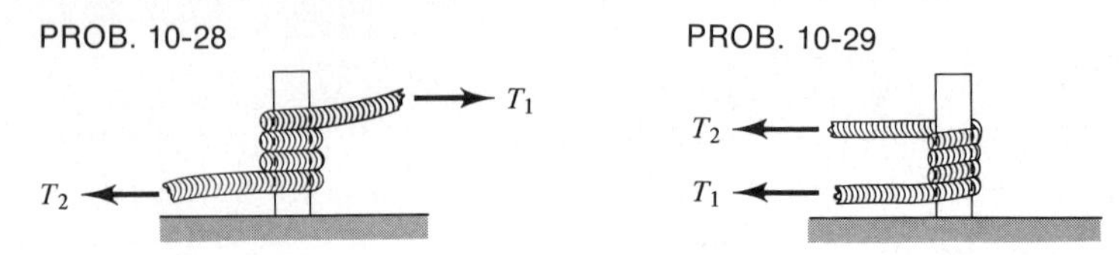

10-30 Determine the range of values of P for which the cable shown is in equilibrium.

PROB. 10-30

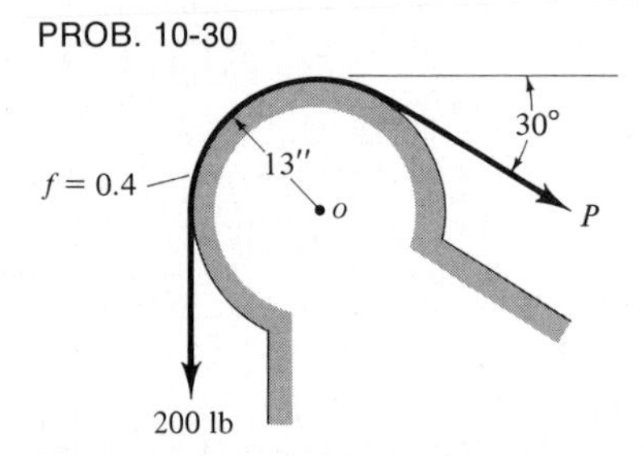

10-31 A 100-ft-long heavy rope hangs in two equal lengths from a large tree bough 40 ft high. How many times should the rope be wrapped around the bough if a 180-lb man wishes to climb one of the ends of the rope up to the bough? The rope weighs 0.15 lb/ft and the coefficient of static friction is 0.55.

10-32 The cable is wrapped once around the pulley as shown. If a 200-in-lb couple acts on the pulley, is the system in equilibrium if (*a*) $f = 0.35$ and (*b*) $f = 0.25$?

PROB. 10-32 PROB. 10-33

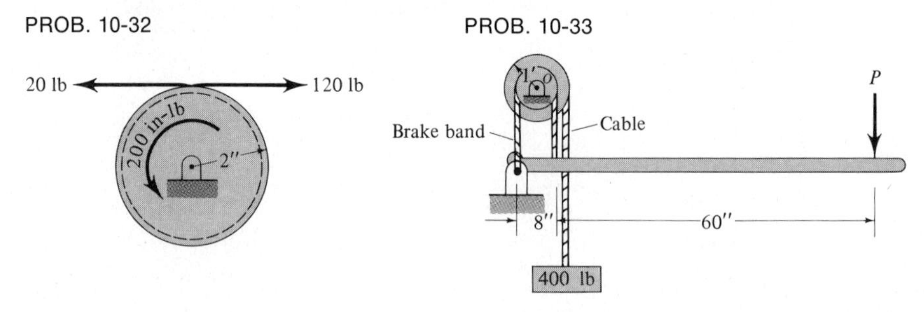

10-33 For the brake-drum configuration shown, find the minimum force P required to hold the weight in statical equilibrium. Use $f = 0.4$. Compare the solution with Example 10-4. Why is this a better braking configuration than that of Example 10-4?

10-34 For the brake-drum configuration shown, find the minimum required tension T to hold the weight in equilibrium. Use $f = 0.4$.

10-35 Same as Prob. 10-34 except the weight is hanging from the left side of the 6-in-radius pulley rather than the right side.

PROBS. 10-34 and 10-35

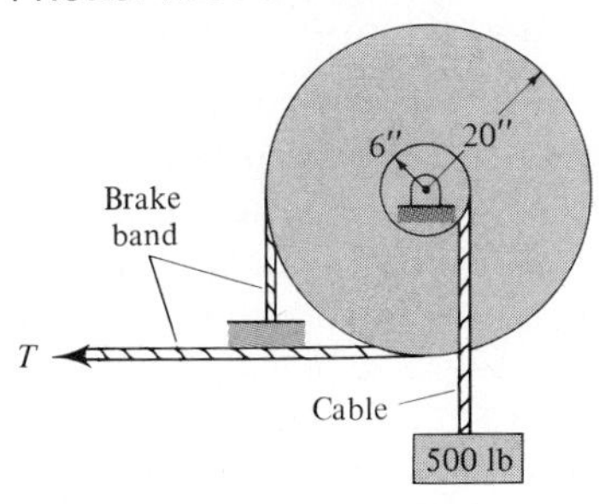

10-36 and 10-37 Determine the smallest force P required to move the 500-lb machine.

PROB. 10-36

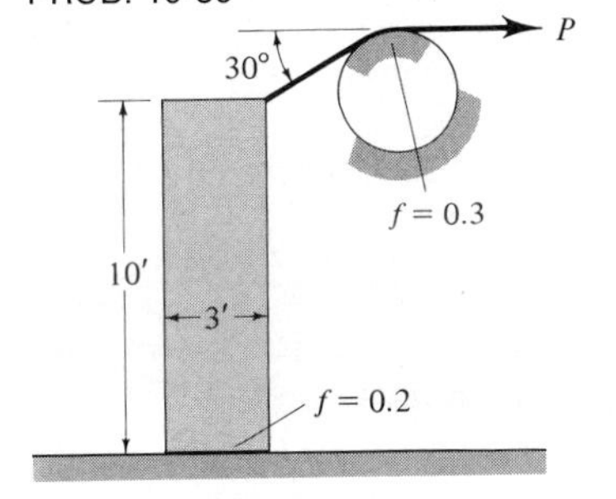

PROB. 10-37

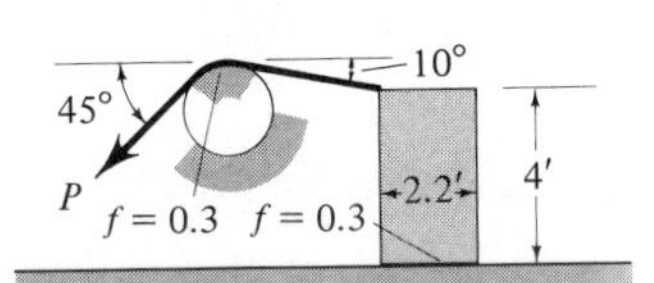

10-38 In the spring-loaded brake-shoe mechanism shown, each spring can exert a maximum spreading force of P lb on the shoes. Assume that the normal pressure between the shoes and brake lining is uniform around the entire contact surface. Derive a relation between the force P and the maximum braking torque which the brake can develop between the housing and the shoes. Use $f = 0.6$.

PROB. 10-38

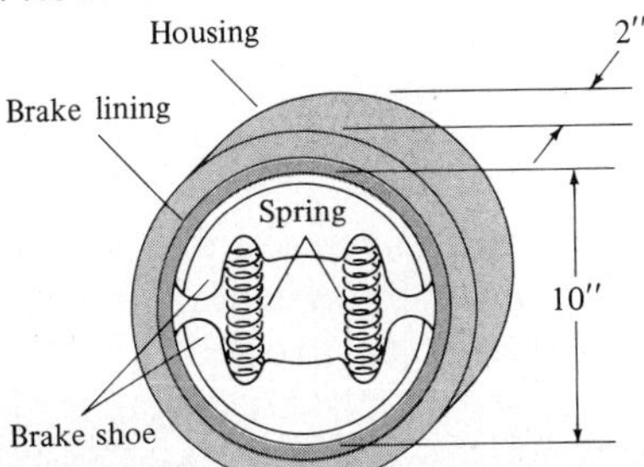

10-39 Rederive an equation similar to Eq. (10-13) for an annular disk with an inside radius of R_i and an outside radius of R_o.

10-40 A solid 9-in-diameter disk-type clutch is lined with a material whose coefficient of static friction is 0.7. What is the ratio of the maximum torque capacity of the clutch to the pushing clutch force P?

10-41 Same as Prob. 10-40 except the clutch disks are annular with an inside diameter of 3 in.

10-42 A flat leather belt passes over an elliptical-shaped cam as shown. If the coefficient of static friction is 0.6, for what values of T will slipping impend?

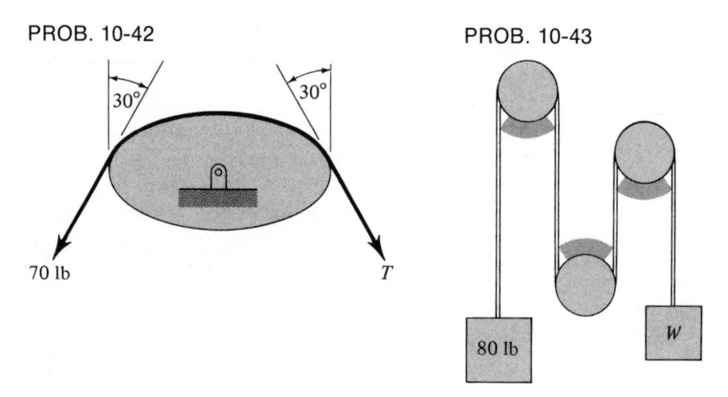

10-43 Determine the range of values of W for which the system shown will be in statical equilibrium. The coefficient of friction between the belt and stationary spools is 0.6.

10-44 The pulley-belt arrangement shown has a coefficient of friction of 0.7. What is the largest torque this system can transmit if the maximum allowable belt tension is 140 lb?

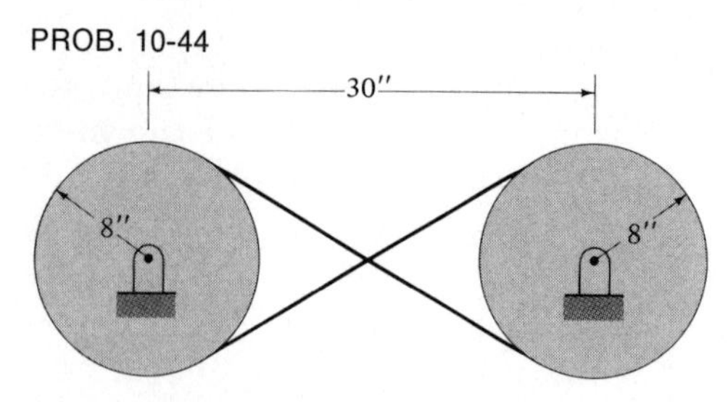

10-45 Body B weighs 1,200 lb and C weighs 680 lb. The coefficient of friction is $2/\pi$ between the rope and the fixed drum, 0.40 between A and B, and 0.30 between B and the ground. Determine the minimum weight A necessary to prevent C from moving downward.

PROB. 10-45

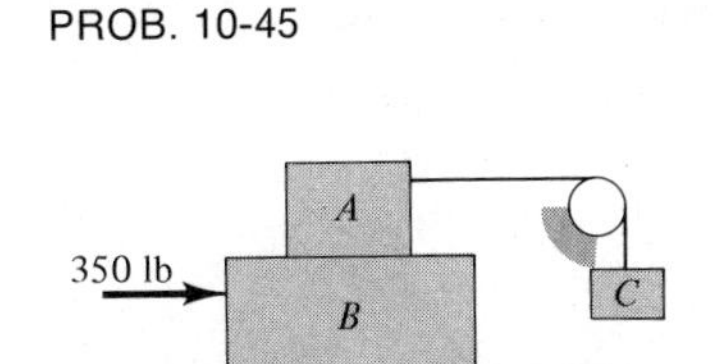

PROB. 10-46

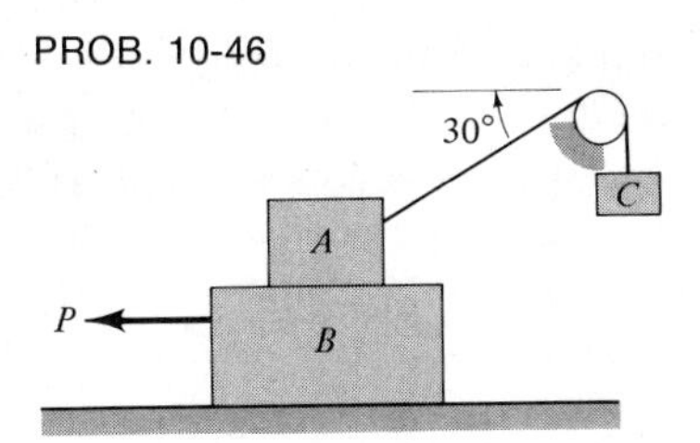

10-46 Block A weighs 100 lb, B weighs 200 lb, and C weighs 25 lb. The coefficient of friction between A and B and between B and the plane is 0.30, and that between the cord and the fixed drum is 0.12. Find the minimum force P necessary to move block B.

10-4 DISCRETE ELASTIC SYSTEMS

We now turn to the type of stability problem similar to the pendulum problems previously indicated in Fig. 10-1. Consider the device shown in Fig. 10-12(a), which consists of a rather substantial weight W mounted atop a long slender deformable elastic column embedded in the ground. Although this type of structure will be dealt with in the next article, suppose, for the time being, we idealize this structure by either of the schemes shown in Fig. 10-12(b) and (c). In Fig. 10-12(b) the rod is considered to be weightless, rigid, and pivoted with linear elastic springs attached to the rod. In Fig. 10-12(c) the rod is again considered to be weightless and rigid, and the pivot has a linear torsion spring which exerts a resistance to rotation proportional to the angular motion. How well either of these idealizations represents the original system depends upon suitable choices of the spring constants k and m, the distance d, and the effective weights W' and W''.

Each of these systems in Fig. 10-12(a) to (c) can be disturbed from its vertical equilibrium position as indicated by the dashed

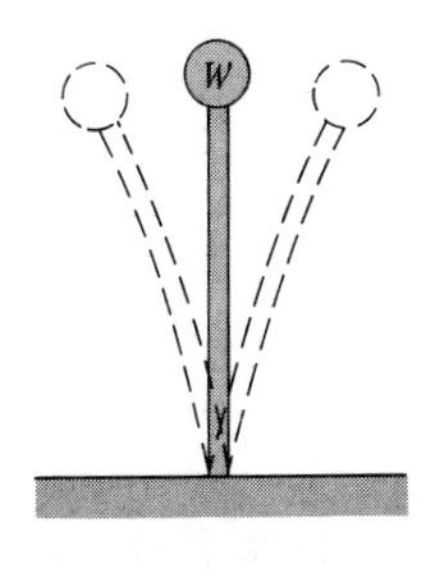

(a)

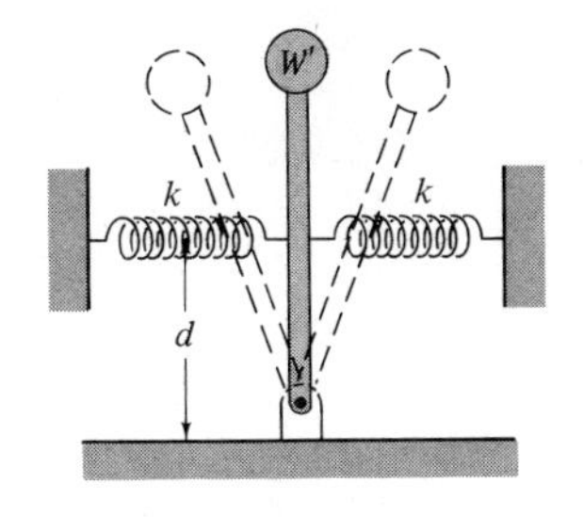

(b)

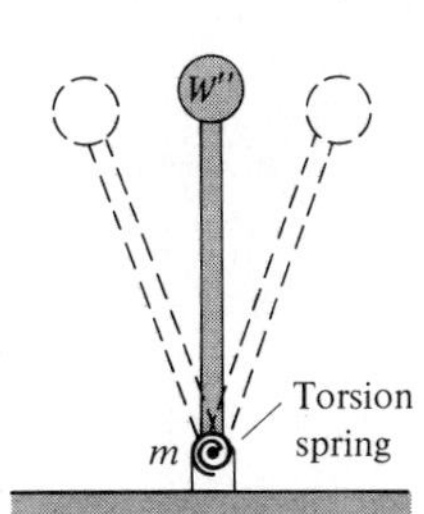

(c)

Fig. 10-12

lines. In each case we can pose the question: Will the system tend to return to its original vertical equilibrium position; i.e., is the vertical equilibrium configuration a *stable* one?

The column structure in Fig. 10-12(a) is typical of a *continuous elastic system*, while those in Fig. 10-12(b) and (c) are called *discrete elastic systems* because the elastic restoring forces (or couples) are concentrated at discrete locations rather than distributed throughout the system as in Fig. 10-12(a). In this section we will examine the stability of some discrete systems.

Consider the discrete system depicted in Fig. 10-13(a) composed of two linear elastic springs with spring constants k attached to a horizontal guide bar, and a vertical slotted rod pivoted at o and acted upon by a downward vertical force P. For convenience we will assume all frictional effects to be negligible. Figure 10-13(b) shows the rod disturbed some slight arbitrary amount θ from its vertical equilibrium configuration, and Fig. 10-13(c) shows the corresponding free-body diagram of the rod with the horizontal spring force F_s and the guide force F_g indicated as components of the resultant force R perpendicular to the smooth slot. In this case the springs have been deformed by an amount $d = (L/2) \tan \theta$, so that the total spring force F_s is given by

$$F_s = 2kd = kL \tan \theta$$

and, therefore,

$$R = \frac{F_s}{\cos \theta} = kL \frac{\sin \theta}{\cos^2 \theta} \tag{10-14}$$

Physically the force R tends to restore the rod to its original vertical position, while the force P tends to increase the disturbance as measured by angle θ. The question is: Which of these tendencies is the dominant one? To answer this question, we can sum moments

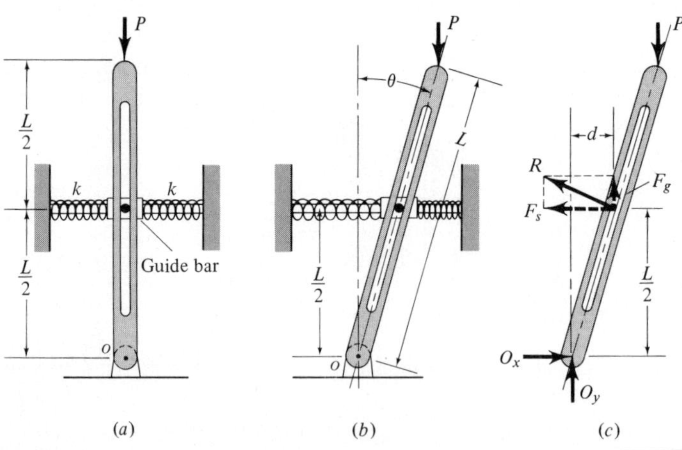

Fig. 10-13

with respect to point o and determine whether the resultant moment is in the same direction as θ (clockwise, in this case) or opposite to θ. Thus, taking clockwise moments to be positive, the resultant moment with respect to point o is

$$M_o = PL \sin \theta - R \frac{L/2}{\cos \theta}$$

To determine if this moment is in the same direction as θ, we can examine the equation

$$PL \sin \theta - R \frac{L/2}{\cos \theta} = \lambda \theta \tag{10-15}$$

where λ is some multiplier. If $\lambda > 0$, the resultant moment is in the same direction as θ and, therefore, tends to increase θ. If $\lambda < 0$, the resultant moment is opposite to θ and tends to decrease θ and restore the rod to the vertical position. Hence we conclude

$$\begin{aligned} \lambda > 0 \quad & \text{implies instability} \\ \lambda < 0 \quad & \text{implies stability} \end{aligned} \tag{10-16}$$

Remark: The case $\lambda = 0$ means that there is no resultant moment, so that the bar is in rotational equilibrium for this particular value of θ, and no judgment can be made regarding stability.

Substituting Eq. (10-14) into (10-15) yields

$$PL \sin \theta - \frac{kL^2}{2} \frac{\sin \theta}{\cos^3 \theta} = \lambda \theta$$

or

$$\frac{\sin \theta}{\theta} L \left(P - \frac{kL}{2} \frac{1}{\cos^3 \theta} \right) = \lambda \tag{10-17}$$

Now

$$0 < \frac{\sin \theta}{\theta} \leqslant 1 \qquad \text{for} \qquad -\frac{\pi}{2} \leqslant \theta \leqslant \frac{\pi}{2}$$

Hence, the algebraic sign of λ is governed by the algebraic sign of the term

$$P - \frac{kL}{2} \frac{1}{\cos^3 \theta}$$

Thus we have, by the inequalities (10-16),

$P - \dfrac{kL}{2 \cos^3 \theta} > 0$ implies instability, i.e., the disturbance tends to increase

$P - \dfrac{kL}{2 \cos^3 \theta} < 0$ implies stability, i.e., the rod tends to return to its vertical equilibrium position

Rewriting these inequalities

$$\cos^3 \theta > \frac{kL}{2P} \qquad \text{implies instability}$$

$$\cos^3 \theta < \frac{kL}{2P} \qquad \text{implies stability}$$

Now

$$0 < \cos \theta \leq 1 \qquad \text{for} \qquad -\frac{\pi}{2} < \theta < \frac{\pi}{2} \qquad (10\text{-}18)$$

so that the second of the inequalities (10-18) means that

$$1 < \frac{kL}{2P} \qquad \text{implies stability}$$

On the other hand, since θ can be *any* arbitrary disturbance, including very small ones, the first of inequalities (10-18) implies that

$$1 > \frac{kL}{2P} \qquad \text{implies instability}$$

Thus, for this particular system, we see that when $P < kL/2$, the vertical equilibrium configuration will be a stable one, while for $P > kL/2$, the vertical configuration will be unstable. The critical value of the load P is

$$P_c = kL/2$$

which is the *limiting value* of the load P for which the vertical equilibrium configuration will be stable. That is,

$$\text{stable } P < P_c < P \text{ unstable}$$

In the special case for which $P = P_c$, the equilibrium configuration is usually referred to as *neutral.*

This same sort of analysis can be applied to most discrete systems to ascertain the stability of a particular equilibrium configuration. The equilibrium configuration need not be a vertical one, such as the case just considered. The main idea is to determine whether or not the forces and/or moments tend to reduce or enlarge a small disturbance.

EXERCISE PROBLEMS

10-47 to 10-49 The systems shown are composed of rods which are considered to be rigid and weightless and linear torsion springs which have a modulus of m in-lb of torque per radian of angular distortion. For what range of values of P will the vertical position be a stable one? Note that $\theta/\sin\theta \geqslant 1$ for $\pi/2 > \theta \geqslant 0$.

PROBS. 10-47 and 10-51

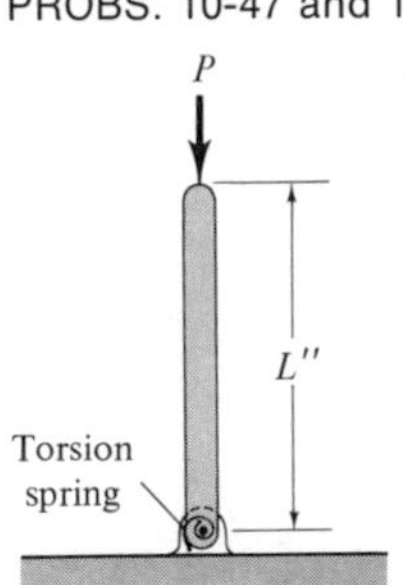

PROBS. 10-48 and 10-52

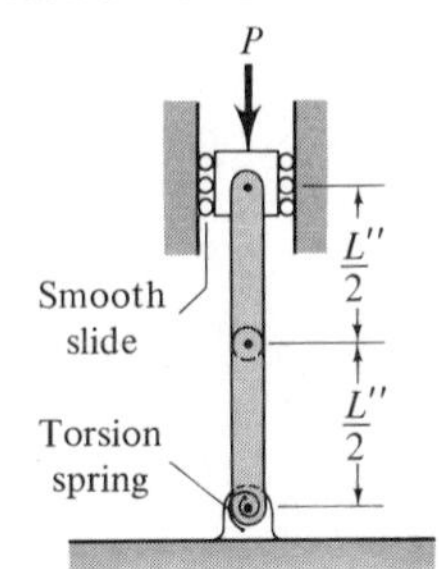

PROBS. 10-49 and 10-53

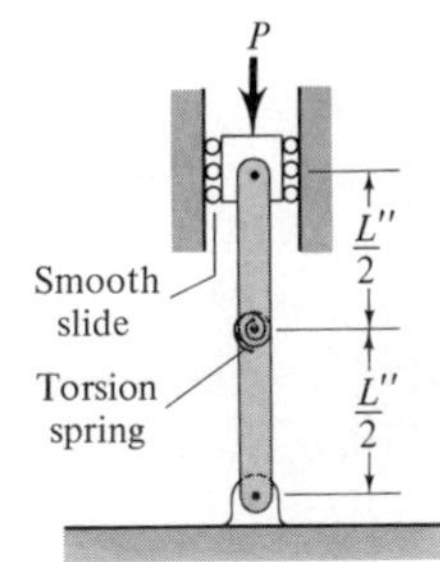

10-50 Consider the mechanism previously shown in Fig. 10-13(a). If the rod is considered rigid but not weightless, what is the value for the critical load P_c? Assume the rod is of weight W and the slot is in the middle of the rod.

10-51 to 10-53 Rework Probs. 10-47 to 10-49 except consider each of the rods to be rigid and homogeneous with weight W.

10-54 In the system shown, the rod is rigid and weightless. The linear spring with modulus k will always remain horizontal. For what value of P does this configuration become unstable?

PROB. 10-54

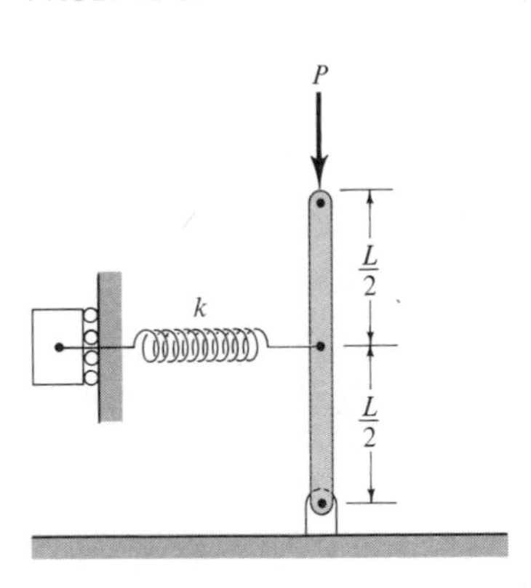

PROB. 10-55

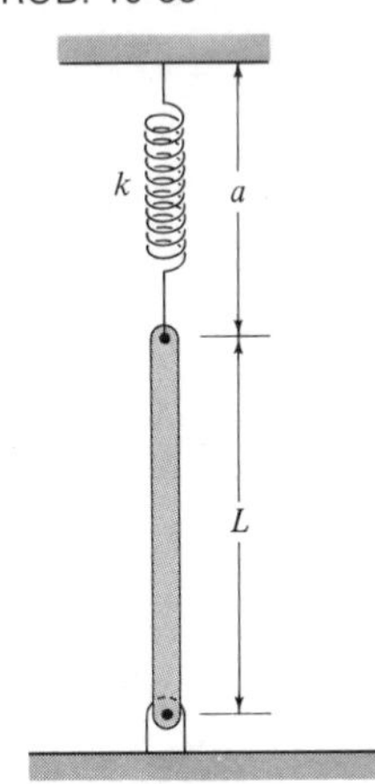

10-55 The homogeneous rigid rod of weight W is attached to an elastic spring and pivoted as shown. What is the required tension T in the spring for the vertical position to be stable? Use the small-angle approximations $\sin\theta \approx \theta$, $\cos\theta \approx 1$.

10-56 Consider the solid homogeneous cone-hemisphere-shaped body shown. For what ratio of cone height h to hemisphere radius R will this configuration become unstable? *Note:* The cg of a cone is at $h/4$, and for a hemisphere, at $3R/8$.

PROB. 10-56

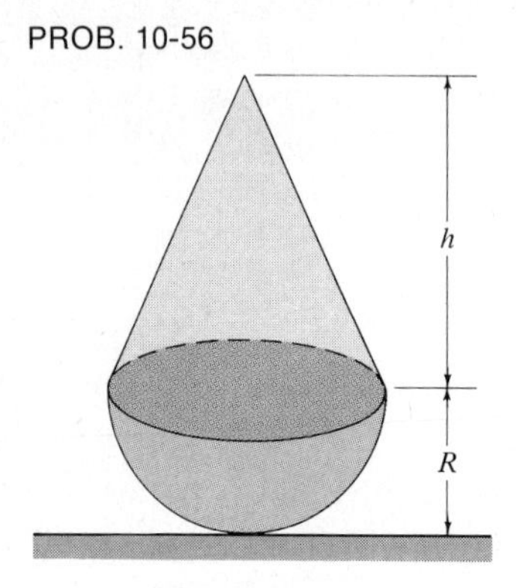

10-5 ELASTIC COLUMNS

A simple column is a long, slender *deformable* structural member subjected to an axial compressive load. Column-like members can be found in almost any type of structure or machine. For example, a compression member of a pin-connected truss, such as the one in Fig. 10-14(a), is a column. Similarly the connecting rod between the piston and crankshaft of an internal-combustion engine is also a column-like member. Of course, we have not as yet given a quantitative meaning to the term "long, slender member" as distinguished from a short, stocky member, although our intuition would lead us to believe that the member in Fig. 10-14(b) would be classified as slender in comparison to the one in Fig. 10-14(c).

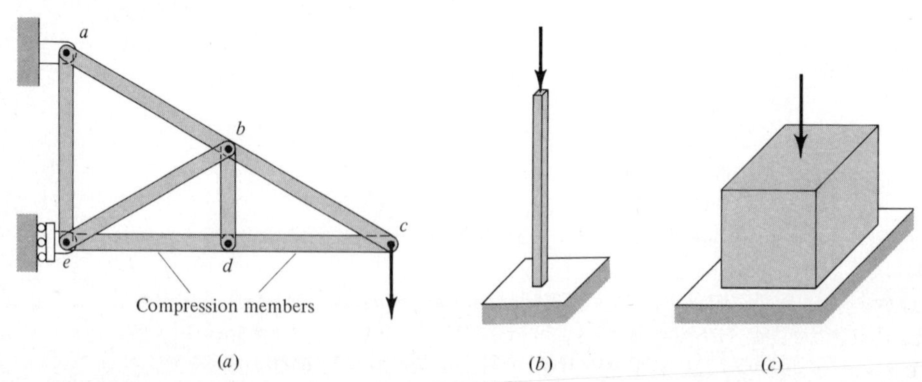

Fig. 10-14

As far as load analysis and load-carrying capacity are concerned, the primary feature that distinguishes a column from an ordinary deformable, axially loaded member is its *mode of failure.* In the tension, torsion, and beam members studied in Chaps. 5 to 9, the equilibrated loading configurations were assumed to be *stable,* and the mode of failure was assumed to be fracture, yielding, or excessive deformation. However, for a column, the loading configuration can become unstable to such an extent that a slight disturbance could cause a dramatic and possibly catastrophic loss of equilibrium. If the structure is unable to reestablish an equilibrated configuration, complete collapse is usually the end result. As a simple experiment, apply a compressive force to the ends of a yardstick and observe the rather dramatic bowing effect which occurs as you increase the load.

To study the stability of columns, let us consider the long, slender compression member of negligible weight pivoted at its ends as shown in Fig. 10-15(a). This idealized situation is actually quite similar to that which exists in member cd or ed of the truss structure in Fig. 10-14(a).

The straight horizontal configuration is a possible equilibrium configuration, as is evident from the free-body diagram of Fig. 10-15(b). Is this configuration stable? That is, if the column is disturbed slightly from this straight configuration, as indicated in Fig. 10-15(c), will the elasticity of the material be able to return the column to its initial straight configuration?

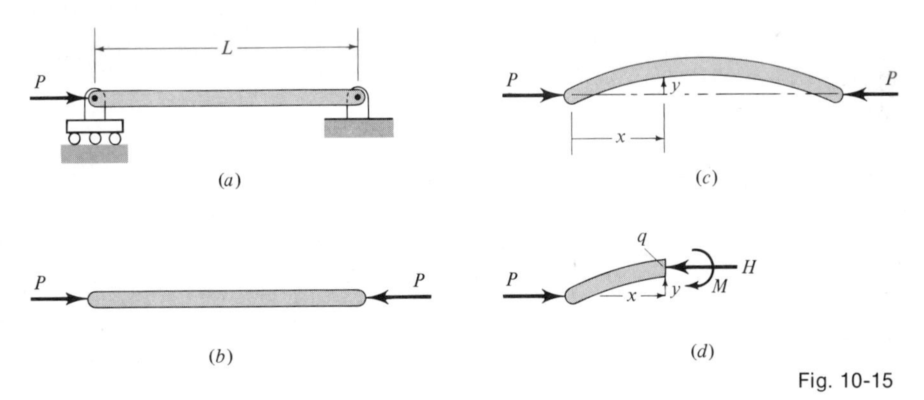

Fig. 10-15

Figure 10-15(d) shows a free-body diagram of a portion of the column with the external load P and the internal reactions H and M – a compressive force and bending couple, respectively. Essentially the situation is similar to that which we encountered in dealing with the discrete systems of the previous section. If the internal restoring couple M due to the elasticity of the column is large enough to overcome the external bending effect of the force P, the column will tend to return to its initial straight position. Otherwise, the external force will overcome the internal resistance and cause the column to continue to bend until either a new bent equilibrium configuration is established or the column completely collapses and ruptures. In each case, from a structural viewpoint, the loss of stability constitutes failure of the column and is commonly referred to as a *buckling failure*. Thus, we seek to determine the critical load P_c for which the straight equilibrium configuration becomes unstable.

Taking moments with respect to point q in Fig. 10-15(d) shows the resultant moment to be

$$M_q = Py - M \tag{10-19}$$

Hence, by the discussion of the previous paragraph, if this resultant moment tends to increase (decrease) the disturbance y, the situation is unstable (stable). To determine which of the terms on the right side of Eq. (10-19) is dominant, let us write

$$Py - M = \lambda y \tag{10-20}$$

where λ is some multiplier such that a positive λ would indicate that the resultant moment was tending to increase the positive y, that is, the term Py was dominant. Hence

$$\begin{aligned} \lambda > 0 \quad & \text{implies instability} \\ \lambda < 0 \quad & \text{implies stability} \end{aligned} \tag{10-21}$$

We must now relate the internal resisting moment M to the elasticity of the column. For the present, let us assume that the material of the column is behaving *elastically*. Then the bent column in Fig. 10-15 is actually an elastic beam loaded in a special manner. Thus, for small disturbances, a positive internal bending moment M is related to a positive deflection y according to the Bernoulli–Euler equation (9-35)

$$\frac{d^2y}{dx^2} = \frac{M}{EI} \tag{9-35}$$

Noting that the bending couple in Fig. 10-15(d) is a negative moment in accordance with the sign convention of Chap. 9, Eq. (10-20) takes the form

$$Py + EI\frac{d^2y}{dx^2} = \lambda y$$

or

$$EI\frac{d^2y}{dx^2} + (P - \lambda)y = 0 \tag{10-22}$$

which is a linear second-order ordinary differential equation whose general solution is of the form

$$y = A \sin Cx + B \cos Cx \tag{10-23}$$

Referring to Fig. 10-15(*c*), the boundary conditions are

$$\begin{aligned} y = 0 \qquad &\text{at} \qquad x = 0 \\ y = 0 \qquad &\text{at} \qquad x = L \end{aligned} \tag{10-24}$$

Substituting these values into Eq. (10-23)

$$0 = A \sin C0 + B \cos C0$$

$$B = 0$$

$$0 = A \sin CL$$

Since $A = 0$ or $C = 0$ would reduce the solution (10-23) to the trivial solution $y = 0$, we see that CL must be some multiple of π.

$$CL = n\pi \qquad n = 1,2,3, \ . \ . \ .$$

Hence, the general solution of Eq. (10-22) is now in the form

$$y = A \sin \frac{n\pi}{L} x \qquad n = 1,2,3, \ . \ . \ . \tag{10-25}$$

Substituting this into Eq. (10-22) yields

$$-EIA\left(\frac{n\pi}{L}\right)^2 \sin \frac{n\pi}{L} x + (P - \lambda)A \sin \frac{n\pi}{L} x = 0$$

$$A \sin\left(\frac{n\pi}{L} x\right)\left[-EI\left(\frac{n\pi}{L}\right)^2 + (P - \lambda)\right] = 0$$

Therefore, since $A \neq 0$, we must have

$$-EI\left(\frac{n\pi}{L}\right)^2 + P - \lambda = 0$$

$$P - EI\left(\frac{n\pi}{L}\right)^2 = \lambda$$

Then, by Eq. (10-21)

$$\begin{aligned} P - EI\left(\frac{n\pi}{L}\right)^2 > 0 \qquad &\text{implies instability} \\ P - EI\left(\frac{n\pi}{L}\right)^2 < 0 \qquad &\text{implies stability} \end{aligned} \tag{10-26}$$

Thus, the critical value P_c at which the original straight column becomes unstable is

$$P_c = EI\left(\frac{n\pi}{L}\right)^2$$

with the smallest value given for $n = 1$ as

$$P_c = \frac{EI\pi^2}{L^2} \tag{10-27}$$

which is called the *Euler buckling load* for an elastic column with pivoted ends. In this equation, E is the modulus of elasticity, usually given in pounds per square inch, I is the second moment of the cross-sectional area about its centroidal axis, usually given in inch^4, and L is the length of the column between the pivots, usually given in inches.

Since we have assumed the column to be long, slender, and elastic, it is often useful to rewrite Eq. (10-27) in a slightly different form. Noting that we can write

$$I = Ar^2$$

where A is the cross-sectional area and r is the radius of gyration (see Appendix A), Eq. (10-27) becomes

$$P_c = \frac{EAr^2\pi^2}{L^2}$$

or

$$\frac{P_c}{A} = \frac{E\pi^2}{(L/r)^2} \tag{10-28}$$

The term L/r is called the *slenderness ratio* of the column and is a quantitative measure of the slenderness by which we can distinguish columns from stocky compression blocks. As a ball-park figure, slenderness ratios between 30 and 200 are common for real structural columns, although these figures vary depending upon the material of the column. Members with slenderness ratios greater than 200 are usually too slender to carry any significant compressive loads.

It is also important to note in Eq. (10-28) the term P_c/A. This term, called the unit buckling load, gives a nominal value for the compressive stress in the column. Since we have assumed the columns to be elastic, *this value should be below the elastic-limit compressive stress for the material.* If this value exceeds the elastic limit, the applicability of the Euler buckling formula [Eq. (10-27) or (10-28)] becomes quite questionable because of the decrease in stiffness which usually accompanies inelastic behavior.

EXAMPLE 10-5

Check the stability of the members of the pin-connected steel truss in Fig. 10-14(a) if the vertical load P is 3 tons. The horizontal members are 10 ft long and the vertical member bd is 5 ft long. All members have a solid $1\frac{1}{4}$-in-diameter circular cross section.

Solution: By the method of joints for trusses (Sec. 4-7), the forces in each of the members of the truss are found to be

$$P_{ab} = P_{bc} = \sqrt{5}P = 13{,}450 \text{ lb} \qquad \text{(tension)}$$
$$P_{ed} = P_{dc} = 2P = 12{,}000 \text{ lb} \qquad \text{(compression)}$$
$$P_{ae} = P_{eb} = P_{bd} = 0 \text{ lb}$$

Hence, only members *ed* and *dc* are carrying compressive loads and are, therefore, columns. The nominal compressive stress in these members is

$$\sigma = \frac{12{,}000}{(\pi/4)(5/4)^2} = 9{,}780 \text{ psi}$$

which is well below the elastic limit of 30,000 psi for mild steel. Therefore, we can anticipate that the Euler buckling formula will be applicable to this situation. With I for a circular cross section given by

$$I = \frac{\pi D^4}{64} = \frac{\pi(5/4)^4}{64} = 0.12 \text{ in}^4$$

the Euler buckling load is

$$P_c = \frac{\pi^2 EI}{L^2}$$
$$= \frac{\pi^2(30 \times 10^6)(0.12)}{(120)^2} = 2{,}470 \text{ lb}$$

Thus, since the actual load in members *ed* and *dc* is 12,000 lb, the structure is *unstable.*

Remark: Although it was not necessary to determine the slenderness ratio, such calculation might be informative. For a solid circular cross section, the radius of gyration is

$$r = \sqrt{\frac{I}{A}} = \sqrt{\frac{\pi D^4/64}{\pi D^2/4}} = \frac{D}{4}$$

Thus, for members *ed* and *dc*, the slenderness ratio is

$$\frac{L}{r} = \frac{120}{5/16} = 384$$

which is much too slender for a structural column.

10-6 DISCUSSION OF ELASTIC BUCKLING

The stability analysis of the previous article was for an idealized column, with the idealization being primarily the weightlessness, the perfectly aligned axial loading, and the frictionless pivots. To initiate buckling of such a column after the critical load has been reached, some sort of external disturbance is needed; otherwise the column will remain straight although unstable. On the other hand, most real columns made of real materials have enough imperfections in fabrication and load alignment, so that no external disturbance is needed to initiate the buckling process once the buckling load is reached. In fact, because of these imperfections, in most cases the buckling load for a *real* column will be somewhat *smaller* than that predicted by the Euler buckling formula [Eq. (10-27)]. Figure 10-16 is indicative of the difference in the behavior between an ideal column and a real column. Figure 10-16(*b*) is a

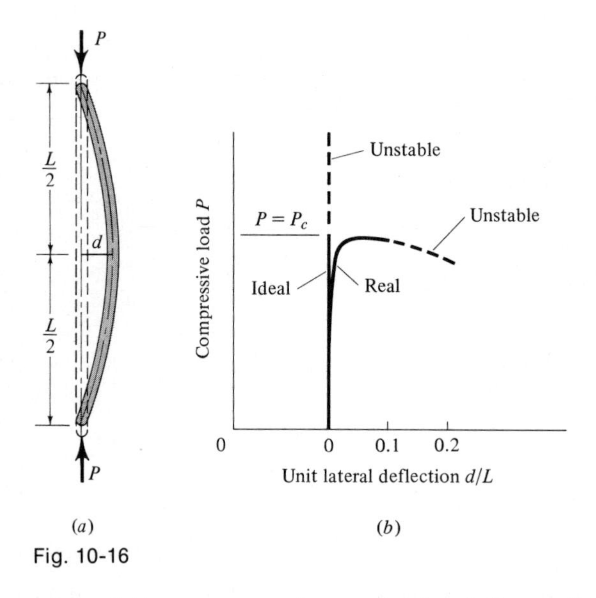

Fig. 10-16

plot of the compressive axial load P vs. the unit lateral deflection d/L. The ideal column remains straight and stable for loads below P_c but becomes unstable (as indicated by the dashed lines) for loads above P_c. The real column initially is straight but gradually bows (bends) a slight amount as the load is increased. This bowing begins to increase significantly as the load nears P_c and would increase dramatically as the column begins to lose its structural stability. If the load P were increased further, total collapse would occur in all probability. For columns made of most structural materials the Euler buckling formula does a reasonably good job of predicting the actual buckling load for elastic behavior.

Figure 10-17 is a plot of Eq. (10-28) for structural steel, the solid line indicating the range of applicability of the Euler formula below the elastic (proportional) limit stress of the material. From this figure it is apparent that, for columns of structural steel, the Euler buckling formula would be applicable for slenderness ratios larger than 100 but would not be reliable for slenderness ratios

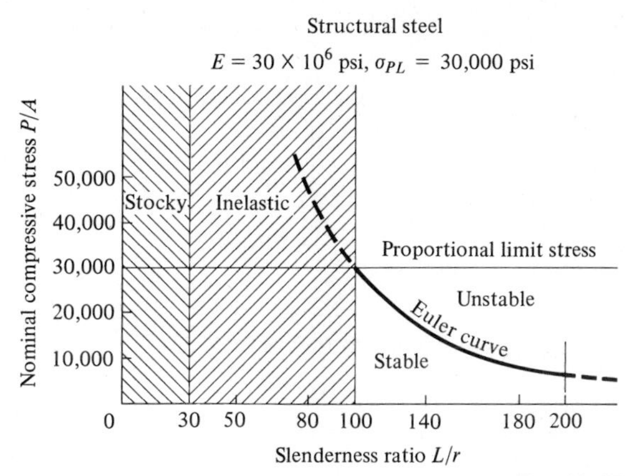

Fig. 10-17

under 100. In all likelihood, a real column of structural steel with a slenderness ratio under 100 would buckle at a load well below that predicted by the dashed line of the extended Euler curve in Fig. 10-17. This is due primarily to the pronounced loss of stiffness at stress levels above the elastic limit [recall the stress-strain curve for mild steel, Fig. 6-4(*a*)]. The next section will discuss a method for predicting the buckling load for inelastic behavior of columns.

Returning to elastic behavior and the Euler buckling formula, the derivation of Eq. (10-27) or (10-28) was for an elastic column pivoted at both ends. Real columns have various types of end conditions or fixities, most of which can be idealized by one of the schemes shown in Fig. 10-18. Recall that, in obtaining the Euler buckling formula, n in Eq. (10-25) was taken to be 1, so that the shape of the disturbed column was half of a sine wave. Accordingly, the length denoted by L_e in Fig. 10-18 corresponds to the effective length of the column whose disturbed shape is (or approximates) a half sine wave. Thus, if we rewrite the Euler buckling formula (10-27) in the form

$$P_c = \frac{\pi^2 EI}{L_e^2} \tag{10-29}$$

we can apply this result to the various end conditions depicted in Fig. 10-18.

(*a*) For pivoted ends, $L_e = L$ and

$$P_c = \frac{\pi^2 EI}{L^2}$$

(*b*) For two fixed ends, $L_e = L/2$ and

$$P_c = \frac{\pi^2 EI}{(L/2)^2} = \frac{4\pi^2 EI}{L^2}$$

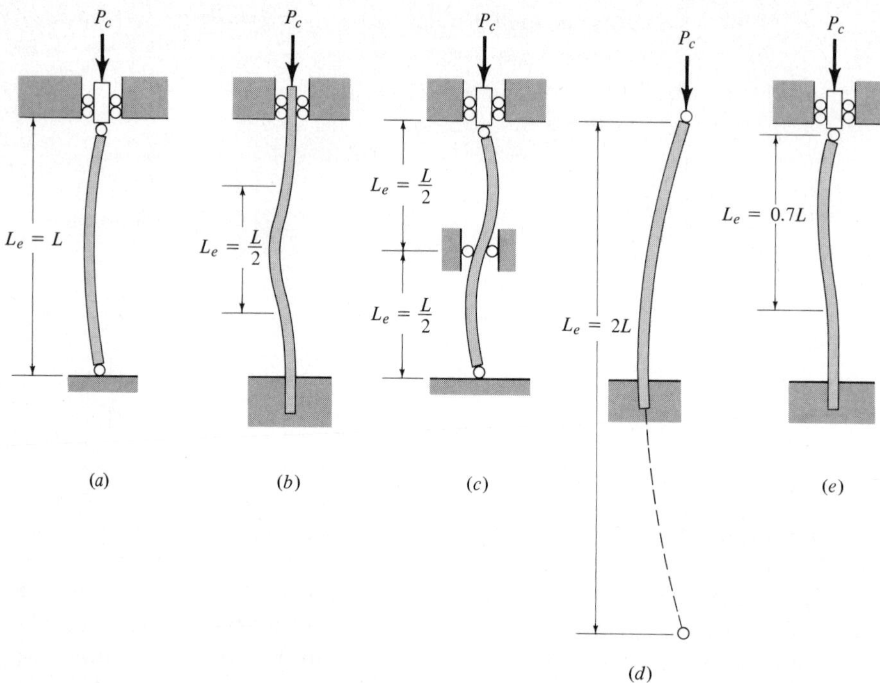

Fig. 10-18

(c) For a pivot at the midpoint, $L_e = L/2$ and

$$P_c = \frac{\pi^2 EI}{(L/2)^2} = \frac{4\pi^2 EI}{L^2}$$

(d) For one fixed end, $L_e = 2L$ and

$$P_c = \frac{\pi^2 EI}{(2L)^2} = \frac{\pi^2 EI}{4L^2}$$

(e) For one fixed and one pivoted end, $L_e = 0.7L$ and

$$P_c = \frac{\pi^2 EI}{(0.7L)^2} = \frac{2\pi^2 EI}{L^2}$$

For other types of fixities, suitable choices of the effective length would give a reasonable value for the critical buckling load.

It is worthwhile to observe that, of those fixities depicted in Fig. 10-18, case (d) having only one fixed end has by far the lowest critical load and is, therefore, the most unstable.

Finally, a comment regarding the calculation of I (or the radius of gyration r). It has been discussed earlier that I is the second moment of the cross section with respect to the centroidal axis of the cross section. However, most cross sections, except circular or square ones, have more than one distinct centroidal axis as, for example, the rectangular and T sections in Fig. 10-19. Therefore, unless such a column was restrained to buckle in a particular manner, the column would tend to buckle about the centroidal axis of *least* I (or r) since this axis would offer the least resistance to bending. Thus a column with the rectangular-shaped cross section of Fig. 10-19(a) would tend to buckle about the indicated horizontal centroidal axis, whereas one with the T shape in Fig. 10-19(b) would probably tend to buckle about the vertical centroidal axis.

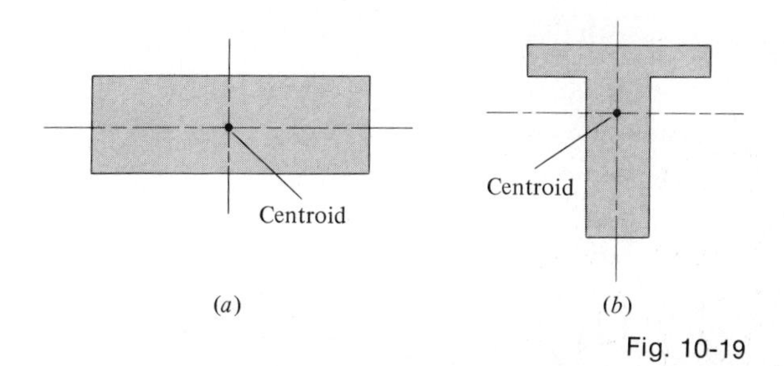

Fig. 10-19

EXAMPLE 10-6

A carpenter intends to use a piece of standard 2×4 spruce timber to temporarily support an overhang on a construction job. The 8-ft timber will be nailed securely at its top end to the overhang, and its bottom end will rest on the foundation. Estimate the maximum safe weight this timber can support.

Solution: Referring to the table of timber sizes in Appendix B, a nominal 2×4 has the cross-sectional properties

$$I_x = 5.65 \text{ in}^4 \qquad I_y = 1.00 \text{ in}^4 \qquad A = 5.34 \text{ in}^2$$

Thus, its least radius of gyration will be

$$r = \sqrt{\frac{I_y}{A}} = 0.433 \text{ in}$$

Since the timber is nailed securely at the top, we will treat the top end as fixed. Strictly speaking, the bottom end is not pivoted, but neither is it fixed nor free. However, the pivoted end more nearly reflects the actual case, so that in effect we have the fixity condition of case (e) in Fig. 10-18 with an effective length of $0.7L$. Hence, the slenderness ratio is effectively

$$\frac{L_e}{r} = \frac{(0.7)(8)(12)}{0.433} = 155$$

By Eq. (10-28), and using 1.6×10^6 psi as an average value for the elastic modulus of wood, we have

$$\frac{P_c}{A} = \frac{\pi^2 E}{(L/r)^2}$$
$$= \frac{\pi^2(1.6 \times 10^6)}{(155)^2} = 656 \text{ psi}$$

so that elastic behavior appears to be a reasonable assumption. Finally, the buckling load will be

$$P_c = (656)A = 656(5.34)$$
$$= 3{,}500 \text{ lb}$$

which will be the maximum load for stability. With no assumed factor of safety, the safeness of this load is somewhat doubtful. What would you suggest as the safe load?

EXERCISE PROBLEMS

10-57 to 10-60 A 12-ft-long column-like member has the cross section shown in the figure. Determine the slenderness ratio of this column if (*a*) the column has pivoted ends and (*b*) the column has two fixed ends.

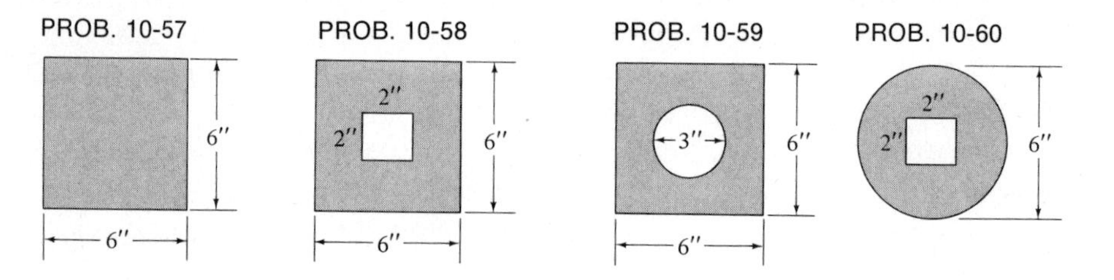

10-61 to 10-64 A 10-ft-long column-like member has the cross section shown in the figure. Determine the maximum and minimum slenderness ratios for this column if (*a*) the column has pivoted ends and (*b*) the column has one fixed end and one pivoted end.

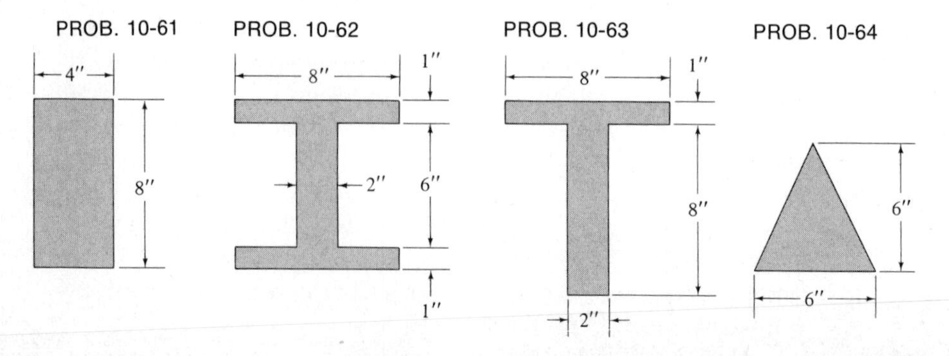

10-65 A common household yardstick has a $1\frac{1}{4} \times \frac{1}{8}$-in retangular cross section. What is the value of the buckling load if the modulus of the wood is 1.5×10^6 psi? Assume pivoted ends.

10-66 A 20-ft-long column of structural steel has a standard W8 × 40 cross section (refer to Appendix B). If this column is to be installed with fixed ends, what will be the slenderness ratio for this column? Will it buckle elastically?

10-67 A mast for a television antenna is to be made of a standard $1\frac{1}{4}$-in steel pipe (refer to Appendix B). If the rotor, antenna, and effective wind loads are estimated to be equivalent to a 50-lb vertical force, how high can this mast safely be if it is to be fixed at the bottom but otherwise free? Use a factor of safety of 3. Is your answer reasonable?

10-68 Same as Prob. 10-67 except the mast is guyed at the top with three guy wires which exert an additional net compressive force on the column of 50 lb.

10-69 The ends of a solid circular $\frac{3}{4}$-in-diameter aluminum bar are embedded in two immovable supports 40 in apart while at room temperature of 65° F. If the bar is heated uniformly, at what temperature will it buckle? Use $E = 10 \times 10^6$ psi and $\alpha = 13 \times 10^{-6}$ in/in/°F.

10-70 An aluminum ($E = 10.5 \times 10^6$) strut is 20 in long and has an elliptical cross-sectional area with a major and minor diameter of 1 and $\frac{1}{2}$ in, respectively. The strut is pinned at each end with the pin running parallel to the minor axis. What is the slenderness ratio? What is the value of the buckling load?

10-71 A piece of 10-ft-long birch timber is to be used to carry a compressive load of 4 tons. The timber has a square cross section, and one of its ends is fixed and the other pivoted. Assuming elastic behavior with an E of 1.8×10^6 psi, find the minimum nominal size timber which is necessary. Use a safety factor of 3.

10-72 A 12-ft-long piece of thin-walled aluminum ($E = 10.5 \times 10^6$) tubing is to carry an axial compressive load of 800 lb. The wall thickness is one-tenth the radius, so that thin-wall approximations can be used. If a safety factor of 2 is required, what minimum size tube is necessary? The ends can be considered to be pivoted.

10-73 A 50-ft boom is supported by two cables as shown. The boom is a truss-like steel structure with the main longitudinal tube-like elements having the configuration shown. What is the slenderness ratio for the boom? Using a factor of safety of 2, find the allowable load P.

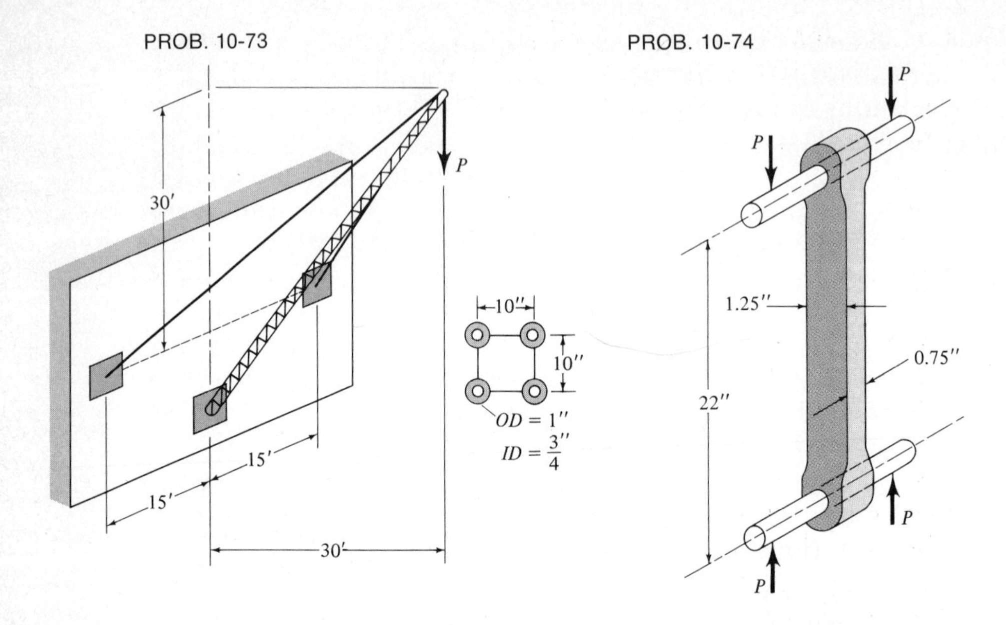

10-74 The connecting rod shown is made of aluminum with a modulus of elasticity of 10×10^6 psi and a proportional limit of 28,000 psi. Determine the maximum allowable value of P.

10-75 An 800-lb platform is to be placed on a 100-ft-high fiberglass pole with a 6 in OD and a 5 in ID. The modulus of elasticity is 10×10^6 psi. The bottom end of the pole is held in a ball-and-socket joint, and the pole is guyed at the platform height. Can the platform be placed at the top of the pole without buckling the pole? If not, how high up can the platform be placed? Neglect the compressive downward pull of the guy wires and the weight of the pole itself.

10-76 Same as Prob. 10-75 except the bottom end is fixed firmly into a foundation.

10-7 INELASTIC BUCKLING: THE ENGESSER EQUATION

In the discussion of columns in Sec. 10-5 leading to the derivation of the Euler buckling formula [Eq. (10-27)], the assumption of elastic behavior was not made until *after* the obtainment of the resultant-moment equation (10-19) and the buckling-criterion equation (10-21). The elasticity assumption was embodied in the use of Eq. (9-35), the Bernoulli–Euler equation, in relating the internal moment M and the deflection y. Therefore, as we now turn to the case of inelastic buckling, we have no need to repeat the arguments and results leading to Eq. (10-21), although it would be well for you to refresh your memory. Thus, the stability of the column is still described by the equation

$$Py - M = \lambda y \tag{10-20}$$

where
$$\begin{aligned} \lambda > 0 \quad & \text{implies instability} \\ \lambda < 0 \quad & \text{implies stability} \end{aligned} \tag{10-21}$$

and our problem is to relate M to y for inelastic behavior.

When the column is in its straight configuration, the primary stress is the axial compressive stress indicated in Fig. 10-20(*a*):

$$\sigma = \frac{P}{A} \tag{10-30}$$

Let us assume that this value is above the proportional limit of the material, so that we are dealing with inelastic behavior. When this column is disturbed slightly, as exaggerated in Fig. 10-20(*b*), a relatively small secondary incremental bending stress $\Delta\sigma$ will be superposed onto the primary axial compressive stress σ. If the disturbance y is sufficiently small, there will be no significant unloading of the primary axial compressive stress, and, therefore, the small incremental bending strain $\Delta\epsilon$ will be related to the small incremental bending stress $\Delta\sigma$ by

$$\Delta\sigma = E_t\,\Delta\epsilon \tag{10-31}$$

where E_t is the tangent modulus at the stress level of the primary compressive stress given by Eq. (10-30). Refer to Fig. 10-20(*c*).

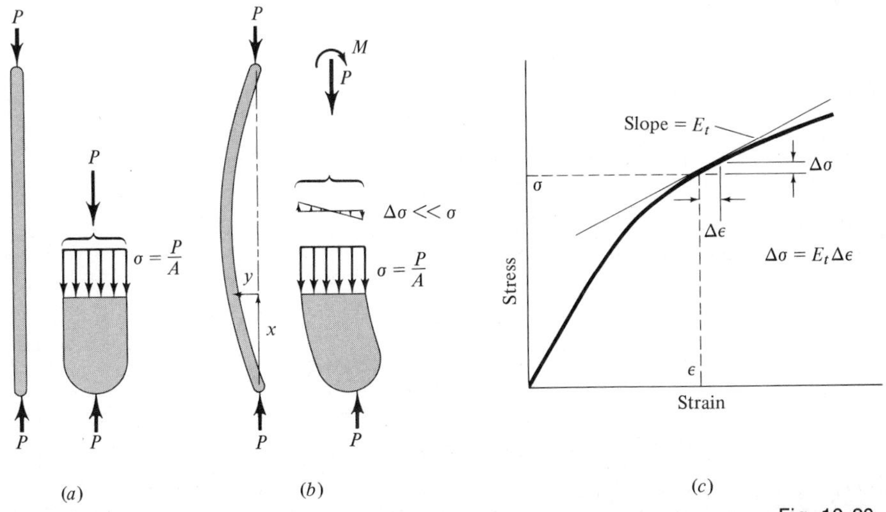

Fig. 10-20

Also, for small deflections, the geometric approximation developed in Chap. 9, namely, Eq. (9-34),

$$\frac{dy^2}{dx^2} = \frac{1}{R} \tag{9-34}$$

where R is the radius of curvature of the bent column, is applicable to the present situation. Now, Eq. (10-31) is similar to the bending stress–bending strain relation of Eq. (9-12) for elastic beams except for E_t in place of E. Hence, except for this difference in moduli, Eqs. (10-31) and (9-34) represent the same basic assumptions used to derive the Bernoulli–Euler equation (9-35):

$$\frac{dy^2}{dx^2} = \frac{M}{EI} \tag{9-35}$$

Consequently, for the case now under consideration, the internal bending moment of the column will be related to the disturbance y by a relation of the form

$$\frac{dy^2}{dx^2} = \frac{M}{E_t I} \tag{10-32}$$

Substituting this into Eq. (10-20) and analyzing the stability of the column would result in a formula for the critical buckling load similar to the Euler formula (10-27), namely,

$$P_c = \frac{E_t I \pi^2}{L^2} \tag{10-33}$$

in which the tangent modulus E_t has replaced the elastic modulus E. This formula is called the Engesser equation, after the man who first developed it. This relation can be rewritten in the equivalent form

$$\frac{P_c}{A} = \frac{E_t \pi^2}{(L/r)^2} \tag{10-34}$$

where L/r is once again the slenderness ratio of the column, and P_c/A is the unit buckling load.

> ***Remark:*** Although there are several other inelastic buckling theories and empirical formulas, the Engesser equation has been shown to predict the buckling load of real inelastic columns with a high degree of reliability. Accordingly, we will not delve into the different arguments and techniques used to develop these other results, which can be found in texts dealing more extensively with stability theory.

Although Eq. (10-34) is quite similar to the Euler formula (10-28), its application to solving problems is considerably more difficult because of the variable value of the tangent modulus. Thus, in solving problems in inelastic buckling, it is often convenient to have a plot of stress σ vs. tangent modulus E_t. With such plots, trial-and-error solutions of the Engesser equation become much more straightforward. Figure (10-21) shows the elastic and inelastic buckling theory of Euler and Engesser for some sample types of materials. The dashed portion of the Euler curve is shown merely to illustrate the difference between the elastic and inelastic theory.

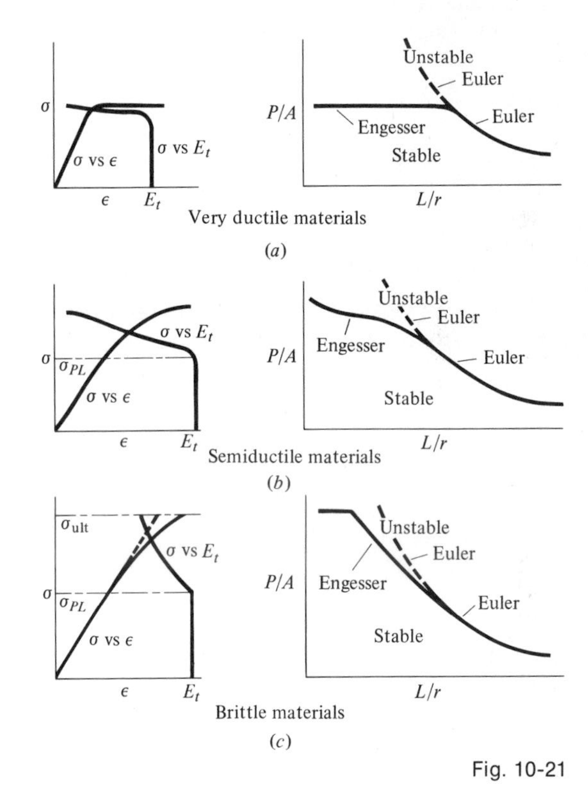

Fig. 10-21

Finally, the same discussion of fixity and end conditions which was made in conjunction with the Euler buckling formula is directly applicable to the Engesser equation for inelastic behavior.

EXAMPLE 10-7

An extruded 2024-T4 aluminum member is to be used as a fixed-ended column. The member is 10 ft long and has an equilateral triangular cross section shown in Fig. 10-22(*a*). Determine the theoretical buckling load for this column. The stress-strain and stress-tangent modulus curves are shown in Fig. 10-22(*b*).

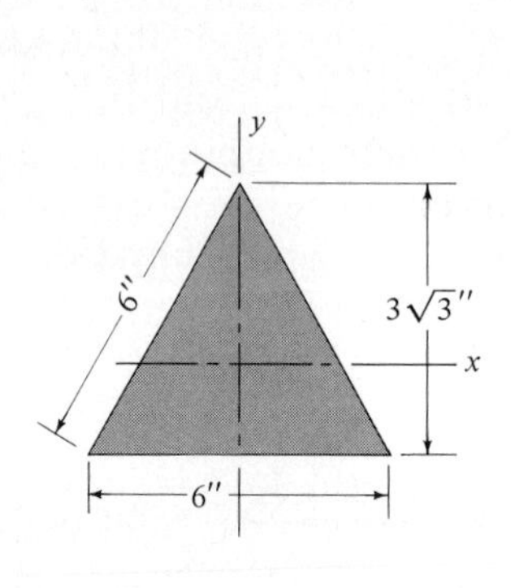

(a)

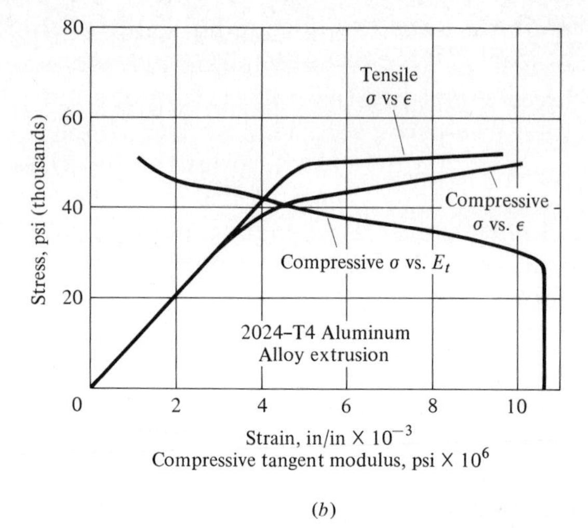

(b)

Fig. 10-22

Solution: We must first determine the least radius of gyration of the cross section. Using Table A-1 in the Appendix for I of a triangular section, we have

$$I_x = \frac{bh^3}{36} = \frac{(6)(3\sqrt{3})^3}{36} = 13.5\sqrt{3}\ \text{in}^4$$

For finding I_y, we will treat the cross section as two right triangles. Therefore,

$$\begin{aligned} I_y &= 2\left[\frac{(b/2)^3(h)}{12}\right] \\ &= 2\,\frac{(3)^3 3\sqrt{3}}{12} = 13.5\sqrt{3}\ \text{in}^4 \end{aligned}$$

Thus, the columns will tend to buckle equally about all centroidal axes with a radius of gyration of

$$\begin{aligned} r &= \sqrt{\frac{I}{A}} = \sqrt{\frac{13.5\sqrt{3}}{1/2(6)(3\sqrt{3})}} \\ &= 1.225\ \text{in} \end{aligned}$$

Since the column is fixed-ended, its effective length is

$$L_e = \tfrac{1}{2}L = (5)(12) = 60\ \text{in}$$

and its slenderness ratio is

$$\frac{L_e}{r} = \frac{60}{1.225} = 49$$

With this slenderness ratio, inelastic behavior is likely. Thus, the Engesser equation (10-34) yields

$$\frac{P_c}{A} = \frac{\pi^2 E_t}{(L/r)^2} = (41.1 \times 10^{-4})E_t \qquad (a)$$

Our problem now is to find some point on the stress–tangent modulus curve in Fig. 10-22(*b*) whose combination of values satisfies the above equation. For example, at a stress level of 40,000 psi, E_t has a value of approximately 4.5×10^6 psi. Thus, from Eq. (*a*),

$$(41.1 \times 10^{-4})(4.5 \times 10^6) = 18{,}500$$
$$\neq 40{,}000$$

Hence, this value of E_t is too small. As a second trial, for a value of $E_t = 8 \times 10^6$, the stress level is about 35,000 psi. Thus,

$$(41.1 \times 10^{-4})(8 \times 10^6) = 32{,}880$$
$$\neq 35{,}000$$

This is much closer and, in fact, is probably close enough. We estimate the unit buckling load to be

$$\frac{P_c}{A} = 34{,}000 \text{ psi}$$

so that the buckling load is

$$\begin{aligned} P_c &= 34{,}000A \\ &= (34{,}000)(\tfrac{1}{2})(6)(3\sqrt{3}) \\ &= 530{,}000 \text{ lb} \end{aligned}$$

10-8 CLOSURE

This concludes our discussion of stability. We have looked at a few relatively simple situations in which the question of stability can be the deciding factor as to the performance of a structure or machine. Simple columns, although fairly common structures, are not the only type of structural member in which stability is a major design consideration. A thin-walled tube used as a torsion member can buckle due to the compressive stresses developed at 45° to the longitudinal axis (recall Mohr's circle of stress). Similarly, a thin member used as a beam, such as a drapery rod, can buckle due to the large compressive bending stresses which may be developed. A shell-type structure such as the hull of a submarine or the fuselage of an airplane can buckle if the external pressures or forces become too large and cause instability.

Generally speaking, the buckling mode of failure is more difficult to predict and analyze, and its subsequent effect more disastrous than other types of failure. Consequently, in designing structures in which buckling is a possible or likely mode of failure, a larger factor of safety is often used than might ordinarily be used.

The discussions in this chapter, although quite introductory in nature, should make you aware of situations in which buckling might occur, and cause you to pause before rendering judgment on the structural performance of a member.

EXERCISE PROBLEMS

10-77 A 17-in-long 2024-T4 aluminum strut is to be pinned at each end and carry a compressive load of 10,000 lb. If the cross-sectional area is a circular thin wall with a radius and thickness of $\frac{3}{4}$ in and $\frac{1}{16}$ in, respectively, is this strut stable? See Fig. 10-22(*b*).

10-78 Same as Prob. 10-77 except the strut has one fixed end and the other end can be considered to be free.

10-79 A piece of 8-ft-long 4 × 4 (nominal) birch timber is to be used as a compressive member with effectively two fixed ends. The stress-strain and stress–tangent modulus curves are shown in the figure. Will this column buckle elastically? Determine the theoretical buckling load.

PROB. 10-79

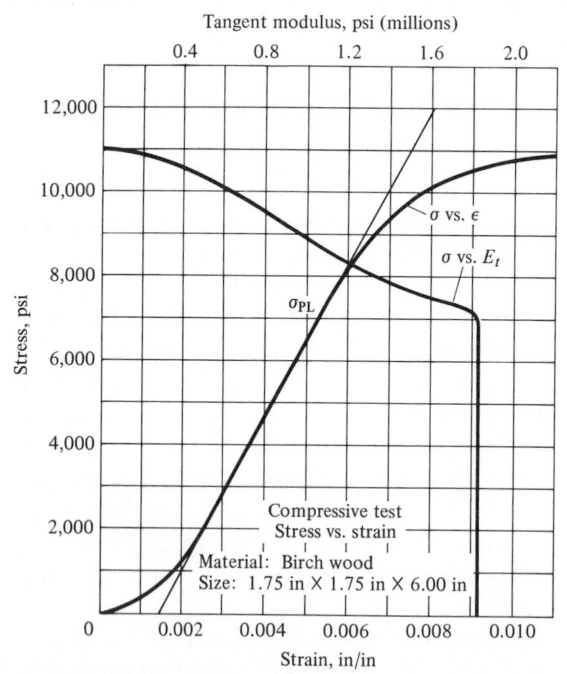

10-80 A column-like member is made of an aluminum-magnesium alloy whose compressive stress-strain curve is approximated in the manner shown in the figure. The member has a solid

1-in-diameter cross section and is to carry a load of 2 tons. Find the maximum allowable length if (*a*) the column has pivoted ends and (*b*) the column has fixed ends.

PROB. 10-80

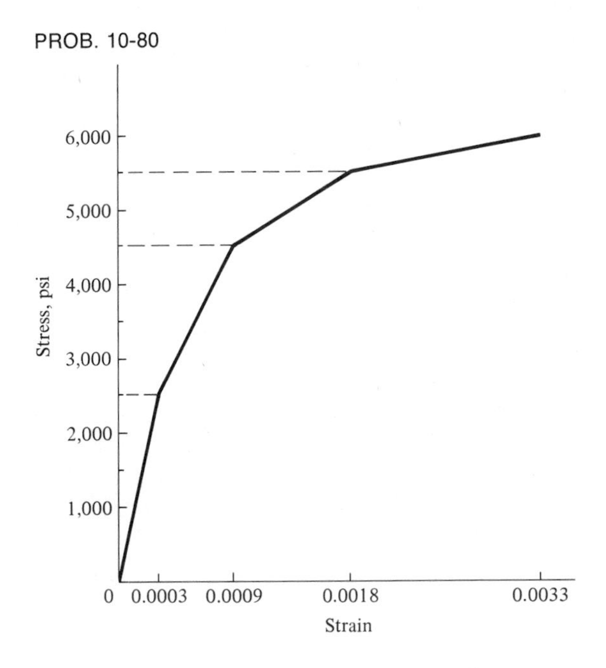

10-81 The ends of an aluminum bar are embedded in two immovable supports 60 in apart while at a temperature of 70° F. If the bar is heated uniformly, at what temperature will it buckle? The bar has a rectangular 6 by 4-in cross section, and the stress-strain curve is approximated by that shown in the figure. Use $\alpha = 13 \times 10^{-6}$ in/in/°F.

PROB. 10-81

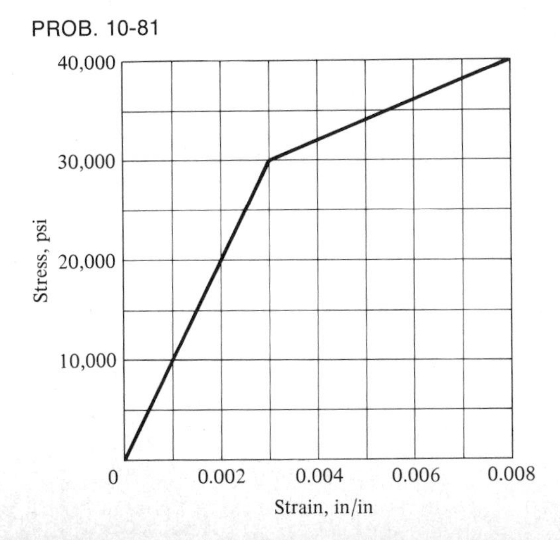

10-82 A strut on a small aircraft is made of a glass-resin laminate whose stress-strain and stress–tangent modulus curves are shown in the figure. The warp is parallel to the load. The strut is 10 in long, has a $1\frac{1}{4}$ by $\frac{1}{2}$-in rectangular cross section, and its ends are pinned with the pin running parallel to the short side. What is the slenderness ratio for the strut? What is the buckling load?

PROB. 10-82

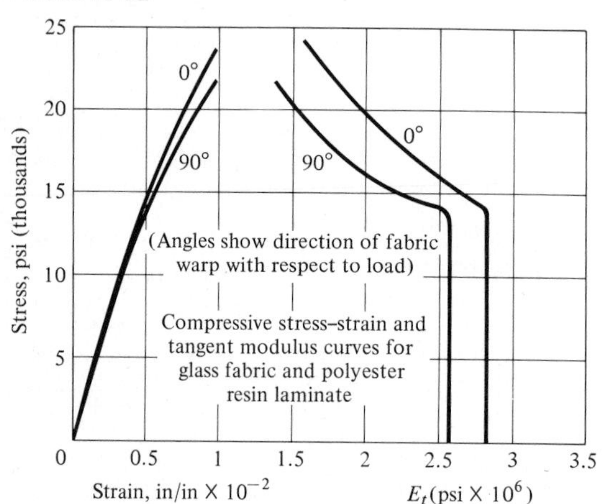

10-83 Recall the boom of Prob. 10-73. In this case the truss-like boom is made of four $4 \times 4 \times \frac{1}{2}$ standard steel angles arranged as shown in the figure. What is the slenderness ratio for this boom? Using a factor of safety of 2.5, find the allowable load P.

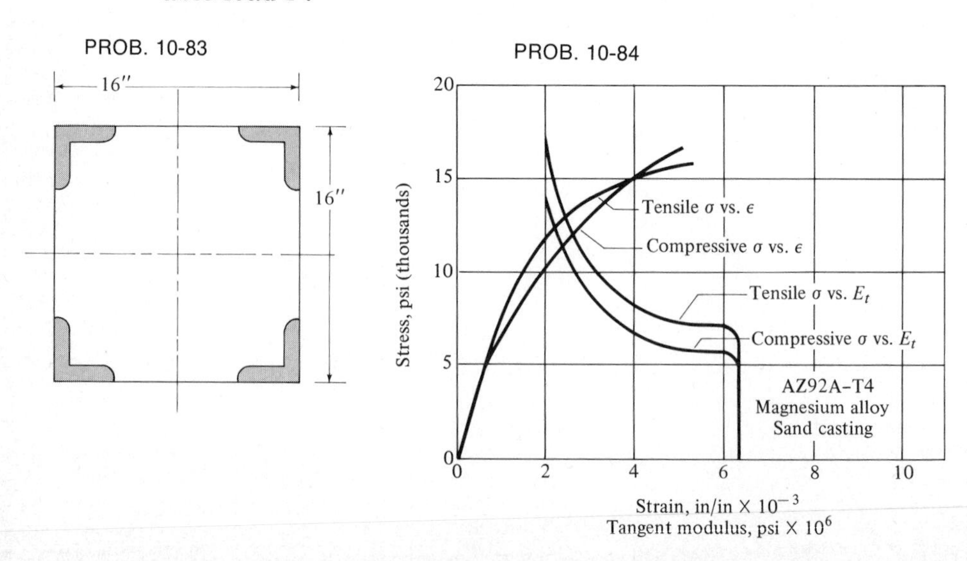

10-84 A column-like member is made of a magnesium alloy whose tensile and compressive properties are indicated by the curves in the figure. The member is 18 in long, has an elliptical cross section 2 by 4 in, and effectively has one free end and one fixed end. Determine the buckling load.

10-85 A $\frac{1}{2}$ by $1\frac{1}{2}$-in steel bar is pinned between two relatively rigid supports 24 in apart, with the pins running parallel to the short side of the bar. The bar is then heated uniformly throughout. If the initial temperature is 70°F, at what temperature will the bar buckle? Use $\alpha = 6.5 \times 10^{-6}$ in/in/°F.

appendix A PROPERTIES OF AREAS AND VOLUMES

A-1 INTRODUCTION

The most common property of a plane area is its size, i.e., the amount of area enclosed within its boundaries. However, the size of an area does not convey any information regarding the shape or orientation of the area, as evidenced by the fact that all the areas in Fig. A-1(*a*) to (*c*) have the same size, but Fig. A-1(*b*) is of different shape from Fig. A-1(*a*), and Fig. A-1(*c*) has a different orientation from (*a*). It is desirable to quantify these geometric differences in terms of certain geometrical properties which are dependent upon the shape and orientation of the area as well as its size. Although many such properties might be defined, two particular properties find wide application in the study of mechanics, namely, the first and second moments of area.

The study of the geometry of areas and volumes preceded the development of calculus by several thousand years. However, it is convenient to formally define the geometric properties in terms of certain mathematical integrals because it is usually through such integrals that these properties arise in the study of mechanics, e.g., resultants of distributed loads, bending stresses in beams, torsion stresses in shafts, etc.

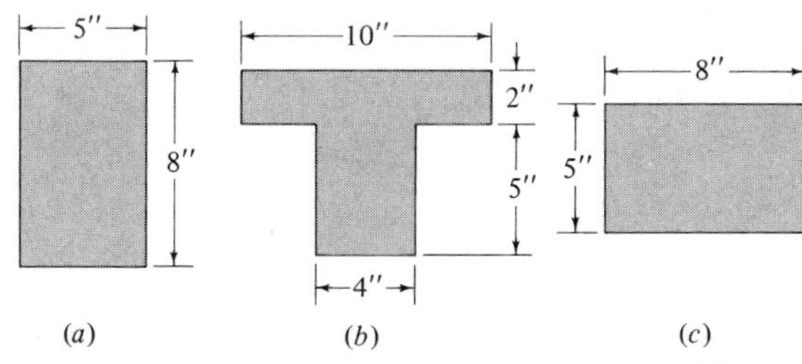

Fig. A-1

A-2 FIRST MOMENT OF AREA: CENTROIDS

Consider the plane area of Fig. A-2(a). Formally, the amount of area, or simply the area, is given by

$$A = \iint_{\text{area}} dA$$

where the double integral is used to indicate that the elemental area dA is usually expressed in terms of differentials of some surface coordinates, such as dx and dy, and the corresponding integration becomes the ordinary Riemann integration of elementary calculus.

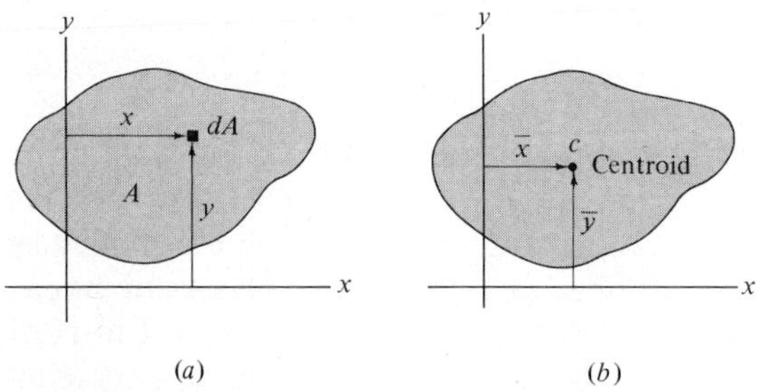

Fig. A-2

Definition A-1 The first moment of an area with respect to an axis is the integral

$$M = \iint_{\text{area}} p\, dA$$

where p is the perpendicular distance from the axis to the elemental area dA. Thus, referring to Fig. A-2(a),

$$M_x = \iint_{\text{area}} y\, dA \quad \text{and} \quad M_y = \iint_{\text{area}} x\, dA \tag{A-1}$$

Definition A-2 The *centroid* of the area is the point whose coordinates $(\bar{x},\bar{y})$ are given by

$$\bar{x} = \frac{M_y}{A} = \frac{\iint_{\text{area}} x\, dA}{\iint_{\text{area}} dA}$$

$$\bar{y} = \frac{M_x}{A} = \frac{\iint_{\text{area}} y\, dA}{\iint_{\text{area}} dA} \tag{A-2}$$

where $\bar{x}$ and $\bar{y}$ are called the centroidal distances or centroidal coordinates relative to the y and x axes, respectively.

Remark: Area, being a physical size, is inherently a positive quantity, whereas a first moment can be algebraically positive, negative, or zero depending upon the location of the area relative to the moment axis. Thus, centroidal distances can be positive, negative, or zero.

Definition A-3 Any axis passing through the centroid of an area is called a *centroidal axis*.

Since by Definition A-3 the centroidal distance between a centroidal axis and the centroid is zero, we have the following:

Theorem A-1 The first moment with respect to any centroidal axis is zero. Conversely, if the first moment with respect to some axis is zero, that axis *must* be a centroidal axis.

As a consequence of the above definitions and theorem, the centroid is sometimes visualized as the place where all the area would have to be concentrated to give the correct first moment with respect to any axis.

A very useful observation is the fact that any axis of symmetry of an area is automatically a centroidal axis. To see why this is true, consider some area with an axis of symmetry which is denoted by the s axis in Fig. A-3(a). Then, for every dA with a positive p, there will be a corresponding dA with a negative p, so that the net value of the integral

$$M_s = \iint_{\text{area}} p \, dA$$

over the entire area will be zero. Thus, by Theorem A-1, the s axis must be a centroidal axis. This means that if an area has two axes

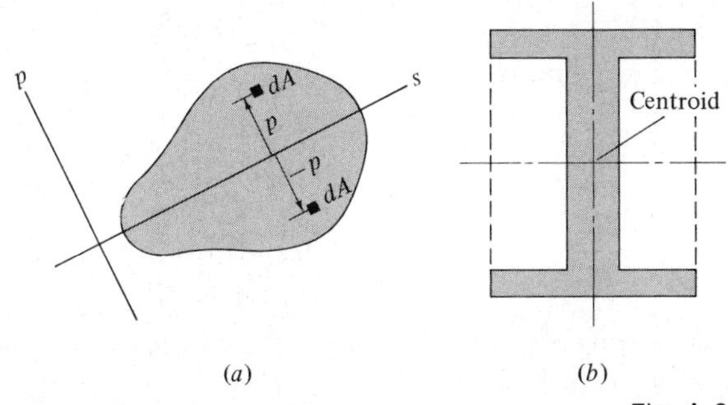

Fig. A-3

of symmetry, its centroid can be located immediately at the intersection of these axes, as, for example, the area in Fig. A-3(*b*).

Composite Areas

The definitions of area, first moment, and centroid are quite general in that they apply to any plane surface area. Quite often a rather complex-shaped area can be visualized as being made up of several simple-shaped areas. For example, the I-shaped area of Fig. A-3(*b*) can be thought of as being composed of three rectangles—the two horizontal flanges and the vertical web. Or, we could visualize this same I shape as being composed of a large solid rectangle with two smaller rectangles removed or cut out as indicated by the dashed lines in Fig. A-3(*b*). In either case the total composite area is the sum of the solid component areas minus those areas that are removed. In general,

$$A_{\text{total}} = \pm A_1 \pm A_2 \pm \cdots \pm A_n \tag{A-3}$$

where n is the number of component areas.

Similarly, in calculating the first moment of some composite area, by Definition A-1,

$$\begin{aligned} M &= \iint_{\text{total area}} p\, dA = \iint_{\pm A_1 \pm A_2 \pm \cdots \pm A_n} p\, dA \\ &= \pm \iint_{A_1} p\, dA \pm \iint_{A_2} p\, dA \pm \cdots \pm \iint_{A_n} p\, dA \end{aligned} \tag{A-4}$$

where the plus or minus sign merely indicates whether or not that particular component area exists or has been removed, and is not indicative of the algebraic sign of the first moment. Thus, Eq. (A-4) is simply the statement that the first moment of a composite area is the algebraic sum of the first moments of the component areas; that is

$$M_{\text{total}} = \pm M_1 \pm M_2 \pm \cdots \pm M_n \tag{A-5}$$

Finally, using Eqs. (A-5) and (A-3) for the centroid, the centroid for a composite area becomes

$$\begin{aligned} \bar{x} &= \frac{\pm M_{1y} \pm M_{2y} \pm \cdots \pm M_{ny}}{\pm A_1 \pm A_2 \pm \cdots \pm A_n} \\ \bar{y} &= \frac{\pm M_{1x} \pm M_{2x} \pm \cdots \pm M_{nx}}{\pm A_1 \pm A_2 \pm \cdots \pm A_n} \end{aligned} \tag{A-6}$$

The following examples will illustrate some of the operational techniques involved in the determination of centroids of plane areas.

EXAMPLE A-1

Find the centroid of the area bounded by the straight line $y = x$ and the parabola $y + 3 = \frac{1}{3}(x - 3)^2$, where x and y are in inches.
Solution: The area under consideration is shown in Fig. A-4(*a*), with the points of intersection given by

$$\begin{gathered} y_{\text{line}} = y_{\text{parabola}} \\ x = \tfrac{1}{3}(x-3)^2 - 3 \\ x^2 - 6x + 9 - 9 - 3x = 0 \\ x^2 - 9x = 0 \\ x = 0,\ 9 \\ y = 0,\ 9 \end{gathered}$$

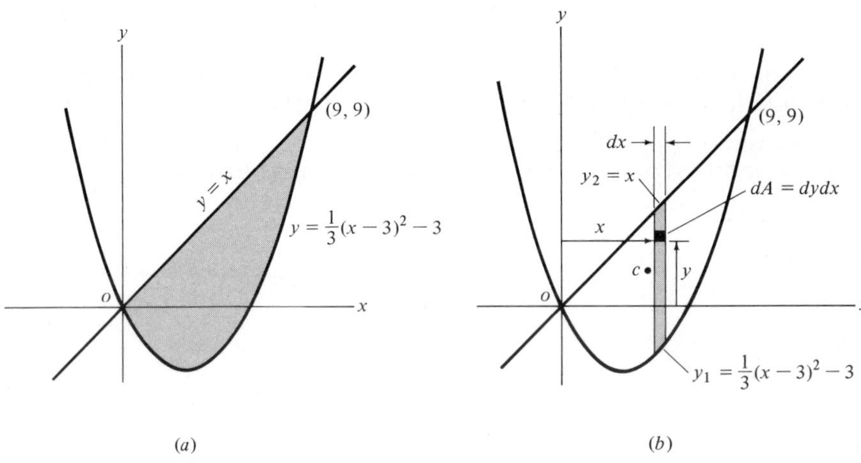

Fig. A-4

The enclosed area is the integral

$$A = \int_0^9 \int_{y_1}^{y_2} dy\, dx \qquad (a)$$

where y_1 is given by the equation of the parabola, and y_2 is the equation of the line. Integrating with respect to y yields

$$A = \int_0^9 \Big[y \Big]_{y_1}^{y_2} dx = \int_0^9 (y_2 - y_1)\, dx$$

$$= \int_0^9 [x - \tfrac{1}{3}(x-3)^2 + 3]\, dx \qquad (b)$$

Remark: This first integration can be interpreted as the area of the vertical strip of width dx indicated in Fig. A-4(*b*).

Integrating Eq. (*b*) yields

$$\left[\frac{x^2}{2} - \tfrac{1}{9}(x-3)^3 + 3x\right]_0^9 = \frac{81}{2} - \frac{(9-3)^3}{9} + 3(9) + \tfrac{1}{9}(0-3)^3$$

$$A = \tfrac{81}{2} = 40.5 \text{ in}^2$$

By Definition A-1, the first moment with respect to the y axis is

$$M_y = \iint x\, dA = \int_0^9 \int_{y_1}^{y_2} x\, dy\, dx \tag{c}$$

Integrating with respect to y yields

$$M_y = \int_0^9 \left[y\right]_{y_1}^{y_2} x\, dx = \int_0^9 (y_2 - y_1)x\, dx$$

$$= \int_0^9 [x - \tfrac{1}{3}(x-3)^2 + 3]x\, dx \tag{d}$$

Remark: This result can be interpreted as the first moment of the thin vertical strip of width dx in Fig. A-4(*b*).

Expanding the integrand and then integrating yields

$$\left[\frac{x^3}{3} - \frac{x^4}{12} + \frac{2x^3}{3} - \frac{3x^2}{2} + \frac{3x^2}{2}\right]_0^9 = 9^2\left[\frac{9}{3} - \frac{9^2}{12} + \frac{(2)(9)}{3}\right]$$

$$M_y = (81)(\tfrac{9}{4}) \text{ in}^3$$

Hence, by Definition A-2, the x centroidal distance is

$$\bar{x} = \frac{M_y}{A} = \frac{(81)(9/4)}{81/2} = 4.5 \text{ in}$$

Similarly, the first moment with respect to the x axis is

$$M_x = \iint y\, dA = \int_0^9 \int_{y_1}^{y_2} y\, dy\, dx$$

$$= \left[\int_0^9 \frac{y^2}{2}\right]_{y_1}^{y_2} dx = \int_0^9 \frac{y_2^2 - y_1^2}{2}\, dx$$

$$= \int_0^9 \left\{\frac{x^2 - [(1/3)(x-3)^2 - 3]^2}{2}\right\} dx \tag{e}$$

Expanding the integrand and then integrating

$$M_x = \left[\frac{-x^5/5 + 3x^4 - 9x^3}{18}\right]_0^9 = \frac{(18)(9)}{10} \text{ in}^3$$

Hence, by Definition A-2, the y centroidal distance is

$$\bar{y} = \frac{M_x}{A} = \frac{(81)(9)/10}{81/2} = 1.8 \text{ in}$$

Then the coordinates of the centroid c are (4.5,1.8), which appear to be physically reasonable for the given area.

EXAMPLE A-2

Find the centroid of the area shown in Fig. A-5(a).

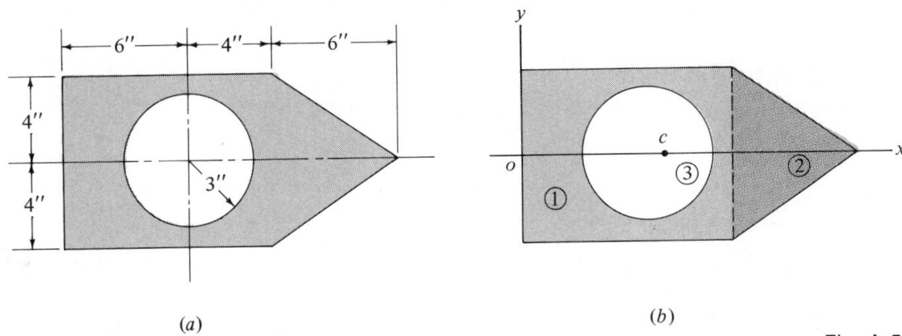

Fig. A-5

Solution: Since the horizontal centerline is an axis of symmetry, this line is a centroidal axis, and the centroid is somewhere on this axis. Thus we need only to locate its horizontal position. Accordingly, let us choose x and y axes as shown in Fig. A-5(b) so that $\bar{y}=0$. To determine $\bar{x}$, let us visualize the total composite area as being made up of the component areas indicated in Fig. A-5(b), where ① is a solid rectangle, ② is a triangle, and ③ is a circular hole. It is usually convenient to set up a table as below.

Component	Area	$\bar{x}$	$M_y = A\bar{x}$
1	$(10)(8) = 80$	5	400
2	$\frac{1}{2}(8)(6) = 24$	$10 + \frac{1}{3}(6) = 12$	288
3	$-\pi(3)^2 = -9\pi$	6	-54π
Total	75.73 in^2		518.35

In this table we have used some results available from Table A-1, namely, that the centroid of a right triangle is at one-third its height (or base). Satisfy yourself as to how each of the entries in the table was obtained. Note also the use of the minus sign to indicate the fact that the circular area is removed from the solid rectangle.

With the results of this table and by Eq. (A-6), we have

$$\bar{x} = \frac{M_y}{A} = \frac{518.35}{75.73} = 6.84 \text{ in}$$

Hence, the centroid of this composite area is located at the coordinate (6.84,0), which appears to be physically reasonable.

EXERCISE PROBLEMS

A-1 By integration, show that the centroid of a right triangle of base b and height h is at $(b/3,h/3)$ as measured from the right angle.

A-2 Using the result of Problem A-1, show by the method of composite areas that the centroid of any triangle is at one-third the perpendicular height measured from any base.

A-3 to A-7 Locate the centroid of the shaded area shown.

PROB. A-3

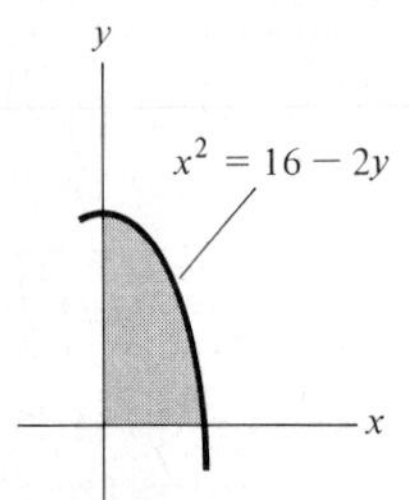

PROB. A-4

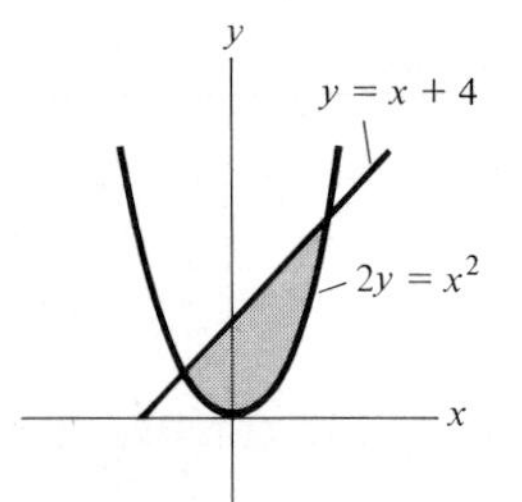

PROB. A-5

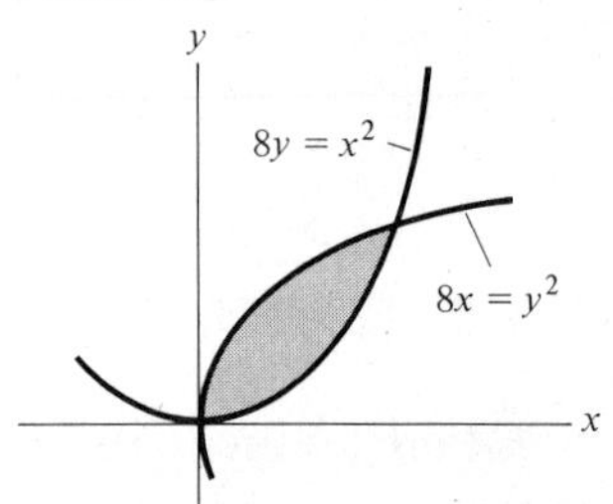

PROB. A-6

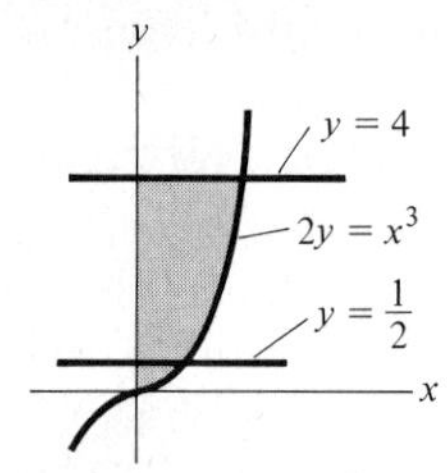

PROB. A-7

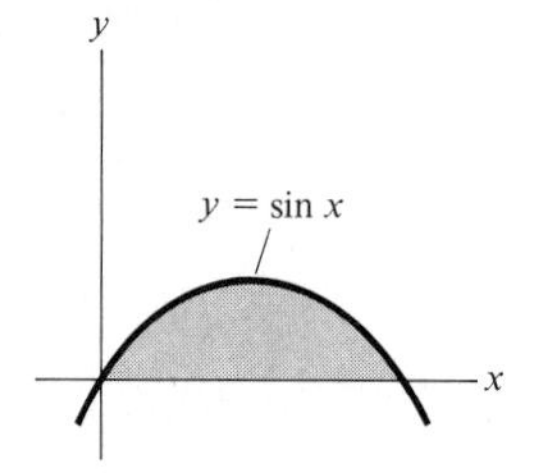

PROB. A-8

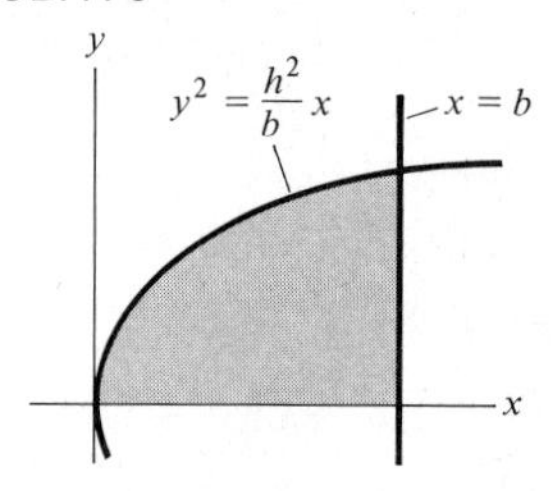

A-8 Locate the centroid for the quadrant of the parabola shown.

A-9 Locate the centroid for the semielliptical area shown.

PROB. A-9

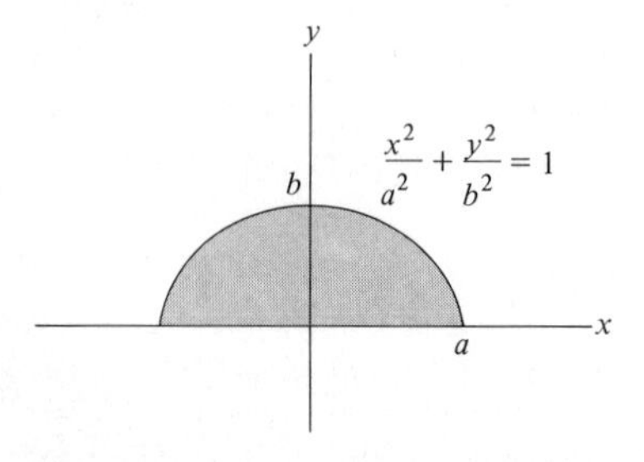

PROB. A-10

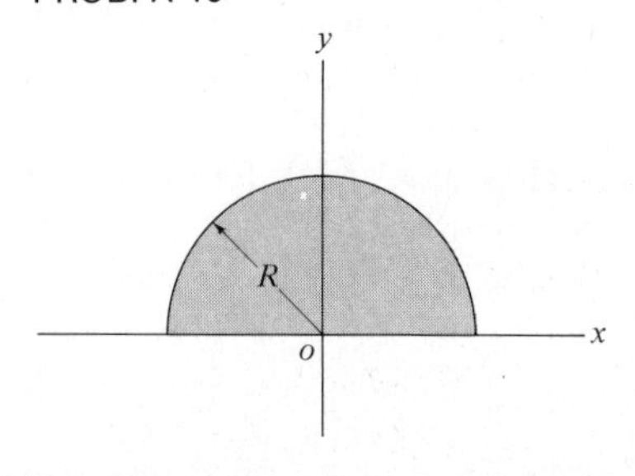

A-10 Locate the centroid for the semicircular area shown.

A-11 to A-16 Locate the centroid of the composite area shown. Refer to the tables in the Appendix as necessary.

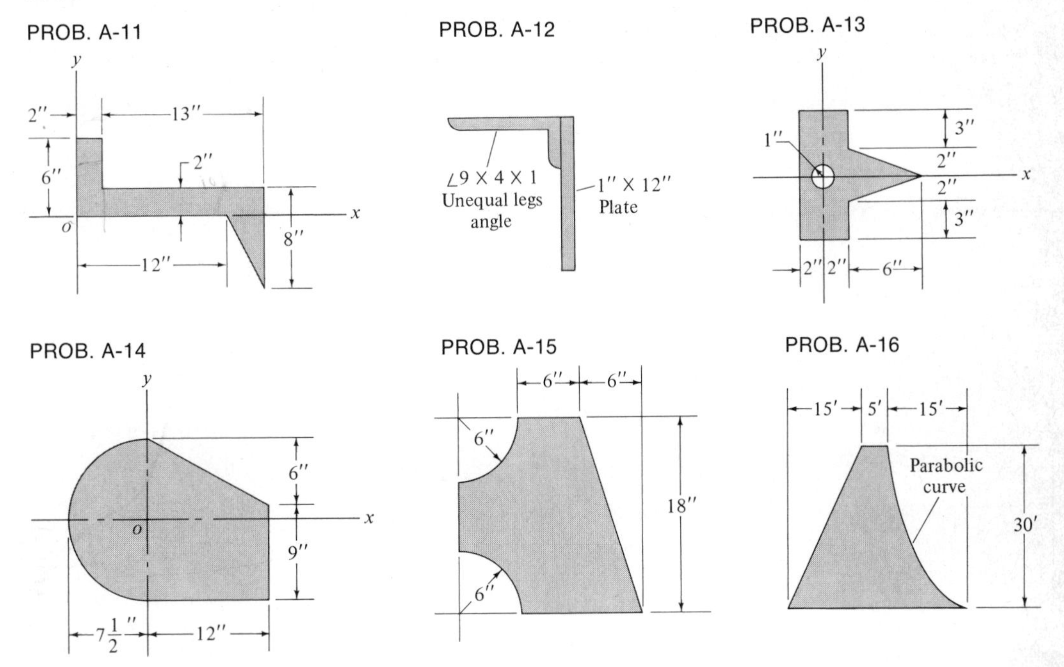

A-3 SECOND MOMENTS OF AREA

In the study of elastic stresses in beams and shafts, an area integral arises which represents a geometric property of the cross-sectional area. Accordingly, it is convenient to define this property by the following:

Definition A-4 The second moment of area (or moment of inertia) with respect to an axis is the integral

$$I = \iint_{\text{area}} p^2 \, dA$$

where p is the perpendicular distance from the axis to the elemental area dA. Thus, referring to Fig. A-6,

$$I_x = \iint_{\text{area}} y^2 \, dA \quad \text{and} \quad I_y = \iint_{\text{area}} x^2 \, dA \qquad \text{(A-7)}$$

where I_x and I_y are the *rectangular* second moments (or just second moments) with respect to the x and y axes, respectively.

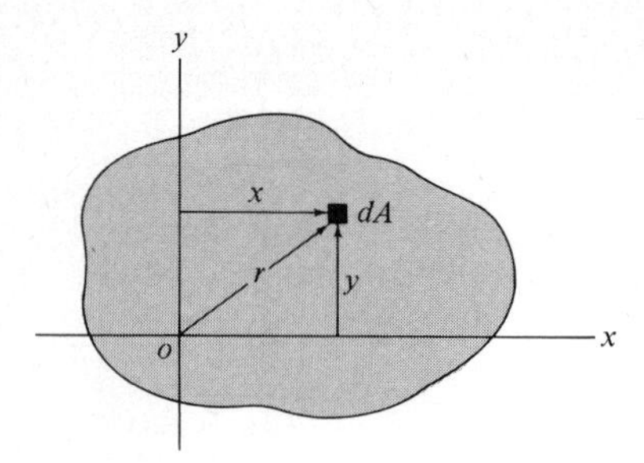

Fig. A-6

Definition A-5 The *polar* second moment of area with respect to a *point* is the integral

$$J = \iint_{\text{area}} p^2 \, dA$$

where p is the distance from the point to the elemental area dA. Referring to Fig. A-6,

$$J_o = \iint_{\text{area}} r^2 \, dA \tag{A-8}$$

Combining Definition A-4 with Definition A-5 we obtain the following:

Theorem A-2 The polar second moment J_o is the sum of the corresponding rectangular second moments I_x and I_y, where the x and y axes are the rectangular axes which intersect at point o; that is,

$$\begin{aligned} J_o &= \iint_{\text{area}} r^2 \, dA = \iint_{\text{area}} (x^2 + y^2) \, dA \\ &= \iint_{\text{area}} x^2 \, dA + \iint_{\text{area}} y^2 \, dA = I_y + I_x \end{aligned} \tag{A-9}$$

The dimensions of both I and J are usually in inch^4 or some other dimension of length raised to the fourth power. Also, since area is inherently a positive value, the second moment is inherently a positive value, with J_o always being larger than the corresponding I_x and I_y. (Recall that the first moment could be either positive or negative.)

Physically, the second moment can be interpreted as a measure of how far the area is distributed away from the particular axis. Thus, referring to Fig. A-7, although both areas are the same size, the one in Fig. A-7(b) would have a considerably larger second moment with respect to the horizontal x axis because more area is located at a greater distance from the x axis. On the other hand, it

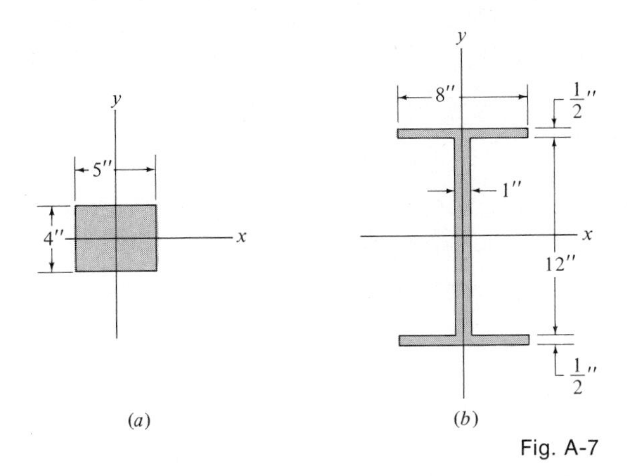

Fig. A-7

is difficult to say which of the two areas would have the larger second moment with respect to the vertical y axis. This idea of how far an area is distributed from a given axis is quantified by the following:

Definition A-6 The radius of gyration of an area with respect to a given axis (or point) is the positive number k such that

$$Ak^2 = I \text{ (or } J\text{)}$$

where I (or J) is the second moment of the area with respect to the given axis (or point).

Thus, for a given area, a large I corresponds to a large k, and vice versa. Generally the radius of gyration is not a readily identifiable physical dimension of the area such as its width, length, centroidal distance, etc. Values of the radius of gyration for standard structural cross sections are given in some of the tables in Appendix B.

> ***Remark:*** We have used the symbol k in this discussion to denote the radius of gyration to avoid confusing this number with the radial distance r in Fig. A-6. However, in structural mechanics it is not uncommon for r to be used to denote radius of gyration, as, for example, in some of the tables in Appendix B and in some of the discussions earlier in this text.

While the definitions of I and J are applicable to any axis or point, the centroidal axes play an important role in the study of elastic bending of beams and the twisting of shafts. Consequently, let us consider the area in Fig. A-8, where the x axis is any arbitrary horizontal axis and the x_c axis is the horizontal centroidal axis, with the distance between these axes denoted by d. By Definition A-4,

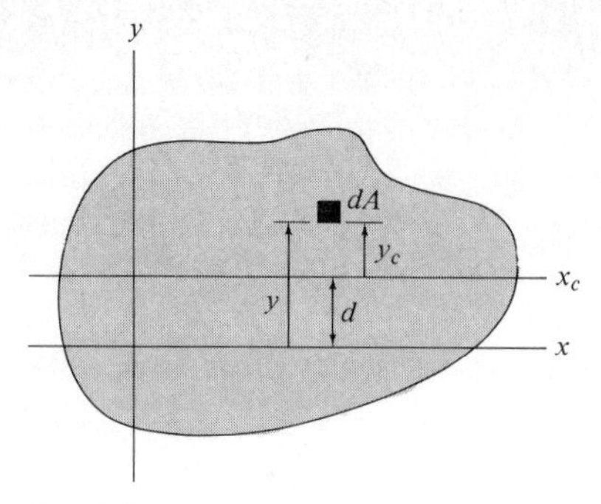

Fig. A-8

$$I_x = \iint_{\text{area}} y^2 \, dA$$

while
$$I_{x_c} = \iint_{\text{area}} y_c^2 \, dA$$

where
$$y = y_c + d$$

Substituting this last expression into the above integral for I_x yields

$$I_x = \iint_{\text{area}} (y_c + d)^2 \, dA$$

$$= \iint_{\text{area}} (y_c^2 + 2y_c d + d^2) \, dA$$

$$= \iint_{\text{area}} y_c^2 \, dA + 2d \iint_{\text{area}} y_c \, dA + d^2 \iint_{\text{area}} dA \qquad \text{(A-10)}$$

The first integral can be identified as I_x and the last term as the product d^2A. Recalling Theorem A-1 and the fact that the x_c axis is a centroidal axis, the second integral in Eq. (A-10) has a value of zero. Hence, Eq. (A-10) becomes

$$I_x = I_{x_c} + Ad^2 \qquad \text{(A-11)}$$

which is the basis for the following theorem:

Theorem A-3 *The Parallel-Axis Transfer Theorem* The second moment of area with respect to an axis is equal to the second moment with respect to the parallel centroidal axis plus the product of the area and the square of the distance between the two axes.

Furthermore, since the area A and the square of the distance d are inherently positive quantities, we have the following:

Theorem A-4 Of all axes *parallel* to a given axis, the parallel centroidal axis has the *least* second moment of area.

Essentially, Theorem A-3 enables us to calculate the second moment with respect to some desired axis if we know the second moment with respect to the centroidal axis, and vice versa.

Just as in the case of first moments of area, it is often convenient to calculate second moments of composite areas by treating the composite area as a combination of several simple areas. Thus the area in question can be visualized as a combination of solid shapes and holes, so that

$$A = \pm A_1 \pm A_2 \pm \cdots \pm A_n \tag{A-12}$$

where the minus sign corresponds to those component areas that are removed. Then the second moment with respect to some x axis can be found by

$$\begin{aligned} I_x &= \iint_{\pm A_1 \pm A_2 \pm \cdots \pm A_n} y^2 \, dA \\ &= \pm \iint_{A_1} y^2 \, dA \pm \iint_{A_2} y^2 \, dA \pm \cdots \pm \iint_{A_n} y^2 \, dA \\ &= \pm I_{1_x} \pm I_{2_x} \pm \cdots \pm I_{n_x} \end{aligned} \tag{A-13}$$

where the choice of plus or minus sign depends upon whether the component area is added or subtracted in Eq. (A-12). Equation (A-13) is simply the statement that the second moment of a composite area is the sum of the second moments of the component areas. Expressions similar to Eq. (A-13) can be obtained for I_y, J, or any other second moment of a composite area.

EXAMPLE A-3

Find the second moment of a solid circular area of radius R with respect to a diametral axis.

Solution: A solid circular area is shown in Fig. A-9(a), and we observe that both the x and y axes are diametral axes. By Definition A-4,

$$I_x = \iint_{\text{area}} y^2 \, dA \qquad \text{and} \qquad I_y = \iint_{\text{area}} x^2 \, dA$$

and, since the x and y axes are both diametral axes,

$$I_x = I_y \tag{a}$$

While we could proceed to evaluate either of the above integrals, it would soon become apparent that the circular boundary is

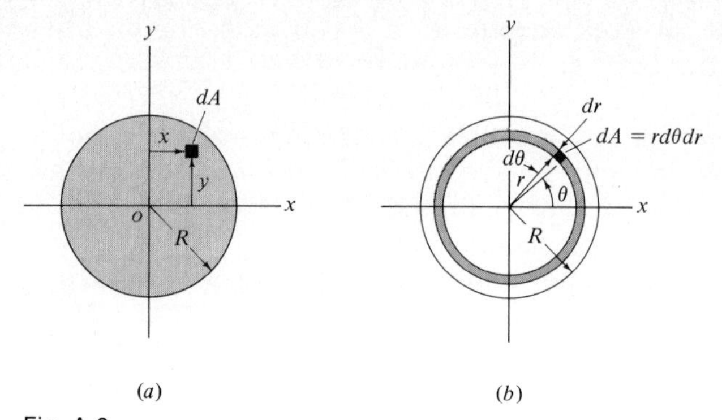

Fig. A-9

not very compatible with the rectangular x,y coordinate system. Therefore, we will proceed in a different manner.

Observe from Definition A-5 and Theorem A-2 that the polar second moment J_o will be related to the second moments I_x and I_y by

$$J_o = I_x + I_y = 2I_x = 2I_y \tag{b}$$

where we have used Eq. (a). Hence

$$I_x = I_y = \tfrac{1}{2}J_o = \tfrac{1}{2}\iint_{\text{area}} r^2\, dA \tag{c}$$

Now, using the polar coordinates indicated in Fig. A-9(b), we obtain

$$\begin{aligned} J_o &= \int_0^R \int_0^{2\pi} r^2(r\, d\theta\, dr) \\ &= \int_0^R r^3 \Big[\theta\Big]_0^{2\pi} dr = \int_0^R r^2(2\pi r\, dr) \end{aligned} \tag{d}$$

Remark: This result can be interpreted as the polar second moment of the thin ring-like area of radius r and thickness dr indicated in Fig. A-9(b).

Integrating Eq. (d) yields

$$J_o = 2\pi \int_0^R r^3\, dr = 2\pi \left[\frac{r^4}{4}\right]_0^R = \frac{\pi R^4}{2}$$

Finally, by Eq. (c) the second moment of a solid circular cross section with respect to a diametral axis is

$$I_x = I_y = \tfrac{1}{2}J_o = \frac{\pi R^4}{4}$$

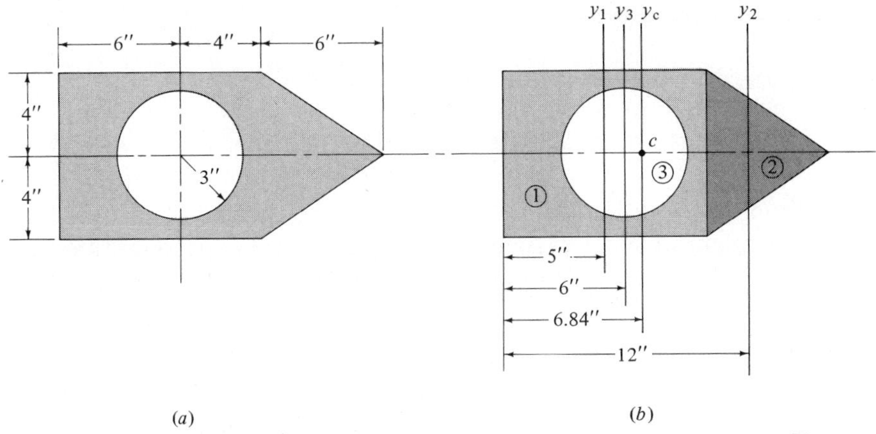

Fig. A-10

EXAMPLE A-4

For the area shown in Fig. A-10(*a*), find the second moment with respect to the vertical centroidal axis.

Solution: The centroid of this composite area was found in Example A-2, and the location of the vertical centroidal axis y_c is indicated in Fig. A-10(*b*). Also in this figure, the y_1, y_2, and y_3 axes are the centroidal axes of the component areas ①, ②, and ③, respectively. Once again it is convenient to set up a table.

Component	I_c	A	d	Ad^2	$I_{y_c} = I_c + Ad^2$
1	$\frac{1}{12}(8)(10)^3 = 667$	80	1.84	271	938
2	$\frac{1}{36}(8)(6)^3 = 48$	24	5.16	639	687
3	$-\frac{\pi(3)^4}{4} = -63.6$	-9π	0.84	-20	-83.6
Total	651.4			890	1,541.4

In this table we have used several results from Table A-1, namely,

$$I_c\ (\text{rectangle}) = \tfrac{1}{12}bh^3$$

$$I_c\ (\text{triangle}) = \tfrac{1}{36}bh^3$$

$$I_c\ (\text{circle}) = \frac{\pi R^4}{4}$$

Satisfy yourself as to how the remaining entries in the above table were obtained, and note the use of the minus sign to indicate the removal of area ③.

Thus, I_{y_c} for the composite area is

$$I_{y_c} = 1{,}541.4 \text{ in}^4$$

EXERCISE PROBLEMS

A-17 Show that for a regular rectangle of height h and base b, the second moment with respect to the horizontal centroid axis is $bh^3/12$. Also show that the second moment with respect to the base axis is $bh^3/3$.

A-18 Show that for a right triangle of height h and base b, the second moment with respect to the horizontal centroidal axis is $bh^3/36$. Also show that the second moment with respect to the base axis is $bh^3/12$.

A-19 For the quadrant of the parabola shown, determine I_x, I_y, and J_o.

PROBS. A-19 and A-20

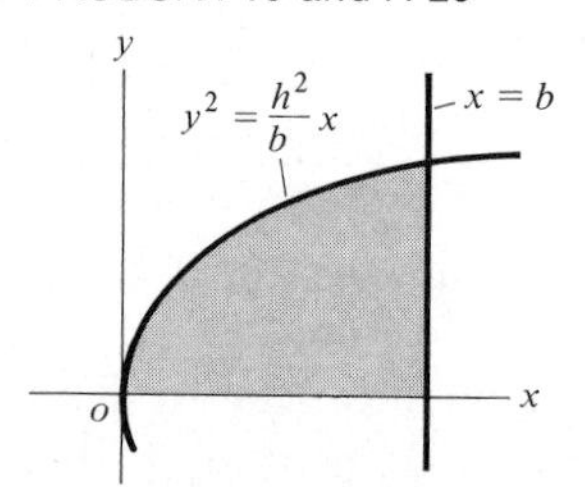

PROB. A-21

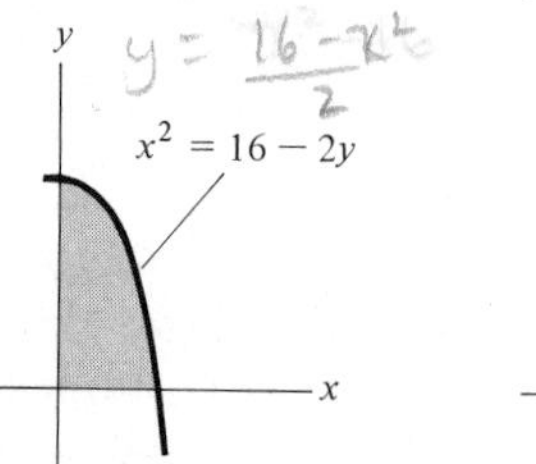

PROBS. A-22 and A-23

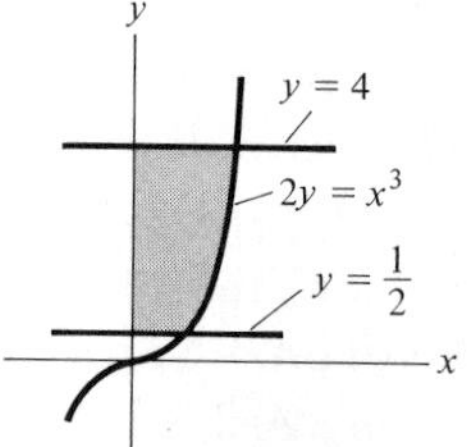

A-20 Using the result of Prob. A-19 and the parallel transfer theorem, determine I_{x_c}, I_{y_c}, and J_c, where c is the centroid.

A-21 For the shaded area shown, find I_y and I_{y_c}.

A-22 For the shaded area shown, find I_x and I_{x_c}.

A-23 For the shaded area shown, find I_y and I_{y_c}.

A-24 For the semicircular area shown, find the second moment with respect to the horizontal centroidal axis.

PROB. A-24

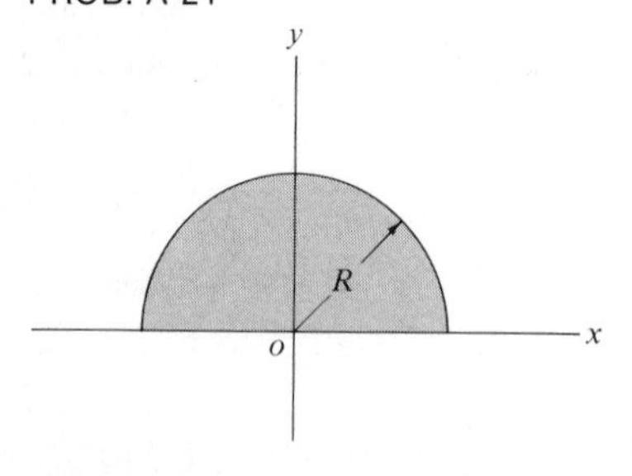

PROB. A-25

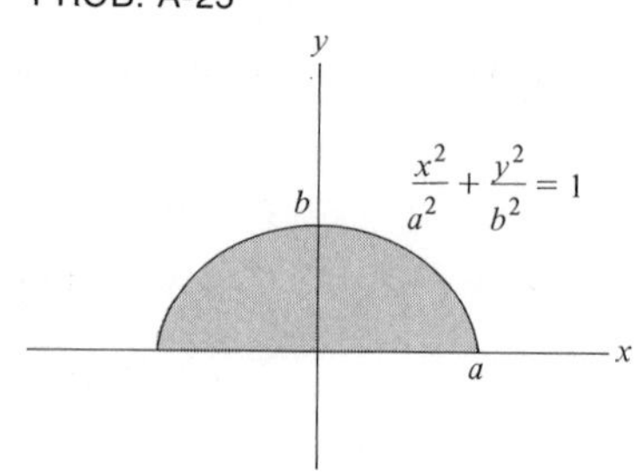

A-25 For the semiellipse shown, find I_x.

A-26 Consider a circular ring-like area of inside radius R_i and outside radius R_o. First show that $J_c = (\pi/2)(R_o^4 - R_i^4)$. Then show that as $R_i \rightarrow R_o$, $2\pi R_o^3 t$ becomes a good approximation for J_c, where $t = R_o - R_i$.

A-27 and A-28 For the shaded area shown, find the radius of gyration with respect to the x axis.

PROBS. A-27 and A-29

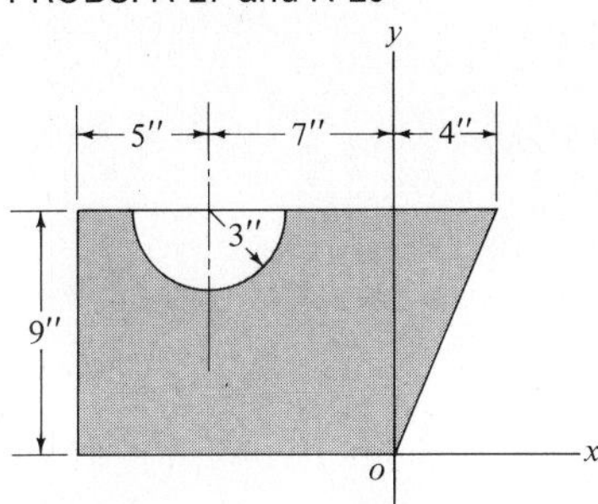

PROBS. A-28 and A-30

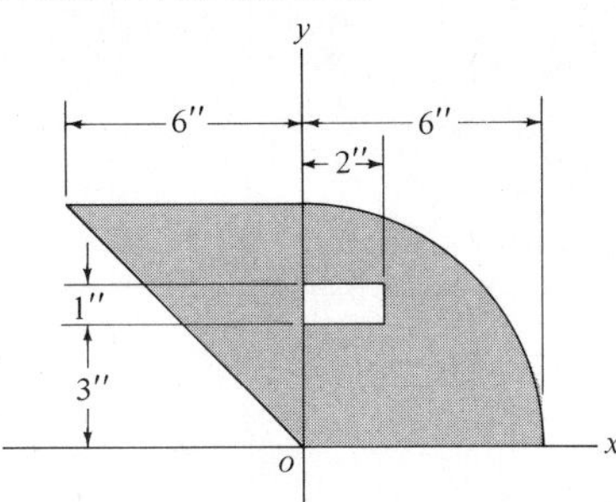

A-29 to A-31 For the shaded area shown, find the radius of gyration with respect to the y axis.

PROB. A-31

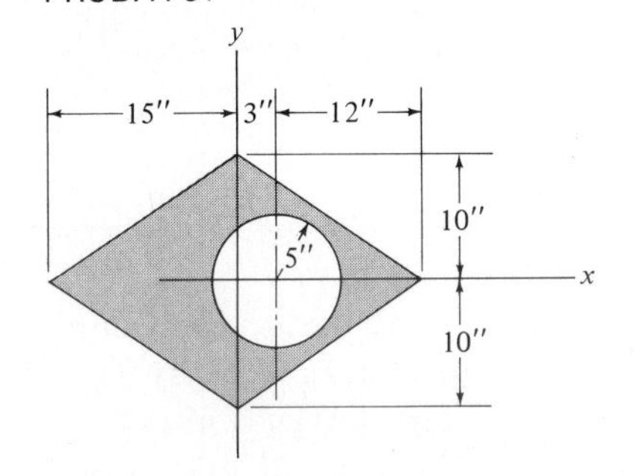

PROB. A-32

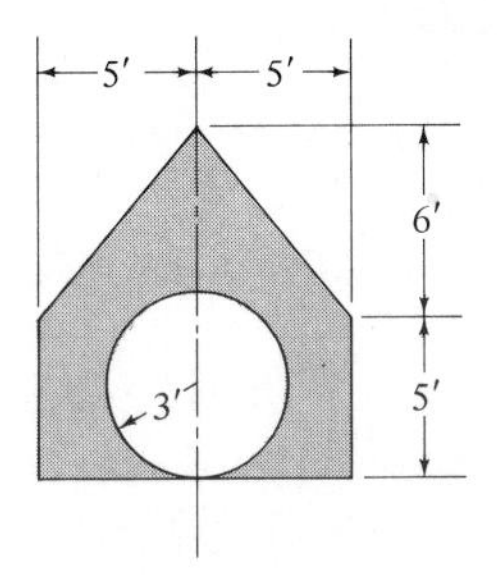

A-32 and A-33 For the shaded area shown, find the second moment with respect to the horizontal centroidal axis.

PROBS. A-33 and A-34

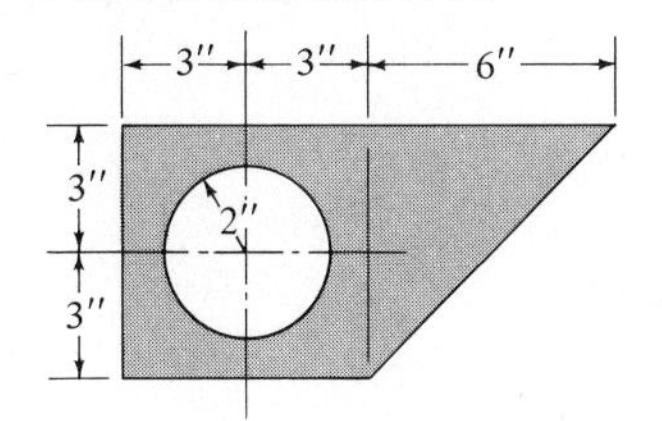

PROB. A-35

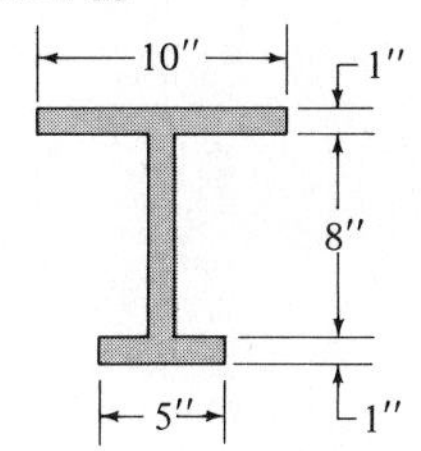

A-34 For the shaded area shown, find the second moment with respect to the vertical centroidal axis.

A-35 to A-40 For the shaded area shown, find the second moment with respect to the horizontal centroidal axis. Refer to the tables in Appendix B as necessary.

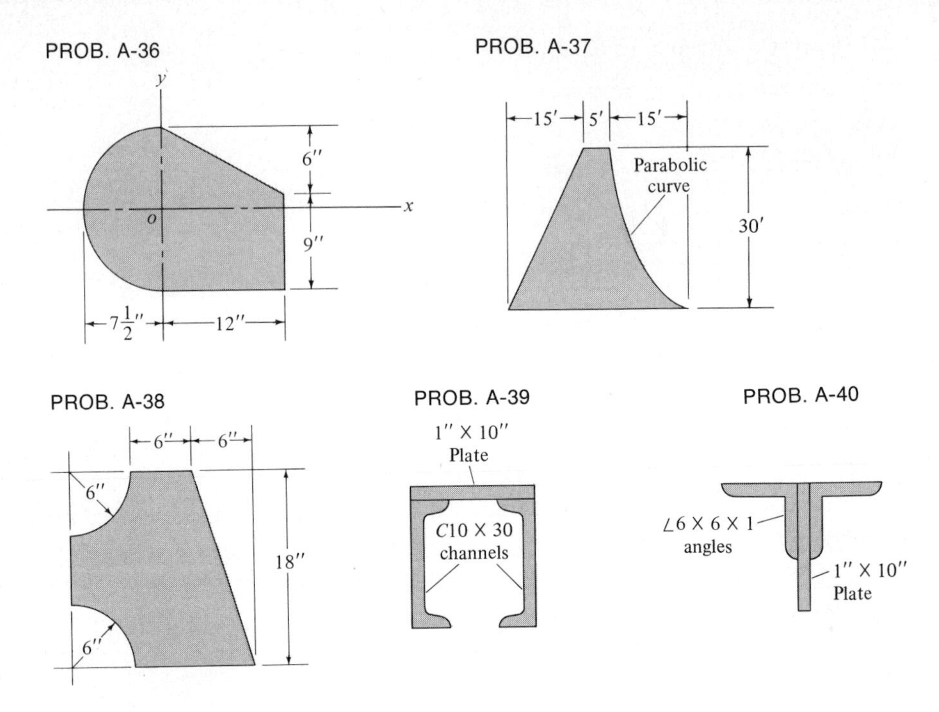

A-4 MAXIMUM AND MINIMUM (PRINCIPAL) SECOND MOMENTS

The second moment of area with respect to any axis has been defined by Definition A-4. Accordingly, I_x and I_y as given by Eq. (A-7) are the second moments with respect to two particular axes, namely, the x and y axes. But what about the second moments with respect to some other axis (or axes), such as the u or v axes in Fig. A-11, which are oriented at some angle θ relative to the x and y axes? According to Definitions A-4 and A-5

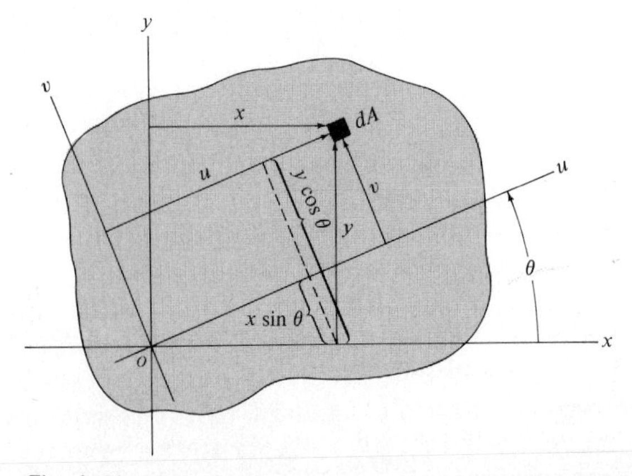

Fig. A-11

$$I_u = \iint_{\text{area}} v^2 \, dA \qquad I_v = \iint_{\text{area}} u^2 \, dA \tag{A-14}$$

and

$$J_o = I_u + I_v = I_x + I_y$$

We now pose the question: Of all possible axes emanating from point o and oriented relative to a given set of rectangular x, y axes, which axis (or axes) will have the largest (or smallest) second moment of area? To answer this question, we observe from Fig. A-11 that

$$v = y \cos \theta - x \sin \theta$$

Substituting this into the expression for I_u in Eq. (A-14) yields

$$\begin{aligned} I_u &= \iint_{\text{area}} (y \cos \theta - x \sin \theta)^2 \, dA \\ &= \iint_{\text{area}} y^2 \cos^2 \theta \, dA - \iint_{\text{area}} 2\, xy \sin \theta \cos \theta \, da \\ &\quad + \iint_{\text{area}} x^2 \sin^2 \theta \, dA \\ &= I_x \cos^2 \theta + I_y \sin^2 \theta - 2 \sin \theta \cos \theta \, P_{xy} \end{aligned} \tag{A-15}$$

where

$$P_{xy} = \iint_{\text{area}} xy \, dA \tag{A-16}$$

is called the *product second moment of area* (or product of inertia) with respect to the x, y axes. Thus, to answer the question posed earlier, we need only to maximize (or minimize) Eq. (A-15) relative to the angle θ. Differentiating Eq. (A-15) with respect to θ yields

$$\frac{dI_u}{d\theta} = -2I_x \sin \theta \cos \theta + 2I_y \sin \theta \cos \theta + 2P_{xy}(\sin^2 \theta - \cos^2 \theta)$$

Setting this equal to zero and using the double-angle identities yields

$$\tan 2\theta = \frac{-2P_{xy}}{I_x - I_y} \tag{A-17}$$

for the angle θ which will maximize (or minimize) the second moment of area relative to all axes passing through point o. A considerable amount of useful information is contained in Eq. (A-17)

Observation 1 In general, Eq. (A-17) yields two values for θ which will differ by 90°. When substituted into Eq. (A-15), one of these will yield the maximum value of I, while the other will yield the minimum. These maximum and minimum values are called the *principal values* of the second moment, and the corresponding orthogonol axes are called the *principal axes* relative to point o.

Observation 2 When P_{xy} is zero, Eq. (A-17) yields the values $\theta = 0$ and $\theta = 90°$, which means that in this case the given x and y axes are the principal axes and I_x and I_y are the principal values.

From this discussion and Eqs. (A-15) and (A-17), it is apparent that to determine the principal values of the second moment relative to some point o, it is sufficient to calculate I_x, I_y, and P_{xy}, where the x and y axes are any pair of mutually orthogonal axes passing through point o. The previous section discussed the calculation of the rectangular second moments I and I_y, while the product second moment P_{xy} can be calculated by Eq. (A-16). By using the same type of argument that was used in conjunction with Fig. A-3 to show that an axis of symmetry was necessarily a centroidal axis, it can be shown that if either the x or y axis is an axis of symmetry, the product moment P_{xy} will be zero. This result along with Observations 1 and 2 above is the basis for:

Theorem A-5 Any axis of symmetry of an area is a *principal centroidal* axis corresponding to either the maximum or minimum centroidal second moment. Furthermore, the centroidal axis perpendicular to this axis of symmetry will also be a principal centroidal axis corresponding to either the maximum or minimum centroidal second moment.

This theorem is extremely useful in problems dealing with the bending of beams and the buckling of columns having symmetrical cross sections such as those discussed in Chaps. 9 and 10. On the other hand, if the area in question does not possess an axis of symmetry, then Eqs. (A-15) to (A-17) are used to determine the principal second moments and principal axes.

Remark: Equation (A-15) is identical in form to Eq. (8-7) dealing with the transformation of normal stresses. Consequently, Eq. (A-15) can be handled by the same graphical technique developed in Sec. 8-4, namely, the Mohr circle. The development of the Mohr circle for second moments is analogous to that for the Mohr circle of stress in Sec. 8-4 and will not be presented here. You simply replace σ_x, σ_y, and τ_{xy} with I_x, I_y, and P_{xy}, respectively.

Just as it was often necessary to transfer rectangular second moments from one axis to another parallel axis by the use of the parallel-transfer theorem (Theorem A-3), it is often necessary to

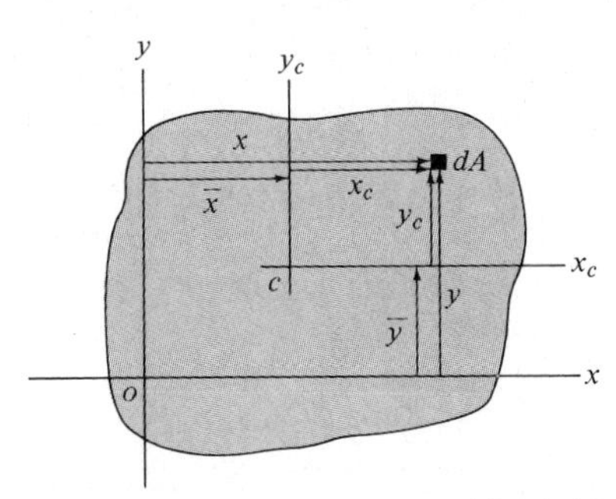

Fig. A-12

transfer product second moments. Referring to Fig. A-12, the x_c and y_c axes are centroidal axes, and the x and y axes are a parallel set of axes with centroidal distances from the x and y axes to the x_c and y_c axes, being $\bar{y}$ and $\bar{x}$, respectively. Then

$$x = x_c + \bar{x} \qquad \text{and} \qquad y = y_c + \bar{y}$$

and
$$P_{xy} = \iint_{\text{area}} xy\, dA = \iint_{\text{area}} (x_c + \bar{x})(y_c + \bar{y})\, dA$$

$$= \iint_{\text{area}} x_c y_c\, dA + \iint_{\text{area}} x_c \bar{y}\, dA + \iint_{\text{area}} y_c \bar{x}\, dA + \iint_{\text{area}} \bar{x}\bar{y}\, dA$$

$$= P_{x_c y_c} + \bar{y} \iint_{\text{area}} x_c\, dA + \bar{x} \iint_{\text{area}} y_c\, dA + \bar{x}\bar{y}A$$

Now, the two middle terms vanish by virtue of the fact that the first moment with respect to a centroidal axis is zero. Thus we arrive at the following:

Theorem A-6 *Product-Transfer Theorem* The product second moment of an area with respect to a pair of orthogonal axes is equal to the product second moment with respect to the parallel pair of centroidal axes plus the product of the area and the centroidal distances between the corresponding axes.

$$P_{xy} = P_{x_c y_c} + A\bar{x}\bar{y} \tag{A-18}$$

Remark: In this theorem the centroidal distances may be positive or negative. Care must be exercised to see that the algebraic signs are correct since the product second moments can be positive or negative (whereas the rectangular second moments could only be positive).

The following example will illustrate the calculation of the product second moments and the determination of the principal second moments.

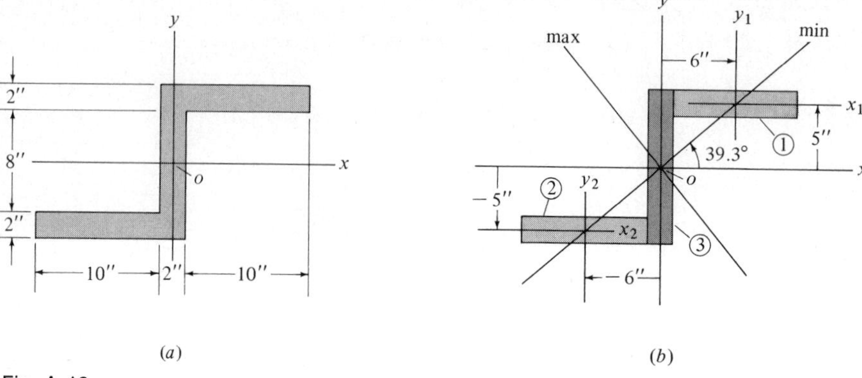

Fig. A-13

EXAMPLE A-5

Find the principal centroidal axes and the corresponding principal values for the area shown in Fig. A-13(*a*).

Solution: Although neither the x nor y axis is an axis of symmetry, the point o is nevertheless the centroid of the area because the area is symmetrical with respect to point o; that is, it is balanced about point o. The total area will be treated as a composite area made up of three rectangles ①, ②, and ③ with corresponding centroidal axes (x_1,y_1), (x_2,y_2), and (x,y), respectively, as indicated in Fig. A-13(*b*). For ①

$$I_x = I_{x_1} + A\bar{y}_1^{\,2} = \tfrac{1}{12}(10)(2)^3 + (20)(5)^5$$
$$= 506 \text{ in}^4$$

$$I_y = I_{y_1} + A\bar{x}_1^{\,2} = \tfrac{1}{12}(2)(10)^3 + (20)(6)^2$$
$$= 886 \text{ in}^4$$

$$P_{xy} = P_{x_1y_1} + A\bar{x}_1\bar{y}_1 = 0 \text{ (because of symmetry)} + (20)(6)(5)$$
$$= 600 \text{ in}^4$$

For ②

$$I_x = \text{same as for ①} = 506 \text{ in}^4$$
$$I_y = \text{same as for ①} = 886 \text{ in}^4$$
$$P_{xy} = P_{x_1y_1} + Ax_2y_2 = 0 + (20)(-6)(-5)$$
$$= +600 \text{ in}^4$$

For ③

$$I_x = \tfrac{1}{12}(2)(12)^3 = 288 \text{ in}^4$$
$$I_y = \tfrac{1}{12}(12)(2)^3 = 8 \text{ in}^4$$
$$P_{xy} = 0 \qquad \text{(because of symmetry)}$$

Thus, for the composite area

$$I_x = 506 + 506 + 288 = 1{,}300 \text{ in}^4$$
$$I_y = 886 + 886 + 8 = 1{,}780 \text{ in}^4$$
$$P_{xy} = 600 + 600 + 0 = 1{,}200 \text{ in}^4$$

Substituting these values into Eq. (A-17) yields

$$\tan 2\theta = \frac{-2(1{,}200)}{1{,}300 - 1{,}780} = \frac{-2{,}400}{-480}$$
$$2\theta = 78.6° \text{ and } 258.6°$$
$$\theta = 39.3° \text{ and } 129.3°$$

as the orientations of the principal axes. These axes are indicated in Fig. A-13(*b*). Substituting these angles into Eq. (A-15) yields

$$\begin{aligned} I_{39.3°} &= 1{,}300 \cos^2 (39.3) + (1{,}780) \sin^2 (39.3) \\ &\quad -2(1{,}200) \sin (39.3) \cos (39.3) \\ &= 778 + 714 - 1{,}176 = 316 \text{ in}^4 \\ I_{129.3°} &= 1{,}300 \cos^2 (129.3) + (1{,}780) \sin^2 (129.3) \\ &\quad -2(1200) \sin (129.3) \cos (129.3) \\ &= 521 + 1{,}066 + 1{,}176 = 2{,}763 \text{ in}^4 \end{aligned}$$

Hence

$$I_{max} = 2{,}763 \text{ in}^4 \qquad \text{and} \qquad I_{min} = 316 \text{ in}^4$$

EXERCISE PROBLEMS

A-41 to A-44 For the area shown, find P_{xy} and $P_{x_c y_c}$.

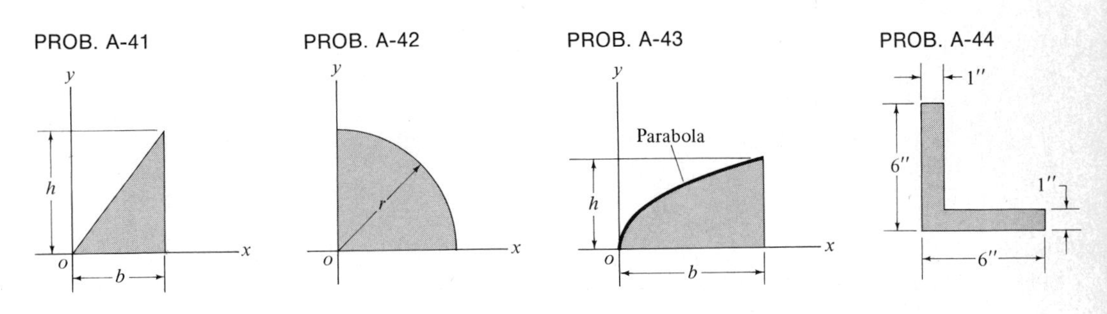

A-45 to A-48 For the areas shown, find the principal centroidal second moments and the corresponding principal axes.

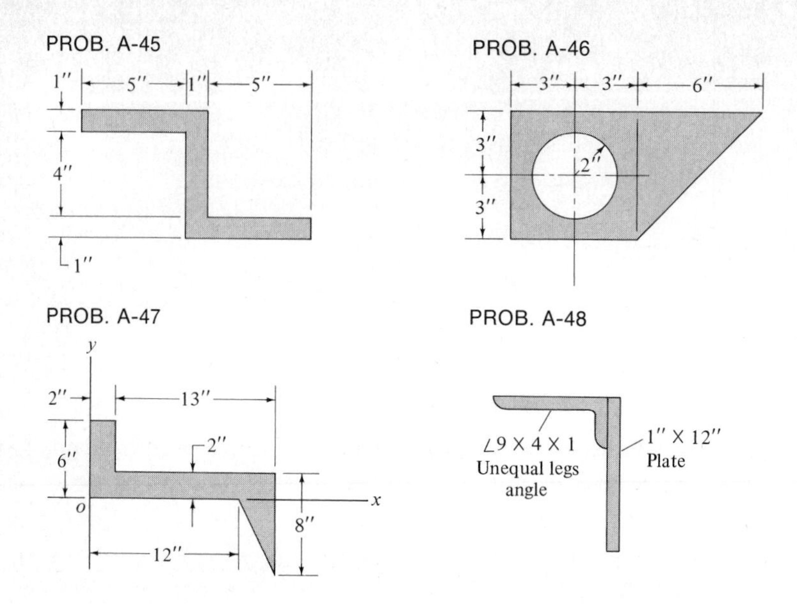

A-5 CENTROIDS OF VOLUMES

The physical concept of mass plays a primary role in the study of dynamics. Consequently, various geometric properties of volumes arise in the formulation of the dynamical equations of real bodies. Several of these properties are analogous to the properties of areas discussed in the previous sections and will be formally defined in terms of integrals over volumes.

Consider the volume in Fig. A-14(a). Formally, the amount of volume, or simply the volume, is given by

$$V = \iiint_{\text{vol}} dV$$

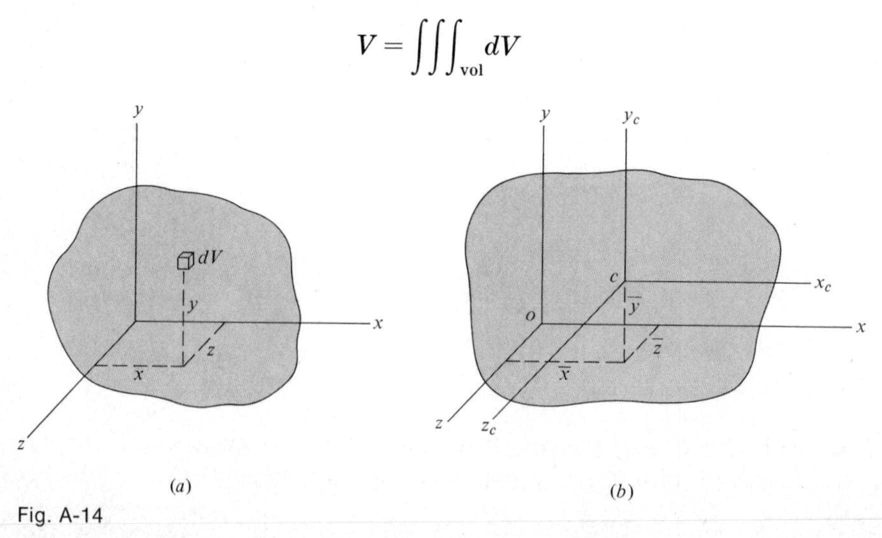

Fig. A-14

where the triple integral is used to indicate that the elemental volume dV is usually expressed in terms of differentials of some three-dimensional coordinates such as dx, dy, and dz, and the corresponding integration operation becomes the ordinary Riemann integration of elementary calculus.

Definition A-7 The first moment of volume with respect to a *plane* is the integral

$$M = \iiint_{\text{vol}} p \, dV$$

where p is the perpendicular distance from the plane to the elemental volume dV. Thus, referring to Fig. A-14(a),

$$M_{xy} = \iiint_{\text{vol}} z \, dV \qquad M_{xz} = \iiint_{\text{vol}} y \, dV \qquad M_{yz} = \iiint_{\text{vol}} x \, dV \tag{A-19}$$

Definition A-8 The centroid of the volume is the point c whose coordinates $(\bar{x},\bar{y},\bar{z})$ are given by

$$\bar{x} = \frac{M_{yz}}{V} \qquad \bar{y} = \frac{M_{xz}}{V} \qquad \bar{z} = \frac{M_{xy}}{V} \tag{A-20}$$

where $\bar{x}$, $\bar{y}$, and $\bar{z}$ are called the centroidal distances from the y–x, x–z, and x–y planes, respectively.

Definition A-9 Any plane passing through the centroid of a volume is called a *centroidal plane.* The intersection of two centroidal planes defines a *centroidal axis* of the volume.

For example, in Fig. A-14(b), the intersection of the x_c–y_c plane with the x_c–z_c plane defines the x_c axis.

By Definitions A-8 and A-9 we have the following theorem:

Theorem A-7 The first moment of volume with respect to any centroidal plane is zero. Conversely, if the first moment of volume with respect to some plane is zero, the plane is a centroidal plane. Furthermore, any plane of symmetry for a volume is necessarily a centroidal plane.

The last statement in Theorem A-7 follows from the same type of argument that was used in conjunction with axes of symmetry for plane areas (see Fig. A-3).

Just as complex-shaped areas could often be visualized as being composed of several simple component areas, complex-shaped volumes can often be visualized as being made up of several simple-shaped volumes, such as rectangular prisms, triangular prisms, spheres, circular cylinders, etc. Accordingly, the total volume becomes

$$V_{\text{total}} = \pm V_1 \pm V_2 \pm \cdots \pm V_n$$

and the total first moment with respect to some plane becomes

$$M_{\text{total}} = \pm M_1 \pm M_2 \pm \cdots \pm M_n$$

where n is the number of component volumes and the plus or minus sign indicates whether the volume is added or removed. Thus, the centroid of a composite volume is given by

$$\bar{x} = \frac{M_{yz,\text{total}}}{V_{\text{total}}} \qquad \bar{y} = \frac{M_{xz,\text{total}}}{V_{\text{total}}} \qquad \bar{z} = \frac{M_{xy,\text{total}}}{V_{\text{total}}} \tag{A-21}$$

Table A-1 gives the centroid of several simple volumes.

EXAMPLE A-6

Find the centroid of a solid hemisphere of radius R.

Solution: Referring to Fig. A-15, the x–y and y-z planes are planes of symmetry, so that the y axis is a centroidal axis. Thus we need only find the $\bar{y}$ centroidal distance to locate the centroid of the hemisphere.

Although the hemisphere is naturally suited to spherical coordinates, it is perhaps more instructive to handle this problem in a different manner.

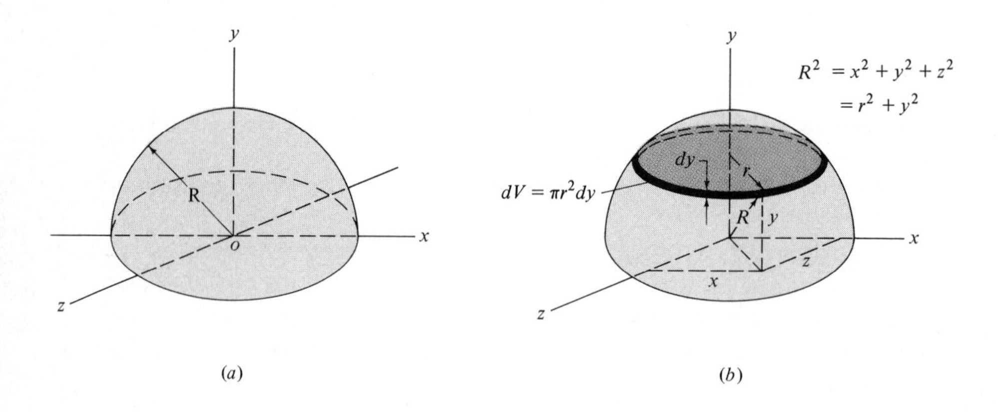

Recalling Eqs. (A-19) and (A-20)

$$M_{xz} = \iiint_{\text{vol}} y \, dV$$

$$\bar{y} = \frac{M_{xz}}{V} \tag{a}$$

Consequently, in Fig. A-15(*b*) we choose as the elemental volume dV the thin disk-like horizontal slice of radius r and thickness dy, so that

$$dV = \pi r^2\, dy$$

where
$$r^2 = R^2 - y^2$$

Notice that since this slice is horizontal, all parts of it are the same distance y from the x–z plane, so that

$$M_{xz} = \int_0^R y[\pi r^2]\, dy = \int_0^R y[\pi(R^2 - y^2)]\, dy$$

$$= \pi \left[R^2 \frac{y^2}{2} - \frac{y^4}{4}\right]_0^R = \frac{\pi R^4}{4} \qquad (b)$$

The volume of the hemisphere is

$$V = \iiint_{\text{vol}} dV = \int_0^R \pi r^2\, dy$$

$$= \int_0^R \pi[R^2 - y^2]\, dy = \pi \left[R^2 y - \frac{y^3}{3}\right]_0^R$$

$$= \pi \tfrac{2}{3} R^3 \qquad (c)$$

Combining Eqs. (*a*) to (*c*) yields

$$\bar{y} = \frac{\pi R^4/4}{2\pi/3R^3} = \tfrac{3}{8} R$$

A-6 SECOND MOMENTS OF VOLUMES

As mentioned earlier, in the study of the dynamical motion of bodies, integrals arise which are analogous to the second moment of area. These integrals are formally defined by the following:

Definition A-10 The second moment of volume (or moment of inertia) with respect to a *plane* is the integral

$$I = \iiint_{\text{vol}} p^2\, dV$$

where p is the perpendicular distance from the *plane* to the elemental volume dV. Thus, referring to Fig. A-16(*a*),

$$I_{xy} = \iiint_{\text{vol}} z^2\, dV \qquad I_{xz} = \iiint_{\text{vol}} y^2\, dV \qquad I_{yz} = \iiint_{\text{vol}} x^2\, dV \qquad \text{(A-22)}$$

where I_{xy} is the second moment with respect to the x–y plane, etc.

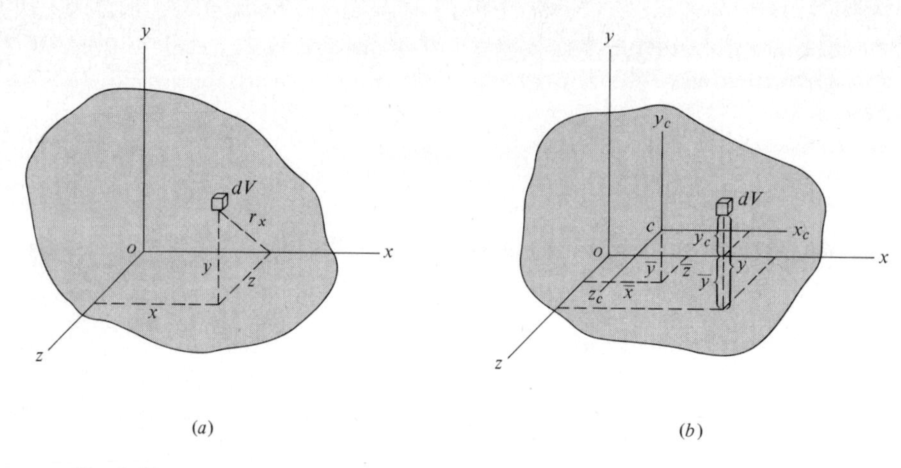

Fig. A-16

Definition A-11 The second moment of volume with respect to an *axis* is the integral

$$I = \iiint_{\text{vol}} p^2 \, dV$$

where p is the perpendicular distance from the *axis* to the elemental volume dV. Referring to Fig. A-16(a), we have

$$I_x = \iiint_{\text{vol}} r_x^2 \, dV \tag{A-23}$$

and similar expressions can be obtained for I_y and I_z.

Noting that

$$r_x^2 = y^2 + z^2$$

we see that

$$\begin{aligned} I_x &= \iiint_{\text{vol}} (y^2 + z^2) \, dV \\ &= I_{xz} + I_{xy} \end{aligned} \tag{A-24}$$

which is the basis for the following theorem:

Theorem A-8 The second moment of volume with respect to an axis is the sum of the second moments with respect to two orthogonal planes which intersect along the axis; e.g., the x–z and x–y planes intersect along the x axis.

This theorem is often quite useful in calculating various second moments of volumes.

Second moments of volumes are usually expressed in inch^5 or some other dimension of length raised to the fifth power. Since volume and distance squared are inherently positive quantities, the second moment of volume is inherently a positive value.

Definition A-12 The radius of gyration of a volume with respect to a given plane (or axis) is the positive number k such that

$$Vk^2 = I$$

where I is the appropriate second moment with respect to the plane (or axis).

Just as with areas, the centroidal axes of volumes play an important role in dynamics. Consequently, a transfer theorem for second moments of volumes is quite useful. Referring to Fig. A-16(*b*), c is the centroid with centroidal axes x_c, y_c, and z_c. The x, y, and z axes are parallel to the corresponding centroidal axes, and the centroidal distances are denoted by $\bar{x}$, $\bar{y}$, and $\bar{z}$. By Definition A-10 and Fig. A-16(*b*), we have

$$I_{xz} = \iiint_{\text{vol}} y^2 \, dV$$

$$I_{x_c z_c} = \iiint_{\text{vol}} y_c^2 \, dV$$

where

$$y = \bar{y} + y_c$$

Substituting this last expression into the above integral for I_{xz} yields

$$I_{xz} = \iiint_V (\bar{y} + y_c)^2 \, dV = \iiint_V \bar{y}^2 \, dV + 2 \iiint_V \bar{y} y_c \, dV + \iiint_V y_c^2 \, dV$$

$$= \bar{y}^2 V + 2\bar{y} \iiint_V y_c \, dV + I_{x_c z_c}$$

$$= I_{x_c z_c} + V\bar{y}^2 \qquad \text{(A-25)}$$

where we have used the fact that the first moment with respect to a centroidal plane is zero. This result is the basis for the following theorem:

Theorem A-9 *The Transfer Theorem* The second moment of volume with respect to a plane (or axis) is equal to the second moment with respect to the parallel centroidal plane (or axis) plus the product of the volume and the square of the centroidal distance between the two planes (or axes).

Since I, V, and the square of the centroidal distance are all inherently positive quantities, we have the following:

Theorem A-10 Of all planes (or axes) *parallel* to a given plane (or axis), the parallel centroidal plane (or axis) has the least second moment of volume.

Finally, second moments of composite volumes can be found by calculating the appropriate second moment for each of the individual component volumes, so that

$$I_{\text{total}} = \pm I_1 \pm I_2 \pm \cdots \pm I_n \qquad \text{(A-26)}$$

where n is the number of component volumes and the minus sign corresponds to those component volumes which are removed rather than added. In utilizing Eq. (A-26), care must be exercised to be sure that second moments with respect to planes are added to (or subtracted from) second moments with respect to the same plane. Furthermore, second moments with respect to planes should not be added to (or subtracted from) second moments with respect to axes.

The following example illustrates some of the ideas regarding second moments of volumes.

EXAMPLE A-7

For the right circular cone shown in Fig. A-17(a), determine I_y, I_{xz}, and I_x.

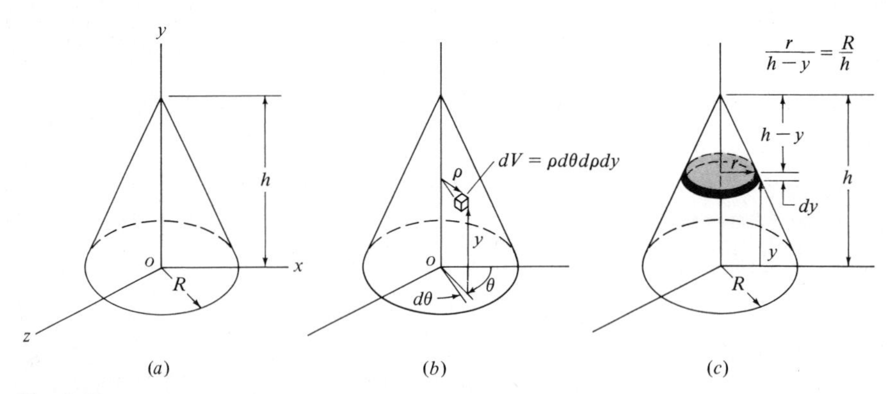

Fig. A-17

Solution: Since the y axis is an axis of symmetry, it is convenient to use the cylindrical polar coordinates ρ, θ, y indicated in Fig. A-17(b). Then, by Definition A-11,

$$I_y = \iiint_{\text{vol}} \rho^2 \, dV \qquad (a)$$

where the elemental volume is given by

$$dV = \rho \, d\theta \, d\rho \, dy \tag{b}$$

Substituting Eq. (b) into Eq. (a) and integrating with respect to θ and then with respect to ρ yields

$$\begin{aligned} I_y &= \int_0^h \int_0^r \int_0^{2\pi} \rho^2 \rho \, d\theta \, d\rho \, dy \\ &= \int_0^h \int_0^r 2\pi\rho^3 \, d\rho \, dy = \int_0^h 2\pi \left[\frac{\rho^4}{4}\right]_0^r dy \\ &= \int_0^h \frac{\pi r^4}{2} \, dy \end{aligned} \tag{c}$$

The integrand in Eq. (c) has a very useful interpretation as the second moment with respect to the y axis of the thin disk-like slice of radius r and thickness dy indicated in Fig. A-17(c). That is, for this thin disk-like volume,

$$dI_y = \frac{\pi r^4}{2} \, dy$$

For the cone, the value of r depends upon the value of y according to

$$r = \frac{R}{h}(h - y) \tag{d}$$

so that at $y = 0$, $r = R$, and at $y = h$, $r = 0$.

Substituting Eq. (d) into Eq. (c) and integrating yields

$$\begin{aligned} I_y &= \frac{\pi}{2} \int_0^h \left[\frac{R}{h}(h - y)\right]^4 dy \\ &= -\frac{\pi}{2} \left[\left(\frac{R}{h}\right)^4 \frac{(h - y)^5}{5}\right]_0^h = \frac{\pi R^4 h}{10} \end{aligned} \tag{e}$$

which agrees with the result given in Table A-1.

To determine I_{xz}, recall Definition A-10:

$$I_{xz} = \iiint y^2 \, dV$$

For this case, we can take as the elemental volume dV the thin disk-like slice previously referred to in Fig. A-17(c). Note that all parts of this slice are the same distance y from the x–z plane. Thus,

$$dV = \pi r^2 \, dy$$

and

$$I_{xz} = \int_0^h y^2(\pi r^2 \, dy)$$

Again, we utilize Eq. (d) relating r and y to obtain

$$
\begin{aligned}
I_{xz} &= \pi \int_0^h y^2 \left[\frac{R}{h}(h-y)\right]^2 dy \\
&= \pi \left(\frac{R}{h}\right)^2 \int_0^h y^2(h^2 - 2hy + y^2)\, dy \\
&= \pi \left(\frac{R}{h}\right)^2 \left[\frac{h^2y^3}{3} - \frac{2hy^4}{4} - \frac{y^5}{5}\right]_0^h \\
&= \frac{\pi R^2 h^3}{30} \qquad (f)
\end{aligned}
$$

Finally, to determine I_x we will utilize Theorem A-8 to write

$$I_x = I_{xz} + I_{xy} \qquad (g)$$

We have already determined I_{xz} in Eq. (f). Also, by Theorem A-8,

$$I_y = I_{xy} + I_{zy} = 2I_{xy}$$

where we have utilized the fact that the x–y and z–y planes are the same so far as the cone is concerned. Thus, by Eq. (e),

$$I_{xy} = \tfrac{1}{2} I_y = \frac{\pi R^4 h}{20} \qquad (h)$$

Substituting Eqs. (h) and (f) into (g) yields

$$I_x = \frac{\pi R^4 h}{20} + \frac{\pi R^2 h^3}{30}$$

which also agrees with the result in Table A-1.

EXERCISE PROBLEMS

A-49 A solid volume is generated by revolving the shaded parabolic area shown about the x axis. Determine the centroid of this generated volume.

A-50 Same as Prob. A-49 except the area is revolved about the vertical axis $x = b$.

PROBS. A-49 and A-50

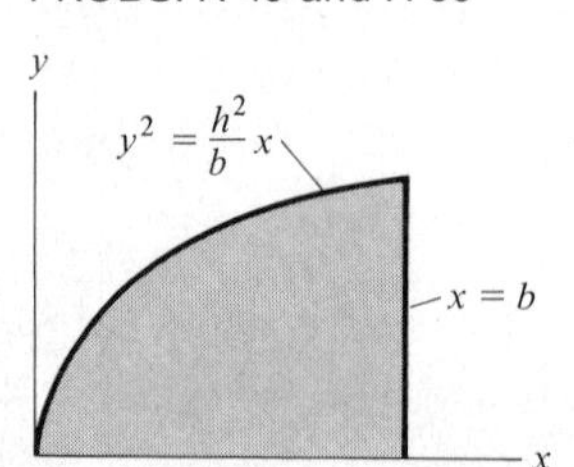

PROBS. A-51 and A-52

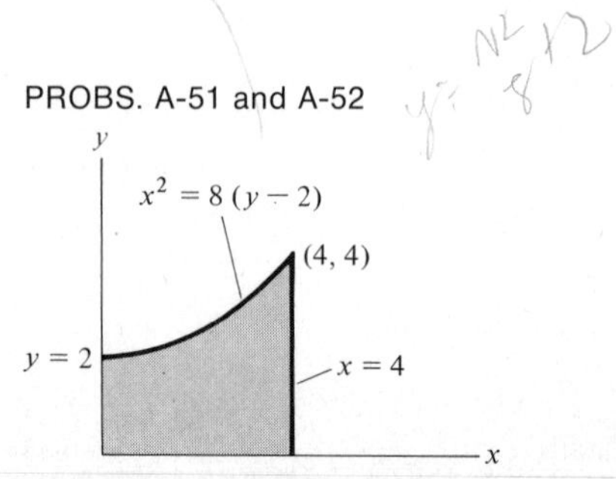

A-51 A dish-like volume is generated by revolving the shaded area shown about the y axis. Determine the centroid of this generated volume.

A-52 A solid volume is generated by revolving the shaded area shown about the x axis. Determine the centroid of this generated volume.

A-53 Determine the vertical centroidal distance of a solid regular right pyramid with a base $b \times b$ and a vertical height of h.

A-54 Determine the centroid of a semiellipsoid of revolution whose surface equation is $x^2/a^2 + y^2/b^2 + z^2/a^2 = 1$. Refer to the figure for Prob. A-64.

A-55 to A-58 Determine the centroid of the volume shown. Use the tables in Appendix B as necessary.

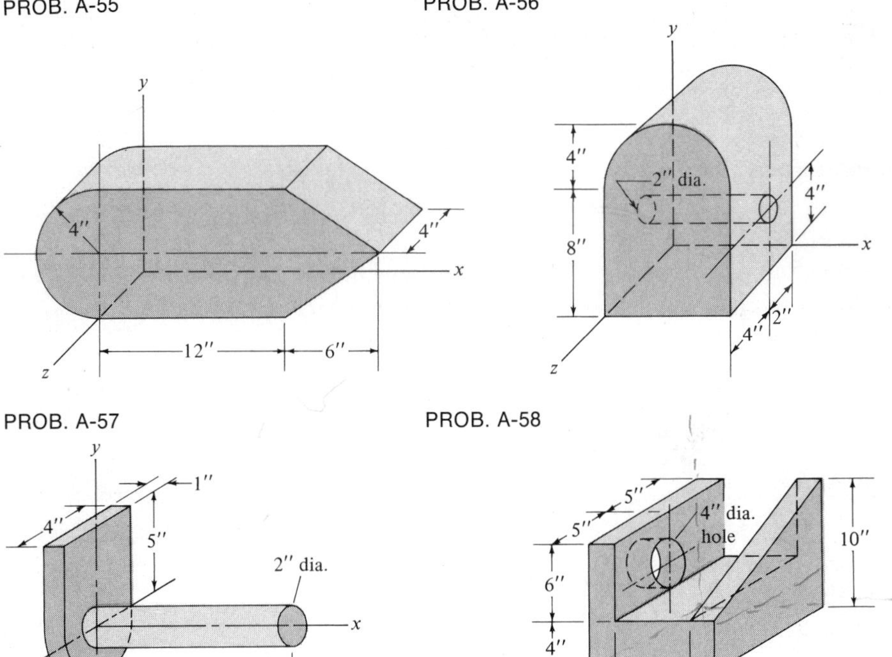

A-59 For the solid hemisphere in Fig. A-15, determine I_{xz}, I_y, and I_x.

A-60 For the solid circular cylinder shown, determine I_x, I_{yz}, and I_y.

A-61 For the solid rectangular prism shown, determine I_{yz}, I_{yx}, and I_y.

PROB. A-60

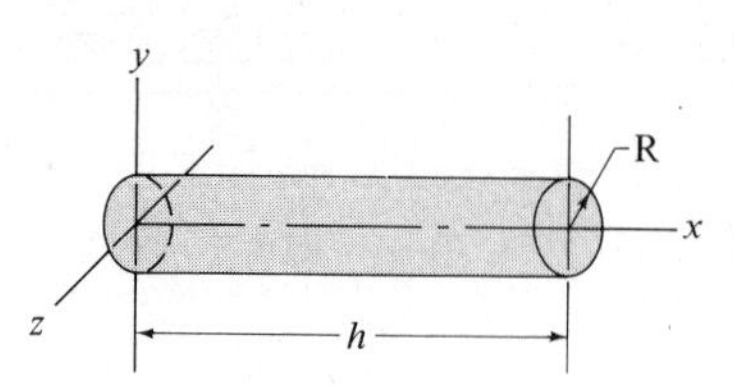

PROB. A-61

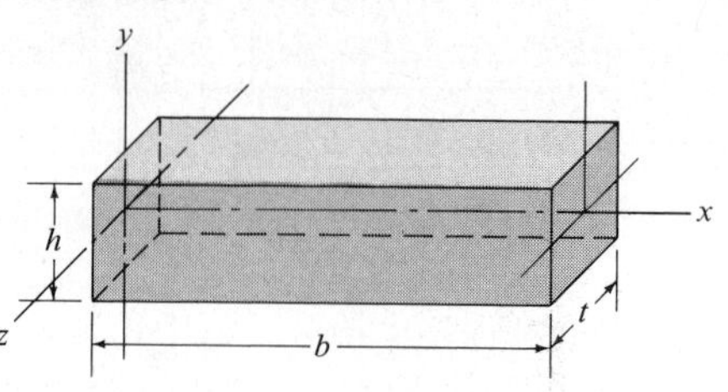

A-62 For the parabolic solid of revolution in Prob. A-49, determine I_x and I_{yz}.

A-63 For the solid of revolution in Prob. A-52, determine I_x and I_{yz}.

A-64 For the semiellipsoid shown, determine I_x and I_y.

PROB. A-64

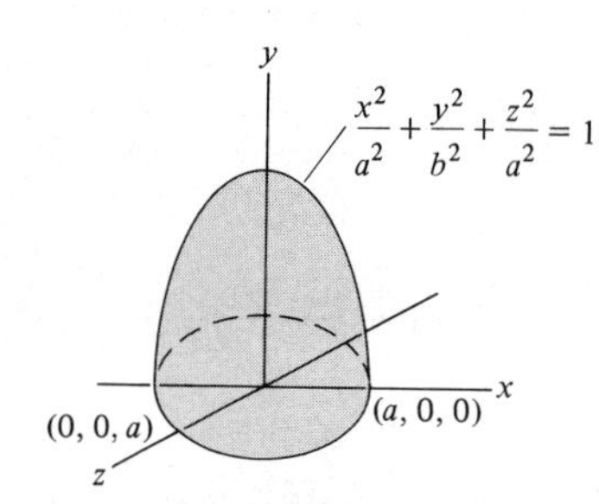

PROB. A-65

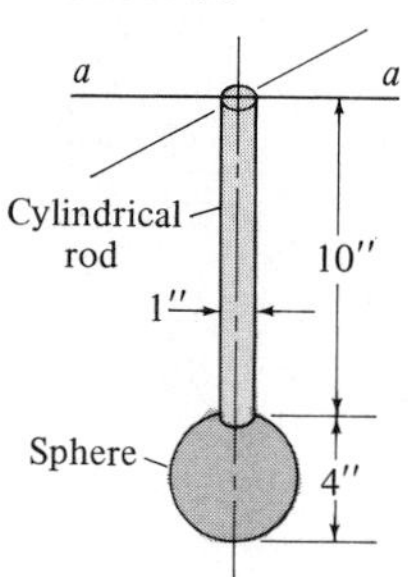

PROB. A-66

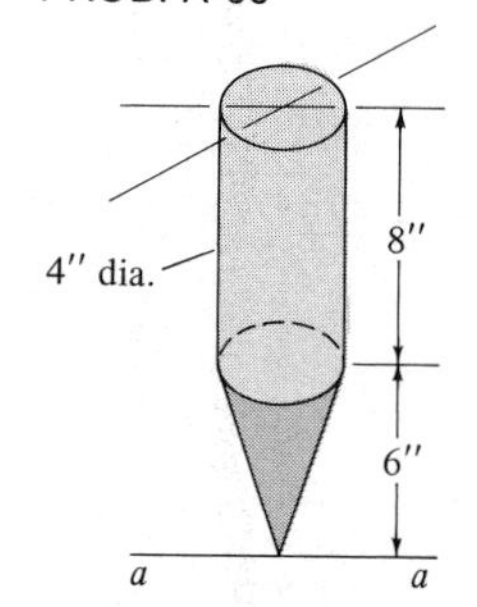

A-65 to A-70 For the composite volume shown, determine I with respect to the indicated a axis. Use the tables in Appendix B as necessary.

PROB. A-67

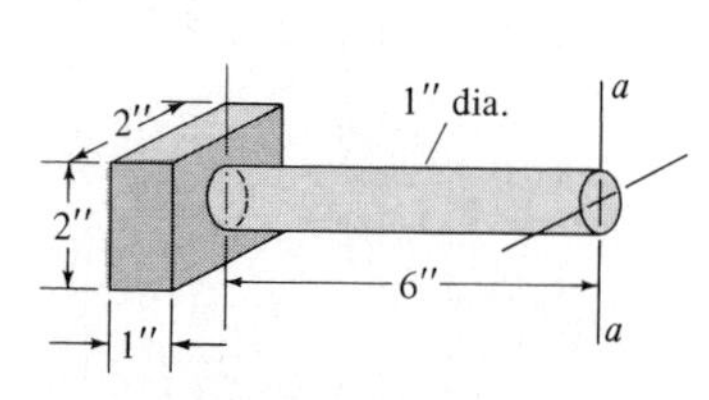

PROB. A-68

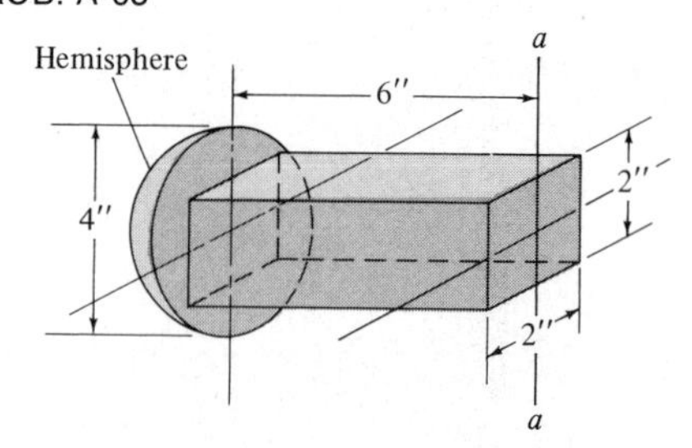

PROB. A-69

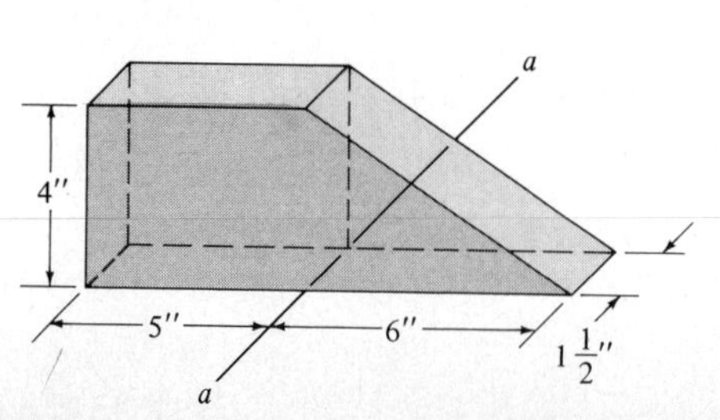

PROB. A-70

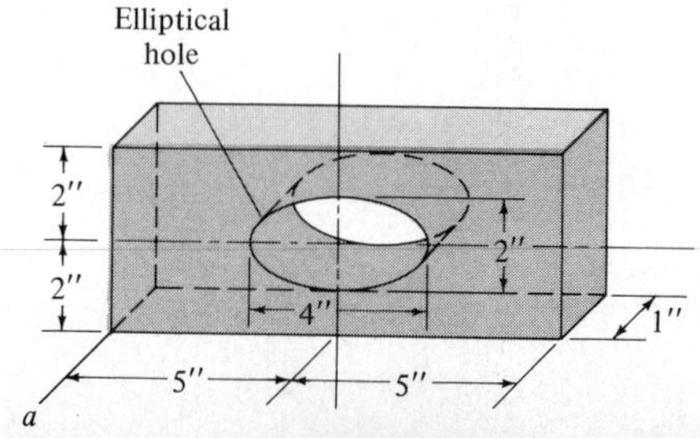

TABLE A-1
PROPERTIES OF AREAS AND VOLUMES

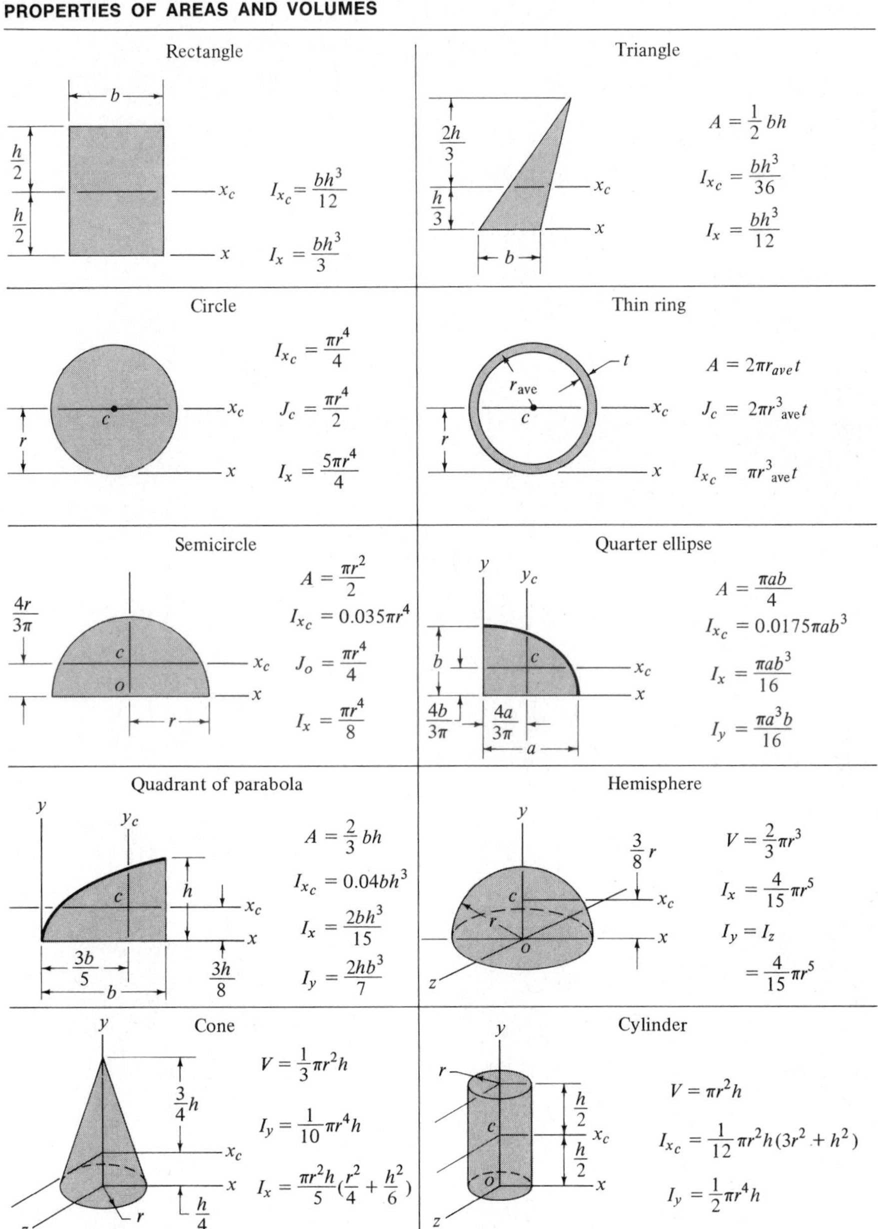

Rectangular prism

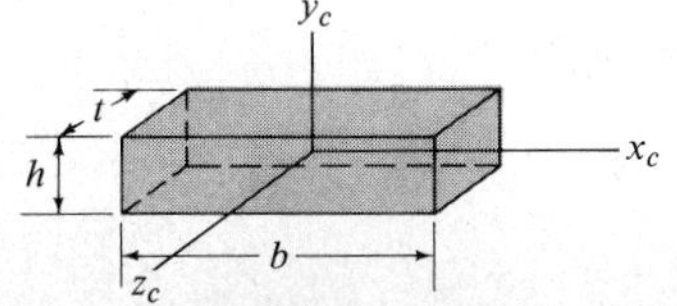

$$I_{x_c} = \frac{1}{12}bht(h^2 + t^2)$$

$$I_{y_c} = \frac{1}{12}bht(b^2 + t^2)$$

$$I_{z_c} = \frac{1}{12}bht(b^2 + h^2)$$

appendix B
USEFUL TABLES

TABLE 1: AVERAGE MECHANICAL PROPERTIES OF SELECTED ENGINEERING MATERIALS

Material	E 10^6 psi	G 10^6 psi	μ	0.2 Percent Yield Strength (tension) 10^3 psi	Ultimate Strength 10^3 psi	Elongation at Rupture in 2 in. Percent	Weight lb/in.^{3}	Coefficient of Thermal Expansion 10^{-6} per °F
Hot rolled steel (SAE 1020)	30	12	0.27	36	65 (ten)	30	0.283	6.5
Structural steel (A-7)	30	12	0.27	35	60-72 (ten)	30	0.283	6.5
High-carbon steel (SAE 1090)	30	12	0.27	67	122 (ten)	10	0.283	6.5
Alloy steel (SAE 4130) (heat treated)	30	12	0.30	100	125 (ten)		0.283	6.5
Stainless steel (18-8)	28	9.5	0.30	80	120 (ten)		0.284	9.6
Gray cast iron (ASTM Class 30)	14.7	5.9	0.20		31 (ten)	<1	0.260	6.7
					124 (comp)			
Cast iron (pearlitic malleable)	26.4	10		80	100 (ten)	7	0.266	6.6
					300 (comp)			
Aluminum 1100-0 (annealed)	10.0	3.8	0.33	3.5	11 (ten)	25	0.098	13.1
Aluminum alloy 2024-T3 (sheet and plate)	10.6	4.0	0.33	50	70 (ten)	18	0.100	12.6
Aluminum alloy 6061-T6 (extruded)	10.0	3.8	0.33	35	38 (ten)	10	0.098	13.1
Aluminum alloy 7075-T6 (sheet and plate)	10.4	3.9	0.33	70	80 (ten)	5	0.101	12.9
Magnesium alloy (H K31A-H24 sheet)	6.5	2.4	0.35	23	34 (ten)	4	0.0647	15
Titanium alloy (6A*l*-4V sheet)	15.9	6.2	0.34	120	130 (ten)	10	0.160	4.6
Brass, hard yellow	15	5.6	0.35	60	74 (ten)	10	0.306	10.5
Copper (DHP. (Hard Temper) (Pipe)	17	6.4		45	50 (ten)	10	0.323	18
Douglas fir timber (air dry; parallel to grain)	1.7				8.1 (ten)		0.020	3.0
					7.4 (comp)			
Red oak timber (air dry; parallel to grain)	1.8				6.9 (comp)		0.025	1.9
Lead (rolled)	2	0.7	0.43	2	2.5 (ten)	50	0.410	16.4
Tungsten carbide (Carboloy, Grade 999)	100		0.24		600 (comp)			2.2
					1.3 (ten)			
Glass (fused silica)	10.0		0.17		13 (comp)	Nil	0.15	4.0
Concrete (low strength)	2		0.15		2 (comp)		0.087	6.0
Concrete (high strength)	3		0.15		5 (comp)		0.087	6.0
Polystyrene (average)	0.5				14 (comp)	2		70
Polyethylene (average)	1.8		0.45		2 (ten)	350	0.033	150
Epoxy (cast; average)	0.65				7 (ten)	4		33
					30 (comp)			
Rubber (natural; molded)			0.50		3 (ten)	800		90

TABLE 2

W SHAPES

Properties for designing

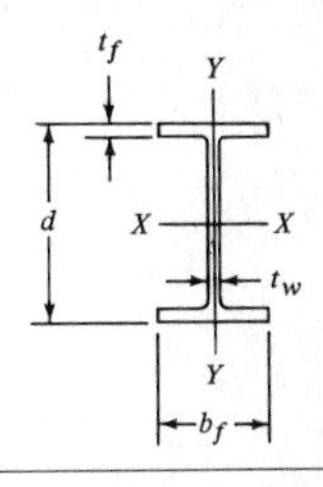

Designation	Area A In.²	Depth d In.	Flange Width b_f In.	Flange Thickness t_f In.	Web Thickness t_w In.	Axis X-X I In.⁴	Axis X-X S In.³	Axis X-X r In.	Axis Y-Y I In.⁴	Axis Y-Y S In.³	Axis Y-Y r In.
W 12×190	55.9	14.38	12.670	1.736	1.060	1890	263	5.82	590	93.1	3.25
×161	47.4	13.88	12.515	1.486	0.905	1540	222	5.70	486	77.7	3.20
×133	39.1	13.38	12.365	1.236	0.755	1220	183	5.59	390	63.1	3.16
×120	35.3	13.12	12.320	1.106	0.710	1070	163	5.51	345	56.0	3.13
×106	31.2	12.88	12.230	0.986	0.620	931	145	5.46	301	49.2	3.11
× 99	29.1	12.75	12.192	0.921	0.582	859	135	5.43	278	45.7	3.09
W 10×112	32.9	11.38	10.415	1.248	0.755	719	126	4.67	235	45.2	2.67
×100	29.4	11.12	10.345	1.118	0.685	625	112	4.61	207	39.9	2.65
× 89	26.2	10.88	10.275	0.998	0.615	542	99.7	4.55	181	35.2	2.63
× 77	22.7	10.62	10.195	0.868	0.535	457	86.1	4.49	153	30.1	2.60
× 72	21.2	10.50	10.170	0.808	0.510	421	80.1	4.46	142	27.9	2.59
× 66	19.4	10.38	10.117	0.748	0.457	382	73.7	4.44	129	25.5	2.58
× 60	17.7	10.25	10.075	0.683	0.415	344	67.1	4.41	116	23.1	2.57
× 54	15.9	10.12	10.028	0.618	0.368	306	60.4	4.39	104	20.7	2.56
× 49	14.4	10.00	10.000	0.558	0.340	273	54.6	4.35	93.0	18.6	2.54
W 8× 67	19.7	9.00	8.287	0.933	0.575	272	60.4	3.71	88.6	21.4	2.12
× 58	17.1	8.75	8.222	0.808	0.510	227	52.0	3.65	74.9	18.2	2.10
× 48	14.1	8.50	8.117	0.683	0.405	184	43.2	3.61	60.9	15.0	2.08
× 40	11.8	8.25	8.077	0.558	0.365	146	35.5	3.53	49.0	12.1	2.04
× 35	10.3	8.12	8.027	0.493	0.315	126	31.1	3.50	42.5	10.6	2.03
× 31	9.12	8.00	8.000	0.433	0.288	110	27.4	3.47	37.0	9.24	2.01
W 8× 28	8.23	8.06	6.540	0.463	0.285	97.8	24.3	3.45	21.6	6.61	1.62
× 24	7.06	7.93	6.500	0.398	0.245	82.5	20.8	3.42	18.2	5.61	1.61
W 8× 20	5.89	8.14	5.286	0.378	0.248	69.4	17.0	3.43	9.22	3.50	1.25
× 17	5.01	8.00	5.250	0.308	0.230	56.6	14.1	3.36	7.44	2.83	1.22
W 8× 15	4.43	8.12	4.015	0.314	0.245	48.1	11.8	3.29	3.40	1.69	0.876
× 13	3.83	8.00	4.000	0.254	0.230	39.6	9.90	3.21	2.72	1.36	0.842
× 10	2.96	7.90	3.940	0.204	0.170	30.8	7.80	3.23	2.08	1.06	0.839
W 6× 25	7.35	6.37	6.080	0.456	0.320	53.3	16.7	2.69	17.1	5.62	1.53
× 20	5.88	6.20	6.018	0.367	0.258	41.5	13.4	2.66	13.3	4.43	1.51
×15.5	4.56	6.00	5.995	0.269	0.235	30.1	10.0	2.57	9.67	3.23	1.46

American Institute of Steel Construction

TABLE 3
S SHAPES
Properties for designing

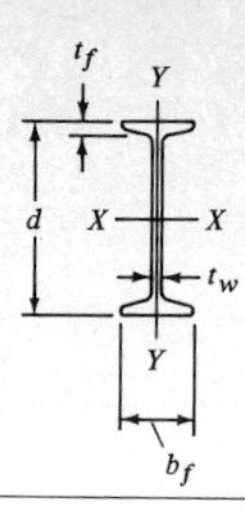

			Flange		Web	Elastic Properties					
						Axis X-X			Axis Y-Y		
	Area A	Depth d	Width b_f	Thickness t_f	Thickness t_w	I	S	r	I	S	r
Designation	In.^{2}	In.	In.	In.	In.	In.^{4}	In.^{3}	In.	In.^{4}	In.^{3}	In.
S 24×120	35.3	24.00	8.048	1.102	0.798	3030	252	9.26	84.2	20.9	1.54
×105.9	31.1	24.00	7.875	1.102	0.625	2830	236	9.53	78.2	19.8	1.58
S 24×100	29.4	24.00	7.247	0.871	0.747	2390	199	9.01	47.8	13.2	1.27
× 90	26.5	24.00	7.124	0.871	0.624	2250	187	9.22	44.9	12.6	1.30
× 79.9	23.5	24.00	7.001	0.871	0.501	2110	175	9.47	42.3	12.1	1.34
S 20× 95	27.9	20.00	7.200	0.916	0.800	1610	161	7.60	49.7	13.8	1.33
× 85	25.0	20.00	7.053	0.916	0.653	1520	152	7.79	46.2	13.1	1.36
S 20× 75	22.1	20.00	6.391	0.789	0.641	1280	128	7.60	29.6	9.28	1.16
× 65.4	19.2	20.00	6.250	0.789	0.500	1180	118	7.84	27.4	8.77	1.19
S 18× 70	20.6	18.00	6.251	0.691	0.711	926	103	6.71	24.1	7.72	1.08
× 54.7	16.1	18.00	6.001	0.691	0.461	804	89.4	7.07	20.8	6.94	1.14
S 15× 50	14.7	15.00	5.640	0.622	0.550	486	64.8	5.75	15.7	5.57	1.03
× 42.9	12.6	15.00	5.501	0.622	0.411	447	59.6	5.95	14.4	5.23	1.07
S 12× 50	14.7	12.00	5.477	0.659	0.687	305	50.8	4.55	15.7	5.74	1.03
× 40.8	12.0	12.00	5.252	0.659	0.472	272	45.4	4.77	13.6	5.16	1.06
S 12× 35	10.3	12.00	5.078	0.544	0.428	229	38.2	4.72	9.87	3.89	0.980
× 31.8	9.35	12.00	5.000	0.544	0.350	218	36.4	4.83	9.36	3.74	1.00
S 10× 35	10.3	10.00	4.944	0.491	0.594	147	29.4	3.78	8.36	3.38	0.901
× 25.4	7.46	10.00	4.661	0.491	0.311	124	24.7	4.07	6.79	2.91	0.954
S 8× 23	6.77	8.00	4.171	0.425	0.441	64.9	16.2	3.10	4.31	2.07	0.798
× 18.4	5.41	8.00	4.001	0.425	0.271	57.6	14.4	3.26	3.73	1.86	0.831
S 7× 20	5.88	7.00	3.860	0.392	0.450	42.4	12.1	2.69	3.17	1.64	0.734
× 15.3	4.50	7.00	3.662	0.392	0.252	36.7	10.5	2.86	2.64	1.44	0.766
S 6× 17.25	5.07	6.00	3.565	0.359	0.465	26.3	8.77	2.28	2.31	1.30	0.675
× 12.5	3.67	6.00	3.332	0.359	0.232	22.1	7.37	2.45	1.82	1.09	0.705
S 5× 14.75	4.34	5.00	3.284	0.326	0.494	15.2	6.09	1.87	1.67	1.01	0.620
× 10	2.94	5.00	3.004	0.326	0.214	12.3	4.92	2.05	1.22	0.809	0.643
S 4× 9.5	2.79	4.00	2.796	0.293	0.326	6.79	3.39	1.56	0.903	0.646	0.569
× 7.7	2.26	4.00	2.663	0.293	0.193	6.08	3.04	1.64	0.764	0.574	0.581
S 3× 7.5	2.21	3.00	2.509	0.260	0.349	2.93	1.95	1.15	0.586	0.468	0.516
× 5.7	1.67	3.00	2.330	0.260	0.170	2.52	1.68	1.23	0.455	0.390	0.522

TABLE 4

CHANNELS
AMERICAN STANDARD
Properties for designing

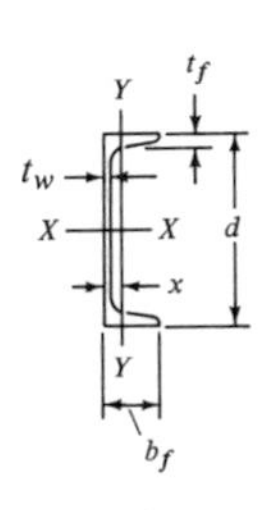

Designation	Area A In.²	Depth d In.	Flange Width b_f In.	Flange Average Thichness t_f In.	Web Thickness t_w In.	$\frac{d}{A_f}$	Axis X-X I In.⁴	Axis X-X S In.³	Axis X-X r In.
C 15×50	14.7	15.00	3.716	0.650	0.716	6.21	404	53.8	5.24
×40	11.8	15.00	3.520	0.650	0.520	6.56	349	46.5	5.44
×33.9	9.96	15.00	3.400	0.650	0.400	6.79	315	42.0	5.62
C 12×30	8.82	12.00	3.170	0.501	0.510	7.55	162	27.0	4.29
×25	7.35	12.00	3.047	0.501	0.387	7.85	144	24.1	4.43
×20.7	6.09	12.00	2.942	0.501	0.282	8.13	129	21.5	4.61
C 10×30	8.82	10.00	3.033	0.436	0.673	7.55	103	20.7	3.42
×25	7.35	10.00	2.886	0.436	0.526	7.94	91.2	18.2	3.52
×20	5.88	10.00	2.739	0.436	0.379	8.36	78.9	15.8	3.66
×15.3	4.49	10.00	2.600	0.436	0.240	8.81	67.4	13.5	3.87
C 9×20	5.88	9.00	2.648	0.413	0.448	8.22	60.9	13.5	3.22
×15	4.41	9.00	2.485	0.413	0.285	8.76	51.0	11.3	3.40
×13.4	3.94	9.00	2.433	0.413	0.233	8.95	47.9	10.6	3.48
C 8×18.75	5.51	8.00	2.527	0.390	0.487	8.12	44.0	11.0	2.82
×13.75	4.04	8.00	2.343	0.390	0.303	8.75	36.1	9.03	2.99
×11.5	3.38	8.00	2.260	0.390	0.220	9.08	32.6	8.14	3.11
C 7×14.75	4.33	7.00	2.299	0.366	0.419	8.31	27.2	7.78	2.51
×12.25	3.60	7.00	2.194	0.366	0.314	8.71	24.2	6.93	2.60
× 9.8	2.87	7.00	2.090	0.366	0.210	9.14	21.3	6.08	2.72
C 6×13	3.83	6.00	2.157	0.343	0.437	8.10	17.4	5.80	2.13
×10.5	3.09	6.00	2.034	0.343	0.314	8.59	15.2	5.06	2.22
× 8.2	2.40	6.00	1.920	0.343	0.200	9.10	13.1	4.38	2.34
C 5× 9	2.64	5.00	1.885	0.320	0.325	8.29	8.90	3.56	1.83
× 6.7	1.97	5.00	1.750	0.320	0.190	8.93	7.49	3.00	1.95
C 4× 7.25	2.13	4.00	1.721	0.296	0.321	7.84	4.59	2.29	1.47
× 5.4	1.59	4.00	1.584	0.296	0.184	8.52	3.85	1.93	1.56
C 3× 6	1.76	3.00	1.596	0.273	0.356	6.87	2.07	1.38	1.08
× 5	1.47	3.00	1.498	0.273	0.258	7.32	1.85	1.24	1.12
× 4.1	1.21	3.00	1.410	0.273	0.170	7.78	1.66	1.10	1.17

American Institute of Steel Construction

TABLE 5
ANGLES
Equal legs
Properties for designing

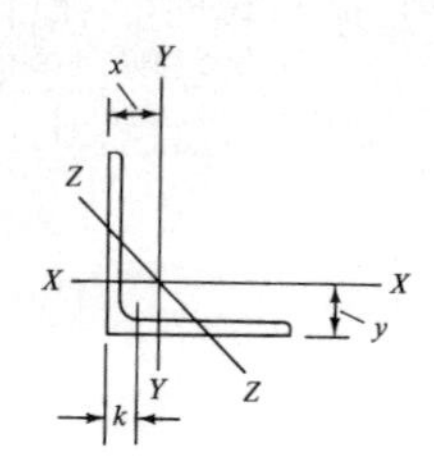

Size and Thickness	k	Weight per Foot	Area	Axis X-X and Axis Y-Y				Axis Z-Z
				I	S	r	x or y	r
In.	In.	Lb.	In.^{2}	In.^{4}	In.^{3}	In.	In.	In.
L 8×8×$1\frac{1}{8}$	$1\frac{3}{4}$	56.9	16.7	98.0	17.5	2.42	2.41	1.56
1	$1\frac{5}{8}$	51.0	15.0	89.0	15.8	2.44	2.37	1.56
$\frac{7}{8}$	$1\frac{1}{2}$	45.0	13.2	79.6	14.0	2.45	2.32	1.57
$\frac{3}{4}$	$1\frac{3}{8}$	38.9	11.4	69.7	12.2	2.47	2.28	1.58
$\frac{5}{8}$	$1\frac{1}{4}$	32.7	9.61	59.4	10.3	2.49	2.23	1.58
$\frac{9}{16}$	$1\frac{3}{16}$	29.6	8.68	54.1	9.34	2.50	2.21	1.59
$\frac{1}{2}$	$1\frac{1}{8}$	26.4	7.75	48.6	8.36	2.50	2.19	1.59
L 6×6×1	$1\frac{1}{2}$	37.4	11.0	35.5	8.57	1.80	1.86	1.17
$\frac{7}{8}$	$1\frac{3}{8}$	33.1	9.73	31.9	7.63	1.81	1.82	1.17
$\frac{3}{4}$	$1\frac{1}{4}$	28.7	8.44	28.2	6.66	1.83	1.78	1.17
$\frac{5}{8}$	$1\frac{1}{8}$	24.2	7.11	24.2	5.66	1.84	1.73	1.18
$\frac{9}{16}$	$1\frac{1}{16}$	21.9	6.43	22.1	5.14	1.85	1.71	1.18
$\frac{1}{2}$	1	19.6	5.75	19.9	4.61	1.86	1.68	1.18
$\frac{7}{16}$	$\frac{15}{16}$	17.2	5.06	17.7	4.08	1.87	1.66	1.19
$\frac{3}{8}$	$\frac{7}{8}$	14.9	4.36	15.4	3.53	1.88	1.64	1.19
$\frac{5}{16}$	$\frac{13}{16}$	12.4	3.65	13.0	2.97	1.89	1.62	1.20
L 5×5× $\frac{7}{8}$	$1\frac{3}{8}$	27.2	7.98	17.8	5.17	1.49	1.57	.973
$\frac{3}{4}$	$1\frac{1}{4}$	23.6	6.94	15.7	4.53	1.51	1.52	.975
$\frac{5}{8}$	$1\frac{1}{8}$	20.0	5.86	13.6	3.86	1.52	1.48	.978
$\frac{1}{2}$	1	16.2	4.75	11.3	3.16	1.54	1.43	.983
$\frac{7}{16}$	$\frac{15}{16}$	14.3	4.18	10.0	2.79	1.55	1.41	.986
$\frac{3}{8}$	$\frac{7}{8}$	12.3	3.61	8.74	2.42	1.56	1.39	.990
$\frac{5}{16}$	$\frac{13}{16}$	10.3	3.03	7.42	2.04	1.57	1.37	.994
L 4×4× $\frac{3}{4}$	$1\frac{1}{8}$	18.5	5.44	7.67	2.81	1.19	1.27	.778
$\frac{5}{8}$	1	15.7	4.61	6.66	2.40	1.20	1.23	.779
$\frac{1}{2}$	$\frac{7}{8}$	12.8	3.75	5.56	1.97	1.22	1.18	.782
$\frac{7}{16}$	$\frac{13}{16}$	11.3	3.31	4.97	1.75	1.23	1.16	.785
$\frac{3}{8}$	$\frac{3}{4}$	9.8	2.86	4.36	1.52	1.23	1.14	.788
$\frac{5}{16}$	$\frac{11}{16}$	8.2	2.40	3.71	1.29	1.24	1.12	.791
$\frac{1}{4}$	$\frac{5}{8}$	6.6	1.94	3.04	1.05	1.25	1.09	.795

American Institute of Steel Construction

TABLE 6
ANGLES
Unequal legs

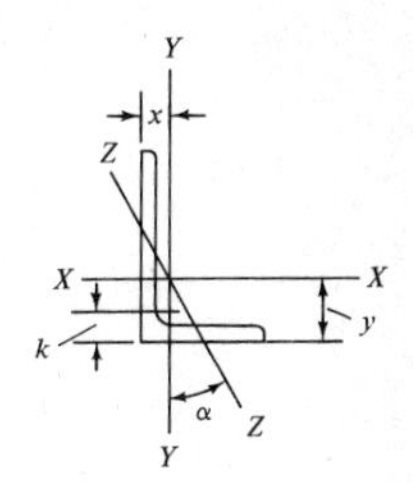

Properties for designing

Size and Thickness	k	Weight per Foot	Area	Axis X-X				Axis Y-Y				Axis Z-Z	
				I	S	r	y	I	S	r	x	r	Tan
In.	In.	Lb.	In.²	In.⁴	In.³	In.	In.	In.⁴	In.³	In.	In.	In.	α
L 9×4× 1	$1\frac{1}{2}$	40.8	12.0	97.0	17.6	2.84	3.50	12.0	4.00	1.00	1.00	.834	.203
$\frac{7}{8}$	$1\frac{3}{8}$	36.1	10.6	86.8	15.7	2.86	3.45	10.8	3.56	1.01	.953	.836	.208
$\frac{3}{4}$	$1\frac{1}{4}$	31.3	9.19	76.1	13.6	2.88	3.41	9.63	3.11	1.02	.906	.841	.212
$\frac{5}{8}$	$1\frac{1}{8}$	26.3	7.73	64.9	11.5	2.90	3.36	8.32	2.65	1.04	.858	.847	.216
$\frac{9}{16}$	$1\frac{1}{16}$	23.8	7.00	59.1	10.4	2.91	3.33	7.63	2.41	1.04	.834	.850	.218
$\frac{1}{2}$	1	21.3	6.25	53.2	9.34	2.92	3.31	6.92	2.17	1.05	.810	.854	.220
L 8×6× 1	$1\frac{1}{2}$	44.2	13.0	80.8	15.1	2.49	2.65	38.8	8.92	1.73	1.65	1.28	.543
$\frac{7}{8}$	$1\frac{3}{8}$	39.1	11.5	72.3	13.4	2.51	2.61	34.9	7.94	1.74	1.61	1.28	.547
$\frac{3}{4}$	$1\frac{1}{4}$	33.8	9.94	63.4	11.7	2.53	2.56	30.7	6.92	1.76	1.56	1.29	.551
$\frac{5}{8}$	$1\frac{1}{8}$	28.5	8.36	54.1	9.87	2.54	2.52	26.3	5.88	1.77	1.52	1.29	.554
$\frac{9}{16}$	$1\frac{1}{16}$	25.7	7.56	49.3	8.95	2.55	2.50	24.0	5.34	1.78	1.50	1.30	.556
$\frac{1}{2}$	1	23.0	6.75	44.3	8.02	2.56	2.47	21.7	4.79	1.79	1.47	1.30	.558
L 6×4× $\frac{7}{8}$	$1\frac{3}{8}$	27.2	7.98	27.7	7.15	1.86	2.12	9.75	3.39	1.11	1.12	.857	.421
$\frac{3}{4}$	$1\frac{1}{4}$	23.6	6.94	24.5	6.25	1.88	2.08	8.68	2.97	1.12	1.08	.860	.428
$\frac{5}{8}$	$1\frac{1}{8}$	20.0	5.86	21.1	5.31	1.90	2.03	7.52	2.54	1.13	1.03	.864	.435
$\frac{9}{16}$	$1\frac{1}{16}$	18.1	5.31	19.3	4.83	1.90	2.01	6.91	2.31	1.14	1.01	.866	.438
$\frac{1}{2}$	1	16.2	4.75	17.4	4.33	1.91	1.99	6.27	2.08	1.15	.987	.870	.440
L 4×3½× $\frac{5}{8}$	$1\frac{1}{16}$	14.7	4.30	6.37	2.35	1.22	1.29	4.52	1.84	1.03	1.04	.719	.745
$\frac{1}{2}$	$\frac{15}{16}$	11.9	3.50	5.32	1.94	1.23	1.25	3.79	1.52	1.04	1.00	.722	.750
$\frac{7}{16}$	$\frac{7}{8}$	10.6	3.09	4.76	1.72	1.24	1.23	3.40	1.35	1.05	.978	.724	.753
$\frac{3}{8}$	$\frac{13}{16}$	9.1	2.67	4.18	1.49	1.25	1.21	2.95	1.17	1.06	.955	.727	.755
$\frac{5}{16}$	$\frac{3}{4}$	7.7	2.25	3.56	1.26	1.26	1.18	2.55	.994	1.07	.932	.730	.757
$\frac{1}{4}$	$\frac{11}{16}$	6.2	1.81	2.91	1.03	1.27	1.16	2.09	.808	1.07	.909	.734	.759
L 4×3× $\frac{5}{8}$	$1\frac{1}{16}$	13.6	3.98	6.03	2.30	1.23	1.37	2.87	1.35	.849	.871	.637	.534
$\frac{1}{2}$	$\frac{15}{16}$	11.1	3.25	5.05	1.89	1.25	1.33	2.42	1.12	.864	.827	.639	.543
$\frac{7}{16}$	$\frac{7}{8}$	9.8	2.87	4.52	1.68	1.25	1.30	2.18	.992	.871	.804	.641	.547
$\frac{3}{8}$	$\frac{13}{16}$	8.5	2.48	3.96	1.46	1.26	1.28	1.92	.866	.879	.782	.644	.551
$\frac{5}{16}$	$\frac{3}{4}$	7.2	2.09	3.38	1.23	1.27	1.26	1.65	.734	.887	.759	.647	.554
$\frac{1}{4}$	$\frac{11}{16}$	5.8	1.69	2.77	1.00	1.28	1.24	1.36	.599	.896	.736	.651	.558
L 3½×3× $\frac{1}{2}$	$\frac{15}{16}$	10.2	3.00	3.45	1.45	1.07	1.13	2.33	1.10	.881	.875	.621	.714
$\frac{7}{16}$	$\frac{7}{8}$	9.1	2.65	3.10	1.29	1.08	1.10	2.09	.975	.889	.853	.622	.718
$\frac{3}{8}$	$\frac{13}{16}$	7.9	2.30	2.72	1.13	1.09	1.08	1.85	.851	.897	.830	.625	.721
$\frac{5}{16}$	$\frac{3}{4}$	6.6	1.93	2.33	.954	1.10	1.06	1.58	.722	.905	.808	.627	.724
$\frac{1}{4}$	$\frac{11}{16}$	5.4	1.56	1.91	.776	1.11	1.04	1.30	.589	.914	.785	.631	.727

American Institute of Steel Construction

TABLE 7
PIPE
Dimensions and properties

Dimension				Weight per Foot Lbs. Plain Ends	Properties			
Nominal Diameter In.	Outside Diameter In.	Inside Diameter In.	Wall Thickness In.		*A* In.[2]	*I* In.[4]	*S* In.[3]	*r* In.
Standard Weight								
$\frac{1}{2}$	.840	.622	.109	.85	.250	.017	.041	.261
$\frac{3}{4}$	1.050	.824	.113	1.13	.333	.037	.071	.334
1	1.315	1.049	.140	1.68	.494	.087	.235	.421
$1\frac{1}{4}$	1.660	1.380	.140	2.27	.669	.195	.235	.540
$1\frac{1}{2}$	1.900	1.610	.145	2.72	.799	.310	.326	.623
2	2.375	2.067	.154	3.65	1.07	.666	.561	.787
$2\frac{1}{2}$	2.875	2.469	.203	5.79	1.70	1.53	1.06	.947
3	3.500	3.068	.216	7.58	2.23	3.02	1.72	1.16
$3\frac{1}{2}$	4.000	3.548	.226	9.11	2.68	4.79	2.39	1.34
4	4.500	4.026	.237	10.79	3.17	7.23	3.21	1.51
5	5.563	5.047	.258	14.62	4.30	15.2	5.45	1.88
6	6.625	6.065	.280	18.97	5.58	28.1	8.50	2.25
8	8.625	7.981	.322	28.55	8.40	72.5	16.8	2.94
10	10.750	10.020	.365	40.48	11.9	161	29.9	3.67
12	12.750	12.000	.375	49.56	14.6	279	43.8	4.38

American Institute of Steel Construction

TABLE 8
SELECTED TIMBER STANDARD SIZES PROPERTIES OF SECTIONS

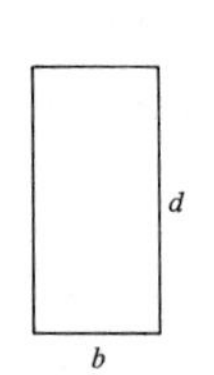

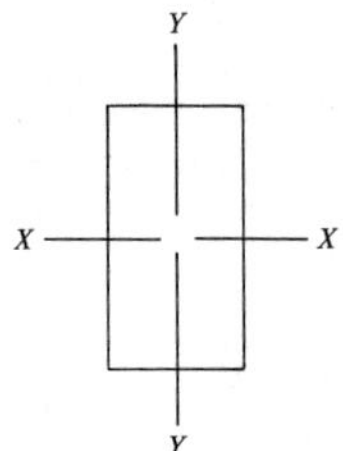

Nominal Size Inches *b* *d*	Actual Size Inches *b* *d*	Area In.[2]	Axis X-X *S* In.[3]	Axis X-X *I* In.[4]	Axis Y-Y *S* In.[3]	Axis Y-Y *I* In.[4]	Board Measure per Lineal Foot	Weight per Lineal Foot Lbs.
2× 2	$1\frac{1}{2}$× $1\frac{1}{2}$	2.25	.56	.42	.56	.42	.33	.63
3	$2\frac{9}{16}$	3.84	1.64	2.10	.96	.72	.50	1.07
4	$3\frac{9}{16}$	5.34	3.17	5.65	1.36	1.00	.67	1.48
6	$5\frac{1}{2}$	8.25	7.56	20.80	2.06	1.54	1	2.29
8	$7\frac{1}{2}$	11.25	14.06	52.73	2.81	2.11	1.33	3.12
10	$9\frac{1}{2}$	14.25	22.56	107.17	3.56	2.67	1.67	3.96
12	$11\frac{1}{2}$	17.25	33.06	190.11	4.31	3.23	2	4.79
14	$13\frac{1}{2}$	20.25	45.56	307.55	5.06	3.80	2.33	5.62
4× 4	$3\frac{5}{8}$× $3\frac{5}{8}$	13.14	7.94	14.39	7.94	14.39	1.33	3.64
6	$5\frac{1}{2}$	19.94	18.27	50.26	12.04	21.83	2	5.54
8	$7\frac{1}{2}$	27.19	33.98	127.45	16.42	29.77	2.67	7.55
10	$9\frac{1}{2}$	34.44	54.53	259.02	20.80	37.71	3.33	9.57
12	$11\frac{1}{2}$	41.69	79.90	459.13	25.18	45.65	4	11.58

Southern Pine Association Technical Bulletin No. 2

TABLE 9
SLOPES AND DEFLECTIONS OF SELECTED BEAMS

	Beam	Slope	Deflection (+upward)
1		$\theta_1 = -\dfrac{PL^2}{16EI}$ at $x = 0$ $\theta_2 = +\dfrac{PL^2}{16EI}$ at $x = L$	$y_{\max} = -\dfrac{PL^3}{48EI}$ at $x = L/2$
2		$\theta_1 = -\dfrac{wL^3}{24EI}$ at $x = 0$ $\theta_2 = +\dfrac{wL^3}{24EI}$ at $x = L$	$y_{\max} = -\dfrac{5wL^4}{384EI}$ at $x = L/2$
3		$\theta_1 = -\dfrac{Pb(L^2 - b^2)}{6LEI}$ at $x = 0$ $\theta_2 = +\dfrac{Pa(L^2 - a^2)}{6LEI}$ at $x = L$	$y_{\max} = -\dfrac{Pb(L^2 - b^2)^{3/2}}{9\sqrt{3}LEI}$ at $x = \sqrt{(L^2 - b^2)/3}$ $y_{\text{center not max}} = -\dfrac{Pb(3L^2 - 4b^2)}{48EI}$
4		$\theta = -\dfrac{PL^2}{2EI}$ at $x = L$	$y_{\max} = -\dfrac{PL^3}{3EI}$ at $x = L$
5		$\theta = -\dfrac{wL^3}{6EI}$ at $x = L$	$y_{\max} = -\dfrac{wL^4}{8EI}$ at $x = L$
6		Zero at ends Zero at center	$y_{\max}(\text{at center}) = -\dfrac{wL^4}{384EI}$ Moment at ends is max. $M_{\max} = -\dfrac{wL^2}{12}$
7	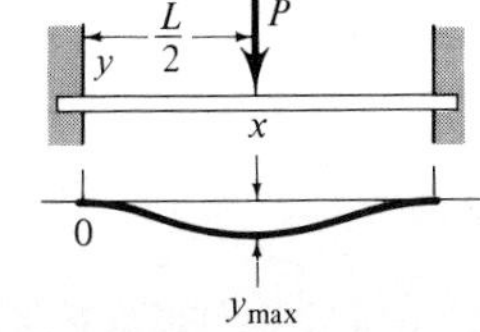	Zero at ends Zero at center	$y_{\max}(\text{at center}) = -\dfrac{PL^3}{192EI}$ Moment at ends is max. $M_{\max} = -\dfrac{PL}{8}$

ANSWERS TO SELECTED PROBLEMS

Chapter 1

1-7 100 mi 53° N of E

1-8 78 mph 50.2° S of W

1-9 25 lb 23°

1-10 13 fps 14.3°

1-12 17 kips 28°

1-16 (*a*) 148 ft 35.8°

(*b*) 89 ft 77°
(*c*) Opposite to (*b*)

1-17 (*a*) 111 fps 5 110, (*b*) 85 fps 40 75

1-20 (*a*) 50 ft, (*b*) 35 ft, (*c*) 3,500 ft^2

1-21 (*a*) 24.7 fps, (*b*) 14.5 fps, (*c*) 1,230 $(fps)^2$

1-23 $\boldsymbol{\alpha} = 78.5$ lb 45°, $\boldsymbol{\beta} = 139$ lb$\longrightarrow$

1-24 $\boldsymbol{\alpha} = 37.5$ in 3 4, $\boldsymbol{\beta} = 74.5$ in$\uparrow$

1-27 $\boldsymbol{\beta} = 10$ ft$\longrightarrow$ or 16°

1-28 $\boldsymbol{\beta} = 10$ ft 23° or 83°

1-30 $\mathbf{F} = (33.6\mathbf{i} + 126\mathbf{j})$ lb

1-33 $\mathbf{F} = (275\mathbf{i} + 55\mathbf{j} - 110\mathbf{k})$ lb

1-35 $\mathbf{F} = (65\mathbf{i} + 40\mathbf{j} - 23.5\mathbf{k})$ lb

1-37 $\boldsymbol{\mu} = \frac{7}{9}\mathbf{i} - \frac{4}{9}\mathbf{j} + \frac{4}{9}\mathbf{k}$

1-39 $\boldsymbol{\mu} = -0.586\mathbf{i} + 0.644\mathbf{j} + 0.493\mathbf{k}$

1-45 (*a*) $169\left(-\frac{\sqrt{2}}{2}\mathbf{i} + \frac{\sqrt{2}}{2}\mathbf{j}\right)$ lb, (*b*) $169\mathbf{j}$ lb

1-49 (*a*) 0, (*b*) 0, (*c*) 0

1-52 (*a*) −288 mi, (*b*) −480 mi, (*c*) −144,000 mi^2

1-55 $\boldsymbol{\beta} = (-4\mathbf{i} - 6.93\mathbf{j})$ in

1-58 (*a*) $(120\mathbf{i} + 86.6\mathbf{k})$ lb, (*b*) −2,500 lb^2, (*c*) 101°

1-59 (*a*) $(90\mathbf{i} + 30\mathbf{j} - 140\mathbf{k})$ lb, (*b*) −4,200 lb^2, (*c*) 106°

1-61 (*a*) $(54.5\mathbf{i} + 86\mathbf{j} - 60\mathbf{k})$ lb, (*b*) −6,020 lb^2, (*c*) 122.5°

1-63 $\boldsymbol{\mu}_{AB} = -0.406\mathbf{i} - 0.820\mathbf{j} + 0.406\mathbf{k}$
$\boldsymbol{\mu}_{AC} = 0.5\mathbf{i} - 0.866\mathbf{j} + 0\mathbf{k}$

$\angle A,B = 59.5°$

1-64 $\boldsymbol{\mu}_{A\text{-}B} = -\frac{12}{13}\mathbf{i} + \frac{4}{13}\mathbf{j} + \frac{3}{13}\mathbf{k}$
$\boldsymbol{\mu}_{A\text{-}C} = -\frac{12}{13}\mathbf{i} - \frac{5}{13}\mathbf{j} + 0\mathbf{k}$

$\angle A,B = 42.8°$

1-67 $\boldsymbol{\alpha} \cdot \boldsymbol{\mu} = 32.1$ lb, $\boldsymbol{\gamma} = (28\mathbf{i} + 16\mathbf{j})$ lb,
$\boldsymbol{\delta} = (12\mathbf{i} - 21\mathbf{j} + 20\mathbf{k})$ lb

1-68 $\boldsymbol{\alpha} \cdot \boldsymbol{\mu} = 28.8$ lb, $\boldsymbol{\gamma} = (10.1\mathbf{i} + 24.8\mathbf{j} - 10.1\mathbf{k})$ lb,
$\boldsymbol{\delta} = (-35.1\mathbf{i} + 18.6\mathbf{j} + 10.1\mathbf{k})$ lb

Chapter 2

2-2 $\mathbf{M}_o = 2{,}710\mathbf{k}$ in-lb

2-4 $\mathbf{M}_o = 10.8$ in-lb 56.3°

2-5 $\mathbf{M}_o = -156\mathbf{k}$ ft-lb

2-7 $\mathbf{M}_o = 238$ in-lb

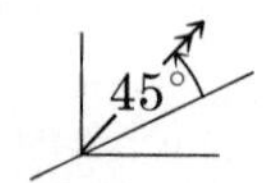

2-9 $\mathbf{M}_o = Fr(1 + \sin\theta)\mathbf{k}$

2-10 $\mathbf{M}_o = -1{,}280\mathbf{k}$ ft-lb

2-15 (*a*) $4\mathbf{i} - 7\mathbf{j} - 6\mathbf{k}$, (*b*) $-1\mathbf{i} + 2\mathbf{j} + 2\mathbf{k}$, (*c*) −1

2-16 (*a*) $8\mathbf{i} - 8\mathbf{j} + 30\mathbf{k}$, (*b*) $-6\mathbf{i} - 32\mathbf{j} + 6\mathbf{k}$, (*c*) 76

2-18 (a) $-6\mathbf{i} - 10\mathbf{j} + 8\mathbf{k}$, ($b$) $\mathbf{0}$, (c) 0

2-20 (a) $-10\mathbf{i} + 4\mathbf{j} - 6\mathbf{k}$, ($b$) $5\mathbf{i} - 2\mathbf{j} + 3\mathbf{k}$, ($c$) 0

2-23 $\mathbf{M}_0 = -(0\mathbf{i} + 3{,}080\mathbf{j} + 1{,}438\mathbf{k})$ in-lb

2-28 $\mathbf{M}_o = (420\mathbf{i} + 0\mathbf{j} - 600\mathbf{k})$ in-lb

2-31 (a) $\mathbf{M}_o = -(141\mathbf{i} + 566\mathbf{j} + 566\mathbf{k})$ in-lb, (b) $d = 8.13$ in

2-33 (a) $\mathbf{M}_o = (36\mathbf{i} - 48\mathbf{j} - 180\mathbf{k})$ ft-lb, (b) $d = 5.43$ ft

2-36 (a) $\mathbf{M}_o = (160\mathbf{i} - 100\mathbf{j} + 180\mathbf{k})$ in-lb, (b) $d = 5.8$ in

2-38 (a) $\mathbf{M}_o = (248\mathbf{i} + 87.5 + 152\mathbf{k})$ in-lb, (b) $d = 6.08$ in

2-39 (a) $\mathbf{M}_o = (-866\mathbf{i} + 500\mathbf{j} + 833\mathbf{k})$ in-lb, (b) $d = 5.2$ in

2-41 $\mathbf{F}_1 = 48$ lb 3 4, $\mathbf{F}_2 = 36$ lb 4 3

2-43 $\mathbf{F}_1 = 379$ lb 1 3, $\mathbf{F}_2 = 402$ lb 1 2

2-44 $\mathbf{F}_1 = 344$ lb 70°, $\mathbf{F}_2 = 491$ lb 20°

2-46 $\mathbf{F}_1 = 15\mathbf{k}$ lb, $\mathbf{F}_2 = (60\mathbf{i} - 20\mathbf{j})$ lb

2-49 (a) No, (b) no, since sums of forces are equal but moments are unequal

2-50 (a) No, (b) yes, by adding a 54.6i-lb force at 14.5 in below lower right-hand corner

2-53 (a) Yes

2-56 (a) $\mathbf{F} = (180\mathbf{i} + 160\mathbf{j})$ lb, $\mathbf{C}_q = 2{,}360\mathbf{k}$ ft-lb
(b) $\mathbf{F} = (180\mathbf{i} + 160\mathbf{j})$ lb, $\mathbf{C}_p = -500\mathbf{k}$ ft-lb

2-58 (a) $\mathbf{F} = -84.5\mathbf{i} - 70.8\mathbf{j}$, $\mathbf{C}_q = 413.5\mathbf{k}$ ft-lb
(b) $\mathbf{F} = -84.5\mathbf{i} - 70.8\mathbf{j}$, $\mathbf{C}_p = -294.5\mathbf{k}$ ft-lb

2-60 (a) $\mathbf{F} = (864\mathbf{i} + 685\mathbf{j})$ lb, $\mathbf{C}_q = 24{,}280\mathbf{k}$ in-lb
(b) $\mathbf{F} = (864\mathbf{i} + 685\mathbf{j})$ lb, $\mathbf{C}_p = 34{,}648\mathbf{k}$ in-lb

2-62 (a) $\mathbf{F} = \mathbf{0}$, $\mathbf{C}_q = -500\mathbf{i} + 1{,}500\mathbf{k}$ (b) same as (a)

2-66 (a) $\mathbf{F} = (-100\mathbf{j} + 75\mathbf{k})$ lb, $\mathbf{C}_q = (125\mathbf{i} - 300\mathbf{j} - 400\mathbf{k})$ in-lb
(b) $\mathbf{F} = (-100\mathbf{j} + 75\mathbf{k})$ lb, $\mathbf{C}_p = (350\mathbf{i} - 300\mathbf{j} - 400\mathbf{k})$ in-lb

2-67 (a) $\mathbf{F} = (-25\mathbf{i} + 40\mathbf{j} - 30\mathbf{k})$ lb, $\mathbf{C}_q = -110\mathbf{j}$ in-lb
(b) $\mathbf{F} = (-25\mathbf{i} + 40\mathbf{j} - 30\mathbf{k})$ lb, $\mathbf{C}_p = (-380\mathbf{j} - 360\mathbf{k})$ in-lb

2-69 $\mathbf{R} = (70\mathbf{i} + 110.7\mathbf{j} - 110.7\mathbf{k})$ lb through o

2-71 $T_A = 133$ lb, $T_B = 167$ lb

2-73 $\mathbf{R} = (47.4\mathbf{i} - 2.5\mathbf{j} + 56.7\mathbf{k})$ lb, $\|\mathbf{R}\| = 74.6$ lb

2-75 $\mathbf{R} = (-38.6\mathbf{i} + 107.6\mathbf{j} + 21.6\mathbf{k})$ lb

2-76 $T_A = 22$ lb, $T_B = 32.7$ lb, $T_C = 48$ lb

2-80 $\mathbf{R} = (60\mathbf{i} + 86\mathbf{j})$ lb, $-$ 15.7 in below o

2-81 $\mathbf{R} = \mathbf{0}$, $\mathbf{C} = -5{,}120\mathbf{k}$ in-lb

2-83 $\mathbf{R} = -1{,}500\mathbf{j}$ lb at $\mathbf{p} = 6.87\mathbf{i}$ ft from o

2-84 $\mathbf{R} = (-300\mathbf{i} - 500\mathbf{j})$ lb, $\mathbf{p} = 0\mathbf{i} + 34.3\mathbf{j}$ in

2-85 $\mathbf{R} = 660\mathbf{i}$ lb, $\mathbf{p} = 2.6\mathbf{j}$ ft

2-87 $\mathbf{R} = (90.7\mathbf{i} - 69.3\mathbf{j})$ lb, $\mathbf{p} = (0\mathbf{i} - 10\mathbf{j})$ in from o

2-89 $\mathbf{R} = 2{,}000$ lb, $\mathbf{F} = 1{,}800\ \mathbf{i} + 4{,}600\mathbf{j}$

2-92 $\mathbf{F}_1 = \frac{1}{3}\mathbf{i}$ kips, $\mathbf{F}_2 = 3\frac{2}{3}\mathbf{i}$ kips

2-94 $\mathbf{R} = \mathbf{0}$, $\mathbf{C} = (120\mathbf{L} - 40\mathbf{k})$ in-lb

2-96 $\mathbf{R} = \mathbf{0}$, $\mathbf{C} = (-120\mathbf{i} + 310\mathbf{k})$ ft-lb

2-98 $\mathbf{F} = -350\mathbf{j}$ lb, $\mathbf{p} = (-9.85\mathbf{i} - 2\mathbf{k})$ in

2-99 $\mathbf{R} = \mathbf{0}$, $\mathbf{C} = (-3\mathbf{i} - 13.5\mathbf{j})$ kip-ft

2-102 $\mathbf{R} = (40\mathbf{i} - 20\mathbf{j} + 80\mathbf{k})$ lb, $\mathbf{C}_o = (400\mathbf{i} - 400\mathbf{j} + 0\mathbf{k})$ in-lb, no

2-103 $\mathbf{R} = \mathbf{0}$, $\mathbf{C} = (30\mathbf{i} - 40\mathbf{j} + 120\mathbf{k})$ ft-lb, no

2-105 $\mathbf{R} = 86.6\mathbf{k}$ lb, $\mathbf{C}_o = (250\mathbf{i} - 693\mathbf{j})$ in-lb, yes

2-107 $\mathbf{R} = 86.6\mathbf{k}$ lb, $\mathbf{p} = (8\mathbf{i} + 2.88\mathbf{j})$ in

2-109 $\mathbf{R} = (-120\mathbf{i} - 160\mathbf{j})$ lb, $\mathbf{p} = 4.24\mathbf{i}$ ft from o

2-110 $\mathbf{F} = (800\mathbf{j} - 200\mathbf{k})$ lb, $\mathbf{C} = (-2{,}400\mathbf{i} - 600\mathbf{k})$ ft-lb

Chapter 3

3-2 63,500 lb

3-4 1,076 lb

3-6 $F = 13{,}500$ lb, $p = 6.24 + 0.728x$ lb/ft^2, x measured from left side

3-8 $F_x = 20{,}750$ lb, $F_y = 8{,}650$ lb

3-11 9,980,000 lb

3-12 2,808,000 lb

3-14 1,248,000 lb

3-17 1,450 tons

3-18 68.6 lb

3-19 $\bar{x} = \dfrac{\ell^2/2 + (\ell + 2r/\pi)(\pi r/2)}{(\ell + \pi r/2)}$, $\bar{z} = \dfrac{r\ell + r^2}{(\ell + \pi r/2)}$

3-23 $\bar{x} = 5$ ft, $\bar{y} = 3$ ft, $\bar{z} = 5$ ft

3-24 $\bar{x} = 3.2$ ft, $\bar{y} = 0.53$ ft, $\bar{z} = 4.16$ ft

3-25 $\bar{x} = 4r/3\pi$, $\bar{y} = 0$, $\bar{z} = 0$

3-26 $\bar{x} = \frac{2}{5}a$, $\bar{y} = \bar{z} = 0$

3-28 $\bar{x} = 2.90$ in, $\bar{y} = \bar{z} = 0$

3-31 $\bar{x} = \bar{z} = 0$, $\bar{y} = h/4$

3-32 $\bar{x} = \bar{z} = 0$, $\bar{y} = 3R/8$

3-33 $\bar{x} = 6$ in, $\bar{y} = 2.75$ in, $\bar{z} = -3$ in

3-35 $\bar{x} = 10$ in, $\bar{y} = 11.53$ in, $\bar{z} = -6$ in

3-37 $\bar{x} = 1.4$ ft, $\bar{y} = 1.81$ ft, $\bar{z} = -1.34$ ft

3-39 $\bar{x} = 4$ in, $\bar{y} = 1\frac{1}{4}$ in, $\bar{z} = 1\frac{2}{3}$ in

3-41 (*a*) $F = 323{,}500$ lb at middle of floor
(*b*) $F = 80{,}870$ lb at 4 ft above floor at middle of wall
(*c*) $F = 107{,}800$ lb at 4 ft above floor at middle of wall

3-44 (*a*) $F = 8{,}420$ lb, $\bar{y} = 2$ ft from top
(*b*) $F = 135{,}000$ lb, $\bar{y} = 8$ ft from top
(*c*) $\mathbf{F} = -(126{,}000\mathbf{i} + 700{,}000\mathbf{j})$ lb, $\bar{x} = 30$ ft from left end, $\bar{y} = 8.4$ ft from top

3-45 $\mathbf{F} = (-2{,}510{,}000\mathbf{i} - 838{,}000\mathbf{j})$ lb, $\bar{x} = -2\frac{2}{3}$ ft and $\bar{y} = 8$ ft, both measured from the lower right corner

3-47 $\mathbf{R} = 17{,}850$ lb (slope 5 : 9) acting at middle of left side

3-48 $\mathbf{R} = (-20{,}736\mathbf{i} - 20{,}736\mathbf{j})$ lb acting at middle of the right circular side of roof

3-51 $\mathbf{R} = -1{,}800\mathbf{j}$ lb, $\bar{x} = 10.1$ ft from left end

3-54 $\mathbf{R} = -5{,}600\mathbf{j}$ lb, $\bar{x} = 8.86$ ft from left end

3-55 $\mathbf{R} = 98{,}500\mathbf{k}$ lb, $\bar{x} = 33.8$ ft, $\bar{y} = 6.06$ ft from top

3-56 $\mathbf{R} = 4{,}800{,}000\mathbf{i}$ lb, $\bar{y} = 53.3$ ft from top

3-58 $\mathbf{R} = -15{,}600\mathbf{j}$ lb, $\bar{x} = 10.75$ ft from left end

Chapter 4

Prob. 4-2

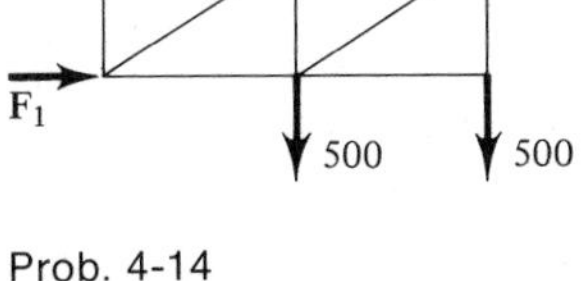

Prob. 4-6 Prob. 4-7

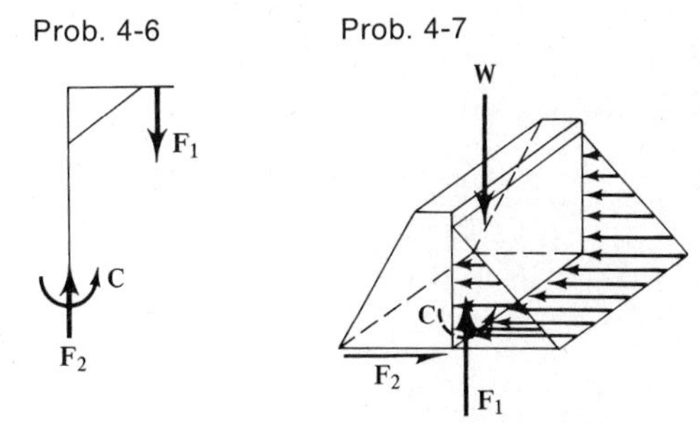

Prob. 4-14

(Maybe)

Prob. 4-15

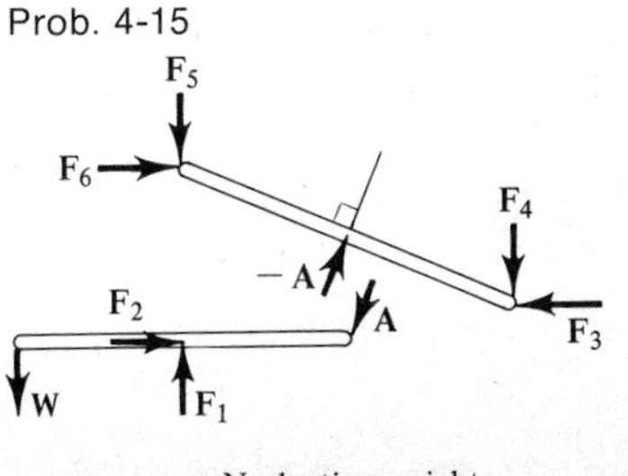

Neglecting weights

Prob. 4-19

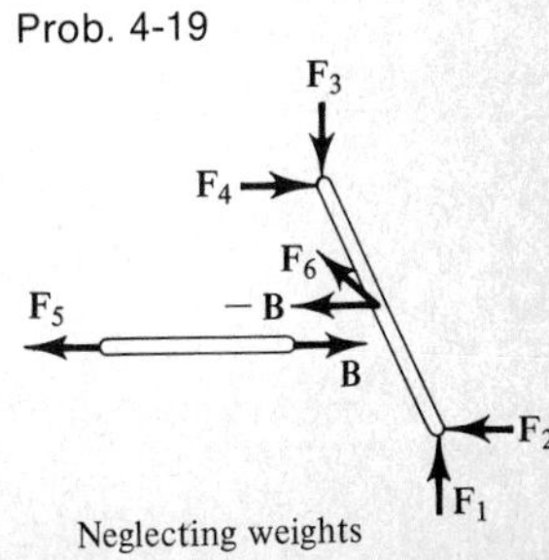

Neglecting weights

4-22 $\mathbf{F}_1 = 14.6\mathbf{i}$ lb, $\mathbf{F}_2 = (-14.6\mathbf{i} + 40\mathbf{j})$ lb, $\|\mathbf{F}_2\| = 42.6$ lb

4-24 $T = 1{,}520$ lb, $F_b = 2{,}200$ lb

4-25 $T_A = 1{,}080$ lb, $T_B = 1{,}080$ lb, $T_C = 660$ lb

4-27 $T_A = 311$ lb, $T_B = T_C = 220$ lb

4-28 $cd = bd = 2{,}800$ lb (compression), $ad = 3{,}000$ lb (tension)

4-31 $\mathbf{F}_g = (-49.5\mathbf{i} + 160\mathbf{j})$ lb, $\|\mathbf{F}_g\| = 167.5$ lb

4-33 $\mathbf{A} = (-562\mathbf{i} + 1{,}000\mathbf{j})$ lb, $\mathbf{B} = 1{,}062\mathbf{i}$ lb

4-35 $T_1 = T_2 = 40.2$ lb, $\mathbf{B} = (62.5\mathbf{i} + 14.2\mathbf{j})$ lb

4-36 $T_1 = T_2 = 100$ lb, $\mathbf{R}_0 = (160\mathbf{i} + 60\mathbf{j})$ lb

4-38 $\mathbf{R}_a = 2{,}200\mathbf{j}$ lb, $\mathbf{R}_b = 1{,}300\mathbf{j}$ lb

4-39 $F_s = 504$ lb, $\mathbf{F}_p = 438\mathbf{i}$ lb

4-42 $\mathbf{F} = 120\mathbf{i}$ lb, $\mathbf{N} = 510\mathbf{j}$ lb, $x = 3.3$ ft from left corner

4-44 $h = 50$ in

4-46 $W = 80$ lb, $\mathbf{F} = 20$ lb $\nwarrow$, $\mathbf{N} = 69.3$ $\nearrow$

4-47 $F = 122$ lb, $\mathbf{F}_f = 22\mathbf{i}$ lb, $\mathbf{N} = 123\mathbf{j}$ lb

4-49 $\mathbf{P} = (84.5\mathbf{i} + 70.8\mathbf{j})$ lb, $\mathbf{C}_p = 294.5\mathbf{k}$ ft-lb

4-50 $\mathbf{P} = (-25\mathbf{i} - 40\mathbf{j} + 30\mathbf{k})$ lb, $\mathbf{C}_p = (380\mathbf{j} + 360\mathbf{k})$ in-lb

4-51 $\mathbf{C} = -248\mathbf{i}$ in-lb, $\mathbf{R}_a = \mathbf{R}_b = (21.6\mathbf{j} - 12.5\mathbf{k})$ lb

4-54 $T_A = T_C = 17.3$ lb, $T_B = 25.4$ lb

4-55 $N = 2{,}600$ lb

4-58 $\mathbf{F}_H = 1.8 \times 10^6\mathbf{i}$ lb, $\mathbf{F}_V = 5.63 \times 10^6\mathbf{j}$ lb, 9.56 ft from right corner

4-60 $F = 16.3$ lb, $\mathbf{A}_y = -0.214\mathbf{j}$ lb, $\mathbf{B}_y = 18.9\mathbf{j}$ lb

4-61 $T_A = 125$ lb, $T_B = 248$ lb, $T_C = 200$ lb, $T_D = 173$ lb

4-63 $W_2 = 30$ lb

4-65 $F = 66$ lb

4-67 $\mathbf{B} = (-2{,}000\mathbf{i} - 250\mathbf{j})$ lb, $\mathbf{C} = (2{,}000\mathbf{i} + 1{,}750\mathbf{j})$ lb

4-69 $A_x = B_x = C_x = D_x = E_x = 0$, $A_y = 150$ lb, $B_y = 300$ lb, $C_y = 450$ lb, $D_y = 900$ lb, $E_y = 450$ lb

4-71 $\mathbf{B}_x = -30\mathbf{i}$ lb, $\mathbf{B}_y = -200\mathbf{j}$ lb, $\mathbf{D}_x = 130\mathbf{i}$ lb, $\mathbf{D}_y = 440\mathbf{j}$ lb

4-73 $T = 25$ lb, $N = 15$ lb, $\mathbf{F} = (20\mathbf{i} + 20\mathbf{j})$ lb

4-75 $\mathbf{D} = (500\mathbf{i} - 1{,}000\mathbf{j})$ lb

4-76 $\mathbf{C} = (-40\mathbf{i} + 130\mathbf{j})$ lb, $\mathbf{D} = (80\mathbf{i} - 160\mathbf{j})$, $\mathbf{E} = (-40\mathbf{i} + 30\mathbf{j})$ lb

4-77 $A_x = 900$ lb, $A_y = 10{,}800$ lb, $B_x = 900$ lb, $B_y = 7{,}200$ lb, $C_x = 2{,}700$ lb, $C_y = 10{,}800$ lb

4-80 $\mathbf{A} = (414\mathbf{i} + 516\mathbf{j})$ lb, $\mathbf{E} = (912\mathbf{i} + 984\mathbf{j})$ lb

4-81 $\mathbf{A} = (-75\mathbf{i} + 1{,}100\mathbf{j})$ lb, $\mathbf{E} = (75\mathbf{i} + 100\mathbf{j})$ lb

4-83 $\mathbf{F} = -57.7\mathbf{j}$ lb, $\mathbf{R}_{AB} = 157.7\mathbf{j}$ lb, $\mathbf{C}_{AB} = 300\mathbf{i}$ in-lb

4-85 ab = 4,950 lb (compression), bc = 2,828 lb (compression), cd = 2,000 lb (tension), de = 2,000 lb (tension), eb = 2,121 lb (tension), db = 3,000 lb (compression), ae = 3,500 lb (tension)

4-86 ab = 2,880 lb (compression), bc = 577 lb (tension), cd = 577 lb (compression), bc = 1,730 lb (compression), de = 4,040 lb (compression), ac = 1,440 lb (tension), ce = 2,020 lb (tension)

4-89 ec = 8 kips (tension), ed = 6 kips (tension), $ef = \sqrt{2}$ kips (compression)

4-91 $cf = 0$, ef = 51.5 kips (compression), $bf = 0$

4-93 cj = 2.83 kips (compression), cd = 3.73 kips (compression), dj = 3.33 kips (tension)

4-95 db = 12 kips (compression), cf = 12 kips (tension), ce = 3.9 kips (compression), eb = 3.9 kips (tension)

4-97 $gf = 0$, af = 50 kips (compression), bf = 50 kips (tension), eb = 30 kips (compression)

4-99 eb = 12,500 lb (tension), bd = 10,000 lb (compression), $ed = ef = 0$

4-101 ab = 50 lb (tension), df = 167 lb (tension)

4-102 $ab = 0$, df = 1,630 lb (tension)

4-104 $ad = ac = ab$ = 978 lb (compression), $bc = bd = cd$ = 326 lb (tension)

Chapter 5

5-2 ① R_N = 600 lb (compression), R_S = 250 lb, C_B = 1,250 ft-lb
② R_N = 250 lb (tension), R_S = 93.7 lb, C_B = 750 ft-lb

5-4 ① $R_N = 0$, R_S = 320 lb, C_B = 640 ft-lb
② $R_N = 0$, R_S = 640 lb, C_B = 5,120 ft-lb

5-7 (*a*) $R_N = 0$, R_S = 25 lb, C_{B_y} = 100 in-lb
(*b*) R_N = 25 lb (tension), R_S = 50 lb, C_{B_y} = 260 in-lb, C_{B_z} = 200 in-lb

5-8 (*a*) R_N = 86.6 lb (tension), R_S = 50 lb, C_B = 150 in-lb
(*b*) $R_N = 0$, R_S = 86.6, C_T = 250 in-lb, C_{B_y} = 520 in-lb

5-10 ① $R_N = 0$, R_{S_y} = 3.5 lb, R_{S_z} = 4.5 lb, C_T = 18 in-lb, C_{B_y} = 9 ft-lb, C_{B_z} = 7 ft-lb
② $R_N = 0$, R_{S_y} = 3.5 lb, R_{S_z} = 13.5 lb, $C_T = 0$, C_{B_y} = 54 in-lb, C_{B_z} = 14 in-lb

5-11 $R_N = 1{,}440\pi$ lb (tension)

5-14 ① $R_N = 166$ lb (tension), ② $R_N = 833$ lb (compression), ③ $R_N = 499$ lb (compression)

5-15 ① $R_N = 336$ lb (tension), $R_S = 300$ lb, $C_B = 1{,}200$ ft-lb
② $R_N = 0$, $R_S = 300$ lb, $C_B = 1{,}200$ ft-lb

5-16 ① $R_N = 0$, $R_S = W$, $C_B = 3W$
② $R_N = 1.89W$ (compression), $R_S = 0.38W$, $C_B = 1.13W$ ft

5-20 $\alpha_{\text{max}} = 37.7°$

5-22 $\sigma_N = 3{,}310$ psi, $\sigma_S = 3{,}940$ psi

5-25 $\sigma_N = 894$ psi, $\sigma_S = 447$ psi

5-26 $\sigma_S = 2{,}529$ psi, $\sigma_B = 2{,}393$ psi

5-27 $\sigma_N = 13.1$ psi

5-28 $\sigma_S = 510$ psi

5-30 $\sigma_B = 6{,}030$ psi, $\sigma_S = 3{,}300$ psi

5-31 $\sigma_S = 40{,}700$ psi, $\sigma_N = 1{,}920$ psi

5-32 OD = 1.89 in, use 2 in OD

5-33 $T = 0.175$ in

5-34 $T = 0.267$ in, $\sigma_S = 3{,}620$ psi

5-35 8 in of weld

5-36 $D = 3$ in

5-39 $\sigma_B = 18{,}400$ psi, $\sigma_N = 12{,}250$ psi

5-41 $\sigma_S = 5{,}100$ psi, $\sigma_N = 4{,}000$ psi, $\sigma_N = 4{,}571$

Chapter 6

6-2 $\sigma_{p\ell} = 4{,}500$ psi, $\sigma_{ys} = 5{,}500$ psi, $E = 450{,}000$ psi

6-3 $E_t = 90{,}000$ psi, $E_s = 250{,}000$ psi

6-8 $\sigma_{p\ell} = 4{,}000$ psi, $\sigma_{ys} = 12{,}500$ psi, $E = 17.4 \times 10^6$ psi

6-9 $E_t = 1.2 \times 10^6$ psi, $E_s = 3.1 \times 10^6$ psi

6-10 $E = 3.8 \times 10^6$ psi, $\sigma_{\text{ult}} = 63{,}800$ psi

6-12 $E = 10 \times 10^6$ psi, $\mu = 0.29$

6-13 $\% = \left(-\frac{2}{3}\epsilon + \frac{\epsilon^2}{9}\right) 100$

6-15 $R = 22.5$ in-lb/in^3, $T = 300$ in-lb/in^3

6-17 $R = 0.46$ in-lb/in^3, $T = 140$ in-lb/in^3

6-18 $R = 81$ in-lb/in^3, $T = 700$ in-lb/in^3

6-19 $R = 15$ in-lb/in^3, $T = 11{,}300$ in-lb/in^3

6-21 $R = 20$ in-lb/in^3, $T = 75$ in-lb/ in^3

6-24 $A = 15$ in^2

6-26 $P = 28{,}000$ lb

6-28 $e = 1.28$ in, $\Delta V = 1.17$ in^3

6-29 $\sigma = 15{,}000$ psi

6-31 $\sigma_a = 5{,}880$ psi, $\sigma_s = 9{,}350$ psi, $e = 0.0723$ in

6-35 $R_T = 8{,}000$ lb (tension), $R_B = 4{,}000$ lb (compression), $e =$ 0.00636 in

6-36 $\sigma_b = 13{,}300$ psi, $\sigma_s = 0$

6-37 $R_T = 4{,}000$ lb (tension), $R_B = 12{,}000$ lb (compression)

6-38 $\sigma_b = 10{,}600$ psi, yes

6-40 $P = 86{,}400$ psi

6-41 $\sigma_b = 12{,}500$ psi, $\sigma_c = 18{,}800$ psi

6-42 $\sigma_s = 41{,}600$, $\sigma_b = 6{,}800$ psi, elastic assumption for steel may be questionable

6-43 $\sigma_{\max} = 18{,}400$ psi

6-45 (*a*) $\sigma_{\max} = 30{,}000$, (*b*) $\sigma_{\max} = 41{,}200$ psi

6-47 $d = 1$ in

6-48 radius $= \frac{5}{16}$ in

6-49 radius $= \frac{1}{2}$ in

6-50 $P = 581{,}000$ lb

Chapter 7

7-1 $J = 0.497$ in^4, $\tau_{\max} = 9{,}050$ psi, nothing

7-3 (*a*) decreases 8 times, (*b*) doubles, (*c*) increases 8 times

7-5 $T = \frac{2}{3}\pi R^3 \tau_{\max}$

7-6 $T = \dfrac{2}{k+3}\pi R^3 \tau_{\max}$

7-7 $T = \frac{2}{3}\pi R^3 \tau_{\max}\left(1 - \dfrac{c^3}{4}\right)$

7-10 $T = \dfrac{\pi}{2} R^3 \tau_{\max}(1 - c^4)$

7-12 25% reduction in torque

7-14 (*a*) $\tau_{\max} = 12{,}200$ psi, (*b*) $\tau_{\max} = 9{,}150$ psi

7-16 The solid shaft

7-17 $\tau_{\max} = 10{,}000$ psi, $T = 15{,}700$ in-lb

7-18 $\tau_{\max} = 10{,}000$ psi, $T = 14{,}700$ in-lb

7-20 $T = 8{,}200\pi$ in-lb (shaft), hp = 245

7-21 (*a*) $G = 4.08 \times 10^6$ psi, (*b*) $\tau_{e\iota} = 20{,}400$ psi, (*c*) $\tau_{\text{ult}} = 30{,}600$ psi

7-23 $T = 2{,}750$ in-lb, $\tau = 14{,}000$ psi

7-24 $T = 5{,}240$ in-lb, $\tau = 20{,}000$ psi

7-25 hp = 2,800

7-26 hp = 1,162

7-28 (*a*) $R_i = 1$ in, (*b*) $\theta = 0.10$ rad, (*c*) $T = 102{,}000$ in-lb

7-29 (*a*) $R_i = 1$ in, (*b*) $\theta = 0.10$ rad, (*c*) $T = 179{,}000$ in-lb

7-30 (*a*) zero, (*b*) $\phi_{\text{res}} = 1.3°$

7-35 $\tau_{\text{steel}} = 10{,}700$ psi, $\tau_{\text{al}} = 5{,}650$ psi, $\phi = 0.143$ rad

7-36 $L_b = 34$ in, $L_s = 16$ in

7-37 $\tau_{\max} = 9{,}350$ psi, $\phi_{A/B} = 0.0025$ rad $= 0.143°$

7-39 hp = 684

7-40 $T_L = 7{,}700$ in-lb, $T_R = 15{,}400$ in-lb

7-41 $T_L = T_R = 33{,}000$ in-lb

Chapter 8

8-1 $\tau = 3{,}183$ psi, $\sigma = 3{,}820$

8-3

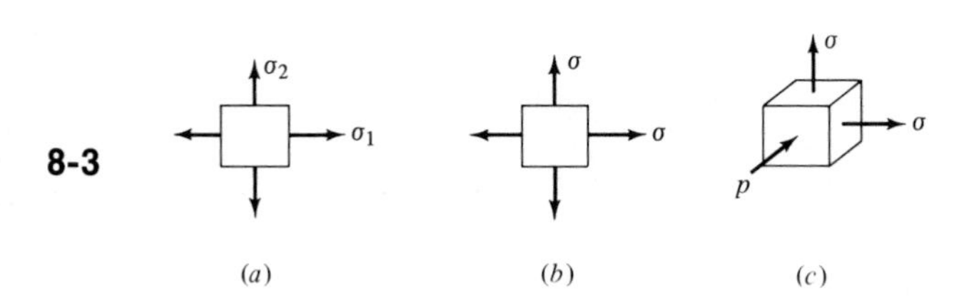

8-5

(*a*) (*b*) (*c*)

8-7

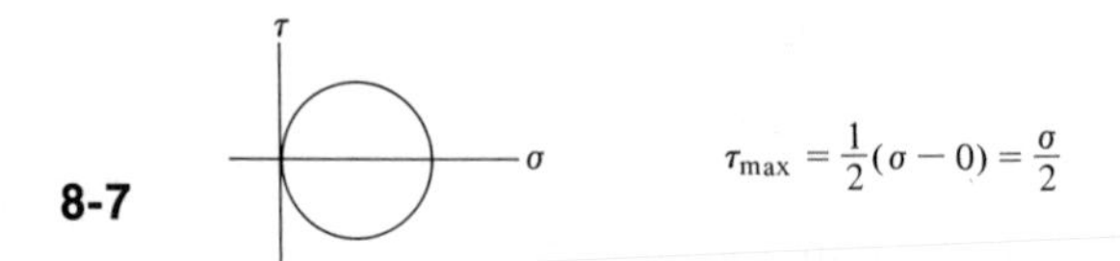

$\tau_{\max} = \frac{1}{2}(\sigma - 0) = \frac{\sigma}{2}$

8-9 τ_{max}, $-\sigma_x$, σ_x $\qquad \tau_{max} = \dfrac{\sigma_x - (-\sigma_x)}{2} = \sigma_x \qquad \theta = 45°$

8-13 $\sigma_{max} = 18{,}000$ psi, $\sigma_{min} = 9{,}000$ psi,
$\tau_{max} = 4{,}500$ psi and $\theta = 45°$

8-14 $\sigma_{max} = 24{,}000$ psi, $\sigma_{min} = -2{,}000$ psi and $\theta = 45°$,
$\tau_{max} = 13{,}000$ psi

8-15 $\sigma_{max} = 17{,}500$ psi, $\sigma_{min} = -7{,}500$ psi and $\theta = -18\frac{1}{2}°$,
$\tau_{max} = 12{,}500$ psi

8-17 $\sigma_{max} = 17{,}000$ psi and $\sigma = -12.6°$, $\sigma_{min} = -17{,}000$ psi,
$\tau_{max} = 17{,}000$ psi

8-19 $\sigma = -298$ psi, $\tau = -39$ psi

8-21 $\sigma_x = 24{,}000$ psi

8-23 $\sigma_x = \sigma_{max} = 2{,}230$ psi, $\sigma_{min} = -1{,}770$ psi, $\tau_{xy} = 0$,
$\tau_{max} = 2{,}000$ psi

8-24 $\sigma_x = \sigma_{max} = 1{,}866$ psi, $\sigma_{min} = 712$ psi, $\tau_{xy} = 0$, $\tau_{max} = 577$ psi

8-25 $T = 663$ in-lb

8-27 $\sigma_{max} = 939$ psi and $\theta = 39.1°$, $\sigma_{min} = -621$ psi, $\tau_{max} = 780$

8-28 No, $\sigma_{max} = 25{,}000 > 24{,}000$ psi

8-29 $Q = 694$ psi

8-30 $Q = 587$ psi

8-31 $T = 424$ in-lb

8-33 $\sigma = \dfrac{pD}{4t}$

8-35 $\sigma = 12{,}500$ psi

8-37 $\sigma = 6{,}856$ psi, $\tau = 3{,}513$ psi

8-38 $\tau = 9{,}770$ psi, $\sigma = 16{,}300$ psi, $\sigma_{max} = 20{,}900$ psi

8-39 $p = 416$ psi based on tensile stress

8-40 $p = 600$ psi

8-41 $\epsilon = 0.571$ in/in

8-43 (*a*) $\epsilon = 0.00125$, (*b*) $\epsilon = 0.05$ in/in,
(*c*) $\epsilon = 0.0511$ in/in

8-44 $\epsilon = \dfrac{e}{L}\dfrac{d}{\sqrt{L_R{}^2 - e^2}}$, $\epsilon \approx \dfrac{ed}{LL_R}$

8-45 $\epsilon_{ac} = \dfrac{\sqrt{(L/\sqrt{2}+X)^2 + (L/\sqrt{2}+Y)^2} - L}{L} \approx \dfrac{X+Y}{\sqrt{2}L}$

$\epsilon_{bc} = \dfrac{\sqrt{(L/\sqrt{2}-X)^2 + (L/\sqrt{2}+Y)^2} - L}{L} \approx \dfrac{Y-X}{\sqrt{2}L}$

8-47 $\Delta V = 0.576 \text{ in}^3$

8-49 $\epsilon_{\max} = 560$, $\epsilon_{\min} = -1{,}560$ and $\theta = -24.4°$

8-51 $\epsilon_{\max} = 1{,}700$ and $\theta = -26.6°$, $\epsilon_{\min} = -800$

8-53 $\epsilon_{\max} = 940$ and $\theta = 19.3°$, $\epsilon_{\min} = -340$

8-55 $\epsilon_{45°} = 0.000900$ in/in

8-57 $\epsilon_{\max} = 320$ and $\theta = -45°$, $\epsilon_{\min} = 120$

8-59 $\epsilon_{\max} = 303$, $\epsilon_{\min} = -983$ and $\theta = -26.9°$

8-62 $\epsilon = 0.00275$ in/in

8-63 $\epsilon_{\max} = 175 \times 10^{-6}$ in/in, $\gamma_{\max} = 250 \times 10^{-6}$ in/in

8-64 $\sigma_{\max} = 16{,}000$ psi, $\sigma_{\min} = 4{,}000$ psi, $\tau_{\max} = 8{,}000$ psi

8-65 $\sigma_{\max} = 62{,}000$ psi, $\sigma_{\min} = -6{,}000$ psi, $\tau_{\max} = 34{,}000$

8-67 $\sigma_{\max} = 11{,}200$ psi, $\sigma_{\min} = 6{,}400$ psi, $\tau_{\max} = 5{,}600$ psi

8-69 $\sigma_{\max} = 1{,}825$ psi, $\sigma_{\min} = -29{,}000$ psi, $\tau_{\max} = 15{,}413$ psi

8-71 $\Delta V = 0.0080 \text{ in}^3$

8-73 $\phi = 144\epsilon$ rad, $T = 25.8\epsilon \times 10^6$ in-lb

8-74 (*a*) $\epsilon = -167 \times 10^{-6}$ in/in, (*b*) $\epsilon = 500 \times 10^6$ in/in,
(*c*) $\epsilon = -46 \times 10^{-6}$ in/in

8-75 $\epsilon_1 + \epsilon_2 = 1.31 \times 10^{-6}p$

Chapter 9

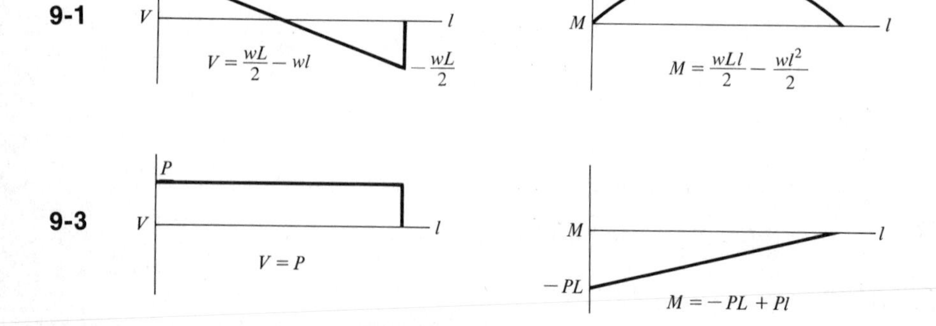

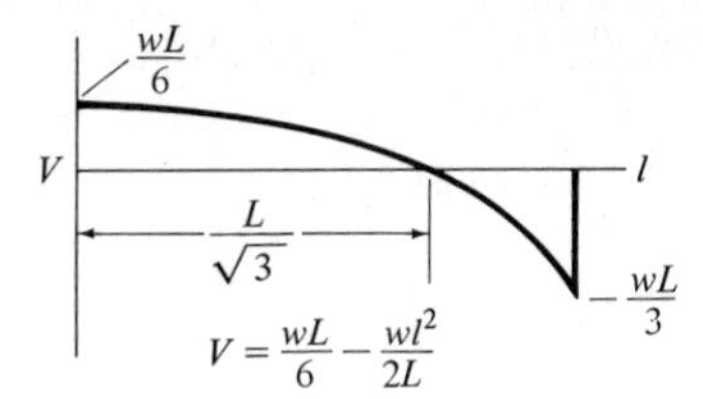

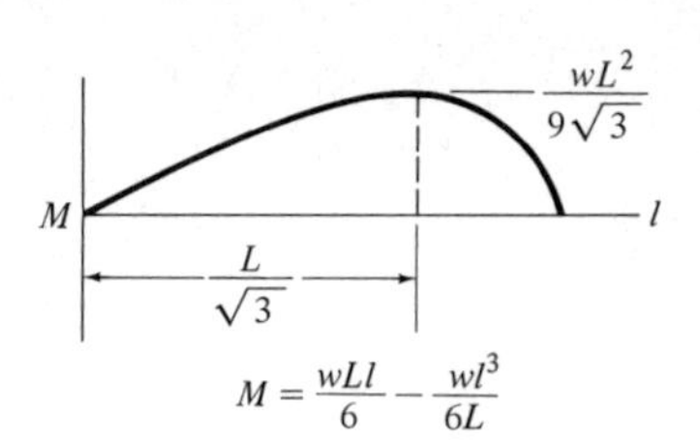

9-7

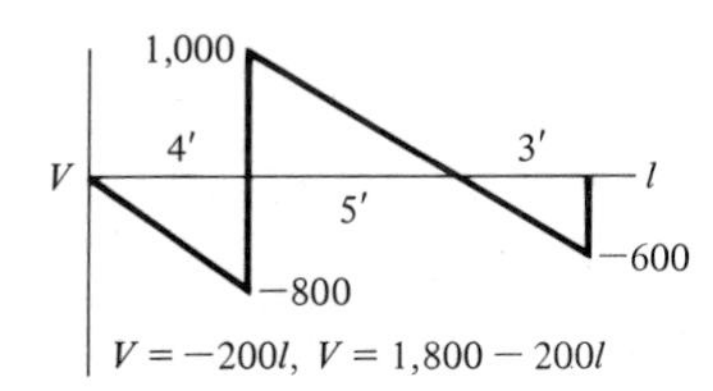

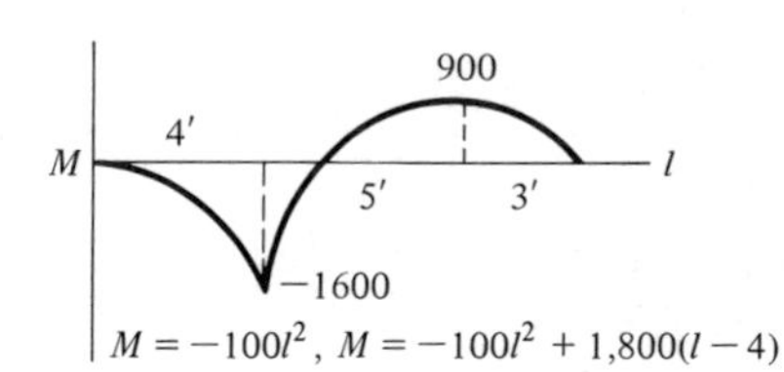

9-9

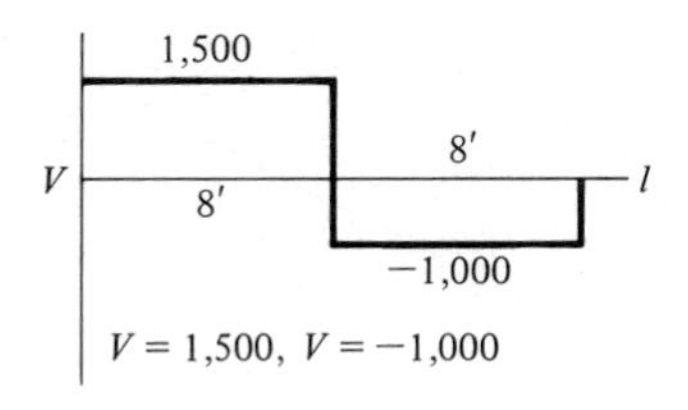

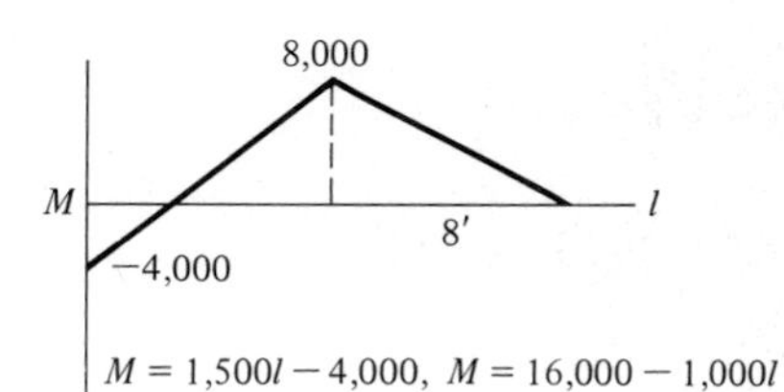

9-11

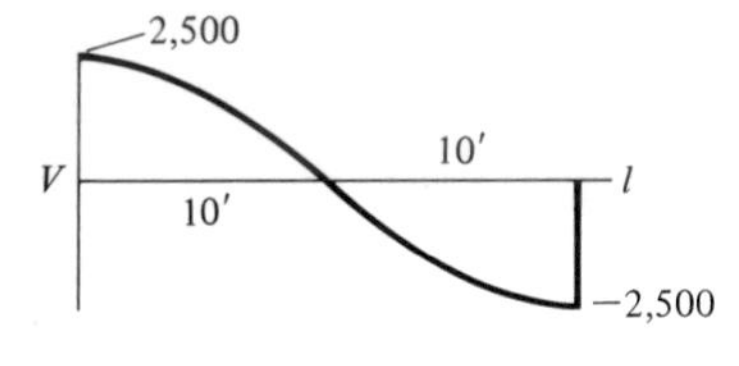

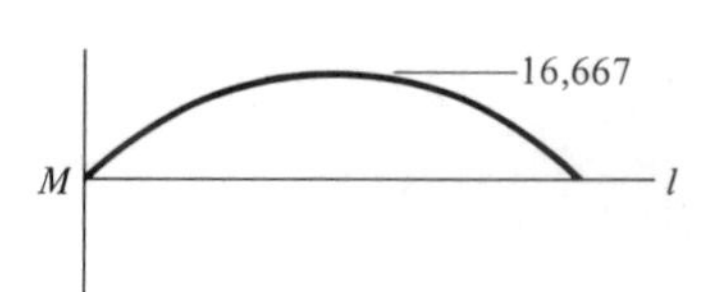

9-13

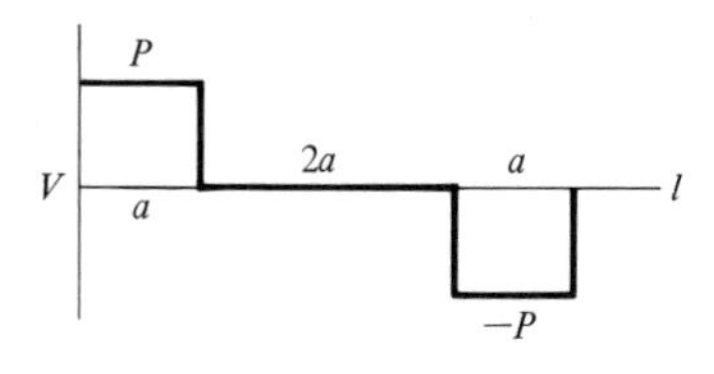

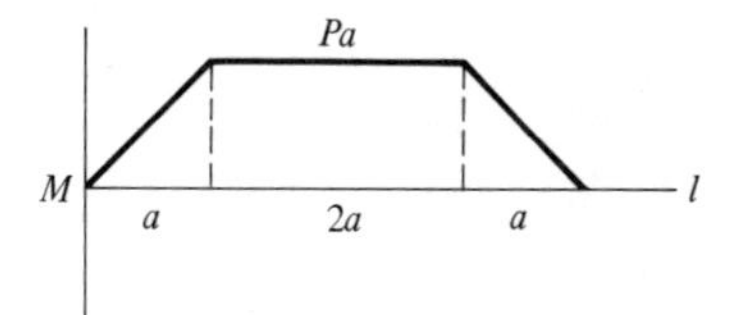

9-15

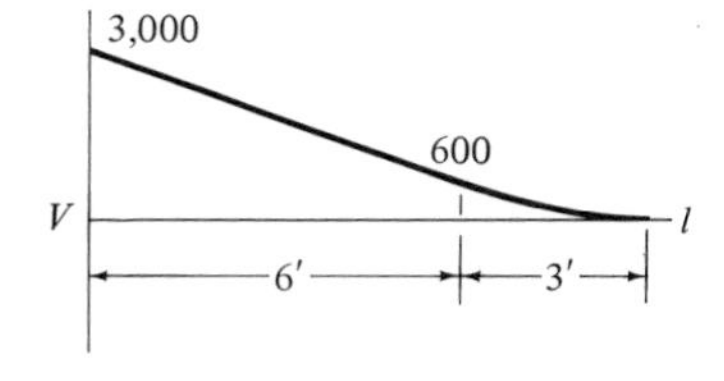

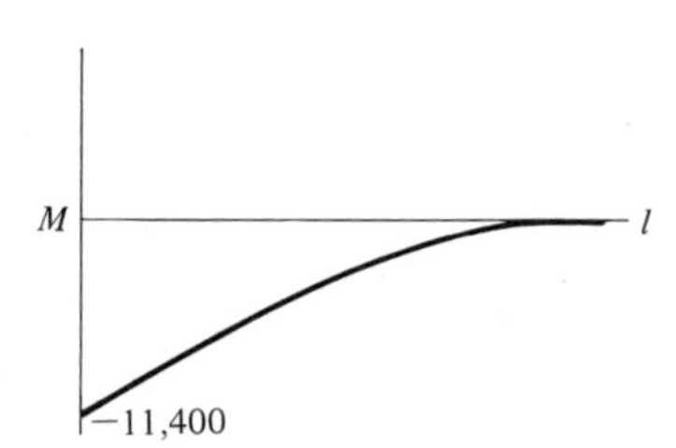

9-17

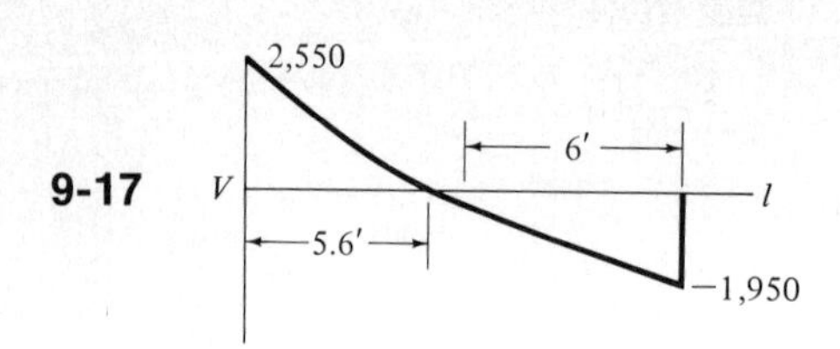

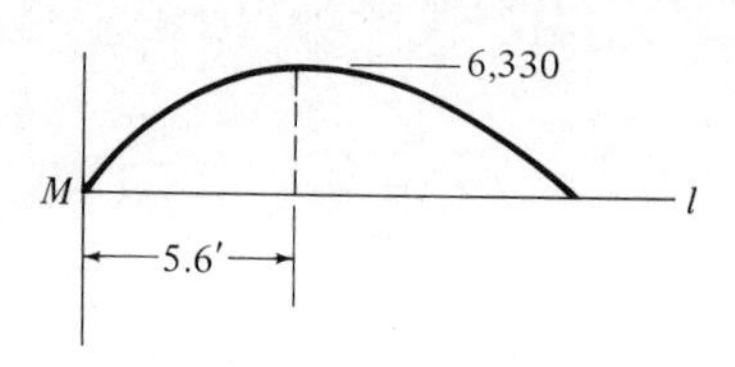

9-19

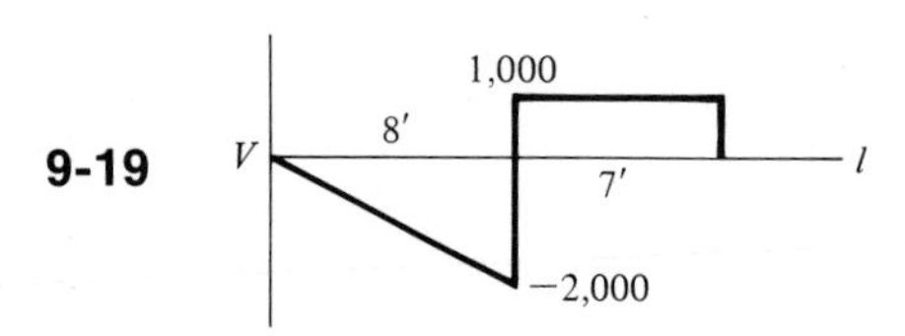

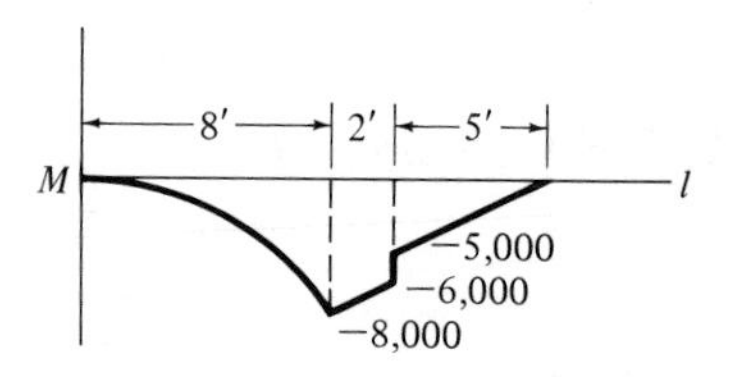

9-21

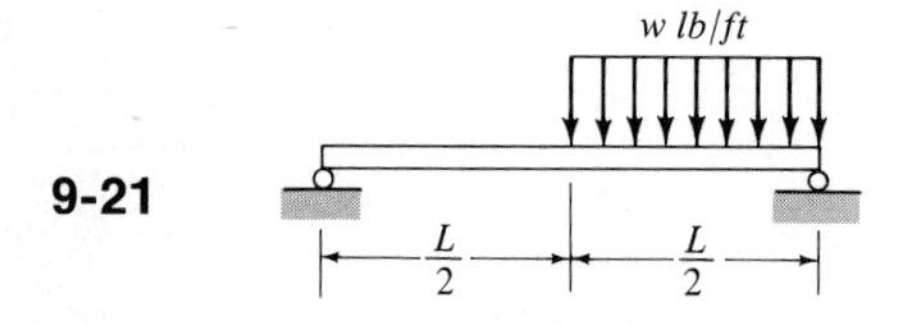

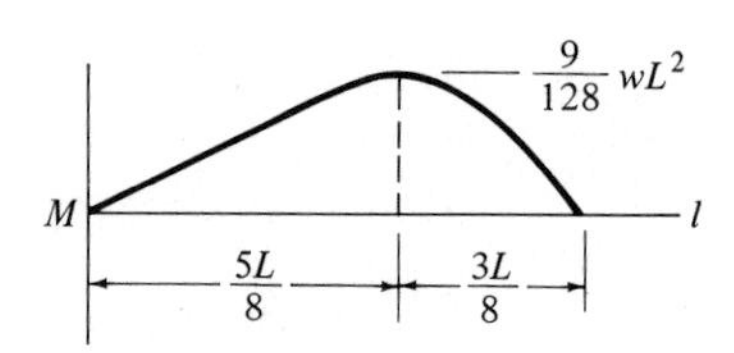

9-24

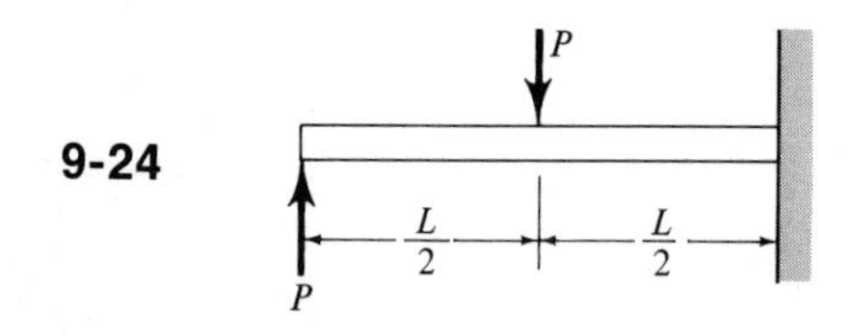

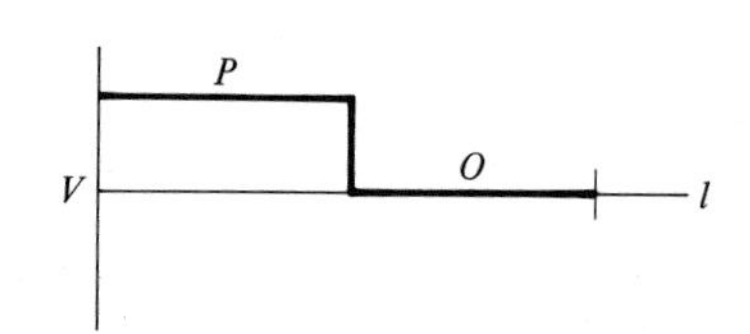

9-27 (*a*) $\bar{y} = 5.98$ in ≈ 6 in, (*b*) $y_p = 7.5$ in

9-29 (*a*) $\bar{y} = 4.81$ in, (*b*) $y_p = 4.25$ in

9-30 $L = 190$ in

9-32 $w = 1{,}625$ lb/ft

9-33 $w = 3{,}010$ lb/ft

9-34 C8 × 13.75

9-35 S18 × 54.7

9-36 $\sigma_t = 13{,}880$ psi at $x = 6.48$ ft,
$\sigma_c = 11{,}912$ psi at $x = 14$ ft

9-39 $W = 23{,}000$ lb

9-40 (*a*) $M = 72{,}000$ in-lb, (*b*) $M = 495{,}000$ in-lb,
(*c*) $M = 540{,}000$ in-lb

9-42 $M_{fp}/M_{e\ell} = 16/3\pi$

9-44 $M_{fp}/M_{e\ell} = 2$

9-46 $M_{ult}/M_{e\ell} = 80/9\pi$

9-48 $M_{ult}/M_{e\ell} = 10/3$

9-50 $P = 6{,}770$ lb

9-54 ① $Q/It = 0.1075\ in^{-2}$, ② $Q/It = 0.0846\ in^{-2}$,
③ $Q/It = 0.1213\ in^{-2}$

9-56 $\tau = 380$ psi

9-58 $L = 7$ ft or 84 in

9-59 $P = 735$

9-60 4×12 in nominal size

9-61 (*a*) $\sigma_t = 827$ psi, (*b*) $\sigma_c = 1{,}240$ psi, (*c*) $\tau_{max} = 146$ psi

9-64 $\sigma_{max} = 4{,}880$ psi

9-65 $P = 43{,}500$ lb for T

9-66 $P = 40{,}320$ lb for T

9-67 Yes, $\sigma_{max} = 25{,}500\ psi < 32{,}000\ psi$

9-68 $\sigma_{max} = 12{,}850$ psi, $\tau_{max} = 10{,}600$ psi

9-69 $\sigma_{max} = 13{,}230$ psi, $\tau_{max} = 6{,}980$ psi

9-71 $\tau_{15^\circ} = 400$ psi, increase by 62.5 psi

9-79 $y_{max} = -wL^4/30EI$

9-80 $y_{max} = -0.432$ in

9-82 $\delta = WL^3/3EI$

9-83 $y_{max} = -11Pa^3/6EI$, $\theta_a = -Pa^2/EI$

9-85 Tensile stress reduced 52.5%, compressive stress 37.4%, deflection 52.6%

9-86 $\sigma = 533$ psi, $y_{max} = 0.308$ in

9-87 $\sigma = 408$ psi, $y_{max} = 0.194$ in

9-88 No

9-89 $y = 0.333$ in

9-92 $R_1 = 3wL/16$, $R_2 = 5wL/8$, $R_3 = 3wL/16$

9-93 $R_1 = 3wL/8$, $R_2 = 5wL/8$, $M_2 = wL^2/8$

9-94 $R_1 = 11P/16$, $M_1 = 3PL/16$, $R_2 = 5P/16$

9-97 $R_1 = 3P/2$, $M_1 = PL/4$, $R_2 = 5P/2$

9-98 $\delta = 3wL^4/8(3EI + kL^3)$

9-99 $M_o = PL/2$

9-102 $\sigma_{max} = 23{,}400$, $\boldsymbol{\delta} = -0.591\mathbf{j} - 0.546\mathbf{k}$

Chapter 10

10-2	63.3 lb $\leqslant P \leqslant$ 635 lb
10-4	$P \geqslant 90$ slipping down $P \leqslant 187.5$ tipping up
10-6	(*a*) $T = 56$ lb for slipping, (*b*) $T = 77.6$ lb for tipping
10-8	$P = 75$ lb, $R_R = 131.25$ lb, $R_L = 18.75$ lb
10-10	$P = 56$ lb, $h = 57.1$ in
10-12	No, it tips for $P = 148$ lb
10-14	(*a*) $\theta = 19.5°$, (*b*) $\theta = 21.4°$, (*c*) $\theta = 36.8°$
10-16	$\theta = 19.5°$, $f = 0.354$
10-17	Yes
10-19	$P = 510$ lb
10-20	Yes, $P = 113$ lb
10-22	$P = 31.9$ lb
10-25	(*a*) Yes, (*b*) yes
10-26	$f = \tan 7.5° \doteq 0.132$
10-27	$P = 75.6$ lb
10-29	$T_1/T_2 = 81$
10-31	Once fully around
10-32	(*a*) Yes, (*b*) no
10-34	$T = 26.9$ lb
10-36	$P = 101$ lb
10-37	$P = 186$ lb
10-38	$T = \pi f D P = \pi 6P$ in-lb
10-40	$T/P = 2.1$
10-42	$T = 20$ lb or 245 lb
10-45	$W_A = 800$ lb
10-46	$P = 111.5$ lb
10-47	$P < m/L$
10-48	$P < m/L$
10-49	$P < 4m/L$
10-52	$P < m/L - W$
10-54	$P_c = kL/4$
10-55	$T > \dfrac{Wa}{2(a+L)}$
10-56	$h > \sqrt{3}R$

10-58 (*a*) $L/r = 78.8$, (*b*) $L_e/r = 39.4$

10-60 (*a*) $L/r = 90$, (*b*) $L_e/r = 45$

10-62 (*a*) $L/r = 41.6$, $L/r = 67.3$, (*b*) $L_e/r = 29.1$, $L_e/r = 47.1$

10-64 (*a*) $L/r = 84.9$, $L/r = 98$, (*b*) $L_e/r = 59.4$, $L_e/r = 68.6$

10-66 $L_e/r = 59$, no

10-68 $L = 627$ in, which is probably unrealistic

10-69 $T = 132°F$

10-70 $L/r = 80$ for both axes, $P_c = 6{,}380$ lb

10-72 $R = 1$ in

10-74 $P = 6{,}222$ lb

10-75 Yes

10-76 Yes

10-77 Yes

10-79 No, $P_c = 98{,}000$ lb

10-80 (*a*) $L = 11.5$ in, (*b*) $L = 23$ in

10-81 $T = 300°F$ based on inelastic theory

10-82 $L/r = 34.64$, $R_c = 11{,}250$ lb

10-84 $P_c = 44{,}000$ lb

Appendix A

A-3 $x = 1\frac{1}{2}$ in, $y = 3.2$ in

A-5 $x = y = 3.6$ in

A-6 $x = 0.83$ in, $y = 2.42$ in

A-11 $x = 7.64$ in, $y = 0.94$ in

A-13 $x = 0.99$ in, $y = 0$

A-14 $x = 2.22$ in, $y = -0.83$ in

A-21 $I_y = 68.3$ in^4, $I_{y_c} = 20.3$ in^4

A-22 $I_x = 38.2$ in^4, $I_{x_c} = 5.3$ in^4

A-27 $k_x = 5.0$ in

A-29 $k_y = 6.34$ in

A-31 $k_y = 6.74$ in

A-32 $k_x = 2.84$ ft

A-34 $I_{y_c} = 386$ in^4

A-36 $I_{x_c} = 3{,}287 \text{ in}^4$

A-39 $I_{x_c} = 400 \text{ in}^4$

A-41 $P_{xy} = \dfrac{b^2h^2}{8}, P_{x_c y_c} = \dfrac{b^2h^2}{72}$

A-43 $P_{xy} = \dfrac{a^2b^2}{6}, P_{x_c y_c} = \dfrac{a^2b^2}{60}$

A-45 $I^{\max} = 173 \text{ in}^4, I_{\min} = 19.7 \text{ in}^4, \theta = 50.7°$

A-47 $I_{\max} = 1{,}370 \text{ in}^4, I_{\min} = 100 \text{ in}^4, \theta = -16°$

A-49 $\bar{x} = \frac{2}{3}b, \bar{y} = \bar{x} = 0$

A-51 $\bar{y} = 1.55 \text{ in}, \bar{x} = \bar{z} = 0$

A-53 $\bar{y} = \dfrac{h}{4}$

A-54 $\bar{x} = \bar{z} = 0, \bar{y} = \frac{3}{8}b$

A-56 $\bar{x} = 4 \text{ in}, \bar{y} = 5.70 \text{ in}, \bar{z} = 3.05 \text{ in}$

A-58 $\bar{x} = 4.75 \text{ in}, \bar{y} = 3.33 \text{ in}, \bar{z} = -5.19 \text{ in}$

A-60 $I_x = \dfrac{\pi R^4 h}{2}, I_{yz} = \dfrac{\pi R^2 h^3}{3}, I_y = \dfrac{\pi R^2 h}{12}(4h^2 + 3R^2)$

A-62 $I_x = \frac{1}{6}\pi h^4 b, I_{yz} = \frac{1}{4}\pi h^2 b^3$

A-64 $I_x = \frac{2}{15}\pi a^2 b(a^2 + b^2), I_y = \frac{4}{15}\pi a^4 b$

A-66 $I_a = 11{,}258 \text{ in}^5$

A-68 $I_a = 1{,}076 \text{ in}^5$

A-69 $I_a = 566 \text{ in}^5$

A-70 $I_a = 1{,}357 \text{ in}^5$

INDEX